What the Critics Say About *Wizard*

It was Seifer's...intention to bring...Tesla's...major scientific and technological contributions...together. He has done a very impressive job collecting a massive amount of documentation of original and secondary sources....The letters in the J.P. Morgan collection, in particular, shed considerable new light on Tesla's connection with Morgan and his contemporaries.

—*Nature*

The best chapter in Seifer's book describes late 19th century science fiction and locates Tesla's projects among other predictions of the future....Seifer is [also] good at describing Tesla's lack of practical, economic and personal judgment and the way his enormous ego invited unscrupulous partners....

—*Washington Post Book World*

Seifer paints a picture of Tesla that anyone familiar with the life of someone such as Orson Welles will recognize. Here was a man who peaked early, traveled in famous company...and started believing his own press hype. That made him spend the rest of his life trying to score another universe-changing coup....*Wizard* does a pretty good job of placing Tesla within the firmament of inventors, thinkers, and futurists. With Seifer's scholarship to build on, anyone reconstructing those dizzy years of invention and litigation at the turn of the century would be foolish to try and leave out Nikola Tesla.

—*Winston-Salem NC Journal*

Despite Tesla's impact on electricity, history does not regard him as highly as many of his inventive contemporaries....As Seifer shows in great detail...Tesla's story is complicated and tests our definition of science....Where does someone like Tesla fit it?

—*MIT's Technology Review*

Wizard...presents a much more accurate...picture of Tesla....[It] is thorough, informative, entertaining and a valuable addition to electrotechnological history, past and future.

—*Electronic Engineering Times*

Here is a deep and comprehensive biography of a great engineer of early electrical science. Indeed, it is likely to become the definitive biography of the Serbian-American inventor Nikola Tesla. The book brings together, into a cohesive whole, the many complex facets of the personal and technical life of the "wizard" who stands alongside Thomas Edison and George Westinghouse as another great implementor and inventor.... Highly recommended.

—*American Academy for the Advancement of Science*

In modern times, Tesla may be enjoying a comeback thanks to books like "Wizard."

—*New York Times*

The next time you dial your cellular phone, there's a guy you should think of. If you don't use a cell phone, then look at the light by which you're reading this newspaper. He helped make it brighter....WIZARD chronicles Tesla's contributions to alternating current, or AC, the electrical system used in most homes today. Though today he is almost forgotten, written out of history by the people he once worked for, Tesla lived in New York's Waldorf Astoria and was world famous.... "His notebooks are filled with mathematics," said Seifer. "He predates Einstein and Bohr with his description of the atom. He was one of the forefathers of quantum physics."

—*Narragansett Times*

[Wizard] brings the many complex facets of [Tesla's] personal and technical life together into a cohesive whole....The book contains excellent discussions of the controversies, fury of activity, and lawsuits surrounding the development of new hardware technology. In many ways, they are similar to the later legal battles in the development of computers....I highly recommend this biography of a great technologist. A.A. Mullin, U.S. Army Space and Strategic Defense Command.

—*Computing Reviews*

Rare insight on a great mind.

—*New Bedford Standard Times*

WIZARD

THE LIFE AND TIMES OF NIKOLA TESLA
BIOGRAPHY OF A GENIUS

Marc J. Seifer

Citadel Press
Kensington Publishing Corp.
www.kensingtonbooks.com

In memory of my Dad, Stanley Seifer

CITADEL PRESS BOOKS are published by

Kensington Publishing Corp.
119 West 40th Street
New York, NY 10018

All Kensington titles, imprints, and distributed lines are available at special quantity discounts for bulk purchases for sales promotions, premiums, fund-raising, educational, or institutional use. Special book excerpts or customized printings can also be created to fit specific needs. For details, write or phone the office of the Kensington sales manager: Kensington Publishing Corp., 119 West 40th Street, New York, NY 10018, attn: Sales Department; phone 1-800-221-2647.

ISBN-13: 978-0-8065-3996-6
ISBN-10: 0-8065-3996-8

First printing: May 1998

28 27 26 25 24 23 22 21 20

Printed in the United States of America

Library of Congress Control Number: 95-49919

First electronic edition: May 1998

ISBN-13: 978-0-8065-3556-2
ISBN-10: 0-8065-3556-3

CONTENTS

FOREWORD

Nikola Tesla was my father's uncle, and as such he was treated by our family much as any uncle might have been who lived at a considerable distance and was advanced in years. But there were stronger bonds between my father and Tesla than might otherwise have been the case. They came from identical social backgrounds, sons of Serbian Orthodox priests, born and raised a few miles apart in the Austro-Hungarian military frontier district county of Lika in the Province of Croatia (my grandmother was Tesla's sister Angelina); they were the only members of a relatively limited extended family to emigrate to America; and they were the only members to undertake science and technology as their life's work.

My father, Nicholas J. (John) Terbo (Nikola Jovo Trbojevich), was thirty years younger than his uncle, came to America thirty years after him, and died thirty years after Tesla. Tesla was already famous as my father was growing up, and he became a model for my father's technical career. Father held about 175 U.S. and foreign patents, the most important of which was his 1923 basic patent on the Hypoid gear, used on the vast majority of the world's cars since 1930. The Hypoid gear introduced advanced mathematics to the art of gear design, much as Tesla's work united electrical theory and electrical engineering. Tesla henceforth proudly referred to my father as "my nephew, the mathematician." (That these patents brought considerable financial as well as professional recognition to my father was also not lost on the often cash-poor Tesla.)

Because the ethnic and professional similarities between Nikola Tesla and my father were so striking, I feel that I have been granted a special privilege through this comparison in understanding Tesla's private personality, including his well-developed sense of humor and his often cavalier disregard for money. Once, when Tesla was visiting us in the early 1930s, my father took him to lunch at the Book Cadillac Hotel, then the finest in Detroit. They arrived late, only a few minutes before a cover charge of $2 or $3 would end. (This was equivalent to $20 or $30 by today's standards.) My father suggested waiting, but Tesla would hear none of it. They sat down amid a flurry of waiters and Tesla ordered a chafing dish, bread, and milk and proceeded to prepare his own lunch to his own specifications (to my father's amusement and the unease of the maître d').

I had not yet reached thirteen when Tesla died in January 1943, and I did not have the sense of the ending of an epoch marked by his passing, both for our family and for an era of individualism in scientific discovery.

I may have reflected with some uneasiness that I had had the opportunity to meet Tesla some three or four years earlier and that no further meetings would ever happen again. I remembered my reluctance to be dragged to the meeting in his suite at the Hotel New Yorker when my mother and I were spending a few days in New York before returning to Detroit after our summer vacation at the Jersey shore. (I would have preferred spending more time at Radio City Music Hall or at the docks, watching the ocean liners.)

I was shy (rather, overwhelmed) and spoke hardly a word to this very tall, very gaunt old man. I would have been repelled—as any young "all-American boy" should have been—to be hugged and kissed by this stranger if my father hadn't often done the same. (This is the way my mother's women friends often acted, but my American mother's brother would have only given me a firm handshake.) Little did I realize that Tesla's hugging, kissing, and patting my head would belie his famous idiosyncrasy of an overriding phobia of germs. Surely, a young boy would have been teeming with "germs"! One could therefore speculate that this "idiosyncrasy" was possibly an affectation designed to preserve his "space."

While Tesla lived, some considerable degree of his fame endured—in no small measure because of his ability to stimulate the media. However, after his death the nation and the world were occupied with other more pressing matters—war and reconstruction, international political realignments, an unmatched explosion of new technology, a new consumer society—and Tesla's fame and recognition nearly evaporated. Only a few in the U.S. and international scientific communities and the abiding respect and admiration of Serbs and all Yugoslavs worldwide kept his name alive.

My awareness of a resurgence of interest in the life and works of Nikola Tesla began in the early 1970s, when I moved from Los Angeles (where it seemed no one had ever heard of Tesla) to Washington, D.C., where at least the name was recognized. In February 1975 my mother phoned to tell me that she had read in the *Los Angeles Times* that Uncle Nikola was to be inducted into the National Inventors Hall of Fame and that I should look into it. I chanced to notice on a local TV news program that evening a short segment monitoring the Hall of Fame and an interview with a girl of ten or twelve who had invented a new can opener or some such. I dismissed the Hall as a commercial promotion and went on to something else.

Only later did I read a newspaper account about the induction of Tesla (along with Orville and Wilbur Wright, Samuel F. B. Morse, and Tesla's nemesis, Guglielmo Marconi) and citing the Hall's sponsorship by the U.S. Department of Commerce's Patent and Trademark Office. The closest living relative of each honoree was to receive the induction diploma at an elaborate ceremony. Lacking any "Tesla" (or even any "Trbojevich") to represent the family, an officer of the Institute of Electrical and Electronics Engineers (IEEE) accepted the Tesla diploma. (The IEEE considers Tesla one of the twelve "apostles" of electrical science and continues to offer an annual prize in the field of Power Engineering in his name.) When I presented myself at the Patent Office a few weeks later, they were delighted and made arrangements to make a second presentation to me at the 1976 awards ceremony held that year at Congress Hall in Philadelphia as part of the U.S. Bicentennial Year celebration.

Since then, there has been an earnest revival of interest in the technological accomplishments of Nikola Tesla and in his personality, philosophy, and culture as well. Part of the drama of his life is that he was a man who not only revolutionized the generation and distribution of electrical energy and made basic contributions to many other facets of modern technology but that he did so without the specific aim of amassing great wealth. This altruism, which is often criticized as "poor business sense," imposed a monetary limitation on future experimentation to test his new innovations. Who knows what advances might have been possible if he had been able to validate them through rigorous experimentation. New science is an expensive endeavor, and finding financial support is a frustrating task for even those as focused as Tesla.

Among the associations that have supported the Tesla renaissance are: the Tesla Memorial Society, which I helped found in 1979, and of which I am pleased to be its Honorary Chairman and Chairman of its Executive Board, and the International Tesla Society, founded in 1983, and of which I am a Life Member. It was while speaking at the first ITS biannual Tesla Symposium in 1984 that I first met a fellow speaker, Dr. Marc J. Seifer, in person. His paper "The Lost Wizard" was the seed from which his new Tesla biography has sprung. I have been impressed with Dr. Seifer's dedication and scholarship in developing his early theories into a well-rounded examination of the mystery of Tesla's great genius.

One of the things that has most intrigued me about a new work on this topic is how much new information keeps surfacing. Dr. Seifer has researched minor characters in Tesla's life as well as the many major ones. This has given him additional insight into Tesla's life and allowed the development of new and different interpretations of many important events, such as the failure of the Wardenclyffe tower project.

Dr. Seifer provides a new look at Tesla's college years, the time when many of his epochal ideas were forming. He has uncovered new information on Tesla's relationship with a number of key individuals, such as his editor, Thomas Commerford Martin, and financial backers John Jacob Astor and John Hays Hammond. A great strength of *Wizard* is its adherence, chapter by chapter, to a rather strict chronology, which makes it easy to follow the breadth and scope of Tesla's life and achievements in an orderly fashion.

I congratulate Dr. Seifer on a decade's journey with Nikola Tesla and am pleased to introduce to you *Wizard.*

William H. Terbo
Honorary Chairman
Tesla Memorial Society

PREFACE

In 1976, while involved in research at the New York Public Library, I stumbled upon a strange text entitled *Return of the Dove* which claimed that there was a man not born of this planet who landed as a baby in the mountains of Croatia in 1856. Raised by "earth parents," an *avatar* had arrived for the sole purpose of inaugurating the New Age. By providing humans with a veritable cornucopia of inventions, he had created, in essence, the technological backbone of the modern era.[1]

His name was Nikola Tesla, and his inventions included the induction motor, the electrical-power distribution system, fluorescent and neon lights, wireless communication, remote control, and robotics.

Tesla—who's he? I said to myself. Because my father had been a TV repairman for several years in the early 1950s and I had spent part of my childhood accompanying him on house calls, helping put up antennas, test and buy radio tubes, play with oscilloscopes, and watch him build TVs, I was amazed that I had never heard of this man.

I remember vividly an event from my grade-school years on Long Island that prepared me for my latter-day interest. It was a Saturday, circa 1959, and I was working on a Boy Scout assignment when I came upon a design for a crystal radio set. My father and I gathered a glass jar and a set of headphones, a crystal detector for changing the ambient AC radio waves into audio DC pulses, some thin copper wire to be wrapped around the jar, a metal switch that was scraped across this coil for the "dial," a small plank to hold the contraption together, and a hundred feet of normal rubber-coated wire for the antenna, which we strung out a second-story window. There was no plug; *all energy was derived from the broadcast signals from the nearby radio stations.* However, after hooking it all together, the reception was faint; I became discouraged.

My father paced the room, considering the problem, muttering, "Something's wrong." After a few moments of deep thought, he made a motion which said, "I've got it." Walking over to our radiator and dragging another wire, which he had hooked up to the jar, Dad attached a ground connection. Suddenly, all the stations began to come in loud and clear, and I marked each of them on the jar along the coil. It was apparent to me then that electrical power was being transmitted from these stations by wireless means and that the earth somehow was intrinsically linked to this system.

And now, here I was, nearly twenty years later, two years out of graduate school with a master's degree, well-read and somewhat knowledgeable about electronics, yet I had never heard of the principal inventor of the very device I had spent endless hours with as a kid. This astonished me in a way difficult to describe. Moreover, when I asked my father about Tesla, he barely knew of him.

Because I believe in seeking original sources, I began to research Tesla's life, starting with the two existing biographies, John O'Neill's classic *Prodigal Genius* and Inez Hunt and Wanetta Draper's *Lightning in His Hand*. Soon after, I began tracking down numerous turn-of-the-century references and also the weighty *Nikola Tesla: Lectures, Patents, Articles*, produced by the Tesla Museum in Belgrade, Yugoslavia. Thus, I was able to ascertain, by following his actual patents, that indeed Tesla existed and that his work was fundamental to all these creations.

That Tesla's name was so little known puzzled me, so in 1980, three years after writing my first article on him, I began a doctoral dissertation on his life. My major purpose was to address the question of name obscurity.

During the writing of my dissertation, several notable Tesla works were compiled. These included the comprehensive and encyclopedic *Dr. Nikola Tesla Bibliography*, by Leland Anderson and John Ratzlaff; Tesla's 1919 autobiography, republished by Hart Brothers; Margaret Cheney's biography *Tesla: Man Out of Time*; two compendiums of Tesla's writings by John Ratzlaff; *Colorado Springs Notes*, produced by the Tesla Museum; and most recently, Leland Anderson's edition of Tesla's private testimony to his lawyers on the history of wireless communication.

Even with all this new material, however, no comprehensive, all-embracing treatise had been achieved. In fact, after studying all these texts, *a number of contradictions and glaring mysteries remained*. These included not only Tesla's obscure early years, tenure at college, and relationship to such key people as Thomas Edison, Guglielmo Marconi, George Westinghouse, and J. P. Morgan but also the worth of Tesla's accomplishments and his exact place in the development of these inventions.

This book attempts to solve the mysteries. Because there have been significant gaps in the record, a clear chronology of Tesla's life is presented. Also addressed are such issues as why his name dropped into obscurity after being a page 1 subject in newspapers around the world at the turn of the century, why he never received the Nobel Prize, even though he was nominated for one, what Tesla did during the world wars, and whether his plan for transmitting electrical power by wireless was feasible.

Using a psychohistorical perspective, the text discusses not only those factors that led to Tesla's genius, but also quirks that led to his undoing. In this vein, it delineates Tesla's relationship to many of his well-known associates, such as John Jacob Astor, T. C. Martin, J. P. Morgan Sr. and Jr., John Hays Hammond Sr. and Jr., Michael Pupin, Stanford White, Mark Twain, Rudyard Kipling, Franklin Delano Roosevelt, George Sylvester Viereck, Titus deBobula, and J. Edgar Hoover. Many of these people are barely touched upon or are not discussed at all in other treatises.

Because Tesla's life is so controversial and complex, I also examine such questions as whether Tesla received impulses from outer space, why he ultimately failed in his partnership with J. P. Morgan in constructing a multifunctional global wireless system for distributing power and information, what his exact relationship to Robert and Katharine Johnson was, and what exactly happened to his particle-beam weaponry system and secret papers. Since I've based the text largely on firsthand documents rather than on the existing biographies, this book offers an essentially new view of Tesla's life. The most recent biography, *Tesla*, by Tad Wise, an admittedly fictionalized version of the inventor's life, was not referred to herein, as the goal of the enclosed is to separate out the myth and uncover who Tesla really was. However, one of Wise's most prominent stories, that Tesla was responsible for the peculiar explosion that devastated Tunguska, Siberia in June of 1908, is addressed in a new appendix for this second edition.

I have visited all major Tesla archival centers such as at the Smithsonian Institution in Washington, D.C.; Columbia University, in New York City; and the Tesla Museum, in Belgrade, Yugoslavia. And because I have also utilized the Freedom of Information Act and accessed the arcane network of Tesla researchers, *I have been able to compile hundreds of documents that have never been discussed before by any Tesla biographies*. In addition, because I am a handwriting analyst, I have also utilized that expertise to analyze a number of the key personalities involved. Through this means, and as a total surprise, I have also been able to discover a heretofore unreported emotional collapse that the inventor suffered in 1906, at the time of the failing of his great wireless enterprise.

Since Tesla lived until the age of eighty-six, the story spans nearly a century. Revered as a demigod by some in the New Age community, Tesla has, at the same time, been relegated to virtual nonperson status by influential segments of the corporate and academic communities. Often billed as a wizard from another world who drew thunderbolts from the skies, Tesla himself helped support the supernatural persona by comparing himself to the Almighty and by frequently grabbing headlines with his sensational talk of interplanetary communication. Because his accomplishments are prodigious, fundamental, and documented, the elimination of his name from many history books is not forgivable. Only by understanding why this occurred can we, as a modern people, hope to rectify the historical record for future generations.

Curiously, the further away we have moved from Tesla's death, the more material on his life has come to the fore. In particular, we must thank John O'Neill, the Tesla Memorial Society, the Tesla Museum in Belgrade, and the International Tesla Society (ITS) for this occurrence and the many Tesla researchers who have written so much about him of late and have participated in the various ITS conferences held every other year, since 1984, at the site of some of his most spectacular experiments, in Colorado Springs.

Because Tesla's eye was always on the future, it seems appropriate to conclude this introduction with the opening lines from his autobiography. They are as true today as we enter the twenty-first century as they were three generations ago, when they were written:

> The progressive development of man is vitally dependent on invention. It is the most important product of his creative brain. Its ultimate purpose is the complete mastery of mind over the material world, the harnessing of the forces of nature to human needs. This is the difficult task of the inventor who is often misunderstood and unrewarded. But he finds ample compensation in the pleasing exercises of his powers and in the knowledge of being one of that exceptionally privileged class without whom the race would have long ago perished in the bitter struggle against pitiless elements. Speaking for myself, I have already had more than my full measure of this exquisite enjoyment, so much that for many years my life was little short of continuous rapture.[2]

1

HERITAGE

Hardly is there a nation which has met with a sadder fate than the Servians. From the height of its splendor, when the empire embraced almost the entire northern part of the Balkan peninsula and a large portion of what is now Austria, the Servian nation was plunged into abject slavery, after the fateful battle of 1389 at the Kosovo Polje, against the overwhelming Asian hordes. Europe can never repay the great debt it owes to the Servians for checking, by the sacrifice of its own liberty, that barbarian influx.

NIKOLA TESLA[1]

It was during a crackling summer storm in Smiljan, a small hamlet at the back edge of a plateau set high in the mountains, when Nikola Tesla was born. The Serbian family resided in the province of Lika, a plateau and gentle river valley in Croatia where wild boar and deer still dwell and farmers still travel on ox-drawn wagons. Only a cart ride from the Adriatic, the land is well protected from invasion by sea, by the Velebit ridge to the west, which runs the length of the province and towers over the coastline as a steep cliff, and by the Dinaric Alps to the east, a chain of mountains that emerge from Austria, span the Balkan peninsula and culminate in the south as the isle of Crete.

Though hidden, Smiljan was centrally located, fifteen miles east of the tiny seaport of Karlobag, six miles west of the bustling town of Gospić and forty-five miles southwest of the cascading wonder known as Plitvice Lakes, an interlinking chasm of caves and streams and magnificent waterfalls that lie at the base of the Dinaric chain.[2]

In the early 1800s, having been briefly part of Napoleon's Illyrian provinces, Croatia was now a domain of Austria-Hungary. With its neighboring Slavic countries of Bosnia, Hercegovina, Montenegro, Serbia, and Slovenia, Croatia was sandwiched between the ruling Hapsburg dynasty to the north and the Ottoman Empire to the south.

In ancient times, and for many centuries, much of the coastline along the Adriatic was ruled by the Illyrians, a piratical tribe believed to have descended from regions around Austria. Successfully protecting their borders from such rulers as Alexander the Great, many Illyrians rose into social prominence; some, at the time of Christ, became emperors.

Slavs, traveling in close-knit clans known as *zadrugas*, were first recognized by the Byzantines in the second century A.D. in the areas around what is now Belgrade. Tesla's appearance resembled the characteristic features of the Ghegs, a tribe described as being tall and having convex-shaped noses and flat skulls. Like other Slavs, these people were originally pagans and worshiped nature spirits and a god of thunder and lightning. Tesla's early ancestors were probably born in the Ukraine. They most likely traveled down through Romania into Serbia and lived near Belgrade, along the Danube. After the Battle of Kosovo in the latter part of the fourteenth century, they crossed the Kosovo plains into Montenegro and continued their migration northward into Croatia in the latter part of the eighteenth century.[3]

All Slavs speak the same language. The major distinction between Croats and Serbs stems from the differences in the histories of their respective countries. The Croats adopted the pope as their spiritual leader and followed the Roman form of Catholicism; the Serbs adopted a Byzantine patriarch and the Greek Orthodox view. Whereas Roman priests remain celibate, Greek Orthodox priests may marry.

In the east and central regions Slavs were more successful in maintaining their own control over what came to be called the kingdom of Serbia; whereas in the west, in Croatia, outside rulers, such as Charlemagne in A.D. 800, occupied the region. While Croatia maintained the Christianization policies of the Franks, the Serbs and Bulgarians drove out the papacy and revived their own pagan faith, which included animal sacrifice and pantheism. Many of the ancient pagan gods were made saints and were celebrated in higher esteem than Jesus. Tesla's patron saint Nicholas was a fourth-century god who protected sailors.[4]

To further alienate the two groups, although speaking the same verbal language, Croats adopted the Latin alphabet, whereas Serbs and Bulgarians took on the Cyrillic alphabet used by the Greek Orthodox church.[5]

Before Turkish rule, from the ninth century until the 1300s, Serbia had maintained autonomy. For Serbia, this period was its golden age, as the Byzantines accepted its autonomous status. Due to the philanthropic nature of its kings, a dynamic medieval art flourished, and great monasteries were erected.[6]

Croatia, on the other hand, was in much more turmoil. Influenced by western Europe, the ruling class attempted unsuccessfully to institute a feudal system of lords and serfs. This policy directly opposed the inherent structure of the democratic *zadrugas*, and so Croatia was never able to establish a unified identity. Nevertheless, one independent offshoot of Croatia, Ragusa (Dubrovnik), which had established itself as a port of commerce and a rival of Venice as a major sea power, became a melting pot for south Slavic culture and a symbol for the Illyrian ideal of a unified Yugoslavia.

The identity of Serbia as a nation, however, changed for all time on June 15, 1389, the day 30,000 Turks obliterated the Serbian nation in the Battle of Kosovo. Cruel conquerors, the Turks destroyed Serbian churches or converted them to mosques. Drafting the healthiest male children into their armies, they skewered and tortured the men and forced the women to convert and marry Turks. Many Serbs fled, taking up residence in the craggy mountains of Montenegro or the hidden valleys of Croatia. Some of those that remained became wealthy as Turkish vassals; others, mostly of mixed blood, became pariahs.

The Battle of Kosovo is as important to the Serbs as the Exodus to the Jews or the Crucifixion to the Christians. It is commemorated every year on the anniversary of the tragedy as Vidov Dan, the day "when we shall see."[7] As one Serb told the author, "It follows us always."[8] The massacre and ensuing defilement of the kingdom became the dominant motif of the great epic poems which served to unify the identity of the Serbian people through their centuries of hardship.

Unlike the Croats, who did not have this kind of all-embracing exigency, the Serbs had Kosovo. Combined with their adherence to the Greek Orthodox religion in a twofold way, Serbs, no matter where they lived, felt united.

The century of Tesla's birth was marked by the rise of Napoleon. In 1809 the emperor wrested Croatia from Austro-Hungarian rule and established French occupation. Extending his domain down the Adriatic coast, Napoleon reunited the Illyrian provinces and introduced French libertarian ideals. This philosophy helped dismantle the outmoded feudal system of lords and serfs and reawaken the idea of a unified Balkan nation. At the same time, the occupation created an identity with the French culture. Tesla's paternal grandfather and maternal great-grandfather both served under the French emperor.[9]

With support from the Russians, Serbian bands united in 1804 under the leadership of the flamboyant hog farmer George Petrovich, known as Kara-George (in Turkish, Black George), a man of Montenegrin heritage trained in the Austrian army. However, in 1811, Napoleon invaded Russia; thus, support for Serbia evaporated.

Forty thousand Turks marched against the Serbs, leveling towns and butchering citizens. Serbs were often executed by impalement, their writhing bodies lined up along the roads to the city. All males captured above the age of fifteen were slaughtered, and women and children were sold as slaves. Kara-George fled the country.

Milosh, the new Serbian leader, was a sly and treacherous character, able to walk a thin line between Serbs and the sultan. In 1817, when Kara-George returned, he was decapitated, his head sent by Milosh to Istanbul. A tyrant as terrible as any Turkish pasha, Milosh became the official head of Serbia in 1830.

One of the more sapient figures of the day was the scholar and Serb Vuk Karadjich (1787–1864). Schooled in Vienna and St. Petersburg, Vuk believed "all Yugoslavs were one."[10]

Pleading with Milosh to build schools and to form a constitution, Vuk created, with a student, a Serbo-Croatian dictionary that combined the two written languages. He published the epic folk ballads, which gained the attention of Goethe, and through this means the Serbian plight and also its unique literature were translated and spread to the western world.[11]

In Croatia, the land of Tesla's birth, Emperor Ferdinand of Austria, in 1843, issued a proclamation forbidding any discussion about Illyrianism, thereby helping keep the Serbs and Croats a separate people. In 1867 the Austro-Hungarian Dual Monarchy was created, and Croatia became a semiautonomous province of the new empire. Simultaneously, in Serbia, Michael Obrenovich was finally able to "secure the departure…of the Turkish garrisons from Belgrade" and convert the state into a constitutional monarchy.

Tesla's background was thus a mixture of crossed influences, a monastic environment, a Byzantine legacy of a once great culture, and incessant battles against barbarous invaders. As a Serb growing up in Croatia, Tesla inherited a rich mix of tribal rituals, egalitarian rule, a modified form of Greek Orthodox Catholicism, pantheistic beliefs, and myriad superstitions. Women cloaked their bodies in black garb, and men packed a cross in one pocket and a weapon in another. Living at the edge of civilization, Serbs saw themselves as protectors of Europe from the Asian hordes. They bore that responsibility with their blood for many centuries.

2

CHILDHOOD (1856–74)

Now, I must tell you of a strange experience which bore fruit in my later life. We had a cold [snap] drier than ever observed before. People walking in the snow left a luminous trail. [As I stroked] Mačak's back, [it became] a sheet of light and my hand produced a shower of sparks. My father remarked, this is nothing but electricity, the same thing you see on the trees in a storm. My mother seemed alarmed. Stop playing with the cat, she said, he might start a fire. I was thinking abstractly. Is nature a cat? If so, who strokes its back? It can only be God, I concluded.

I cannot exaggerate the effect of this marvelous sight on my childish imagination. Day after day I asked myself what is electricity and found no answer. Eighty years have gone by since and I still ask the same question, unable to answer it.

NIKOLA TESLA[1]

Nikola Tesla descended from a well established frontier *zadruga* whose original family name had been Draganic.[2] By the mid-1700s the clan had migrated to Croatia, and the Tesla name arose. It was "a trade name like Smith...or Carpenter," which described a woodworking ax that had a "broad cutting blade at right angles to the handle."[3] Supposedly, the Teslas gained the name because their teeth resembled this instrument.

The inventor's grandfather, also named Nikola Tesla, was born about 1789 and became a sergeant in Napoleon's Illyrian army during the years 1809–13. Like other Serbs living in Croatia, Nikola Tesla, the elder, was honored by fighting for an emperor who sought to unify the Balkan states and overthrow the oppressive regime of the Austro-Hungarians. He "came from a region known as the military frontier which stretched from the Adriatic Sea to the plains of the Danube including...the province of Lika [where the inventor was born]. This so-called 'corpus separatum' in the Hapsburg monarchy had its own military administration different from the rest of the country, and therefore [they were] not subjects of the feudal lords."[4] Mostly Serbs, these people were warriors whose responsibility was to protect the territory from the Turks. And in return, unlike the Croats, Serbs were able to own their own land.

Shortly after Napoleon's defeat at Waterloo in 1815, Nikola Tesla married Ana Kalinic, the daughter of a prominent officer. After the collapse of Illyria, the grandfather moved to Gospić, where he and his wife could raise a family in a civilized environment.[5]

On February 3, 1819, Milutin Tesla, the inventor's father, was born. One of five children, Milutin was educated in a German elementary school, the only one available in Gospić. Like his brother Josip, Milutin tried to follow in his father's footsteps. In his late teens he enrolled in an Austro-Hungarian military academy but rebelled against the trivialities of regimented life. He was hypersensitive and dropped out after an officer criticized him for not keeping his brass buttons polished.[6]

Whereas Josip became an officer and later a professor of mathematics, first in Gospić and then at a military academy in Austria,[7] Milutin became politically active, wrote poetry, and entered the priesthood. Influenced by the philosopher Vuk Karadjich, Milutin promulgated the "Yugoslav idea" in editorials published in the local newspapers under the nom de plume Srbin Pravicich, "Man of Justice." Tesla wrote that his father's "style of writing was much admired...pen[ning] sentences...full of wit and satire." He called for social equality among peoples, the need for compulsory education for children, and the creation of Serbian schools in Croatia.[8]

Through these articles, Milutin attracted the attention of the intellectual elite. In 1847 he married Djouka Mandić, a daughter from one of the more prominent Serbian families.[9]

Djouka's maternal grandfather was Toma Budisavljevic (1777–1840), a regal, white-bearded priest who was decorated with the French Medal of Honor by Napoleon himself in 1811 for providing leadership during the French occupation of Croatia. Soka Budisavljevic, one of Toma's seven children, followed the family tradition by marrying a Serbian minister, Nikola Mandić, who himself came from a distinguished clerical and military family. Their daughter, Djouka, who was born in 1821, was Tesla's mother.[10]

Eldest daughter of eight children, Djouka's duties increased rapidly, for her mother was stricken with failing eyesight and eventually became blind.

"My mother...was a truly great woman of rare skill and courage," Tesla wrote. Probably due to the magnitude of her responsibilities, which included, at age sixteen, preparing for burial the bodies of an entire family stricken with cholera, Djouka never learned to read. Instead, she memorized the great epic Serbian poems and also long passages of the Bible.[11]

Tesla could trace his lineage to a segment of the "educated aristocracy" of the Serbian community. On both sides of the family and for generations there could be found clerical and military leaders, many of whom achieved multiple doctorates. One of Djouka's brothers, Pajo Mandić, was a field marshal in the imperial Austro-Hungarian army. Another Mandić ran an Austrian military academy.[12]

Petar Mandić, a third brother and later favorite uncle of Nikola's, met with tragedy as a young man when his wife passed away. In 1850, Petar entered the Gomirje Monastery, where he rose in the clerical hierarchy to become the regional bishop of Bosnia.[13]

In 1848, through the help of the Mandić name, Milutin Tesla obtained a parish at Senj,[14] a northern coastal fortress located just seventy-five miles from the Italian port of Trieste. From the stone church, situated high on austere cliffs, Milutin and his new bride could overlook the blue-green Adriatic Sea and the mountainous islands of Krk, Cres, and Rab.

For eight years the Teslas lived in Senj, where they sired their first three children: Dane (pronounced Dah-nay), born in 1849, the first son, and two daughters, Angelina, born the following year, who would later become the grandmother to the current honorary head of the Tesla Memorial Society, William Terbo, and Milka, who followed two years later. As with her other two sisters and like her mother, Milka would eventually marry a Serbian Orthodox priest.

Djouka was proud of her son, Dane, who used to sit with the fishermen on the shore and bring back stories of great adventure. Like his younger brother, who was yet to be born, Dane was endowed with extraordinary powers of eidetic imagery.[15]

Due to a profound sermon on the subject of "labor," as a result of which Milutin was awarded a special red sash by the archbishop, the minister was promoted to a congregation of forty homes in the pastoral farming village of Smiljan,[16] situated only six miles from Gospić. Milutin was returning home, where his father still lived. In 1855, the young minister, his pregnant wife, and his three children packed their oxcart and made the fifty-mile journey over the Velebit ridge through the Lika valley to their new dwelling.

In 1856, Nikola Tesla was born. He was followed three years later by Marica, mother-to-be of Sava Kasonovic, the first Yugoslavian ambassador to the United States, and the man most responsible for creating the Tesla Museum in Belgrade.

Smiljan was an ideal setting for the young boys to grow up in. Nikola, raised in large measure by his two older sisters, appears to have led an

Arcadian childhood, annoying the servants, playing with the local birds and animals on the farm, and learning inventions from his older brother and mother.

Down to the local creek the boys would go to swim or catch frogs in the spring or summer and to build dams in the autumn and early winter in vain attempts to stop the seasonal flooding of the land.[17] One of their favorite recreations was a smooth waterwheel, a device which contained inherent concepts that would later form the basis of Tesla's innovative bladeless steam turbines.

Other inventions included a cornstalk popgun, which contained principles that Tesla later adapted when he fashioned particle-beam weapons, a special fishing hook for catching frogs, snares for capturing birds, and a parasol used in an unsuccessful attempt to glide off the roof of the barn. Young Niko must have taken quite a leap, because his fall laid him up for six weeks.[18]

Perhaps the boy's most ingenious creation was a propeller driven by sixteen May bugs glued or sewn four abreast onto the wooden blades. "These creatures were remarkably efficient, for once they were started they had no sense to stop and continued whirling for hours and hours.... All went well until a strange boy came to the place. He was the son of a retired officer in the Austrian Army. That urchin ate May-bugs alive and enjoyed them as tho [*sic*] they were the finest blue-point oysters. That disgusting sight terminated my endeavors in this promising field and I have never since been able to touch a May-bug or any other insect for that matter."[19]

Some of the inventor's earliest memories, when he was three, were recalled when he was an octogenarian. Many years before pigeons, Tesla showered his affections on the family cat, Mačak, "the fountain of my enjoyment.... I wish that I could give you an adequate idea of the depth of affection which existed between me and him."

After dinner, Niko and his cat would rush out of the house and frolic by the church. Mačak would "grab me by the trousers. He tried hard to make me believe he would bite, but the instant his needle sharp incisors penetrated the clothing, the pressure ceased and their contact with my skin was as gentle and tender as that of a butterfly alighting on a petal."

Tesla liked best to wallow in the grass with Mačak. "We [would] just roll... and roll... in a delirium of delight."

In this peaceful setting, young Niko was introduced to the barnyard animals. "I would... take one or the other under my arm and hug and pet it," he wrote, "[especially] the grand resplendent cock who liked me."[20] It was also at this time that the boy began to study flight, a topic of interest that caused him in later life to invent a variety of novel flying machines. His relationship with birds, however, was filled with contradictions.

> My childhood...would have passed blissfully if I did not have a powerful enemy,...our gander, a monstrous ugly brute, with a neck of an ostrich, mouth of crocodile and a pair of cunning eyes radiating intelligence and understanding like the human....
>
> One summer day my mother had given me a bath and put me out for a sun warming in Adam's attire. When she stepped in the house, the gander espied me and charged. The brute knew where it would hurt most and seized me by the nape almost pulling out the remnant of my umbilical cord. My mother, who came in time to prevent further injury, said to me: "You must know that you cannot make peace with a gander or a cock whom you have taunted. They will fight you as long as they live."[21]

Tesla had run-ins with other animals as well, such as a local wolf, who fortunately turned and ran from him; the family cow, which Tesla rode and one time fell off of; and the giant ravens, whom he claimed to have snared with his bare hands by hiding under the bushes and leaping out at them, as a cat would.

Tesla also liked to tell the story of his two homely aunts, who often visited his home. One, an Aunt Veva, "had two protruding teeth like the tusks of an elephant. She loved me passionately and buried them deep in my cheek in kissing me. I cried out from pain but she thought it was from pleasure and dug them in still deeper. Nevertheless," Tesla recalled, "I preferred her to the other...[as] she used to glue her lips to mine and suck and suck until by frantic efforts I managed to free myself gasping for breath."

One day when they came over and Tesla was still small enough to be held in the arms of his mother, "they asked me who was the prettier of the two. After examining their faces intently, I answered thoughtfully, pointing to one of them, 'This here is not as ugly as the other.'"[22]

Tesla inherited his sense of humor from his father, who, for instance, once cautioned a cross-eyed employee who was chopping wood near the minister and his son, "For God's sake...do not strike at what you are looking at but what you expect to hit."

Father Tesla was known to talk to himself and even carry on conversations with different tones in his voice, a trait also noted in the inventor, especially in his later years.[23] Milutin also trained his sons in exercises in developing memory and their intuitive faculties. Able to recite at length works in several languages, he often remarked playfully that if some of the classics were lost he could restore them. "My father...spoke fluently a great many languages and also ranked high as a mathematician. He was an omnivorous reader and possessed a large library from which it was my privilege to gather a great deal of information during my years of life spent at home."[24]

Texts from this library included works in German by Goethe and Schiller, encyclopedic works in French by D'Alembert, and other classics, probably in English, from the eighteenth and nineteenth centuries.[25]

The inventor's colorful autobiography remains the primary source of information on his childhood. Although Dane and his parents appear prominently in the work, Tesla's sisters are barely mentioned. Certainly he loved the girls—he exchanged letters with them regularly for the duration of his life—but they seemed not to have overtly influenced his upbringing. It was his mother whose untiring work habits and proclivity toward invention influenced the inventor-to-be.

Whereas Milutin ran the parish and published his articles, Djouka directed the servants and ran the farm. She had the responsibility of growing the crops, sewing all the clothes, and designing needlework, a practice that made her famous in the region. Tesla also ascribes his proclivity to invention to his mother; she conceived of many household appliances, including churns, looms, and "all kinds of [kitchen] tools. [My mother] descended from a long line of inventors." Starting before dawn, she did not quit work until eleven o'clock at night.[26]

In 1863 disaster struck the Tesla household.[27] Dane, who "was gifted to an extraordinary degree," was out riding the family horse, which was "of Arabian breed, possest [*sic*] of almost human intelligence." The horse may have been the type easily spooked. The previous winter, it had thrown Milutin in the middle of the forest after an encounter with wolves and had run home, leaving him unconscious. The horse, however, was smart enough to bring the search party back to the scene of the accident, and thus the father was saved. This animal, which "was cared for and petted by the whole family," threw the brother as well, but Dane died of the injuries. The family never really recovered. "[Dane's] premature death left my parents disconsolate....The recollection of his attainments made every effort of mine seem dull in comparison. Anything I did that was creditable merely caused my parents to feel their loss more keenly. So I grew up with little confidence in myself."[28]

Upset by his brother's death and the rejection by his parents, particularly his mother, the seven-year-old ran away from home and hid in an "inaccessible" mountain chapel which "was visited only once a year." By the time he had reached the place, it was nightfall. There was little the boy could do but force his way in and spend the night "entombed....It was an awful experience."[29]

Shortly after this tragedy, Milutin was promoted and given the parish at the ornate "onion bulb church" in the town of Gospić.[30] The family moved the few miles to Gospić, where father Tesla also took a post teaching religion at the local gymnasium (high school).[31] Niko was of school age, so he began his formal education at this time. However, he had great difficulty adjusting to city life, for he missed the farm and the idyllic existence he had once enjoyed. "This change of residence was like a calamity to me. It almost broke my heart to part from our pigeons, chickens and sheep, and our magnificent flock of geese which used to rise to the clouds in the morning and return from the feeding grounds at sundown in battle formation, so perfect that it would have put a squadron of the best aviators of the present day to shame."[32]

The boy would wake up in the middle of the night with nightmares of Dane's death, which he claimed to have witnessed, and of the funeral, which probably involved an open casket. "A vivid picture of the scene would thrust itself before my eyes and persist despite all efforts to banish it.... To free myself of these tormenting appearances, I tried to concentrate my mind on something else...[by] continuously [conjuring up] new images.... I was opprest [*sic*] by thoughts of pain in life and death and religious fear...swayed by superstitious beliefs and lived in constant dread of the spirit of evil, of ghosts and ogres and other unholy monsters of the dark."[33]

It was at this time that Tesla began to have what today are known as out-of-body experiences, although he never ascribed anything mystical or paranormal to them. "Blurred [at first]...I would [see]...on my journeys...new places, cities and countries—live there, meet people and make friendships...and, however unbelievable, it is a fact that they were just as dear to me as those in actual life and not a bit less intense in their manifestations."[34]

Tesla stated he had such great powers of eidetic imagery that he sometimes needed one of his sisters to help him tell which was hallucination and which was not. Like Dane, his thoughts were often interrupted with annoying flashes of light. These psychoneurological disturbances continued throughout his life. On the positive side, the problem was also attributed to his inventive bent. He could use his powers of visualization to mold his various creations, and even run and modify them in his mind, before committing them to paper and the material world.

While still in grade school, Niko had obtained a post at the local library in Gospić classifying the various books. But he was forbidden to read at night for fear that his eyes would be strained in dim light, and Milutin "would fly into a rage" to stop him.[35] Undeterred, Niko would swipe some household candles, seal up the cracks in his room, and continue reading through the night. The book that Tesla claimed changed his life was *Abafi*, a story, translated into Serbian, about the son of Aba, by the Hungarian author Josika. "Up to the age of eight, my character was weak and vacillating.... This work somehow awakened my dormant powers of will and I began to practice self-control. At first my resolutions faded like snow in April, but in a little while I conquered my weakness and felt a pleasure I never knew before—that of doing as I willed."[36]

Thus, by the age of twelve he was successfully experimenting with acts of self-denial and self-mastery, a paradoxical pattern which played itself out repeatedly throughout his life. Simultaneously, Tesla began to develop peculiarities, probably stemming from the stress associated with his brother's death, strained relationship with his parents, and denial of his sexual desires.[37] At this time he became ill and claimed that a heavy dose of Mark Twain's writings turned his spirit and cured him. "Twenty-five years later, when I met Mr. Clemens and we formed a friendship...I told him of the experience and was amazed to see that a great man of laughter burst into tears.[38] During this period I contracted many strange...habits....I had a violent aversion against the earrings of women but other ornaments, as bracelets, pleased me....The sight of a pearl would almost give me a fit but I was fascinated with the glitter of...objects with sharp edges....I would not touch the hair of other people except, perhaps, at the point of a revolver. I would get a fever by looking at a peach....Even now [at age sixty-one] I am not insensible [*sic*] to some of these upsetting impulses."[39]

The youngster, however, also undertook normal boyhood adventures, including a few near-death experiences: on one occasion "by falling headlong into a huge kettle of boiling milk, just drawn from the paternal herds";[40] on another occasion nearly drowning after swimming under a raft; and on a third occasion by nearly being swept over a waterfall at one of the nearby dams. These were rather unpleasant experiences, but not as bad, according to the inventor, as the following: "There was a wealthy lady in town who used to come to the church gorgeously painted up and attired with an enormous train and attendants. One Sunday I had just finished ringing the bell in the belfry and rushed downstairs when this grand dame was sweeping out and I jumped on her train [and] tore it. Livid, my father gave me a gentle slap on the cheek, the only corporal punishment he ever administered to me but I almost feel it now."[41]

Niko became ostracized and avoided social interaction. Fortunately, he was able to redeem himself through his inventive mind. One day, the local firemen brought out their new engine and started a fire to demonstrate it. To the embarrassment of the officials, the hose, which drew its water from the local river, would not work. Intuitively, Tesla realized that there was a kink in the rigging. Tearing off his Sunday best, he dived into the water, unscrambled the line, and became the hero of the day. This event became a strong inducement for the boy to continue his interest in invention. Simultaneously, this act symbolized a new way to obtain love and admiration not only from his parents but also from society.

Between the ages ten and fourteen, Niko attended the Real Gymnasium, equivalent to junior high school. (It appears that both his father and uncle taught there.) It was a fairly new institution, with a well-equipped physics department. "I was interested in electricity almost from the beginning of my educational career," he wrote. "I read all that I could find on the subject...[and] experimented with batteries and induction coils."[42]

Tesla also began experimenting with water turbines and motors that utilized power derived from differentials in air pressure. His goal, though unattainable due to a flaw in his logic, was a perpetual-motion machine that would work by maintaining a steady-state vacuum and harnessing, like a windmill, the rush of incoming air. This movement, he hoped, would turn a generator endlessly.

After seeing a drawing or photograph of Niagara Falls, Tesla announced to his Uncle Josip that one day he would place a gigantic wheel under the falls and thereby harness it. Most likely, he also visited the magnificent network of waterfalls at Plitvice Lakes for additional inspiration, as they were only a day's journey away.

In 1870, at the age of fourteen, Niko moved from Gospić to Karlovac (Carlstadt), where he saw a locomotive for the first time, to attend the Higher Real Gymnasium, located by a swamp on a tributary of the Sava River near Zagreb. The youth lived with his aunt Stanka, his father's sister, and her husband, Colonel Brankovic, "an old war-horse."[43]

During his stay "at Karlovac, he frequently visited his cousin, Milica Zoric, at the family estate in Tomingaj....[Niko] who would often go there for vacations...[found it to be] a sort of sanctuary...."[44]

At Karlovac he was trained in languages and mathematics. His most influential professor was Martin Sekulic, a physics teacher, who "demonstrated the principles by apparatus of his own invention. [It was]...a freely rotatable bulb, with tinfoil coatings, which was made to spin rapidly when connected to a static machine. It is impossible for me to convey an adequate idea of the intensity of feeling I experienced in witnessing his exhibitions of these mysterious phenomena. Every impression produced a thousand echoes in my mind."[45]

Through hard work, Tesla condensed the four years of schooling into three and began to plan a way to approach his father with his controversial decision not to enter the ministry but to study engineering instead. "It is not humans that I love, but humanity" he tried to tell his father.

During his last year at Karlovac, after a day of exploration by a nearby marsh, he caught a fever which he said was malaria. The seriousness of his condition may have been exacerbated by an inadequate diet. "I was fed [by my aunt] like a canary bird....When the Colonel would put something substantial on my plate she would snatch it away and say excitedly to him: 'Be careful, Niko is very delicate.' I had a voracious appetite and suffered like Tantalus. But I lived in an atmosphere of refinement and artistic taste quite unusual for those times and conditions."[46]

Upon graduation, Tesla received notice from his father that he should go on a hunting expedition, as there was an epidemic in town. The youth returned to Gospić, anyway. The streets were stacked with corpses, the atmosphere thick with smoke, for the people mistakenly thought that cholera was being transmitted through the air rather than by drinking water. Partly due to his weakened condition from his earlier illness, Tesla quickly became a victim. Bedridden for nine months, he nearly died. "In one of the sinking spells which was thought to be the last, my father rushed into the room.... 'Perhaps,' I said, 'I may get well if you will let me study engineering.' 'You will go to the best technical institution in the world,' he solemnly replied, and I knew that he meant it."[47]

The Teslas settled on the Polytechnic School at Graz, Austria, located 175 miles to the north. First, however, the boy would have to serve three years in the army. With a major war breaking out against the Turks, Milutin directed his son to pack his gear and go into the hills to avoid the draft. There the youth could maintain a low profile and at the same time recover his health. "For most of this term I roamed in the mountains, loaded with a hunter's outfit and a bundle of books, and this contact with nature made me stronger in body as well as in mind...but [my] knowledge of principles was very limited."[48]

Misguided inventions of this period included a "submarine tube...[able] to convey letters and packages across the seas...[and] a ring around the equator" for transporting people from one end of the globe to the other."[49] One day, while playing with snowballs on the side of a mountain, however, Tesla discovered the concept of hidden trigger mechanisms able to unleash great reservoirs of energy: "One...found just the right conditions; it rolled until it was a large ball and then spread out rolling up the snow at the sides as if it were a giant carpet, and then suddenly it turned into an avalanche...stripping the mountainside clear of snow, trees, soil and everything else it could carry with it."[50]

But contact with war was unavoidable, and on occasion the youth chanced upon its ravages. Twenty-five years later, he would recall, "I have seen men hung, beaten to death, shot, quartered, stuck on a pointed stick, heads chopped off and children on a bayonette like quails 'en broche' at Delmonico's."[51] Fortunately, Tesla avoided capture, and in 1875 he returned to Gospić. With a new fellowship from the Military Frontier Authority, he began school in Austria the following semester.

3

College Years (1875–82)

It has cost me years of thought to arrive at certain results, by many believed to be unattainable, for which there are now numerous claimants, and the number of these is rapidly increasing, like that of the colonels in the South after the war.

Nikola Tesla[1]

Eighty miles south of Vienna, in the capital of the province of Styria, was the Polytechnic School in Graz. Milutin had chosen the school because it was one of the most advanced of the region. The physicist and philosopher Ernst Mach had taught there a few years earlier, as had the psychophysiologist Gustav Theodor Fechner. Planning on becoming a professor, Tesla undertook courses in arithmetic and geometry from Professor Rogner, a lecturer known for his histrionics; theoretical and experimental physics with the punctilious German professor Poeschl; and integral calculus with Professor Allé. Allé "was the most brilliant lecturer to whom I ever listened. He took a special interest in my progress and would frequently remain for an hour or two in the lecture room, giving me problems to solve, in which I delighted."[2] Other courses taken included analytical chemistry, mineralogy, machinery construction, botany, wave theory, optics, French, and English.[3] To save money, he roomed with Kosta Kulishich, whom he had met at the Student Society of Serbia. Kulishich later became a professor of philosophy in Belgrade.[4]

Tesla plunged into his work with great intensity. Studying upward of twenty hours a day, he changed his major to engineering and extended his curriculum to study other languages—he could speak about nine of them—and the works of such writers as Descartes, Goethe, Spencer, and Shakespeare, many of which he knew entirely by heart. "I had a veritable mania for finishing whatever I began," he recalled, reflecting on his next self-appointed assignment. The collected works of Voltaire comprised "one hundred large volumes in small print which that monster had written while drinking seventy-two cups of black coffee per diem."[5] This task cured him of the compulsion but did not serve to quell the pattern of relentless self-denial and self-determination. Because he was praised by his teachers, the other students became jealous, but at first Tesla remained unperturbed.

Returning home the following summer, having passed his freshman year with all A+'s,[6] the young scholar expected to be praised by his parents. Instead, his father tried to persuade his son to stay in Gospić. Unbeknown to Tesla, his teachers had written Milutin warning that the boy was at risk of injuring his health by obsessively long and intense hours of study. A rift was created between father and son, perhaps in part because the Military Frontier Authority had been abolished and the scholarship was no longer available.

Reacting to the ridicule from other students, who resented Tesla for his monastic study habits and close association with the faculty, Tesla took up gambling. "He began to stay late at the Botanical Garden, the students' favorite coffee house, playing cards, billiards and chess, attracting a large crowd to watch his skillful performances."[7] Tesla's father "led an exemplary life and could not excuse the senseless waste of time and money. . . ." "I can stop whenever I please," he told his father, "but is it worth while to give up that which I would purchase with the joys of Paradise?"[8]

During his sophomore year, a direct-current Gramme dynamo was delivered from Paris to Professor Poeschl's physics class. It was equipped with the customary commutator, a device that transferred the current from the generator to the motor. *Electricity in its natural state is alternating.* This means that its direction of flow changes rapidly. *An analogous situation would be a river that flowed downstream, then upstream, then downstream, and so on many times per second.*[9] One can see the difficulty in harnessing such a river with, for instance, a waterwheel, for the wheel would constantly change its direction as well. The commutator is comprised of a series of wire brushes that serve to transfer the electricity into only one direction of flow, that is, a direct current (DC). It is a cumbersome device and sparks considerably.

When Professor Poeschl displayed this up-to-date equipment, Tesla *intuitively deduced that the commutator was unnecessary* and that alternating current (AC) could be harnessed unencumbered. He voiced this opinion, which appeared utterly fantastic at the time. Poeschl devoted the rest of the lecture to a detailed explanation of how this goal was impossible. Driving the point home, Poeschl embarrassed his student by disconnecting the "superfluous" commutator and noting with feigned surprise that the generator no longer worked.[10] "Mr. Tesla may do many things, but this he can not accomplish. His plan is simply a perpetual motion scheme."[11] Tesla would spend the next four years obsessed with proving the professor wrong.

Another invention Tesla worked on at the same time, but under the tutelage of Professor Allé, was that of a mechanical flying machine. As a child, Tesla had heard stories from his grandfather about Napoleon's employment of hot-air balloons, which were used to observe enemy troop movements and for dropping bombs. No doubt he had also studied the principles involved in school and quite possibly saw such futuristic creations floating in the Austrian skies when he went off to college.

By his third year Tesla was running into difficulties at school. Having surpassed his classmates in his studies, he became bored and frustrated by his inability to find a solution to his AC problem. He began to gamble more heavily, sometimes twenty-four hours at a stretch. Although Tesla tended to return his winnings to heavy losers, reciprocation did not occur, and one semester he lost his entire allowance, including the money for tuition. His father was fuming, but his mother came to him with "a roll of bills" and said, "Go and enjoy yourself. The sooner you lose all we possess the better it will be. I know that you will get over it."[12]

The audacious youth won back his initial losses and returned the balance to his family. "I conquered my passion then and there," he wrote, and "tore it from my heart so as not to leave a trace of desire. Ever since that time I have been as indifferent to any form of gambling as to picking teeth."[13] This statement appears to be an exaggeration, as Tesla gambled quite freely with his future and was known to play billiards when he came to the United States. An Edison employee recalled: "He played a beautiful game. [Tesla] was not a high scorer, but his cushion shots displayed skill equal to that of a professional exponent of this art."[14] It has also been suggested that years later, in the early 1890s, Tesla bilked some of the wealthy socialites in New York by feigning minimal ability in the sport.[15]

Exam time came, and Tesla was unprepared. He asked for an extension to study but was denied. He never graduated from the Austrian Polytechnic School and did not receive any grades for his last semester there. Most likely, he was discharged, in part for gambling and, supposedly, "womanizing."[16] According to his roommate, Tesla's "cousins, who had been sending him money, therefore withdrew their aid." Fearing that his parents would find out, Tesla disappeared without word. "Friends searched everywhere for him and decided that he had drowned in the river."

Clandestinely packing his gear, Tesla traveled south, over the border into Slovenia, where he arrived in Maribor in late spring of 1878 to look for work. He played cards with the local man on the streets, as is still the custom today, and soon gained employment with an engineer "earning 60 florins a month,"[17] but the job was short-lived. Tesla continued traveling, making his way through Zagreb, to the small coastal village of Min-Gag. He would not return home, for he did not want to confront his parents. At the same time, however, Tesla also continued his quest for a solution to the problem of removing the commutator from the DC generator.

His cousin, Dr. Nikola Pribic, recalled a story he had heard as a boy growing up in Yugoslavia in the 1920s: "My mother told us...he would always like to be alone [when Tesla visited us]. In the morning he would go off into the woods and meditate. He would measure one tree to another making notes, experimenting [stringing wires between them and transmitting current]. Peasants passing by would be astonished at such an erratic person....They would approach and say, 'We're sorry; your [cousin] seems to be crazy.'"[18]

Having finally located his son, after word from Kulishich, who had seen Tesla in Maribor, Milutin traveled north to discuss his academic problems. Tesla refused to return to Graz, so Milutin offered a solution: His son would make a fresh start at another university. They returned to Gospić.

Reaccepted into the family, Tesla began once again to attend church to hear his father's sermons. There he met Anna. She was "tall and beautiful [with] extraordinary understandable eyes." For the first and only time in his life, Tesla would say, "I fell in love." Delighting in her company, Nikola would take Anna for strolls by the river or back to Smiljan, where they would talk about the future. He wanted to become an electrical engineer; she wanted to raise a family.[19]

The following year, Milutin passed away, and a few months later, in 1880, Tesla left for Bohemia (now in the Czech Republic) to "carry...out my father's wish and continue my education." He promised to write to Anna, but their romance was doomed, and she would marry shortly thereafter.

Tesla enrolled in the Charles-Ferdinand branch of the University of Prague, one of the foremost institutions in Europe, for the summer term.

According to Ernst Mach, who, a decade earlier, had transferred from Graz to be appointed Rector Magnificus, Prague was a city "rich in talented people," with street signs often appearing in a half-dozen languages. Although the city was filled with majestic buildings, sanitary conditions were severely lacking. To avoid typhoid fever one had to boil water or obtain mineral water from springs to the north.[20]

Just two years after Tesla's stay, Harvard psychologist William James would come to visit, to meet with Mach and Mach's archrival, Carl Stumpf, "Ordinary Professor of Philosophy." Stumpf was a student of the controversial ex-priest Franz Brentano (who also influenced another pupil, Sigmund Freud) and was also Tesla's philosophy teacher. Other courses Tesla undertook included analytical geometry with Heinreich Durege, experimental physics with Karel Domalip, both "Ordinary Professors," and higher mathematics with Anton Puchta, who was an "Extraordinary

Professor" from the German Technical University also in Prague.[21]

With Stumpf, Tesla studied Scottish philosopher David Hume. Raised as a child prodigy of music, the acerbic and "sharp-nosed" Stumpf[22] opposed a number of key psychophysicists, including the famed Wilhelm Wundt as well as Mach, but at the same time he also helped shape the thinking of a number of key students, such as phenomenologist Edmund Husserl and Gestalt psychologist Wolfgang Kohler.[23]

A persuasive advocate of Hume's "radical skepticism," Stumpf argued for the concept of the "tabula rasa." Basing his thinking on Aristotle and John Locke, both of whom repudiated the concept of innate ideas, Stumpf stated that the human mind was born a blank slate, a "tabula rasa"; impinging on it, after birth, were all of the "primary quality of things," that is, true knowledge about the world. Through the sense organs, Tesla learned, the brain mechanically recorded incoming data. The mind, according to Hume, was nothing more than a simple compilation of cause-and-effect sensations. What we call ideas were secondary impressions derived from these primary sensations. The will and "even the soul w[ere] reduced by Hume to impressions and associations of impressions."[24] At this time, Tesla also studied the theories of Descartes, who envisioned animals, including man, as simply "automata incapable of actions other than those characteristic of a machine."[25]

This line of thinking would dominate Tesla's worldview and would ironically serve as the template for a mechanistic paradigm that would lead the inventor to discover his most original creations, even though the whole idea of original discovery appears to be antithetical to this extrinsically motivated Aristotelian premise. According to Tesla and to this view, all of his discoveries were derived from the outside world.

Although Tesla does not overtly refer to Stumpf's perceived adversary, in retrospect, it appears obvious that Stumpf's opposition did not stop Tesla from studying Mach's experiments in wave mechanics. Born in Moravia (now the Czech Republic) in 1838, Mach graduated from the University of Vienna in 1860. By 1864 he was a full professor at Graz, and by 1867 he was head of the department of experimental physics at Prague, with four books and sixty-two articles to his credit. Influenced by the research in psychophysics of Fechner in Graz and Ludwig von Helmholtz in Berlin, Mach studied the workings of the human eye, along with his Prague colleague "famed physiologist and philosopher" Jan Purkyne. Both the eye and ear collected information from the outside world, analyzed it, and transferred it, via electrical impulses in the nerves, to the respective processing centers in the brain. This traditional line of research had been taken by many other well-known scientists, including Isaac Newton, Johann von Goethe, and Herbert Spencer, all favorites of Tesla's.

In his laboratory, Mach had constructed a "famous instrument known as a wave machine. This device could make progressive [and standing] longitudinal [and] transverse waves...." Mach could display a number of mechanical effects with these acoustic waves and "demonstrate the analogy between acoustic and electromagnetic events." By this means, the "mechanical theory of the ether" could also be demonstrated.[26]

By studying acoustical-wave motion in association with mechanical, electrical, and optical phenomena, Mach discovered that when the speed of sound was achieved, the nature of an air flow over an object changed dramatically. This threshold value became known as Mach 1.

Mach also wrote on the structure of the ether and hypothesized that it was inherently linked to a gravitational attraction between all masses in the universe. Influenced overtly by Buddhist writings, which no doubt filtered down to esoteric discussions by the university students, Mach could hypothesize that no event in the universe was separate from any other. "The inertia of a system is reduced to a functional relationship between the system and the rest of the universe."[27] This viewpoint was extended to the relationship of mental events to exterior influences. Like Stumpf, he agreed that every mental event had to have a corresponding physical action.[28]

Since Mach's writings so closely parallel Tesla's later research and philosophical outlook, Mach seems a curious omission from Tesla's published writings.

By the time Tesla left the university at the end of the term, he had made great strides, both theoretical and practical, in solving the AC dilemma. "It was in that city," Tesla said, "that I made a decided advance, which consisted in detaching the commutator from the machine and studying the phenomena in this new aspect."[29]

With the death of his father, he needed to earn his own living. He began an apprenticeship in teaching but did not enjoy it. Uncle Pajo suggested that he move to Hungary, where employment could be obtained through a military friend, Ferenc Puskas, who ran the new "American" telephone exchange with his brother Tivadar.[30] In January 1881, Tesla moved to Budapest, but he found to his dismay that the operation had not yet been launched.

The Puskas brothers were very busy men, running operations in St. Petersburg and overseeing, in Paris, Thomas Edison's incandescent lamp exhibit at the Paris Exposition and fixing the lighting system at the opera house there.[31] Out of funds and without a job, Tesla approached the Engineering Department at the Central Telegraph Office of the Hungarian government and talked his way into a position as a draftsman and designer. Working for a subsistence salary, he utilized what little surplus funds he had to purchase equipment to further his experiments.

Anthony Szigeti, a former classmate and engineer from Hungary, "with the body of Apollo...[and] a big head with a lump on the side...[that] gave him a startling appearance,"[32] became Tesla's friend and confidant. Many a night, when the budding inventor was not enmeshed in his research, the two fellows would meet at the local cafes, where they would discuss the events of the day or compete in such friendly games as determining who could drink the most milk. On one such occasion, Tesla claimed he was beaten after the thirty-eighth bottle![33]

Due to his meager funds and general inability to budget himself, Tesla had but one suit, which had withered from use. It was the time of a religious festival, and Szigeti inquired what Tesla would be wearing. Stuck for an answer, the youthful inventor came upon the clever idea of turning his suit inside out, planning thereby to show up with a seemingly new set of clothes. All night was spent tailoring and ironing. But when one starts with a wrong premise, no amount of patching can right the problem. The outfit looked ridiculous, and Tesla stayed home instead.[34]

In a few months, the American telephone exchange opened there in Budapest and Tesla and Szigeti immediately gained employment. The new enterprise allowed the young engineers to finally learn firsthand how the most modern inventions of the day operated. It was also the first time that Tesla was introduced to the work of Thomas Edison, the "Napoleon of Invention," whose improvements on Bell's telephone helped revolutionize the field of communications. Up the poles Tesla would climb to check the lines and repair equipment. On the ground he worked as a mechanic and mathematician. There he studied the principle of induction, whereby a mass with an electric or electromagnetic charge can provide a corresponding charge or force or magnetism in a second mass without contact. He also studied a number of Edison's inventions, such as his multiplex telegraph, which allowed four Morse-coded messages to be sent in two directions simultaneously, and his new induction-triggered carbon disk speaker, the flat, circular, easily removable device which is still found in the mouthpiece of every telephone today. As was his nature, Tesla took apart the various instruments and conceived ways of improving them. By giving the carbon disk a conical shape, he fashioned an amplifier, which repeated and boosted transmission signals. Tesla had invented a precursor of the loudspeaker. He never bothered to obtain a patent on it.

Except for friendly diversion with Szigeti, Tesla's every spare moment was spent reworking the problem of eliminating the commutator in DC machines and harnessing AC without cumbersome intermediaries. Although a solution seemed imminent, the answer would not be revealed. Hundreds of hours were spent building and rebuilding equipment and discussing his ideas with his friend.[35]

He pored over his calculations and reviewed the work of others. Tesla later wrote: "With me it was a sacred vow, a question of life or death. I knew that I would perish if I failed."[36] Monomaniacal in pursuit of his goal, he gave up sleep, or rest of any kind, straining every fiber to prove once and for all that he was right and Professor Poeschl and the rest of the world were wrong. His body and brain finally gave out, and he suffered a severe nervous collapse, experiencing an illness that "surpasses all belief." Claiming that his pulse raced to 250 beats per minute, his body twitched and quivered incessantly.[37] "I could hear the ticking of a watch...three rooms [away]. A fly alighting on a table...would cause a dull thud in my ear. A carriage passing at a distance...fairly shook my whole body....I had to support my bed on rubber cushions to get any rest at all....The sun's rays, when periodically intercepted, would cause blows of such force on my brain that they would stun me....In the dark I had the sense of a bat and could detect the presence of an object...by a peculiar creepy sensation on the forehead." A respected doctor "pronounced [his] malady unique and incurable." Desperately clinging to life, Tesla was not expected to recover.[38]

Tesla attributes his revival to "a powerful desire to live and to continue the work" and to the assistance of the athletic Szigeti, who forced him outdoors and got him to undertake healthful exercises. Mystics attributed the event to the triggering of his pineal gland and corresponding access to higher mystical states of consciousness.[39] During a walk in the park with Szigeti at sunset, the solution to the problem suddenly became manifest as he was reciting a "glorious passage" from Goethe's *Faust*.

See how the setting sun, with ruddy glow,
The green-embosomed hamlet fires.
He sinks and fades, the day is lived and gone.
He hastens forth new scenes of life to waken.
O for a wing to lift and bear me on,
And on to where his last rays beckon.

"As I uttered these inspiring words," Tesla declared, "the truth was [suddenly] revealed. I drew with a stick on the sand the diagrams shown six years later in my address before the American Institute of Electrical Engineers....Pygmalion seeing his statue come to life could not have been more deeply moved. A thousand secrets of nature which I might have stumbled upon accidentally I would have given for that one which I had wrestled from her against all odds and at the peril of my existence."[40]

Tesla emphasized that his conceptualization involved new principles rather than refinements of preexisting work.

The AC creation came to be known as the rotating magnetic field. Simply stated, Tesla utilized two circuits instead of the customary single circuit to transmit electrical energy and thus generated dual currents ninety degrees out of phase with each other. The net effect was that a receiving magnet (or motor armature), by means of induction, would rotate in space and thereby continually attract a steady stream of electrons whether or not the charge was positive or negative. He also worked out the mechanism to explain the effect.[41]

Motor Schematics Showing Magnetic Field Rotation.

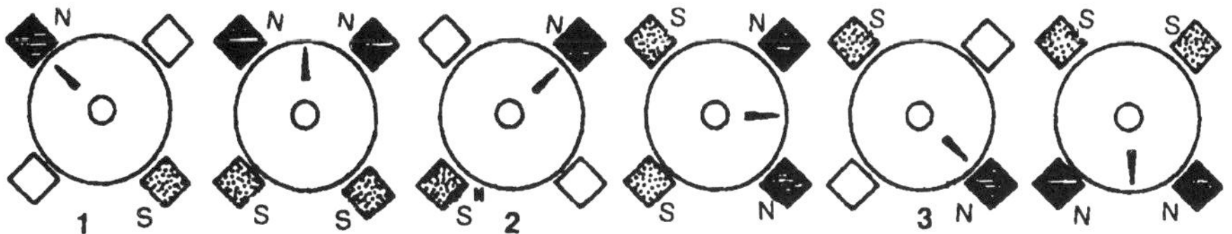

Follow the letter *N* in each of the stages above.

Tesla referred to this diagram (or one quite similar to it) in his lecture before the American Society of Electrical Engineers for the first time in 1888. Each square represents the same armature at different points in its rotation. There are two independent circuits or currents set up diagonally across from one another which are 90 degrees out of phase with each other in both position and timing. So, for instance, in the first position, the armature points to the north pole (of the north/south circuit, which runs from bottom right to top left). The other circuit (running from bottom left to top right) is in the position of changing so that neither pole has a charge. If we look to the next square to the right (which is occurring a fraction of a second later as the currents continue to alternate), we note that the charge is beginning to enter the second circuit (i.e., running from bottom left to top right). At this point in the other circuit, the charge is beginning to reverse itself as well but still has the same polarity. As there are two north poles set up at this fraction of a second, the armature rotates to go between the two of them. In the third square, the bottom right–top left circuit is now neutralized on its way to reversing its polarity, while the bottom left–top right circuit maintains the polarity it has just entered. Therefore, the armature continues its movement to the most northward position, and so on.[42]

"In less than two months," Tesla wrote, "I evolved virtually all the types of motors and modifications of the system now identified with my name.... It was a mental state of happiness as complete as I have ever known in life. Ideas came in an uninterrupted stream and the only difficulty I had was to hold them fast."[43] Tesla invented, at this time, not only single-phase motors whereby the two circuits were 90 degrees out of phase with each other, but also polyphase motors which used three or more circuits of the same frequency in various other degrees out of phase with each other. Motors would be run entirely in his mind; improvements and additions to design were conceived; finally, plans and mathematical calculations would eventually be transferred to a notebook. This step-by-step procedure would become customary.

Ferenc Puskas, who had initially hired Tesla, asked if he wanted to help his brother Tivadar run the new Edison lighting company in Paris. Tesla said, "I gladly accepted."[44] Szigeti was also offered a position, so Tesla was fortunate to have a good friend with whom he could share the new adventure.

Was Tesla the first to conceive of a rotating magnetic field? The answer is no. As far back as 1824, a French astronomer by the name of François Arago experimented with spinning the arm of a magnet by using a copper disc.

The first workable rotating magnetic field similar to Tesla's 1882 revelation was conceived three years before him by Mr. Walter Baily, who demonstrated the principle before the Physical Society of London on June 28, 1879, in a paper entitled "A Mode of Producing Arago's Rotations."[45] The invention comprised two batteries connected to two pairs of electromagnets bolted catty-corner to each other in an *X* pattern, with a commutator used as a switching device. The rotating magnetic field was initiated and maintained by manually cranking the commutator. On this occasion Baily stated, "The disk can be made to rotate by means of intermittent rotation of the field effected by means of electromagnets."[46]

Two years later, at the Paris Exposition of 1881, came the work of Marcel Deprez, who calculated "that a rotating magnetic field could be produced without the aid of a commutator by energizing electromagnets with two out-of-step AC currents."[47] However, Deprez's invention, which won an award at the electrical show, had a major problem: One of the currents was "furnished by the machine itself." Furthermore, the invention was never practically demonstrated.[48]

Other researchers who conceived of a rotating magnetic field analogous to Tesla's but *after* his revelation (in early 1882) were Professor Galileo Ferraris of Turin, Italy (1885–88), and an American engineer, Charles Bradley (1887). Ferraris was influenced by the work of Lucien Gaulard and George Gibbs, who designed AC transformers during the mid 1880s. In 1883 they presented their AC system at the Royal Aquarium in London,[49] and in 1885 they installed an AC system of power distribution in Italy, where they met Ferraris.[50] Purchased by George Westinghouse for $50,000, the system was installed in Great Barrington, Massachusetts, the following year by William Stanley, Westinghouse's head engineer. The Gaulard-Gibbs invention, however, did not do away with the commutator, which was the express purpose of Tesla's design.

In Ferraris's published treatise on his independent discovery of a rotating magnetic field, he wrote, "This principle cannot be of any commercial importance as a motor." After learning of Tesla's work, Ferraris stated, "Tesla developed it much further than [I]...did."[51]

Bradley filed for a patent for an AC polyphase device on May 8, 1887 (no. 390,439) after nine Tesla AC patents had been granted. Haselwander, in the same year, utilized slip rings in place of commutators on DC Thomson-Houston equipment and also designed two- and three-phase windings on DC armatures.[52]

The question of priority concerning Tesla's invention was discussed by Silvanus P. Thompson, a physics professor in London, in his 1897 comprehensive text on AC motors. Thompson (no relationship to Elihu Thomson), considered at the time to be "perhaps the best known writer on electrical subjects now living," said that Tesla's work separated itself clearly from predecessors and contemporaries in his "discovery of a new method of electrical transmission of *power* [emphasis added]."[53]

A question that remains unanswered was whether or not Tesla knew of Baily's work. It is quite possible that he had read Baily's paper, although no one at the time, including Baily, comprehended the importance of the research or understood how to turn it into a practical invention.[54] Tesla stated in the early 1890s, "I am aware that it is not new to produce the rotations of a motor by intermittently changing the poles of one of its elements.... In such cases, however, I imply true alternating currents; and my invention consists in the discovery of the mode or method of utilizing such currents."[55]

A few years later, in a well-publicized case involving patent priorities on what came to be known as the "Tesla Alternating Current Polyphase System," Judge Townsend of the U.S. Circuit Court of Connecticut noted that before Tesla's invention and lecture to the American Institute of Electrical Engineers (AIEE) in 1888, there had been no AC motors; furthermore, no one attending the lecture recognized any priorities. Whereas Baily had dealt with "impractical abstractions, Tesla had created a workable product which initiated a revolution in the art."[56] The Tesla patents were also sustained against individual cases involving Charles Bradley, Mons. Cabanellas and Dumesnil, William Stanley, and Elihu Thomson.[57]

In citing a previous case on a similar issue, Judge Townsend responded to what today is called the "doctrine of obviousness":

> The apparent simplicity of a new device often leads an inexperienced person to think that it would have occurred to anyone familiar with the subject, but the decisive answer is that with dozens and perhaps hundreds of others laboring in the same

> field, it had never occurred to anyone before [*Potta v. Creager*, 155 U.S. 597]....Baily and the others [e.g., Bradley, Ferraris, Stanley] did not discover the Tesla invention; they were discussing electric light machines with commutators....Eminent electricians united in the view that by reason of reversals of direction and rapidity of alternations, an alternating current motor was impracticable, and the future belonged to the commutated continuous current....
>
> It remained for the genius of Tesla to...transform the toy of Arago into an engine of power.[58]

The discovery of how to effectively harness the rotating magnetic field was really only a fraction of Tesla's creation. Before his invention, electricity could be pumped approximately one mile, and then only for illuminating dwellings. After Tesla, electrical *power* could be transmitted hundreds of miles, and then not only for lighting but for running household appliances and industrial machines in factories. Tesla's creation was a leap ahead in a rapidly advancing technological revolution.

4

TESLA MEETS THE WIZARD OF MENLO PARK (1882–85)

O, he's a great talker, and, say, he's a great eater too. I remember the first time I saw him. We were doing some experimenting in a little place outside Paris, and one day a long, lanky lad came in and said he wanted a job. We put him to work thinking he would soon tire of his new occupation for we were putting in 20–24 hours a day, then, but he stuck right to it and after things eased up one of my men said to him: "Well, Tesla, you've worked pretty hard, now I'm going to take you into Paris and give you a splendid supper." So he took him to the most expensive cafe in Paris—a place where they broil an extra thick steak between two thin steaks. Tesla stowed away one of those big fellows without any trouble and my man said to him: "Anything else, my boy? I'm standing treat." "Well, if you don't mind, sir," said my apprentice, "I'll try another steak." After he left me he went into other lines and has accomplished quite a little.

THOMAS EDISON[1]

Taking the advice of Ferenc Puskas, Tesla left Budapest for Paris in April 1882 delighted with the chance to meet the Edison people from America and ready to build his motor and to find investors. Concurrently, he was getting paid for the experience. Paris in the 1880s was a center of modern fashion: men in their cutaway coats and silk top hats, women with braided hair, in long frilled dresses with bustles, and wealthy tourists ready to take back the latest fineries to their respective nations. Tesla was met by Ferenc's brother Tivadar Puskas, a hard driver but also a man known to talk in "air balloons."[2] Tesla, whose head could also soar into the clouds, had met a powerful ally. Mindful of the need for secrecy, they discussed strategies for approaching Charles Batchelor, manager of the newly formed Compagnie Continental Edison, with Tesla's new motor as the young inventor was introduced to operations.

Formerly a resident of Manchester, England, Batchelor, a "master mechanic," had been sent to America a decade earlier to present innovative threadmaking machinery recently created by his employers, the Coates Thread Company.[3] There he met Edison and shortly became his most trusted associate. Batchelor worked on the first phonographs and on perfecting the filament for the lightbulb. He also ran operations in New Jersey and then in Europe, owning a 10 percent share of Edison's many worldwide companies.[4] An open-minded individual, Batchelor was approachable, although also rather busy.

Anthony Szigeti probably emigrated from Budapest at the same time as Tesla, since both were hired by Puskas and "were together almost constantly in Paris." Szigeti wrote, "Tesla [was very]...much excited over the ideas which he then had of operating motors. He talked with me many times about them and told me his plan...[of] constructing and operating motors...[and] dispensing with the commutator."[5]

Having just purchased a large factory at Ivry-sur-Seine, for the construction of generators and manufacture of lightbulbs, Batchelor, as Edison's closest partner, was planning on erecting central lighting stations throughout Europe. He also had plans in England, where the Crystal Palace Exposition was then displaying Edison's new incandescent lamp.[6] Batchelor would need good men to run the concerns and wrote Edison frequently as to the expertise of the various workers. He was particularly impressed with Puskas, who had successfully run the Edison lighting exhibit at the Paris Exposition of 1881. "Puskas...[is the only worker] having any idea of 'push,'" he wrote, "and I think that you should insist on him [becoming a partner]."[7]

Within six months Edison Continental would be producing lamps superior to those from America;[8] the company would erect central stations in most of the major cities of Europe for indoor lighting and also administer the large outdoor arc lamps which were being used to illuminate the urban streets. Tesla, who was working at Ivry-sur-Seine, would be trained with the other workers to travel out and help run these facilities. "I never can forget the deep impression that magic city produced on my mind. For several days after my arrival I roamed thru [*sic*] the streets in utter bewilderment of the new spectacle. The attractions were many and irresistible, but, alas, the income was spent as soon as received. When Mr. Puskas asked me how I was getting along...I [replied] 'the last twenty-nine days of the month are the toughest!'"[9]

In the mornings, before work, Tesla would arise at 5:00 A.M. to swim twenty-seven laps at a bathhouse on the Seine, and in the evenings he would play billiards with the workers and discuss his new AC invention.

"One of them, Mr. D. Cunningham, foreman of the Mechanical Department, offered to form a stock company. The proposal seemed to me comical in the extreme. I did not have the faintest conception of what that meant except that it was the American way of doing things."[10]

T. C. Martin writes: "In fact, but for the solicitations of a few friends in commercial circles who urged him to form a company to exploit the invention, Mr. Tesla, then a youth of little worldly experience, would have sought an immediate opportunity to publish his ideas, believing them to be [a]...radical advance in electrical theory as well as destined to have a profound influence on all dynamo electric machinery."[11]

In his spare time, and as was his custom, Tesla wrote out the specifications and mathematics of his AC invention in a notebook[12] and worked on alternative designs for his flying machine. He probably sought out financial backers, for he received an invitation to go on a shooting expedition from a "prominent French manufacturer."[13] Perhaps the inventor had not totally recovered from the strange illness he had almost succumbed to in Budapest, for after this outing he suffered the "sensation that my brain had caught fire. I saw a light as [though] a small sun was located in it and I [passed] the whole night applying cold compressions to my tortured head." Writing this passage almost forty years later, Tesla claimed that "these luminous phenomena still manifest themselves from time to time, as when a new idea opening up possibilities strikes me."[14]

In the summer he worked on the lighting at the opera house in Paris or went to Bavaria to help in the wiring of a theater; and in the autumn he may have helped in the laying of underground cables for the new central station going up in Paris or traveled to Berlin to install incandescent lighting at the cafes.[15]

At the end of the year Tesla "submitted to one of the administrators of the Company, Mr. Rau, a plan for improving their dynamos, and was given an opportunity." Louis Rau, who was director of the Compagnie Continental Edison in rue Montchanien and had "his beautiful home lit with the Edison system,"[16] allowed Tesla to implement his modernization plan. Shortly thereafter the young inventor's automatic regulators were completed and accepted gratefully.[17] Tesla was probably hoping to be compensated for his new contributions, but he was sent to work in Strasbourg before financial compensation was awarded.

In January 1883, Batchelor shipped twelve hundred lamps to the Strasbourg plant, located at the railroad station.[18] And within three months Tesla arrived to oversee the operations. There he would stay for the next twelve months.

Batchelor had been urging Edison to test the generators coming from America for at least "two or three days with a [full] load," as fires from faulty armatures and poor insulation were becoming too common. The powerhouse at Strasbourg, in particular, had been subject to this type of problem.[19] Since "all our plants are differently constructed,"[20] it would take well-trained and creative engineers to run things smoothly. Batchelor demonstrated confidence in Tesla's abilities by sending him to Strasbourg; however, he seems not to have mentioned Tesla in his correspondence with Edison. In any event, Tesla's account of the situation in Strasbourg corroborates Batchelor's: "The wiring was defective and on the occasion of the opening ceremonies a large part of a wall was blown out thru [*sic*] a short-circuit right in the presence of old Emperor William I. The German Government refused to take the plant and the French Company was facing a serious loss. On account of my knowledge of the German language and past experience, I was entrusted with the difficult task of straightening out matters."[21]

Having anticipated a long stay in the region, Tesla had brought with him from Paris materials for his first AC motor. As soon as he was able, Tesla constructed the motor in secret in a closet "in a mechanical shop opposite the railroad station";[22] however, summer would arrive before this first machine was in operation. Anthony Szigeti, his assistant, forged an iron disk, which Tesla "mounted on a needle," having surrounded it, in part, with a coil.[23] "Finally," Tesla wrote, "[I] had the satisfaction of seeing rotation effected by alternating currents of different phase, and without sliding contacts or commutator, as I had conceived a year before. It was an exquisite pleasure, but not to compare with the delirium of joy following the first revelation."[24]

Tesla presented his new creation to his friend, Mr. Bauzin, the mayor of the town, who tried his best to interest wealthy investors; "but to my mortification...[there was] no response." Upon his return to Paris, he sought promised compensation for achieving a difficult success in Strasbourg. Approaching his employers, "after several days of...circulus vicious, it dawned on me that my reward was a castle in Spain....Mr. Batchelor [pressed] me to go to America with a view of redesigning the Edison machines; I determined to try my fortunes in the land of Golden Promise."[25]

John O'Neill, Tesla's first major biographer, has suggested that Batchelor wrote a note of introduction to Edison which read, "I know two great men and you are one of them; the other is this young man."[26] Evidence for the veracity of this oft-repeated tale is lacking. Batchelor, for instance, had been back in America for at least three months prior to Tesla's arrival;[27] thus, he would not have had to write a letter. Furthermore, there is evidence that Edison had already met Tesla in Paris during a little-known sojourn he took to look over his European operations at that time.[28] O'Neill also refers to Batchelor incorrectly as Edison's "former assistant"[29] when Batchelor was probably Edison's closest lifelong colleague. Edison does, however, substantiate that "Tesla worked for me in New York. He was brought over from Paris by Batchelor, my assistant,"[30] but there is no reference to Batchelor's appreciation of Tesla's genius. On October 28, 1883, fully a year *after* Tesla began working for Edison Continental, while he was stationed in Strasbourg, Batchelor singled out "the names of...two [or three] I can mention as capable as far as their work shows: Mr. Stout—an inspector; Mr. Vissiere—my assistant; Mr. Geoffrey—whose plants are always spoken well of....There are others capable, but I think these are the best."[31] Certainly, had Tesla impressed Batchelor as O'Neill contends, he should have been listed in this letter or in numerous other letters to Edison that I have reviewed.

Before Tesla left for America, he spent time with a scientist who was studying microscopic organisms found in common drinking water. Combined with the scare he had with his bout with cholera a few years earlier, Tesla acquired a phobia which led him to shun unpurified water, scour his plates and utensils before eating, and refrain from frequenting unsavory restaurants. He would later write, "If you would watch only for a few minutes the horrible creatures, hairy and ugly beyond anything you can conceive, tearing each other up with the juices diffusing throughout the water—you would never again drink a drop of unboiled or unsterilized water."[32]

In the spring of 1884, with funds for the journey supplied by Uncles Petar and Pajo,[33] Tesla packed his gear and caught the next boat for America. Although his ticket and money and some of his luggage were stolen, the young man was not deterred. "Resolve, helped by dexterity, won out in the nick of time...[and] I managed to embark for New York with the remnants of my belongings, some poems and articles I had written, and a package of calculations relating to solutions of an unsolvable integral and to my flying machine."[34] The voyage appears not to have been a happy one; a "mutiny" of sorts occurred on board, and Tesla was nearly knocked overboard.[35]

In 1808, Sir Humphry Davy created artificial illuminescence by running an electric current across a small gap between two carbon rods. This simple device evolved into the arc lamp, used in English lighthouses in the 1860s and displayed at the Philadelphia Exposition of 1876 by Moses Farmer. By 1877 numerous investigators were exploring the possibility of placing the incandescent effect within glassed enclosures because they would be much safer this way for marketing to households, and a race developed between such inventors as Charles Brush, Thomas Edison, Moses Farmer, St. George Lane-Fox, Hiram Maxim, William Sawyer, and Joseph Swan.

"I saw the thing had not gone so far but that I had a chance," Edison said.[36] And so he challenged William Wallace, Farmer's partner, to a race as to who would be the first man to create an efficient electric light. Boasting that he would soon light up New York City with 500,000 incandescent lamps, Edison and his business manager, Grosvenor Lowery, were able to secure large amounts of capital from such investors as Henry Villard, owner of the first trans-American railroad, and financier J. Pierpont Morgan.

In November 1878, after three years of research, a hard-drinking telegrapher by the name of William Sawyer and his lawyer-partner Albion Man, applied for a patent for an incandescent lamp with carbon rods (filaments) and filled with nitrogen. They proclaimed that they had beaten Edison. Joseph Swan, another competitor, removed the nitrogen and kept the carbon filament but created a *low*-resistance lamp. Realizing that the amount of power required to send electricity a few hundred feet was prodigious using the low-resistance design, Edison created, in September 1878, a *high*-resistance vacuum lamp that utilized considerably less power. Together with a revolutionary wiring called a feeder line,[37] his success was further augmented by a new Sprengel pump, which William Crookes had been recommending for the creation of vacuums in glass-enclosed tubes. It would be another six months—April 22, 1879—before he would file for a patent, but his new design would lower power requirements and thereby cut copper costs one hundredfold.[38]

The competition was fierce, and Edison's financial backers were running scared. They suggested to Edison that they purchase the Sawyer patents and combine the two companies. Edison had not yet settled on carbon as a filament and was exhausting his working capital in experiments with boron, iridium, magnesium, platinum, silicon, and zirconia. At the same time, he had also sent explorers to the Amazon, Bolivia, Japan, and Sumatra in search of rare forms of bamboo, which he was also considering. It would not be until 1881 that he finally settled upon a form of carbonized paper.

During this time, however, and without Edison's knowledge, Sawyer and Man approached Lowery. Their lamp was superior to Edison's; it was patented, and it worked. Lowery tried to bring Edison in for a four-way discussion; however, Edison sent an emissary who "dared not relay to Edison all Lowery had said. But Edison heard enough to be jolted from his indecision.... Cursing and spraying tobacco juice, he exclaimed it was the old story—lack of confidence!"[39]

Edison was adamant about not joining with Sawyer or Swan or anyone else. He continued rash publicity campaigns which announced "a veritable Aladdin's lamp.... [It is] Edison's light, the great inventor's triumph."[40]

With the backing of Wall Street moguls, Edison began to illuminate Menlo Park and the private homes of the wealthy in New York City. The first was that of J. Pierpont Morgan, at Thirty-Sixth Street and Madison Avenue. The year was 1881.

To run the generator Edison designed a steam engine and boiler and placed the power plant under the stable in a newly dug cellar at the back edge of the property. Wires were connected to the new incandescent lights placed in the gas fixtures of the home via a brick-lined tunnel which ran the length of the yard just beneath the surface. "Of course, there were the frequent short circuits and many breakdowns on the part of the generating plant. Even at the best, it was a source of a good deal of trouble for the family and neighbors. who complained of the noise of the dynamo. Mrs. James M. Brown next door said that its vibrations made her house shake." Morgan had to pile sandbags around the inside of the cellar and place the machinery on heavy rubber pads "to deaden the noise and the vibrations. This final experiment restored quiet and brought peace to the neighborhood until the winter, when all the stray cats in the neighborhood gathered on this warm strip in great numbers and their yowlings gave grounds [from the neighbors] for more complaints."[41]

The following year, on September 4, 1882, the new Central Station at Pearl Street opened. It provided electric lighting to many Wall Street buildings, including Morgan's office.

Tesla's ship dropped anchor in New York in late spring of 1884, just as the monumental decade-long project the Brooklyn Bridge was being completed and the last components of the Statue of Liberty were being hoisted into position. Twenty-eight years old, "tall and spare, [with] thin, refined face"[42] and sporting a mustache, Tesla still had the look of an adolescent.

His first impression of the New World was that it was uncivilized, a hundred years behind the lifestyle of the great European cities. Deferring his planned meeting with Edison one day to look up an old friend, Tesla had the good fortune to pass by "a small machine shop in which the foreman was trying to repair an electric machine.... He had just given up the task as hopeless."[43] One rendition of the story has Tesla agreeing to fix the machine "without a thought for compensation."[44] On a separate occasion, Tesla revealed that "it was a machine I had helped design, but I did not tell them that. I asked... 'what would you give me if I fix it?' 'Twenty dollars' was the reply. I took off my coat and went to work, [and]...had it running perfectly in an hour."[45] The story is important because, depending on the rendition, two different Teslas emerge, one motivated by money and one not.

In either case, Tesla was shocked by the rough character of the New World.[46] He proceeded cautiously to Edison's new laboratory, a former ironworks at Goerck Street, situated only a few blocks from the central lighting station Edison was constructing at Pearl Street.[47] Batchelor probably met Tesla and introduced him to the inventor. "I was thrilled to the marrow by meeting Edison," Tesla said.[48]

Possibly aware of the proximity of Transylvania to Tesla's birthplace and a resurgence of interest in the tales of Vlad Dracula, the fifteenth-century alleged vampire who lived in the region, Edison inquired whether or not the "neophyte...had ever tasted human flesh?"[49]

Aghast at the question and Edison's "utter disregard of the most elementary rules of hygiene,"[50] Tesla replied in the negative and asked what Edison's diet consisted of.

"You mean to make me so all-fired smart?"

Tesla nodded.

"Why, I eat a daily regimen of Welsh rabbit," Edison replied. "It's the only breakfast guaranteed to renew one's mental faculties after the long vigils of toil."

Wanting to emulate the grand wizard, the neophyte took on the peculiar diet, "accepting as true, in spite of a protesting stomach, the jocular suggestion."[51]

Tesla's various accounts of this meeting differ markedly, depending on his mood at the time of the telling and his awareness of the size and shape of the audience. For in his autobiography, published in six installments of Hugo Gernsback's futuristic magazine *Electrical Experimenter,* Tesla wrote that 'the meeting with Edison was a memorable event in my life. I was amazed at this wonderful man who, without early advantages and scientific training, had accomplished so much. I had studied a dozen languages, delved in literature and art, and had spent my best years in libraries...and felt that most of my life had been squandered."[52]

It wasn't long before Tesla realized that his academic training and mathematical skills had given him a great engineering advantage over Edison's plodding strategy of trial and error. In a bitter moment of reminiscence, at the time of Edison's death in 1931, Tesla said: "If he had a needle to find in a haystack he would not stop to reason where it was most likely to be, but would proceed at once with the feverish diligence of a bee, to examine straw after straw until he found the object of his search.... I was almost a sorry witness of his doings, knowing that just a little theory and calculation would have saved him 90 per cent of the labor.... Trusting himself entirely to his inventor's instinct and practical American sense...the truly prodigious amount of his actual accomplishments is little short of a miracle."[53]

It was little wonder that Tesla was completely unsuccessful in describing his new AC invention to Edison and had to settle for Batchelor's suggestion that he redesign the prevailing DC machinery instead. According to Tesla, "the Manager had promised me $50,000 on completion of this task,"[54] and so Tesla set himself to work, "experiment[ing] day and night, holidays not excepted," as was the custom of the factory.[55]

Thomas Alva Edison was an extremely complex fellow. Ornery, ingenious, determined, and unyielding, he was a fierce competitor and the single most important inventive force on the planet. He had descended from a grandfather, John Edison, a Tory who had been tried for treason during the American Revolution and banished to Canada, and a father, Samuel Edison, who had tied his son to a whipping post and beaten him publicly after young Al, as he was called then, had started a fire in a barn which threatened the rest of the buildings in the community.[56] He had scrapped with and outwitted others on his way to Wall Street and had outdistanced competing inventors numerous times. Notches on Edison's belt of "better mousetraps" included the telephone transmitter (microphone), an electrical pen, a musical telephone, and the duplex, an ingenious device which enabled a telegraph to send four messages in two directions simultaneously.

Edison was known to curse and swap jokes with his men at his research and development center, the world's first invention factory. He kept his business free of cockroaches with a protective electric grid lining the edges of the floor and "electrifried larger varmints" with his "rat paralyzer"; he even occasionally wired the washbasin to keep his men on their toes. Edison was a trickster, a storyteller, and a con artist. The use to the consumer and the cost of production or "the market test [was] the sole test of achievement.... Everything he did was directed by [that] realization."[57]

In an entirely different realm of invention, besides being a better technician than anyone else, Edison was a *creator;* his most original work was a machine that talked: the phonograph. With this device, Edison had entered the realm of the immortal; he was the "Wizard of Menlo Park."

Inviting the public to his laboratory on a number of occasions, Edison amazed people at all levels of society with machines that sang and reproduced the sound of birds, artificial lamps that changed the darkness into a light cherry red, and various other mechanical contrivances to make one's workload easier.

The invention of the electric light was to Edison not only a new, clever technology; it contained the seeds of a new industry. His mere presence in the field drove the stocks of the gas-burning companies into the grave. Yet Edison planned to utilize their pipes by channeling copper wire through them instead of dangerous gas and to replace flame by electricity. He moved the center of his operation from New Jersey to New York City. There Edison rented a town house for his wife and family in celebrated Gramercy Park, the abode of such luminaries as authors Mark Twain and Stephen Crane, sculptor Augustus Saint-Gaudens, architect Stanford

White, *Century* editor Richard Watson Gilder, and publisher James Harper.[58] Edison later described his plans for orchestrating a revolution in home illumination: "I had the central station in mind all the time.... I got an insurance map of New York City, laid out a district [bounded by] Wall Street, Canal, Broadway [and the] East River, [and purchased] two old bum buildings down in Pearl Street. They charged us $75,000 a piece. I tell you it made my hair stand on end."[59]

Edison's financial problems were numerous. Not only were there expensive start-up costs, there were also problems with the extreme inefficiency of the DC system and court battles on invention priorities and marketing battles against such competitors as Brush Electric, Consolidated Electric, Sawyer-Man, Swan Incandescent, Thomson-Houston, United States Electric, and the Westinghouse Corporation.

"Tell Westinghouse to stick to air brakes. He knows all about them," Edison complained;[60] but Westinghouse would not listen.

Edison's other major competitor was Elihu Thomson. With Edison embroiled in a legal contest with Sawyer, Thomson used the ambiguity of the moment to appropriate the incandescent lamp Edison had given him and make it the template for ones produced and sold by the Thomson-Houston Electric Company. On October 8, 1883, the patent office ruled that William Sawyer had priority over Edison "for an incandescent lamp with carbon burner."[61] This decision, though later overturned in Edison's favor, enabled Thomson to continue his piracy. Due to Sawyer's priority, Thomson now saw himself as "ethically in the clear"[62] as no clear-cut inventor had supposedly been established.

Edison thus came to vigorously dislike Thomson, a man who had betrayed his trust, and Westinghouse, who was now siding with Sawyer. For safety, aesthetic, and practical reasons, Edison was a proponent of underground cables and DC. "Nobody hoisted water and gas mains into the air on stilts," he said.[63] He publicized the fact that electricians were dying on the dangerous overhead wires of his competitors, but this battle eventually became transformed into DC versus AC; Edison stayed with DC, while Thomson and Westinghouse began to experiment with AC. As AC utilized much higher voltages, Edison warned the public against it. A long legal battle with Westinghouse ensued and ran into the millions of dollars. Thomson again managed quietly to avoid the courts while he expanded his business.

Francis Upton, Edison's mathematician, graduate of Helmholtz's laboratory, and contemporary of Tesla's in terms of European education, had calculated in 1879 that to light 8,640 lamps for only nine city blocks, the cost would be $200,812 for the 803,250 pounds of copper required. Through clever wiring, improvements in lamp design, and "an invention corollary to the parallel circuitry," Edison had cut copper costs almost 90 percent, but no matter what he did, a power station could never reach beyond a radius of one or two miles.[64]

Upton, whom Edison affectionately referred to as "Culture," suggested that they look into the new advances in AC, and so he was sent, in 1884, to Europe to negotiate with Karl Zipernowski, Otto Blathy, and Max Deri, three Hungarians who had greatly improved the Gaulard-Gibbs AC transformer. Edison even paid $5,000 for an option on this "ZBD" system, but it was mostly to placate Culture. The wizard did not trust AC, and if his "damn fool competitors" were in it, he certainly didn't want any part of it. Twenty years of experience, ingenuity, and doing the impossible with DC had to be worth something. The "bugs" could be worked out.

Yet at the same time that Edison constructed DC generators to make the earth tremble[65] and competitors stole his ideas or fashioned other primitive electric-lighting devices, a Serbian genius in his very midst had designed a system which made this prevailing technology *obsolete.*

According to W. L. Dickson, one of Edison's earliest biographers and longtime employee at Menlo Park and Goerck Street, "Nikola Tesla, that effulgent star of the scientific heavens, even then gave strong evidence of the genius that has made him one of the standard authorities of the day." Tesla's "brilliant intellect" had held Dickson and the other workers "spellbound" as he "alternately fired [us] with the rapid sketching of his manifold projects or melted [us] into keenest sympathy by pictures of his Herzogovinian home....But like most holders of God's intrinsic gifts, he was unostentatious in the extreme, and ready to assist with counsel or manual help any perplexed member of the craft."[66]

Although unable to interest Edison in his AC motor, Tesla was able "within a few weeks [to win]...Edison's confidence." Tesla's greatest success came when he fixed a badly broken set of dynamos on Henry Villard's ocean liner, the *Oregon,* the first boat ever to have electric lighting. "At five o'clock in the morning, when passing along 5th Avenue on my way to the shop," Tesla recalled, "I met Edison with Batchelor and a few others who were returning to retire.

"'Here is our Parisian running around at night,' he said. When I told him I was coming from the *Oregon* and had repaired both machines he looked at me in silence....But when he walked some distance I heard him remark: 'Batchelor, this is a damn good man,' and from that time on I had full freedom in directing the work."[67]

Alternately spending time at the Pearl Street Station or the Goerck ironworks, Tesla installed and fixed indoor incandescent lamps and outdoor arc lamps, reassembled many of Edison's DC generators, and designed twenty-four different types of machines that became standards which replaced those being used by Edison.[68] At the same time, he worked on patents on arc lamps, regulators, dynamos, and commutators for DC apparatus, trying to devise a way to approach his boss with his new invention, obtain a raise, and gain compensation for the lump sum he had allegedly been promised.

The atmosphere was informal, Tesla occasionally dining with Edison, Batchelor, and other higher-ups, such as Edward Johnson, president of the Edison Illuminating Company, or Harry Livor, another engineer and small-time entrepreneur in machine-works manufacturing. Their favorite spot was a small restaurant opposite the Edison showroom at 65 Fifth Avenue. There they would swap stories and tell jokes.[69] Afterward, some would retire to a billiard house where Tesla would impress the fellows with his bank shots and vision of the future.[70]

Livor boasted of an agreement with Edison and Batchelor resulting in a company capitalized at $10,000, formed for the manufacture of shafting. Edison and Batchelor provided the machinery and money, Livor, the tools and services.[71] Impressed, Tesla asked for advice, particularly how to obtain a raise from his present modest salary of eighteen dollars per week to a more lucrative twenty-five dollars. "Livor gladly undertook this service... to intercede with Batchelor... but greatly to his surprise was met with an abrupt refusal."

"No," replied Batchelor, "the woods are full of men like [Tesla]. I can get any number of them I want for $18 a week." Tate, who began employment as Edison's secretary shortly after this episode, which Livor related to him, noted that Batchelor "must have been referring to the woods I failed to find in the vicinity of Harlem."[72] Tesla's version of the story is somewhat different: "For nine months my hours [at the Edison Machine Works] were 10:30 A.M. till 5 A.M. the next day. All this time I was getting more and more anxious about the invention [AC induction motor] and was making up my mind to place it before Edison. I still remember an odd incident in this connection. One day in the latter part of 1884 Mr. Batchelor, the manager of the works, took me to Coney Island, where we met Edison in the company of his former wife. The moment that I was waiting for was propitious, and I was just about to speak, when a horrible-looking tramp took hold of Edison and drew him away, preventing me from carrying out my intentions."[73]

In analyzing this story, a discrepancy as to the timing was discovered, for Edison's wife caught typhoid fever in July 1884 and died on August 9. Since Tesla had arrived in May or June, and if Edison's wife was present, then the event took place in late June or early July, only a few weeks after he began his employment. In a close working environment, with the hours as described, even a few weeks could seem like a very long time. One way or another, with the death of Edison's wife and Edison's extreme dislike of such AC men as Elihu Thomson and George Westinghouse, no time for discussing an AC invention may have been "propitious." The "horrible-looking tramp" who grabbed Edison away was probably Edison himself, who was known to dress like a "Bowery bum," Tesla using a euphemism to soften the story. "The manager had promised me fifty thousand dollars [for redesigning equipment], but when I demanded payment, he merely laughed. 'You are still a Parisian,' remarked Edison. 'When you become a full-fledged American, you will appreciate an American joke.'"[74]

If a "completion agreement"[75] had truly been made with Edison, Tesla should have had it put in writing. It seems unlikely that this amount of money for a somewhat ambiguous bargain would be offered, but it was well within Edison's nature to make "expensive if indefinite promises of rewards as a way of getting the men to work for low wages." Edison, who could be more deaf than he actually was, at times, was known to "put on" his college-educated 'sperts, as when he convinced the chemist Martin Rosanoff that his first lightbulb filament was made out of Limburger cheese! Deeply hurt, Tesla left the company and set out on his own.[76]

5

LIBERTY STREET (1886–88)

There were many days when [I] did not know where my next meal was coming from. But I was never afraid to work, I went to where some men were digging a ditch...[and] said I wanted to work. The boss looked at my good clothes and white hands and he laughed to the others...but he said, "All right. Spit on your hands. Get in the ditch." And I worked harder than anybody. At the end of the day I had $2.

NIKOLA TESLA[1]

Although Tesla felt cheated when he departed from the Edison Machine Works in the early months of 1885, his time spent there had enabled him to study the master at work. Simultaneously, it allowed Tesla to begin to organize his own company and write up first drafts in a notebook on advances in arc-lighting design and on the construction of DC commutators. It also enabled him to see that Edison was mortal and fallible and that he, Tesla, had a scheme significantly more advanced. A new confidence began to emerge.

In March 1885, Tesla met with the well-established patent attorney Lemuel Serrell, a former agent of Edison's, and Serrell's patent artist, Raphael Netter.[2] Serrell taught Tesla how to break down complex inventions into individualized improvements, and on the thirtieth of the month they applied for Tesla's first patent (no. 335,786), an improved design of the arc lamp which created a uniform light and prevented flickering. In May and June they applied for other patents on improvements on the commutator for the prevention of sparking and for regulating the current by means of a novel independent circuit coupled with auxiliary brushes. In July yet another arc-lighting patent was filed. This one enabled exhausted lamps to automatically separate themselves from the circuit until such time as the carbon filaments could be replaced. Unfortunately, the design had been anticipated by Elihu Thomson. Although "embarrassed" by having been unaware of the state of the art in America at this time, Tesla was able to create novel refinements, and they were patentable.[3]

During his trips to Serrell's office, the inventor met with B. A. Vail and Robert Lane, two businessmen from New Jersey.[4] With ambiguous assurances that they were also interested in the AC motor, Tesla agreed to form a lighting and manufacturing company with them in Tesla's name in Vail's town of Rahway, New Jersey. There, after nearly a year of toil working with Paul Noyes, from Gordon Press Works, he completed the installation; this, his first and only municipal arc-lighting system, was used to illuminate the streets of a town and some factories.[5] The efficiency and original approach of the system attracted the attention of George Worthington, editor of *Electrical Review,* who "took pleasure" in featuring the company on the front page of the August 14, 1886, issue.

For the next few months, the Tesla Electric Light & Manufacturing Company reciprocated by advertising in the journal. Vail hired the mechanical artist Mr. Wright of New York City to draw the lamp and dynamo and at the same time, along with Tesla's help, created bold copy which claimed: "the most perfect...and entirely new [arc lighting] system of...automatic self-regulat[ion]" In a display ad four times the size of most other electrical concerns, the Tesla system guaranteed "absolute safety and great saving of power...with no flickering or hissing."[6]

Having obtained stock in the company and with a little money in his pocket, Tesla moved into a garden apartment in Manhattan. Decorating the grounds "in the continental fashion with colored glass balls on sticks," the cosmopolitan's delight was short-lived. "Children broke in and stole the balls, so Tesla replaced them with metal ones. The stealing continued, however, so Tesla ordered his gardener to bring them into the house every night."[7]

Unfortunately, neither Vail, who was president of the company, nor Lane, who was vice president and treasurer, cared about Tesla's other creation. To them, an AC motor was a seemingly useless invention. The sensitive inventor became incensed, for he had postponed exploiting the AC system until the Rahway project was completed under the assumption that his backers would support that quest as well. To his shock, Tesla was forced out of his own concern and handed "the hardest blow I ever received."[8] "With no other possession than a beautifully engraved certificate of stock of hypothetical value,"[9] the inventor was bankrupt. Betrayed by men he trusted, the inventor came to consider the winter of 1886–87 a time of "terrible headaches and bitter tears, my suffering being intensified by my material want."[10] He was forced to work as a ditchdigger. The occupation was particularly demeaning for the self-perceived aristocrat. "My high education in various branches of science, mechanics and literature seemed to me like a mockery."[11]

Tesla's crisis abated in the spring. Having interested the foreman in his engineering prowess, he was introduced to Alfred S. Brown, a prominent engineer who worked for Western Union Telegraph Company. Brown, who himself held a number of patents on arc lamps,[12] had probably seen the article and advertisements on Tesla in *Electrical Review*. Well aware of the limitations of the prevailing DC apparatus, he became immediately impressed with the "merits" of Tesla's AC inventions and thereupon contacted Charles F. Peck, "a distinguished lawyer" from Englewood, New Jersey.[13] Peck "knew of the failures in the industrial exploitation of alternating currents and was distinctly prejudiced to a point of not caring even to witness some tests."

"I was discouraged," Tesla recalled, "until I had an inspiration. Do you remember the 'Egg of Columbus?' I asked. The saying goes that at a certain dinner the great explorer asked some scoffers of his project to balance an egg on its end. They tried in vain. He then took it, and cracking the shell slightly by a gentle blow, made it stand upright. This may be a myth, but the fact is that he was granted an audience by Isabella, the Queen of Spain, and won her support."

"And you plan to balance an egg on its end?" Peck inquired.

"Yes, but without cracking the shell. If I should do this, would you admit that I had gone Columbus one better?"

"All right," he said.

Having finally gained the lawyer's attention, Tesla cut to the quick. "And would you be willing to go out of your way as much as Isabella?"

"I have no crown jewels to pawn," Peck retorted, "but there are a few ducats in my buckskins and I might be able to help you to an extent."[14]

After the meeting, Tesla rushed to the local blacksmith with a hard-boiled egg and had a mate cast in iron and brass. When he returned to the lab, he constructed a circular enclosure with polyphase circuits along the perimeter, and when he placed the egg in the center and turned on the current, the egg began to spin. As the egg's speed of rotation increased, its wobbling ceased, and it stood on its end. Not only was Tesla able to "go Columbus one better," he was also easily able to display the principles behind the idea of his rotating magnetic field. Peck was won over, and together the three men formed a new electric company in Tesla's name.

Peck, who had connections with John C. Moore, a banker with connections to J. P. Morgan, provided the bulk of the capital, and Brown provided technical expertise and located the laboratory at 89 Liberty Street, adjacent to what today is the World Trade Center. In return, Tesla agreed to split his patents on a fifty-fifty basis. In actuality, the three equally shared one patent for an AC dynamo, Peck and Tesla split five more patents on commutators, motors, and power transmission, and the balance of inventions conceived during this period were placed in the name of the Tesla Electric Company. Their first patent was filed on April 30, 1887.[15] Finally, Tesla had arrived. He would begin an unprecedented excursion into the field of invention, a flow of intense activity which would continue unabated for fifteen years.

Driven by his wish to maintain priority in a variety of areas and upon the realization that new technologies could influence the course of history, Tesla began a vigorous schedule that frightened those around him. On many occasions, he drove himself until he collapsed, working around the clock, with few breaks. "Tesla produced as rapidly as the machines could be constructed three complete systems of AC machinery—for single-phase, two-phase, and three-phase currents—and made experiments with four- and six-phase currents. In each of the three principal systems he produced the dynamos for generating the currents, the motors for producing power from them, and transformers for raising and reducing the voltages as well as a variety of devices for automatically controlling the machinery. He not only produced the three systems but provided methods by which they could be interconnected and modifications providing a variety of means of using each of the systems."[16] He also calculated, in fundamental fashion, the mathematics behind these inventions.

On May 10, Anthony Szigeti landed in New York, and by the end of the week he was working at Liberty Street. With Tesla as designer, Brown as technical expert, and Szigeti as assistant, they began manufacturing their first AC induction motors. Peck, who along with Brown would be associated with Tesla for the next decade as a quiet backer, helped implement the patent applications by seeing investors in California, Pennsylvania, and New York.

Within a few weeks, *Electrical World* editor T. C. Martin stopped by the shop and coaxed Tesla into writing his first article on the invention. Immediately taken by him, Martin described the long-limbed electrician as having "eyes that recall all the stories one has read of keenness of vision and phenomenal ability to see through things. He is an omnivorous reader, who never forgets; and he possesses the peculiar facility in languages that enables the educated native of eastern Europe to talk and write in at least half a dozen tongues. A more congenial companion cannot be desired...the conversation, dealing at first with things near at hand and next...reaches out and rises to the greater questions of life, and duty, and destiny."[17]

T. C. Martin, with heavy emphasis in his signature on the C, was a complex person who would come to play a significant role in Tesla's life. In 1893 he edited the most important compilation of Tesla's writings assembled during his lifetime. Flamboyantly mustachioed and with large, round, soulful eyes and a shaved head, Martin, now married, had been a former seminary student who had emigrated from England when he was only twenty-one. Born in the same year as Tesla, Martin had worked for the Wizard of Menlo Park in the late 1870s before moving to the island of Jamaica. Returning to New York in 1883, he quickly became editor of *Operator and Electrical World.* Started in 1874 by the well manicured W. J. Johnston from a "little four-page telegraph sheet prepared and issued by Western Union operators in New York City, for circulation among their fellows,"[18] the *Operator* began to gain prominence after Thomas Edison started contributing significant pieces. As soon as Martin was hired, the paper's name was changed simply to *Electrical World.*

The following year, in 1884, T. C. Martin became vice president of the newly formed American Institute of Electrical Engineers (AIEE), and in 1886 his first book appeared, *The Electrical Motor and Its Applications.* A few months later, he was elected president of the AIEE.[19]

With his newfound prominence and very British attitude, T. C. Martin's sense of self-worth rose to the occasion. In very deliberate fashion, he organized a rebellion at *Electrical World,* with his coeditor Joseph Wetzler and a few other workers, against the owner, the proper, pedantic, and overbearing W. J. Johnston.[20] A capable editor in his own right, Johnston was forced to fire his editors and work on the journal himself, "as if Martin had never existed."

Along with Wetzler, Martin gained employment with *Electrical Engineer,* a competing company which gained great prominence when the duo climbed aboard. As a friend of Edison, and with his new base of operations, Martin was prepared to seize the moment. "An industrious writer with graceful style,"[21] T. C. Martin had the capability to cross over into higher social circles. He was a leader, an opportunist, egoist, and charmer. He was also one of the most influential personalities in the glamorous futuristic field of electrical engineering. Having discovered this new volcano of vision in Nikola Tesla, Martin approached him with the idea of helping choreograph Tesla's entrée into the electrical-engineering community.

The Serb was mysterious. He could rebuff lesser mortals and enjoyed the habits of a recluse. But Thomas Commerford Martin had tact and tenacity of purpose. He helped arrange for the esteemed engineering professor William Anthony, of Cornell University, to come to Liberty Street and test the new AC motors for efficiency. And Tesla reciprocated by traveling to Cornell to display his motors to Anthony and three other professors, R. H. Thurston, Edward Nicholas, and William Ryan. Anthony, who was twenty years their senior and a graduate of both Brown and Yale universities, had just retired from Cornell after fifteen years in order to take a position designing electrical measuring instruments for Mather Electrical Company in Manchester, Connecticut. Soon to be president of the AIEE himself, Anthony was pleased with his tests. Along with Martin, he helped coax Tesla into presenting his motor before the newly formed electrical society.

Martin had great difficulty persuading Tesla "to give any paper at all." Martin said that "Tesla stood very much alone, [as] the majority [of the electricians] were entirely unfamiliar with [the motor's] value." In haste, Tesla wrote out his lecture the night before in pencil. It had not been easy for him to construct an efficient machine, but having finally succeeded and having passed all of Professor Anthony's stringent tests for efficiency, "nothing now stood in the way of [its] commercial development...except that they had to be constructed with a view to operating on the circuits then existing which in this country were all of high frequency."[22]

On May 15, 1888, Tesla appeared before the AIEE to read his landmark paper "A New Alternating Current Motor." He had already filed for fourteen of the forty fundamental patents on the AC system, but he was still reluctant to fully announce his work. Realizing that the invention was worth at a minimum hundreds of thousands of dollars, Tesla and company sought investors through the advice of their new patent attorneys, Parker W. Page, Leonard E. Curtis, and Gen. Samuel Duncan, the last a leader of the firm and respected member of the New York Bar Association.[23] By the time of the lecture, Tesla, Peck, and Brown had already been negotiating with prospective buyers, such as Mr. Butterworth, a gas manufacturer from San Francisco, and, through General Duncan, George Westinghouse of Pittsburgh,[24] but nothing as yet was settled.

Westinghouse was already utilizing an alternating current system developed by the "erratic" French inventor Lucien Gaulard and the "sporty" entrepreneur John Dixon Gibbs of England.[25] In 1885 his manager of the electrical division, Guido Pantaleoni, had returned to Turin, Italy, to attend the funeral of his father. By coincidence, through his engineering professor Galileo Ferraris, whom Westinghouse himself had met while visiting Italy in 1882, Pantaleoni was introduced to Lucian Gaulard, who had installed his AC apparatus between Tivoli and Rome. Gaulard and Gibbs had already made headlines two years earlier when they first exhibited their invention at the Royal Aquarium in London; but in Turin the system won a gold medal and a prize of £400 awarded by the Italian government. Westinghouse purchased the American patent rights in late November after receiving a cable request from Pantaleoni.

The Gaulard-Gibbs system, although improved by the Hungarian ZBD system, still had serious problems. For Westinghouse, this was further complicated by the fact that Edison owned the option on the ZBD; and it was probably to block competitors that Edison purchased the system to begin with.[26]

In the same year, in America, after creating the Westinghouse

Electric Company, Westinghouse placed William Stanley in charge of the Gaulard-Gibbs modifications. Simultaneously, he brought to America Reginald Belfield, the engineer who had helped install the Gaulard-Gibbs system at the Inventions Exhibition in London two years earlier. Stanley, a frail, thin-faced temperamental "little man"[27] with piercing eyes, aquiline nose, wispy mustache, and Alfalfa hairdo, was a native of Brooklyn who had worked for Hiram Maxim, inventor of the machine gun. Although it was Westinghouse's idea to place Stanley in charge of the Gaulard-Gibbs apparatus, Stanley would later maintain that Westinghouse never fully understood the system until he got it in working order.[28] This appears unlikely, for it was a private joke among the upper echelon of the Westinghouse Company that Stanley had a penchant for claiming new discoveries when they became such to him. In any case, Westinghouse hedged his bets by establishing numerous DC central stations as well while research on AC was in progress.[29]

"Nervous and agile,"[30] Stanley was a hypersensitive individual who never really got along with Westinghouse. Due to ill health, and on the advice of the general manager, Col. Henry Byllesby, who proposed that success might be more forthcoming if Stanley separated himself from the pressures of the company, the inventor returned to his childhood summer retreat in the Berkshires in Great Barrington, Massachusetts, to work on the Gaulard-Gibbs system in private, taking Reginald Belfield with him. Stanley converted the Gaulard-Gibbs arrangement to parallel circuitry and independent control of separate fixtures and at the same time created a transformer which stepped up the AC from 500 volts to 3,000 when delivered along a transmission line and stepped them back down to original levels when entering households. This invention, although very similar to the ZBD configuration, was nevertheless patentable. It enabled AC to be sent three-quarters of a mile, or approximately one-quarter of a mile farther than the lower voltages of the prevailing DC systems.[31]

On April 6, 1886, George Westinghouse, along with Col. Henry Byllesby, traveled up to New Hampshire to witness the landmark apparatus for themselves. Prior to coming to Westinghouse, Byllesby had been employed at the Edison Machine Works as a mechanical engineer and was one of the designers of the Pearl Street station.[32] "From that time on," Byllesby said, "we progressed with amazing speed."[33] By the time of Tesla's lecture, Westinghouse noted that his company had "sold more central station[s]...on the alternating current system than all of the other electric companies in the country put together on the direct current system,"[34] but few engineers understood the principles involved.

In fierce competition with Westinghouse and a third player, Elihu Thomson of Thomson-Houston Electric Company, Thomas Edison had received a report on his own alternating current ZBD system. His engineers in Berlin indicated that the use of such high voltages was exceedingly dangerous.[35] Thomson, who himself had lectured at the AIEE a year before on the topic of AC, supported Edison's contention that AC was too risky.[36] Thus, at the time Tesla spoke, the battle of the currents had already begun, but the makeup of the contenders was complex. In 1886, fully two years before Tesla's high-voltage AC system became manifest, Edison had written to his manager, "Just as certain as death Westinghouse will kill a customer within six months after he puts in a system of any size. He has got a new thing and it will require a great deal of experimenting to get it working practically. It will never be free from danger."[37]

Tesla's lecture began with a brief description of the "existing diversity of opinion regarding the relative merits of the alternate and continuous current systems. Great importance," Tesla continued, "is attached to the question whether alternate currents can be successfully utilized in the operation of motors." He followed this preamble with a lucid description of the problems of the prevailing technology and his elegant solution, explained in words, diagrams, and simple mathematical calculations. The lecture was so thorough that many engineers, after studying the work, felt that they had known it all along:

> I have the pleasure of bringing to your notice a novel system of electric distribution and transmission of power by means of alternate currents...which I am confident will at once establish the superior adaptability...and will show that many results heretofore unattainable can be reached by their use....
>
> In our dynamo machines, it is well known, we generate alternate currents which we direct by means of a commutator, a complicated device, and the source of most of the troubles experienced....Now, the currents so directed cannot be utilised in the motor, but they must be reconverted into their original state....In reality, therefore, all machines are alternate-current machines, the current appearing continuous only in the external circuit during their transit from generator to motor.[38]

Since Tesla's lecture was dealing with fundamentals, it was easily understood, even though his invention was so revolutionary.

By demonstration in Tesla's laboratory after the lecture, the inventor showed that his synchronous motors could almost instantaneously be reversed. He also described, in precise mathematical calculations, how to determine the number of poles and speed of each motor, how to construct single-phase, two-phase, and three-phase motors, and how his system could be interlinked with DC apparatus. The lecture made use of entirely new principles.[39]

Now electricity could be transported hundreds of miles from a single distribution point, and not just for lighting streets or dwellings but for running appliances in households and industrial machinery in factories.

At the end of the talk, T. C. Martin called upon Professor William Anthony to present his independent tests of the Tesla motors. He had designed dynamos himself, which he had displayed a decade earlier at the Philadelphia Exposition of 1876. Tugging nervously at his scraggly beard, Anthony confirmed that the Tesla motors he had taken back to Cornell had an efficiency comparable to the best DC apparatus. "A little over 60%," he said, for the larger models. Moreover, the reversal of direction that the machines could achieve took place "so quickly that it was almost impossible to tell when the change took place."[40]

Boiling inside at having been anticipated by a newcomer, the persnickety professor Elihu Thomson stepped forward. Wanting to reestablish that his work in AC predated Tesla's, Thomson pointed out how their inventions differed: "I have been very much interested in the description given by Mr. Tesla of his new and admirable little motor," he managed with a pert smile. "I have, as probably you may be aware, worked in somewhat similar directions, and towards the attainment of similar ends. The trials which I have made have been by the use of a single alternating current circuit—not a double alternating circuit—a single circuit supplying a motor constructed to utilize the alternation and produce rotation."[41]

Unbeknown to Thomson, his words at that moment would come back to haunt him, because he had identified precisely the difference between the two creations. Whereas Thomson's single AC circuit had to still make use of a commutator and thus was highly inefficient, Tesla's system utilized two or more circuits out of phase with each other and constructed in such a way as to make the commutator obsolete. Tesla, of course, recognized the importance of Thomson's words and reiterated the point to establish clearly that his invention, just presented, was not analogous to Thomson's prior work:

"Gentlemen," Tesla began, "I wish to say that the testimony of such a man as Professor Thomson flatters me very much." Pausing with a smile and a bow of recognition, Tesla timed his coup de grâce with understated finesse. "I had a motor identically the same as that of Professor Thomson, but I was anticipated by him.... That peculiar motor represents the disadvantage that a pair of brushes [i.e., a commutator] must be employed."[42]

In this brief riposte Tesla claimed the high ground, and created an enemy who would fight him on this and other priority issues (e.g., the Tesla coil) for the rest of their lives.

Now Westinghouse had to move fast. He realized the value of the Tesla patent applications, having had nearly a month to look them over, along with the report from Professor Anthony.[43] A week after the lecture, on May 21, he sent Col. Henry Byllesby to Tesla's laboratory. Byllesby met with fellow engineer Alfred Brown on Cortland Street, where he was introduced to Charles Peck, the lawyer and major financial backer of the Tesla Electric Company. Together with a fourth man, Mr. Humbard, they went over to Liberty Street to meet the inventor and see the machines in operation.

"Mr. Tesla struck me as being a straight-forward, enthusiastic, sort of a party," Byllesby wrote to Westinghouse, "but his description was not of a nature which I was enabled, entirely, to comprehend. However, I saw several points which I think are of interest. In the first place, as near as I can get at it, the underlying principle of this motor is the principle which Mr. Shallenberger is at work on at the present moment. The motors, as far as I could judge...are a success. They start from rest, and the reversion of the direction of rotation is suddenly accomplished without any short-circuiting....In order to avoid giving the impression that the matter was one which excited my curiosity, I made my visit short."[44]

Back at Cortland Street, Brown and Peck informed Byllesby that he had to make a decision "by ten o'clock, Friday of this week," as the company was also negotiating with a Mr. Butterworth from San Francisco. They claimed that Professor Anthony had joined this California syndicate and was backing up Butterworth's offer of $250,000 in short-term notes and a royalty of $2.50 per watt of horsepower. "I told them the terms were monstrous," Byllesby said, "but they replied that they could not possibly hold the matter over longer than the date mentioned. I told them I thought there was no possibility of our considering the matter seriously, but that I would let them know before Friday."

Byllesby suggested that Westinghouse come to New York himself or send Shallenberger and another representative, but Westinghouse, who was familiar with the San Francisco syndicate, told Byllesby instead to stall them and try to secure more favorable terms.[45]

During the six-week interim, Westinghouse conferred with his specialists, Oliver Shallenberger and William Stanley, and his lawyer E. M. Kerr. Just three weeks prior to Tesla's lecture, Shallenberger had discovered "by chance" that a loose spring spun in "a shifting magnetic field. Directly he said to his assistant Stillwell who was also present..., 'There's a meter in that and perhaps a motor.' Within two weeks he designed and built a most successful alternating current meter of the induction type" which became standard for the field; and, like Tesla's creation, his apparatus utilized a rotating magnetic field.[46] Shallenberger, however, did not yet really understand the principles involved, nor had he had time to apply for a patent.

Stanley, on the other hand, claimed that there was nothing new in Tesla's creation. He pointed out that in September 1883 he had put the idea down in a notebook that an induction coil could be excited by AC. "I have built an AC system on basically the same principle which allows electromotive force to be transmitted from power stations to homes for the purpose of illuminating them," he told Westinghouse.[47] But the fact of the matter was, Stanley's system still used a commutator. His ego had gotten in the way of his ability to reason objectively that his scheme was not analogous to Tesla's.

Kerr reminded Westinghouse that unless he had a competing patent of sufficient strength, he would be powerless. Westinghouse was aware that Professor Ferraris of Turin, Italy, had published a paper on the rotating magnetic field one or two months prior to Tesla's lecture. Ferraris had also constructed discs that rotated in AC fields in university presentations as early as 1885. Tesla willingly admitted that "Professor Ferraris not only came independently to the same theoretical results, but in a manner identical almost to the smallest detail,"[48] but Ferraris wrongly concluded that "an apparatus founded upon this principle cannot be of any commercial importance as a motor."[49] Nevertheless, Kerr realized the legal importance of Ferraris's work. He suggested to Westinghouse that they purchase the U.S. patent options, so Pantaleoni was sent to Italy. He paid 5,000 francs, or about $1,000, for the rights.[50] But time was running out; the Tesla people would not wait forever. Westinghouse wrote Kerr:

> I have been thinking over this motor question very considerably, and am of the opinion that if Tesla has a number of applications pending in the Patent Office, he will be able to cover broadly the apparatus that Shallenberger was experimenting with, and that Stanley thought he had invented. It is more than likely that he will be able to carry his date of invention back sufficient time to seriously interfere with Ferraris, and that our investment there will probably prove a bad one.
>
> If the Tesla patents are broad enough to control the alternating motor business, then the Westinghouse Electric Company cannot afford to have others own the patents.[51]

Concerning the sticky point of royalties, which the Tesla syndicate placed at the audacious figure of $2.50 per watt, Westinghouse wrote, "The price seems rather high, but if it is the only method for operating a motor by the alternating current, and if it is applicable to street car work, we can unquestionably easily get from the users of the apparatus whatever tax is put upon it by the inventors."[52] Thus, in no uncertain terms, Westinghouse writes here the portentous statement that royalty payments could be passed on to customers, a concept he would be forced later to conveniently overlook.

6

INDUCTION AT PITTSBURGH (1889)

[My] first impression [was that of a man with] tremendous potential energy of [which]...only part had taken kinetic form. But even to a superficial observer, the latent force was manifest. A powerful frame, well proportioned, with every joint in working order, an eye as clear as crystal, a quick and springy step—he presented a rare example of health and strength. Like a lion in a forest, he breathed deep and with delight the smoky air of his factories.

NIKOLA TESLA ON GEORGE WESTINGHOUSE[1]

Although George Westinghouse had made his fortune with the invention of air brakes for trains, he was not just a railroad man. He was a descendant of the aristocratic Russian von Wistinghousen family; his father was also an inventor, with six fundamental patents of farming machinery. With his brother Henry (who later became his partner), George early on was introduced to such devices as the battery and the sparking leyden jar (a glass jar lined with foil and used for storing an electric charge). Having been a cavalry boy and, later, a navy engineer during the Civil War, George Westinghouse had experience and vision; he knew that the future was in electricity.

In late July 1888, Tesla took a train to Pittsburgh to meet with George Westinghouse and finalize the sale of his patents. It may have been the middle of summer, but oddly, the inventor welcomed the intense heat. He looked forward to the meeting.

Considerable in stature, with a walrus-sized mustache, Chester A. Arthur sideburns, and a remarkable wife of equal proportion who wore a bustle that jutted three feet to the rear, Westinghouse greeted the lanky inventor. A garrulous man, George Westinghouse subsumed those around him with his geniality and unbounded confidence. He took Tesla to his home and then on a tour of the plant. With nearly four hundred employees, Westinghouse's electric company was mainly producing "alternators, transformers and accessories for equipping central stations for supplying incandescent lighting."[2] Barrel-chested and physically expansive, Westinghouse counterbalanced in appearance the spindle-legged foreigner, who walked as "straight as an arrow, [with his] head erect...but with a preoccupied air as if new combinations were crystallizing in his brain."[3]

Tesla said:

> Though past forty then, [Westinghouse] still had the enthusiasm of youth. Always smiling, affable and polite, he stood in marked contrast to the rough and ready men I met. Not one word which would have been objectionable, not a gesture which might have offended—one could imagine him as moving in the atmosphere of a court, so perfect was his bearing in manner and speech. And yet no fiercer adversary than Westinghouse could have been found when he was aroused. An athlete in ordinary life, he was transformed into a giant when confronted with difficulties which seemed unsurmountable. He enjoyed the struggle and never lost confidence. When others would give up in despair he triumphed.[4]

Known for his foresight and courage, Westinghouse had already *quadrupled* the sales of his electric company, from $800,000 in 1887 to over $3 million in 1888, even though he was in the midst of expensive legal and propaganda battles with Edison.[5] Extraordinary in his ability to generate enthusiasm in his workers and a decisive man of action, he immediately gained the respect of those he met, particularly Nikola Tesla.

Westinghouse offered Tesla $5,000 in cash for a sixty-day option, $10,000 at the end of the option if they elected to purchase the patents, three notes of $20,000 at six-month intervals, $2.50 per watt in royalties, and two hundred shares of stock in the Westinghouse Company. Minimum payment on the royalties was calculated at "$5,000 for the first year, $10,000 for the second year, and $15,000 for each succeeding year thereafter during the life of the patents."[6] Westinghouse also agreed to pay for any legal expenses in litigation on priority issues, but a clause for lowering payments was added should any suits be lost. Calculated out, for fifteen years, this figure, minus the stock, came to $75,000 in initial outlays and $180,000 in royalty payments, or approximately $255,000.[7]

Tesla owned four-ninths of his company, the balance shared by Peck and Brown, presumably three-ninths to the former partner and two ninths to the latter.[8] Concerning total amounts paid out by Westinghouse, Tesla also referred to European patents, especially in England and Germany.[9] Thus, it is hard to determine exactly how much Tesla received for his forty patents. Westinghouse was not only getting a simple induction motor but also a variety of synchronous and load-dependent motors as well as armatures, turbines, regulators, and dynamos. Tesla may have sold additional inventions later on in separate agreements; the value of his stock holding is also unclear.

A decade later, Tesla wrote to another financier, John Jacob Astor, that "Mr. Westinghouse agreed to pay for my rotating field patents about $500,000, and, despite...hard times, he has lived up to every cent of his obligation."[10] Since Tesla was trying to raise money from Astor, he may have exaggerated the sum. Two years earlier, *Electrical Review* noted that the Westinghouse annual report listed the purchase of the patents at $216,000,[11] which is a figure that corresponds roughly to the Byllesby memorandum above, minus a few years' worth of royalty payments. If this was the case, then Tesla probably received for himself about half that figure, or $100,000, the entire amount paid in installments during the years 1888–97.[12]

During the negotiations, Tesla agreed to move to Pittsburgh to help develop his motor. It is quite possible that he received no salary for his stay there, for he had a peculiar "principle, ever since I devoted myself to scientific laboratory research, never to accept fees or compensations for professional services."[13] Tesla had been paid for his patents and was receiving royalties (or payments against royalties), so there was an income. Further evidence that no additional daily or weekly compensation was received is implied in a signed agreement by George Westinghouse, dated July 27, 1889, substantiating that Tesla worked in Pittsburgh for one year and that during that time he was paid with "one hundred and fifty (150) shares of Capital Stock." In return, Tesla promised to assign any patents to the Westinghouse Company which were directly related to the development of his induction motor patents. Other compensation was received from Westinghouse, however, for other contributions. For instance, when Tesla discovered that Bessemer steel created a vastly superior transformer than ones made out of soft iron, Tesla was paid approximately $10,000 for the idea.[14]

Tesla gave up his garden apartment in New York and moved into one of several hotels in Pittsburgh, including the Metropolitan, the Duquesne, and the Anderson.[15] Hotel living would become a lifestyle which he never departed from.

His talk, just two months old, had already catapulted him to fame. "About the middle of August 1888 in the Westinghouse testing room at Pittsburgh," Charles Scott, his assistant to be, remembered: "I had just come with the company and was assistant to E. Spooner who was running the dynamos testing room at night. He called me and said, 'There comes Tesla.'

"I had heard of Tesla," Scott continued, having "read [Tesla's] paper on the polyphase induction motor which my former college professor had pronounced as a complete solution of the motor problem. And now I was to see Tesla himself."

Fair-haired, with round, rimless glasses, Scott had only learned "that there was such a thing as alternating current" the summer before, in 1887. "I had...graduated from college two years earlier, and I wondered why I had not heard of such things from my professors." His only introduction was an *Electrical World* article by William Stanley, which was "a fascinating...key to many mysteries."[16] Now, a year later, he was to meet Nikola Tesla, the man who so elegantly solved all the puzzles proposed by Stanley. "There he came, marching down the aisle with head and shoulders erect and with a twinkle in his eye. It was a great moment for me."[17]

Scott, who later became an engineering professor at Yale University, was "Tesla's wireman...in preparing and making tests. It was a splendid opportunity for a beginner, this coming in contact with a man of such eminence, rich in ideas, kindly and friendly in disposition. Tesla's fertile imagination often constructed air castles which seemed prodigious. But, I doubt whether ever his extravagant expectations for the toy motor of those days measured up to actual realization...for the polyphase system which it inaugurated...exceed[ed] the wildest dreams of the early day[s]."[18]

Scott was not only Tesla's assistant, as time went on and against the opinion of many colleagues, he became a champion of Tesla's cause, a bearer of the truth, that is, that Tesla was the inventor of the induction motor. Another staunch supporter was Swiss immigrant Albert Schmid, coauthor of two AC patents with Tesla. Even though Westinghouse himself was also an ally, there was a bevy of other workers who tried seriously to strip Tesla of the crown. Major adversaries of the early period included Oliver Shallenberger, inventor of the AC meter, and his helpmate Lewis B. Stillwell, inventor of the Stillwell booster, which operated somewhat like the Tesla coil. At a later period, the key antagonist was Andrew W. Robertson, Westinghouse's chief executive officer.

Yet another opponent was William Stanley, the first American to have ever successfully instituted an AC system in the country. Stanley had split off from the Westinghouse Corporation (circa 1892–93) in order to sell his own polyphase motors, which were clear patent infringements on the Tesla system. This position was supported by the courts a few years later, and Stanley was forced to purchase the Tesla motors from Westinghouse.[19]

To fathom the depth of hostility that existed within the Westinghouse camp against Tesla, one need only read Lewis B. Stillwell's chapter on the history of alternating current, written forty years after the fact in a text entitled *George Westinghouse Commemoration.* Edited by Charles Scott, the book was widely distributed by the corporation and reissued in 1985. In the introduction to Stillwell's chapter it is recounted

> how Westinghouse brought the Gaulard-Gibbs system to America, how it was modified, and then given practical demonstrations by Stanley...and what has happened since.
>
> In 1888 came Shallenberger's brilliant invention of the induction meter. In the same year Nikola Tesla was granted his United States patents covering the polyphase motor and system. Westinghouse promptly secured the American rights. Tesla came to Pittsburgh to develop his motor. He made vain attempts to adapt it to the existing single phase, 133-cycle circuits....The *obvious advantages* [emphasis added] of direct connection of engines and generators called for a lower frequency....Two were selected as standard, namely 60 cycles for general use and 30 cycles for conversion into direct current.[20]

If we analyze the structure of this Stillwell quote, we note that although the topic sentence refers to Shallenberger, the entire paragraph is about Tesla. The word brilliant is used to describe an accidental discovery that a spring reacted to alternating currents[21] when no adjective is used to describe the inventor of an entire power system!

Tesla refers to the same situation in his autobiography: "My system was based on the use of low frequency currents and the Westinghouse experts had adopted 133 cycles with the object of securing advantages in transformation [because their Gaulard-Gibbs system operated at that frequency]. They did not want to depart from the standard form of apparatus and my efforts had to be concentrated upon adapting the motor to their conditions."[22]

With 120 power plants set up at 133 cycles per second, one can understand the predicament Tesla was placed in. Since Shallenberger's meter was compatible with the prevailing 133-cycle single circuit, it appeared logical that Tesla's polyphase motor could be made compatible as well.

In December 1888, Edison's propaganda battle against Westinghouse peaked when Edison began to allow H. P. Brown (who was not an Edison employee) to come to his Menlo Park laboratory in order to electrocute various animals with AC. A few months earlier, Brown had experimented in electrocuting animals at the School of Mines, a division of Columbia University, in New York City. Brown, an electrical engineer who lived on Fifty-fourth Street, had become upset over the many accidental deaths of his colleagues. He had collected a list of over eighty casualties, and although many of the men died because of DC, Brown decided that AC was the real culprit. Within two years, Brown began to manufacture electric chairs for various prisons which he sold for $1,600. He also planned to get paid to be the executioner. During the summer of 1888 the *New York Times* reported that he "tortured and electrocuted a dog.... First try[ing] continuous currents at a force of 300 volts...when the shock came the dog yelped.... At 700 volts he broke his muzzle and nearly freed himself. He was tied again. At 1,000 his body contorted in pain....'We will have less trouble when we try alternating current,' Mr. Brown said. It was proposed that he put the dog out of its misery at once. This was done on an alternating current of 300 volts killing the beast."[23]

A number of cities had adopted electrocution to rid the streets of unwanted canines, but the state of New York went a step further and set up a commission in 1886 "to report...on the most humane method of capital punishment."[24] Under the auspices of the Medico-Legal Society of New York, Brown arose as chief spokesman.

William Kemmler, a hooligan who had axed his mistress to death, became the test case for the use of electricity as a means of capital punishment.

Ostensibly, because the Westinghouse motors could produce the more deadly frequency, Brown surreptitiously purchased some working models in order to continue his gruesome experiments. Naturally, Westinghouse was upset over the devastating publicity. He and Tesla faced the possibility that the new AC polyphase system might never succeed in competition with existing AC and DC technologies, as both former systems required much lower voltages.

As Brown prepared to experiment with larger animals in order to assure the commission that electricity could kill criminals in a "humane" way, the Kemmler trial proceeded to question various electrical experts on the use of the Westinghouse currents for the electric chair.

Edison saw the controversy as a slick way to capitalize on the campaign against Westinghouse and the new Tesla technology. "Edison's scheme for electrical execution of criminals is the best so far presented. He proposes to manacle the wrists, with chain connections, place...the culprit's hands in a jar of water diluted with caustic potash and connected therein...to a thousand volts of alternating current...place the black cap on the condemned, and at a proper time close the circuit. The shock passes through both arms, the heart and the base of the brain, and death is instantaneous and painless."[25]

Able to fuel his vendetta, Edison provided access to his famous laboratory for Brown to "Westinghouse" twenty-four dogs, which he purchased from the local children at twenty-five cents apiece. Edison also "Westinghoused" two calves and a horse![26]

Perturbed, George Westinghouse wrote a letter of appeal to the *New York Times* which stated that AC was no more dangerous than DC, since people have been shocked and injured by DC as well.[27] Westinghouse assured the public of the safety of his system; so Brown, also in the *Times*, challenged Westinghouse a few days later "to meet me in the presence of the competent electrical experts and take through his body the alternating current while I take through mine a continuous current. The A.C. must have not less than 300 alternations per second."[28]

On July 23, 1889, Edison was questioned under oath by Kemmler's attorney, W. Bourke Cockran, an Irish immigrant schooled in France, in his second term in the House of Representatives. Having gained a reputation locally for fighting Tammany Hall, Cockran had also achieved national recognition as the "Boy Orator" for taking on William Jennings Bryan's presidential opponent, William McKinley in well-covered debates. Now he set his sights on tackling the Wizard of Menlo Park.[29]

QUESTION: Has Harold P. Brown any connection with yourself or the Edison Company?

EDISON: Not that I know of...

QUESTION: What would happen in case Kemmler should be kept on the chair several minutes with the current working on him?...Would he be carbonized?

EDISON: No. He would be mummyized...

QUESTION: This is your belief, not from knowledge?

EDISON: From belief. I never killed anybody...

"Finally, Mr. Cockran alluded to the rivalry between the Edison and the Westinghouse Companies and asked Mr. Edison if he loved Mr. Westinghouse as a brother. There was more than usual stillness, followed by Edison's answer: 'I think Mr. Westinghouse is a very able man.'...Mr. Cockran gave the 'wizard' a light from his cigar stump he had been chewing and dismissed him."[30]

It would be another full year before the actual execution took place, but public opinion continued to run against the dangerous Westinghouse current. Although Edison did not author the electric-chair ideas, he did everything he could to help the cause, providing his staff, especially the ingenious A. E. Kennelly, later a professor at Harvard, to aid Brown; in addition, he lent his name.

Outcries began to appear in various periodicals concerning the "electrical executioners." For example, the following editorial was published in a number of the papers and journals: "It is hard to conceive of a more horrible experiment than that which will be made on Kemmler....In a secret place, he will be compelled to go through a process of mental and moral, if not also, bodily torture and nobody can tell how long it will last."[31]

This ominous passage actually was not severe enough, for the execution of Kemmler became a nightmare. The job was completely bungled when, after electrocution, "to the horror of all present, the chest began to heave, foam issued from the mouth, and the man gave every evidence of reviving."[32]

The execution was likened to the work of barbarians and torturers and to scenes "worthy of the darkest chambers of the Inquisition of the 16th Century." One eyewitness who was completely disgusted was Dr. Jenkins, who told the *New York Times,* "I would rather see ten hangings than one such execution as this." Top electricians were also interviewed.

"I do not care to talk about it," Westinghouse said. "It was a brutal affair. They could have done better with an axe."

Even Edison was affected. "I have merely glanced over an account of Kemmler's death," he said, "and it is not pleasant reading.... One mistake in my opinion was in leaving everything to the doctors.... In the first place the hair on Kemmler's head was non-conductive. Then the top of the head I do not believe a good place to give a shock.... The better way is to place the hands in a jar of water... and to let the current be turned on there.... I think when the next man is placed in the chair to suffer the death penalty that death will be accomplished instantly and without the scenes at Auburn today."[33]

Although Westinghouse tried to distance himself from the nefarious deeds, his company still suffered greater damage than Edison's from a publicity standpoint because it was AC that was used to electrocute Kemmler. Mass hysteria threatened to overpower the attempt to institute the new Tesla AC invention, let alone the prevailing Gaulard-Gibbs AC system.

Tesla realized that eventually the company would have to come around to the lower frequencies if they wanted to use his creation, but to his shock, "in 1890, the induction motor work was abandoned."[34]

Westinghouse let it be known that his hands were tied, that his backers would not continue throwing tens of thousands of dollars away on futile research. They had given Tesla a fair chance to alter his equipment to satisfy the needs of the company. It seemed folly to destroy all prevailing equipment to satisfy the untried requirements of this new technology. Furthermore, they were against the idea of paying royalties should the motor eventually prove profitable. Enough was enough.

In a quandary, Tesla negotiated with Westinghouse a compromise solution. He would abandon the royalty clause of the contract if Westinghouse promised to commit his workers once again to the invention.

Westinghouse was in a corner. He knew that he had to curtail all work on the motor at this time to satisfy the tide of hostility that was rising against Tesla. He also realized that the invention was too important, and he believed that eventually a solution would be found. No one knows for sure exactly what happened, but it appears that Westinghouse made a tacit commitment to Tesla that he would get the company to resume work on the motor if Tesla removed the $2.50-per-watt royalty clause in the contract. If the motor came on the market and the polyphase system was adopted, the yearly figures cited above, as payments against royalties (worth approximately $255,000), would be honored instead.

Tesla was aware of the historical importance of his invention. He realized that it would *alter the world* beneficially in measurable ways. His motor, for instance, would provide an inexpensive replacement for potentially hundreds of thousands of hours of manual labor. At the same time, his creation would carve his name deep into the history books, alongside such heroes as Archimedes and Faraday. Moreover, he knew that his system was the most efficient, that it was fundamental, and that, if adopted, it would prevail. He also wanted very much to resume his preferred path of pioneer inventor.

Tesla was not counting out debits and credits on a balance sheet; rather, he viewed his partnership with Westinghouse in a more flexible way. He was also negotiating in good faith and assumed that if he lightened the potential financial burden, the company would somehow reciprocate. By offering goodwill, he was hoping to reap what he had sown. Speaking of Westinghouse many years later, Tesla said: "George Westinghouse was, in my opinion, the only man on the globe who could take my alternating current system under the circumstances then existing and win the battle against prejudice and money power. He was a pioneer of imposing stature and one of the world's noblemen."[35]

This, however, was a public statement; his private feelings were more complex. It is clear from reading through decades of letters to the corporation that Tesla maintained a close relationship with Westinghouse. Yet often there were undercurrents of resentment due mainly to lack of appreciation by the Westinghouse concern of Tesla's sacrifice and continuing contribution to the company. Tesla was also upset because the full scope of his patents became simplified and implications arose to suggest that he merely invented an induction motor and not an entire power system.

Finally, after nearly two years of inactivity, the Westinghouse people resumed their efforts to make the Tesla motor practicable. In 1891, Benjamin Lamme, a portly, easygoing, but studious youngster, began to reexamine Tesla's patents and Tesla and Scott's experimental motors. After conversing with Tesla in New York and talking over the matter with Scott, Lamme approached his overseers with a plan to resume work on the motor.

Lamme realized that Tesla had "exhausted all the possibilities" of trying to adapt his motor to the higher frequencies and that he was forced to "return to [the] low frequencies...insist[ing] on the superiority of his polyphase system."[36] This idea, as stated above, was rejected—most likely by Shallenberger and Stillwell. Lamme, as the junior engineer, had to proceed carefully. With Scott's aid, he "finally obtained permission" to take up the work on his own, although there is little doubt that a number of officials opposed the idea. "By this time, the 60-cycle system was coming in quite rapidly," Lamme said, so he suggested this frequency to the staff. Shallenberger "lost his temper and talked some plain-language to me." No doubt he said that there would be no possible way that they would utilize the lower frequencies. "This looked pretty serious to me, who as a mere boy in the test room, had got into a row with a chief technical authority of the company. I explained my situation to Mr. Schmid who simply laughed about it....However, somewhat to my surprise, Mr. Shallenberger always took my part, thereafter....This, of course, gave me a larger idea of the man himself; and I have always looked back with the greatest pleasure to my acquaintanceship with such a man."

What happened, most likely, is that Schmid, along with Scott, went behind the scenes and explained to Shallenberger that here was their opportunity to finally make use of the motor without giving any more credit to Tesla. They would simply let it be known that a new and brilliant engineer working at the company had "discovered" the efficiencies of lower frequencies, and so the credit would go to Lamme. No wonder Schmid laughed about it. With a way out, Shallenberger reversed himself and patted Lamme on the back, Lamme somewhat naively concluding that he built "the first induction motor...which bears any close example to the modern type....I [also]...designed the great generators for Niagara which were without precedent. They were marvels of engineering achievement."[37] Having rediscovered what Tesla had suggested all along, Lamme often made it seem as though he were the originator of the idea.

Uneducated readers, left with incomplete source materials, of which there are many, were forced to conclude that when it comes to the AC polyphase system, it was "that versatile genius B. G. Lamme, [who was the]...pillar of the Westinghouse company" who made it possible.[38] But people who read Scott carefully knew the truth: "Strenuous efforts to adapt the Tesla motor to [the prevailing] circuit were in vain. The little motor insisted in getting what it wanted, and the mountain came to Mohamet."[39]

7

BOGUS INVENTORS (1889–90)

Keely has discovered that all sympathetic streams, cerebellic, gravital, magnetic, and electric, are composed of triple flows; this fact governing all the terrestrial and celestial orders of positive and negative radiation.... He has discovered that the range of molecular motion in all quiescent masses is equal to one-third of their diameters, and that all extended range is induced by sound-force, set at chords of the thirds which are antagonistic to the combined chords of the mass of the neutral centres that they represent.

"Who Is the Greatest Genius of Our Age?"
REVIEW OF REVIEWS, 1890[1]

Tesla left Pittsburgh in the autumn of 1889 to return to New York and start his second laboratory, now on Grand Street. There he would begin work on high frequency apparatus, wireless transmission, and theories on the relationship between electromagnetic radiation and light. In particular, the inventor wanted to replicate the findings of the German academician Heinrich Hertz, a student of Hermann Ludwig von Helmholtz's who had recently published his landmark experiments in wave propagation. Tesla said that this work "caused a thrill as had scarcely ever been experienced before."[2] "I was not free at Pittsburgh," he continued. "I was dependent and could not work.... When I [left] that situation ideas and inventions rushed through my brain like a Niagara."[3]

Before he set up shop, the inventor traveled to Paris to attend the Universal Exposition and witness the unveiling of a gargantuan architectural triumph, the Eiffel Tower. Returning to a city filled with fond memories, Tesla could say hello to old friends and tell them how far he had come. There the budding creator could stop once again at the Louvre to gaze upon the "marvels" of Raphael, his self-perceived counterpart in the fine arts.[4] Nevertheless, Tesla also had mixed feelings, as he was traveling in the shadow of his nemesis, Thomas Edison, who not only attended the fair but who also provided a one-acre site for displaying his various inventions. In particular, the phonograph created a sensation, and Edison was received as a veritable demigod.

While Edison, who was accompanied by his new wife, Mina, just twenty-two years old, lunched with Alexandre Eiffel in his apartment at the top of the tower, Tesla met with Prof. Wilhelm Bjerknes at the fair to "witness the beautiful demonstrations [of his] vibrating diaphragms."[5] A Norwegian physicist from the University of Stockholm, Bjerknes had, along with Jules-Henri Poincaré, not only replicated the work of Heinrich Hertz on the propagation of electromagnetic waves through space; they had also, according to Hertz, discovered *multiple resonance* and had worked out the mathematics of the phenomena.[6] Tesla was able to study Bjerknes's oscillator, which provided a variety of electromagnetic waves and a resonator for augmenting them, and also discuss with him theoretical implications concerning the properties of the electromagnetic waves thereby produced.

And while the Wizard of Menlo Park met with Louis Pasteur at his laboratory in Paris and received the French Grand Cross of the Legion of Honor for his achievements, Tesla developed one of his most important discoveries, namely, that so-called Hertzian waves not only produced transverse oscillations, suggested by Bjerknes, but also longitudinal vibrations structured much like sound waves, "that is to say, waves, propagated by alternate compression and expansion...of the ether."[7] This conceptualization would come to play a pivotal role in the wireless transmitters Tesla would construct over the next decade.

As Tesla packed his bags for a brief visit to see his family,[8] Edison continued on his tour. In Italy he was honored by Queen Marguerite of Venice and King Humbert of Rome, in Berlin he met with Helmholtz at his laboratory, and at Heidelberg, Edison displayed his phonograph before a "monster meeting" of fifteen thousand people at which the machine "delivered a speech in good German."[9] Edison's favorite moment, however, was when he attended a large dinner given by Buffalo Bill, who was touring Europe with his Wild West Show. And as Tesla left for New York, Edison continued on to London, where he could visit his central stations but also learn firsthand that the new use of the Tesla AC was here to stay.[10] In Deptford, for instance, in this year, an engineer by the name of Ferranti installed probably the first-ever working single-phase generating station; with a Tesla system, he was able to transmit 11,000 volts seven miles away to London.[11] Although a truly epoch making achievement, this plant, for some unknown reason, received virtually no publicity.

This was a complex time in the history of the electrical industry; those electrical experts in Europe and America who took the time to study the Tesla creation saw its benefits immediately. In Switzerland and Germany polyphase induction motors were being constructed by C. E. L. Brown and Michael von Dolivo-Dobrowolsky, and in America Elihu Thomson of Thomson-Houston and William Stanley did the same. As in any branch of science, it was customary to study and replicate the work of others, but in electrical engineering success ensured not only a place in history but also a substantial profit. Thus, there would be many, such as most of those mentioned above, who would try and claim the polyphase system as their own.

There was another motor invention, however, called the hydro-pneumatic pulsating vacuo engine, which was on much safer ground; nobody could replicate the intricacies of its machinery, and nobody but the inventor, John Ernst Worrell Keely, knew how it worked. Keely had gotten the idea for the motor after reading the treatise "Harmonies of Tones and Colours, Developed by Evolution," by Charles Darwin's niece Mrs. F. J. Hughes, which discussed the structure of the ether and various theoretical harmonic laws of the universe.[12] Hailed as a virtual perpetual-motion machine, the Keely motor never ceased to intoxicate the public, for Keely had the uncanny ability to keep the secret going, but continually achieve abstruse results. "In the opinion of Madame Blavatsky, [Keely] has discovered Vril, the mysterious force of the universe in which Lord Lytton drew attention in his 'Coming Race'.... Keely calls it sympathetic negative attraction."[13]

Rivaling snake-oil salesmen in the ability to "humbug" the public, Keely, a former "circus sleight-of-hand performer," had formed a company in 1874, capitalized with $100,000 of stock in order to sell his motor, and had been successfully doing so for nearly fifteen years, until 1889, when his work was questioned.

Public Opinion wrote that, "Engineers, scientific men, and capitalists made frequent pilgrimages to Keely's Philadelphia laboratory to see the 'Keely motor mote.' Sometimes it moted and sometimes it didn't, but Keely always had a great tale to tell. Keely's chief accomplishment was a ready use of jargon of scientific and unscientific terms. He talked about 'triune currents of polar flow of force,' the 'reflex action of gravity,' 'chords of mass,' 'sympathetic outreaches of distance,' 'depolar etheric waves' and a lot of other things which didn't mean anything, but [he] never told why his motor moted and [why he] never took a patent" (although he did have them drawn up).[14]

T. Carpenter Smith, writing in *Engineering Magazine* described an eyewitness account with the "inventor" at work: "Mr. Keely proceeded to produce the force by striking a large tuning-fork with a fiddle-bow and then touching the generator with the fork. After two or three attempts, which he said failed by reason of not getting the 'chord of the mass,' he turned a small valve on the top of the generator. When a slight hiss was heard, loud cheers greeted his announcement that he had 'got it.'...The state of mind of the audience may be imagined when the shout of one enthusiast 'Keely, you are next to God Almighty!' seemed only a natural expression."[15]

This mountebank inventor headlined the New York dailies with his newest creations and also the accompanying cry of fraud. Factions of the public demanded he be jailed, the court giving Keely sixty days to "divulge his secret,"[16] but Keely held fast. Threatening to stop work on his inventions unless the court dropped its suit,[17] Keely was held in contempt of court and in November of 1888, he was placed in jail.[18]

Shortly thereafter, through his counsel, Keely revealed that the "missing link" was a "copper tube in the form of a hoop," and he was released on bail a few days later. He argued that he had indeed "obeyed all orders of the court" in explaining his invention in detail and the suit against him was overturned![19]

Like Edison, who promised and gave the world the "Lamp of Aladdin," and Tesla, who discovered and harnessed a purported alternating current perpetual-motion machine, Keely billed his own invention as the "greatest scientific discovery of the century."

Apparently vindicated, Keely proceeded with his deception. In 1890 the world celebrity and palmist Count Louis Hamon, better known as Cheiro, visited the Keely lab; and by 1895, John Jacob Astor became an investor.

Unfortunately for Tesla, however, like Keely, he, too, was gaining a reputation for making outlandish claims. For instance, Tesla said that his system could "place 100,000 horse-power on a wire" and send it hundreds of miles with almost no loss of power when the prevailing technology could only send a few hundred volts one mile, and in that case, with power dropping off markedly with distance.[20] Never mind that this prognostication came true just a few years later; Tesla's style was that of a visionary, and the claim seemed ridiculous. To the unimaginative, the uninformed, or people who listened only to the opposition, he was little different from Keely, and so he suffered through guilt by association.[21]

As far back as 1884, *Scientific American* had published an exposé on the Keely motor, speculating that its source of power was a secret chamber of compressed air. This was confirmed in 1898, after Keely's death, in an investigation of his laboratory conducted by Mrs. Bloomfield-Moore's son Clarence. Having waited for his mother, an ardent Keely admirer, to pass on before he tore the place apart, Moore discovered in the cellar a large tank and a series of pipes leading up through the floor above where the demonstrations took place. Keely's "'etheric force' was nothing more than compressed air [released by]...the pressure of his foot tapping on a concealed spring valve."[22]

Other bogus inventors of the day included Gaston Bulmar, who tried to sell General Electric (GE) special pills that turned water into gasoline; Walter Honenau, who derived *free energy* from an H_2O "hydro-atomizer" and "King Con," Victor "the Count" Lustig, who was eventually arrested for devising and selling a special money machine that cranked out crisp twenty-dollar bills from inserted white paper.[23]

In an age of new marvels that almost daily transformed society in irreversible ways, the public was "ripe for picking"; naive investors were often bilked by promises of impossible schemes. Thus, the inventor was perceived in contradictory ways, as executioner or light bearer, con man or wizard.

When Tesla returned from Europe, he wanted nothing better than to have his AC invention placed in the hands of its new rightful owner so that he could proceed with other burgeoning interests. Naturally, he would continue to aid Westinghouse in any way that he could, by continually offering advice to Scott, Schmid, or Lamme or stopping by at Pittsburgh to provide hands-on expertise. Throughout the 1890s, whenever the opportunity presented itself, Tesla would also introduce eminent prospective clients to the Westinghouse concern. As was his custom and style, Tesla never considered a commission for this service, although he did obtain vital equipment for his laboratory, which at first was provided without cost.

And as with Keely, Bulwer-Lytton's electrical-like energy called "vril power" had also been attached to Tesla by this time as well, via a letter from a lady in 1890, who "dreamed that if I [Tesla] would read the 'Coming Race' of Bulwer, I would discover great things which would advance [my work considerably]." But Tesla would not pursue the mystical treatise for a decade, and therefore it appears that although the inventions discussed in the story bore great similarities to some of Tesla's later creations, the reader must not say afterward that "the beautiful things which I shall invent were suggested by Bulwer."[24] Nevertheless, the similarities remain, and one wonders whether or not Tesla actually read the book at the time or knew of its contents.

8

SOUTH FIFTH AVENUE (1890–91)

At one bound [Tesla] placed himself abreast such men as Edison, Brush, Elihu Thomson, and Alexander Graham Bell.... His performance touched on the marvelous.

JOSEPH WETZLER IN *HARPER'S WEEKLY*, JULY 11, 1891[1]

Tesla returned from Paris during the summer of 1889 to his new laboratory near Bleecker Street. Down the road from one of Edison's showrooms, the lab took up the entire fourth floor of a six-story building located at 33-35 South Fifth Avenue (which today is called West Broadway). At the same time, he toured the hotels, moving into the Astor House, a posh five-story establishment situated by a trolley line in the heart of the city.

Over the summer, Tesla's "best friend," Anthony Szegeti, passed away. He wrote home to notify his family. "I feel alienated," he told Uncle Pajo, "and it is difficult [for me to adapt to the American lifestyle]."[2]

Now part of the nouveau riche, and also the star of the family, the inventor began sending money home to his mother and sisters and also to some of the cousins. Addressing the letters mostly to his sisters' husbands, all of whom were priests, Tesla wrote Uncle Pajo, "Somehow it is hard to correspond with the ladies."[3] Although occasionally he did write his sisters, mostly he just sent checks, and each of them would repeatedly write back to try and get a more personal word from "the only brother that we have."[4] Throughout the 1890s, Tesla sent several thousand forints, at 150 forints a clip, which was an amount equivalent to six months' rent at a well-to-do home or six months salary for a Serbian workman. Some of the funds were given as gifts, some to pay back his uncles for their aid in funding his education and sojourn to the New World, and other funds, which were partially obtained through European royalties, were used as investments.[5] To Uncle Petar, who had advanced to become a Metropolitan (Cardinal) in

Bosnia, Tesla revealed that he was receiving many letters from dignitaries and also such respect that it was difficult for him to describe.[6]

Uncle Pajo occasionally shipped European bottles of wine to his finicky nephew, unhappy with the selection in the United States. Impatiently waiting for these bottles was, for Tesla, like "waiting for the messiah."[7]

As Tesla's fame grew, and reports of his successes made the headlines in their local papers, Tesla became a virtual demigod to the Serbian and Croatian people, and a noble, though distant benefactor to his family. "We think about you even in [our] dreams," one of his brothers-in-law wrote.[8]

Except for occasional dinners with such friends as T. C. Martin or necessary trips to Pittsburgh, the inventor spent virtually all of his waking existence at the lab. His partner, Alfred S. Brown, would stop by to help when needed, but mostly Tesla worked with either one or two assistants or alone. As was his custom, he could labor seven days a week and around the clock, stopping only to freshen up at the hotel or for a necessary appointment. Monastic by choice and compelled by an all-consuming desire to be a major player in the burgeoning new age, the wizard preferred working through the night, when distractions could be minimized and concentration could be intensified.

Now free, he began his investigations along a number of separate but interrelated lines. As an experimental physicist, he began to study the difference between electromagnetic and electrostatic phenomena, and also the relationship of the structure of the ether to that of electricity, matter, and light. As an inventor, he began to design equipment for generating extremely high frequencies and voltages and for transforming direct into alternating current, or vice versa, and for creating uniform oscillations. Tesla also wanted to devise ways to manufacture light and to explore the concept of wireless communication. Already concerned about the fragility of the earth's natural resources, the finite supply of timber and coal, Tesla spent endless hours in contemplation, reviewing and replicating the findings of others, criticizing or improving upon their inventions, and also designing completely original creations. His goal was influenced by an evolutionary perspective and pragmatic considerations: He wanted to devise mechanical means for doing away with needless tasks of physical labor so that humans could spend more time in creative endeavors.

Unlike Karl Marx, who saw the worker becoming "an appendage of the machine,"[9] Tesla realized that machines could liberate the worker.

The inventor, in Tesla's eyes, had always been and always would be the light giver of the species, guiding its future through advanced technology. The masses, in turn, would benefit because machines would perform menial tasks so that they could pursue more intellectual occupations. With increased technology, cultural evolution would proceed at ever faster rates. "Conversely," Tesla warned, "everything that is against the teachings of religion and law of hygiene...tend[s] to decrease [human energy]."[10] Impure drinking water, in particular, was one of the greatest dangers.

Within the next eighteen months Tesla initiated most of the inventions that would occupy him for the next half century. During the last weeks of 1889, Martin met with him on several occasions in order to finalize his article on the Serb's heritage and plans for the future. The inventor would talk late into the night about his youth and the incessant struggle of his ancestors to fight off the diabolical Turks. As Martin took notes, Tesla outlined some of his inventions, particularly his work with high frequencies and his original theories on the relationship between electromagnetism and the structure of light. Martin tried to talk the inventor into presenting his ideas before the AIEE, but Tesla evaded a direct response. "Suppose I were to obtain for you Lord Kelvin's lectures? I know they are a bit wearisome, and somewhat beyond solution, but I believe that you, Samson like, can wrestle the honey from this lion's jaw."[11]

"Perhaps" was Tesla's reply.

On January 21, 1890, Professor Anthony took over the presidency of the AIEE from Elihu Thomson (who followed T. C. Martin) and opened the year with his own lecture on new electrical theories.[12] Happy to see the professor again, Tesla attended the seminar and was elected vice president. Participating in the discussion which followed, he was joined by Irish mathematician Arthur Kennelly of the Edison Company and Michael Pupin, a physics teacher and fellow Serb.

Having just returned from Helmholtz's laboratory in Germany, Pupin was unaware of the extent of the animosity that existed between the Edison and Westinghouse camps.

Pupin was from Idvor, a Serbian town north of Belgrade. His father had been a *knez,* or village leader, much like Tesla's father, but unlike Milutin Tesla, Mr. Pupin was an illiterate peasant and not part of the clerical aristocracy. Many of Pupin's relatives, like Tesla's, were war heroes who had fought off the Turks to protect the empire; and, like Tesla, Pupin had avoided military service.

Michael Pupin emigrated to the United States in 1874. After working at odd jobs, he entered Columbia College in New York in 1878. Graduating in 1884 with a keen interest in electrical theory and with honors, Pupin received a fellowship to study abroad. He wanted to go to Cambridge to learn under James Clerk Maxwell, but he found upon his arrival that Maxwell had been dead for four years.[13] This tendency to overlook the obvious appears to be a theme that runs through Pupin's life. After Cambridge, he went on to the University of Berlin, where he received a doctorate in physics. In 1889 he returned to New York to become an instructor at Columbia College.

In February 1890, the full-page Martin article on Tesla was published in *Electrical World* accompanied by a very prominent photograph of the youthful-looking engineer. For Tesla it was excellent publicity, the first major essay to appear portraying the up-and-coming inventor.

A meeting of the AIEE, devoted entirely to the new Tesla AC system, was planned for the following month. In particular, the conference had been sparked by a number of important developments; most notably plans coming from Switzerland and Germany for a proposed long-distance AC power-transmission experiment,[14] the impending success of the Westinghouse Company in instituting a hydroelectric plant utilizing the Tesla AC system at a mining camp in Telluride, Colorado, and the announcement of an International Niagara Commission to look into the best way to harness Niagara Falls.

At the March AIEE meeting, Prof. Louis Duncan was the main speaker; his lecture began with a mathematical dissection of the workings of "the novel and admirable little machine invented by Mr. Tesla." A former officer from the U.S. Naval Academy, Duncan had recently transferred from the South Pacific to Johns Hopkins University, where he stayed on to teach. An important ally, he gave the Tesla invention academic credibility. "The great advantage of the motor," Duncan said, "lies in the fact that it has no commutator and it permits the use of very high voltages. In the future, power will be transmitted electrically at voltages that will make machines with the commutator next to useless." After the lecture, there was a discussion, and Tesla participated.[15]

Pupin, who spoke that summer in Boston and again in New York the following year on "Alternating Current Theory," was fast becoming an admirer of Tesla's work. However, at the same time, Pupin was also becoming embroiled in the controversy as to who was the real author of the AC polyphase system, and from Tesla's vantage point, Pupin made the mistake of befriending the wrong people.

At the Boston meeting, Pupin "noticed that my audience was divided into two distinct groups; one group was cordial and appreciative, but the other was as cold as ice. The famous electrical engineer and inventor, Elihu Thomson, was in the friendly group, and he looked me up after the address and congratulated me cordially. That was a great encouragement and I felt happy." Other prominent individuals, however, tried to have Pupin fired from the electrical engineering department at Columbia because of his adherence to AC,[16] but Pupin overrode the controversy and at the same time increased his friendship with Thomson.

Unbeknown to Pupin, Thomson himself was in a quandary because he now recognized the clear advantage of the Tesla system, but he was locked out of its use because Westinghouse owned the patents.

Although the Thomson-Houston Electric Company was extremely profitable, the concern faced certain doom if it was unable to use efficient AC machinery. As Elihu Thomson had worked with AC for over a decade, he felt fully justified in adapting a Tesla-like system, especially because there were other engineers who also claimed legitimate priorities of aspects of the system, notably Shallenberger and Ferraris. Furthermore, Thomson himself had come close to conceiving a similar workable plan. That Tesla held fundamental patents on a completely revolutionary invention was continually overlooked by Thomson as he sought ways to rationalize his position while working at his company, presiding at the AIEE, and writing in the electrical journals. He had successfully circumvented the Edison lightbulb patents by paying Sawyer for a license to produce the stopper lamp, (an exhausted light bulb similar to Edison's which used a rubber plug—stopper—to hold in the vacuum) and so he sought a similar tack with AC.

During a heated series of articles in *Electrical World* between Thomson and Tesla, it appears that Thomson admits to some of his distaste for his rival when he writes, "I confess that my statement as to the motive of my critical remarks may have been out of place. They were elicited, however, by Mr. Tesla having on a former occasion misunderstood my motives."[17] And so Pupin's feelings for his Serbian brother became undermined as his friendship with Thomson grew.

Tesla agreed to present his work in high-frequency phenomena during a three-day symposium of the AIEE, which was held in May 1891, three months after he first published his research on the topic.[18] A hall was booked at Columbia College, which, at the time, was located between Park and Madison avenues, on Forty-ninth and Fiftieth Streets, and the public was invited.[19]

It is difficult to calculate the enormous impact that this lecture had on the engineers of the day and on the course of Tesla's life, for it is clear that after the event Tesla was perceived in an extraordinary way. Joseph Wetzler, or Josh, as he preferred to be called, covered the talk for *Electrical World.* But Tesla's lecture was too important for a single exposure in a journal of limited circulation, and Wetzler was able to also peddle the piece as a spectacular full-page account in the prestigious *Harper's Weekly.*

"[With] lucid explanations in pure nervous English," Wetzler proclaimed, "this stripling from the dim border-land of Austro-Hungary...[had] not only gone far beyond the two distinguished European scientists Dr. Lodge and Professor Hertz in grasp of electro-magnetic theory of light, but...he had actually made apparatus by which electrostatic waves or 'thrusts' would give light for ordinary every-day uses." Tesla had not only presented a remarkable display of electrical phantasmagoria, he had also outlined new "fundamental and far reaching principles."[20]

Wetzler cogently pointed out that Tesla had gone far beyond Heinrich Geissler and Sir William Crookes in the production of light by use of vacuum tubes, and he had also "eclipsed" the Wizard of Menlo Park with his refinements of the incandescent lamp. "But Mr. Tesla was not satisfied with these results, brilliant as they were. He had set himself no less a task than to create a lamp which, without any external connection to wires...would glow brightly when placed anywhere in an apartment."[21] And if that were not enough, for a conclusion, Tesla passed tens of thousands of volts of AC through his body to light up lightbulbs and shoot sparks off his fingertips and to show the world that it was not at all a killer current when utilized correctly. "Exhausted tubes...held in the hand of Mr. Tesla...appeared like a luminous sword in the hand of an archangel representing justice," said one reporter.[22]

With Gano Dunn, his assistant,[23] Tesla began his talk somewhat nervously but gained momentum as he progressed: "Of all the forms of nature's immeasurable, all-pervading energy, which ever and ever change and move, like a soul animates an innate universe, electricity and magnetism are perhaps the most fascinating.... We know," Tesla continued, "that [electricity] acts like an incompressible fluid; that there must be a constant quantity of it in nature; that it can neither be produced or destroyed...and that electric and ether phenomena are identical." Having set up the premise that our world is immersed in a great sea of electricity, the wizard proceeded to astound the audience with his myriad experiments. And at the end, Tesla's matched his opening in poetic expression: "We are whirling through endless space, with an inconceivable speed," he said. "All around us everything is spinning, everything is moving, everywhere is energy." Based on this premise, Tesla ended with a prophetic supposition, which has often been interpreted by some to suggest that a zero point, or free energy strata, exists. "There must be some way of availing ourselves of this energy directly," he said. "Then, with the light obtained through the medium, with the power derived from it, with every form of energy obtained without effort, from stores forever inexhaustible, humanity will advance with giant strides. The mere contemplation of these magnificent possibilities expands our minds, strengthens our hopes and fills our hearts with supreme delight."[24]

Those who attended the lecture, including Professor Anthony, Alfred S. Brown, Elmer Sperry,[25] William Stanley, Elihu Thomson, and Francis Upton, would remember the historic occasion for the rest of their lives.[26] Robert Millikan, for instance, who later won a Nobel Prize for his work with cosmic rays, was a graduate student at Columbia at the time. He said many years later, "I have done no small fraction of my research work with the aid of the principles I learned that night."[27]

Michael Pupin, who also attended, however, appears not to have been so entranced. "[While] I was lecturing," Tesla told a Serbian reporter, "Mr. Pupin, with his friends, [most likely, Elihu Thomson and Carl Hering] interrupted...by whistling, and I had difficulty quieting down the misled audience."[28] Pupin wrote to Tesla to smooth things over and to set up an appointment to see the motor, as he was scheduled to deliver a number of lectures on polyphase currents, but Tesla avoided him.[29] In Europe rumors began to circulate about the new electrical Svengali in America, and Tesla was quickly invited to speak before European scientific societies.

9

REVISING THE PAST (1891)

Many of the investigations of the book apply to polyphase systems circuits [with chapters] on induction motors, generators, synchronous motors, [etc.]....A part of this book is original, other parts have been published before by other investigators....I have, however, omitted altogether literary references, for the reason that incomplete references would be worse than some, while complete references would entail expenditure of much more time than is at my disposal....I believe that the reader...is more interested in the information than in knowing who first investigated the phenomenon.

CHARLES STEINMETZ[1]

Three months after Tesla's Columbia College lecture, in August 1891, two engineers, Charles Eugene Lancelot Brown, of the Swiss firm of Maschinenfabrik Oerlikon, and Michael von Dolivo-Dobrowolsky, representing the German firm Allgemeine Elektrizitäts-Gesellschaft (AEG), galvanized the engineering community when they successfully transmitted 190 horsepower from a waterfall at a cement factory on the Neckar River in Lauffen to the International Electrical Exposition, which was being held at Frankfurt, Germany, 112 miles away. With the support of three different governments, the lines passed through Wurtemberg, Bavaria, and Prussia before arriving in Frankfurt.[2]

By using oil as an insulator, as Tesla had explained in his Columbia College lecture, Brown was able to generate as much as 40,000 volts on his equipment, 25,000 of which he transported along the wires, stepping them down to usable frequencies when they reached the exposition. The efficiency of 74.5 percent astounded his colleagues. Dobrowolsky, who suggested that the invention was his conceptualization, utilized a three-phase AC with a working frequency of 40 cycles per second (instead of the single-phase current, with a frequency of 133 cycles per second, that the

Westinghouse people kept insisting on). The power was so great at the Frankfurt site that a large advertising sign with a thousand incandescent lamps was lit, and a motor pump powering an artificial waterfall was harnessed.[3]

On December 16, Michael Pupin delivered a talk before the AIEE on polyphase systems. Having given the same talk a week before at the New York Mathematical Society, Pupin was proud to advance abstract theories on this new field of polyphase systems. With his hair swept back, wearing wire-rimmed glasses, a brush mustache tapered at the ends, and a professorial three-piece suit, Pupin was fast readjusting to his return to the States. Now he was beginning to carve a name for himself. Dutifully, he had written Tesla before the talk to discuss his motors, but the inventor had eluded him.

In his opening remarks at the AIEE, with Arthur Kennelly, Elihu Thomson, Charles Bradley, and Charles Steinmetz present, Pupin referred to the "beautiful inventions of Nikola Tesla and the completeness of the success which Dobrowolsky and Brown obtained by the practical applications of these inventions," but he also described the German operation in such a way as to imply that aspects of it were independently conceived.

It appears that Tesla did not attend the lecture. Instead, he wrote Pupin the day after; but it was not to offer congratulations or to extend an invitation to meet. Tesla suggested that Pupin obtain the original patent specifications; the German design was simply a copy of his work.

But Pupin shot back, "I don't think that you ought to find fault with me for not having given your inventions a fuller discussion.... In the first place, it was a little bit too soon to discuss the practical details in a paper which treats of the most general fundamental principles of the polyphasal systems. Secondly, I know of your motors only by hearsay; I have not had the pleasure of being shown one by anybody.... I looked you up twice at your hotel and wrote you once... but all my efforts were in vain."[4] Pupin tried at the end of the letter to set up a personal interview, but Tesla was not one who could easily forgive such naïveté, especially from a Serb who spoke the native tongue so poorly.[5] To the hypersensitive Tesla, Pupin was a man who spread falsehoods. And his continual association with Thomson did not help. As Tesla was about to take a trip to Europe, a meeting of amelioration never took place.

Concerning the dispute over whose invention it really was, it is important to realize that obfuscation of the truth continues to this day.[6] The problem began with Michael von Dolivo-Dobrowolsky himself, who was reluctant to admit that he got the idea from Tesla, and it was perpetuated by his friend Carl Hering, who wrote a prodigious number of articles on the episode in the journals as the event unfolded throughout 1891. Hering had been a professor of engineering at the university in Darmstadt, Germany, in the early 1880s. His protégé Dobrowolsky, a native of St. Petersburg and the son of a Russian nobleman, replaced Hering when he retired from the university at the end of 1883.

C. E. L. Brown, a Swiss native and son of a designer of steam engines, began successfully transmitting electrical power with AC dynamos he had constructed while working in Lucerne. Brown, who was a year younger than Dobrowolsky and seven years younger than Tesla, had received most of his training in Winterthur and Basel, where he worked for the Burgin machine shops. In 1884 he began his employment at Oerlikon and within two years became director of operations.[7] On February 9, 1891, Brown delivered an address in Frankfurt on the subject of long-distance transmisssion of electrical power, and that is where he met Dobrowolsky. A partnership was formed between Oerlikon and AEG, and within seven months, their success between Lauffen and Frankfurt was achieved.[8]

Now, with Dobrowolsky's claims and Hering's one-sided reporting in the electrical journals,[9] factions of the American engineering community that were locked out of the Tesla patents could extol the Lauffen-Frankfurt venture while at the same time continuing to imply that Tesla's work was not intrinsic to its success. Ironically, the Westinghouse side wanted to downplay the event as well, not only because it proved Tesla right and them wrong but also because it dwarfed their success at Telluride. Thus, when perusing the Westinghouse literature, one is hard-pressed to find any mention of Lauffen-Frankfurt at all.

Pupin, in his lectures, did not support Tesla's role, nor did Kennelly, Thomson, or Bradley. Charles Proteus Steinmetz, however, was in another category. Like Pupin, he had just emigrated from Europe, and also like Pupin, he was academically oriented, with no particular economic stake in the invention at that time.

Steinmetz, who had fled Germany in 1889 in order to escape imprisonment for being a revolutionary socialist, was a brilliant student of mathematics from the University of Breslau. A dwarf hunchback, with his head sunk into his shoulders and one leg shorter than the other, Steinmetz had to continually overcome his odd appearance and frail disposition by displaying an advanced intellect. Just twenty-six years old and still attempting to cultivate a mustache and beard, Steinmetz, who was gaining a reputation for his work on the law of hysteresis (which involved a mathematical explanation that explained the lagging of magnetic effects when electromagnetic forces are changed) recognized some flaws in Pupin's talk. As this would be one of his earliest attempts to express himself before his peers in the difficult English language,[10] he carefully bolstered the addendum by bringing along calculations and drawings. Working in Yonkers, Steinmetz had developed a single-phase commutator motor a year before, in the summer of 1890.[11]

With brazenly long shoulder-length hair, the gnome was dressed in a slightly wrinkled three-piece suit adorned with an opulent watch chain and a pair of pince-nez that hung distinctively from a strap attached to his right collar. Standing to his full height of four feet and reaching for his glasses to read off his calculations, Steinmetz noted in his German accent that "Ferraris built only a little toy." He also went on to correct Pupin's implication that the Dobrowolsky creation was the first to use a three-phase system. "I cannot agree with that in the least, for [that] already [exists] in the old Tesla motor." Summing up, Steinmetz concluded, "I really cannot see anything new...in the new...Dobrowolsky system."[12]

It would take a few months for Steinmetz to realize why his colleagues raised their eyebrows as he dashed all hopes for Dobrowolsky's claims. Nevertheless, they were impressed with his analysis and mathematical expertise. Elihu Thomson returned to his firm of Thomson-Houston in Lynn, Massachusetts, with the knowledge that a new mathematical genius had arrived from Europe, and soon thereafter Thomson-Houston offered Steinmetz a job at Lynn.

Meanwhile, in Pittsburgh, unbeknown to Edison, Westinghouse had been surreptitiously meeting with Henry Villard, Edison's financial backer, over a two-year period to discuss a possible merger. Villard, who had recently combined a number of smaller companies with Edison Electric to create Edison General Electric, was well aware that Edison did not get along with Westinghouse. Villard was an immigrant from Germany, the son of a Bavarian judge. Having tried in his earlier days to set up a "free soil" German settlement in Kansas, Villard was the individual who drove the golden spike in the Northern Pacific Railroad to link the West Coast with the East. He conferred with J. Pierpont Morgan, the real power behind the operation, and had Morgan send Edward Dean Adams, a longtime banking associate, to Menlo Park to try and get Edison to align with Westinghouse. Thriving on "beating the other guy," Edison would hear none of it. "Westinghouse," he said, "has gone crazy over sudden accession of wealth or something unknown to me and is flying a kite that will land him in the mud sooner or later."[13]

Legal fees in trying to protect the Edison lightbulb patents had already cost Edison $2 million and Westinghouse the same. The Edison camp had decided to sue Westinghouse rather than Thomson-Houston, because the Pittsburgh company had purchased United States Electric, the concern that held the competing patents from Sawyer-Man and Hiram Maxim, while Thomson-Houston had only a lease agreement. Thus, while two giants fought each other in what Edison called "a suicide of time," Thomson-Houston got rich.

On July 14, 1891, after many years of battles and appeals on priority of invention of the lightbulb, Judge Bradley ruled in favor of Edison. Although Westinghouse was caught with the wrong lightbulb patents, his Tesla AC power system was an asset worth attaining; but Westinghouse was proving difficult to negotiate with. Villard therefore began to make overtures to Tesla directly, but the inventor had to yield to Westinghouse's decisions.

"Dear Sir," Tesla wrote Villard in his tidiest penmanship, "I have approached Mr. Westinghouse in a number of ways and endeavored to get to an understanding...[but] the results have not been very promising....Realizing this, and also considering carefully the chances and probabilities of success, I have concluded that I cannot associate myself with the undertaking you contemplate." Tesla reluctantly concluded the letter by wishing the financier "best success in [his] pioneer enterprise."[14]

Villard switched tactics and approached Thomson-Houston with the thought of buying them out. He had gone up to Lynn, Massachusetts, in February and had continued secret negotiations with Charles Coffin, chief executive officer (CEO) of Thomson-Houston, throughout the summer. In December a meeting was held at 23 Wall Street, in Morgan's office, to finalize plans for a merger. After Morgan looked over the financial records of both companies, he realized that Edison Electric, which was in debt for $3.5 million, had less revenue than the smaller and solvent Thomson-Houston. Morgan thereupon reversed himself and suggested that Thomson-Houston buy out Edison Electric. Either way, he created a monopoly. Simultaneously, Morgan maneuvered Villard out of the company altogether—he had to blame someone for the problems—and Charles Coffin took control of the new consolidation. They named the company General Electric (GE).

Because of the enormous debt of his company and the possibility that he was working with inferior DC equipment, Edison had lost his edge. The thought of working with that patent pirate Elihu Thomson and the removal of his own name from the marquee made the electrical wizard, for the time, a beaten man. Although he stirred the hornets' nest before he left, Edison realized that a new age of electricity had arrived, one that would not countenance his commonsense, trial-and-error approach. Over a year before the actual merger was completed, he wrote Villard, "It is clear that my usefulness is gone....Viewing it from this light you will see how impossible it is for me to spur my mind, under the shadow of possible future affiliations....I would now ask you not to oppose my gradual retirement from the lighting business, which will enable me to enter into fresh and congenial fields of work."[15] And so Edison turned his interests to furthering the work of Edward Muybridge, a pioneer in motion pictures. In 1888 and 1891 he had his first patents on a device he called the kinetograph, and a few years later, he developed a fully working movie camera and projection system. In 1893, Edison could write to the elderly Muybridge[16] that he now had a peep-show device that people would pay five cents to see.[17]

The "Morganization" of GE created an even greater foe for Westinghouse but also a critical problem for GE. Whereas Westinghouse was blocked from using an efficient lightbulb, GE was blocked from generating AC. As the Edison patents were only valid for another two years, certainly Westinghouse was in the better position. But in 1891–92 it was still too early in the game to realize this. From the point of view of the courts, it was still undecided as to who the author of the AC polyphase system really was, even though Westinghouse had the Ferraris patent trump card to back those of Tesla's, and so, over the next few years, Westinghouse was forced to sue not only a number of subsidiaries of GE but also some independents, such as William Stanley, who were now producing polyphase systems on their own.

From GE's point of view, there was a whole host of patents in AC that Thomson owned, but any others that they could obtain would undoubtedly help in the legal arena. Thus, they approached Charles Steinmetz with a scheme to work on improvements on AC designs in such a way that they would obscure Tesla's role. Attracted to the intrigue, Steinmetz accepted the challenge.[18]

The fray between Westinghouse and GE took a new turn in the race to win the bid to light the upcoming Chicago World's Fair and to harness Niagara Falls. In the courts, the suits switched from lightbulbs to power generation, and at their respective plants attention turned toward a way to compete with the success achieved by Brown and Dobrowolsky.

For Westinghouse Corporation, Schmid, Scott, and Lamme could confer with Tesla, while Stillwell and Shallenberger brooded, and the money men reluctantly agreed to dismantle the very lucrative but outmoded Gaulard-Gibbs machinery. For GE, the situation was more complex. They had hoped that someone like Steinmetz or Thomson could come up with a competing design, but they hadn't realized that Tesla held all the fundamental patents. Quite simply, there was no other system. Tesla had understood the foundation. One couldn't proceed without him.

Thomson and Steinmetz were reduced to figuring out ways to somehow bypass the patents by designing "teaser currents"[19] or some other smokescreen device in order to pretend that they had created a separate invention. In a case of industrial espionage, Thomson-Houston apparently paid a janitor to steal the Tesla blueprints from the Westinghouse plant.[20] Embarrassed to explain how the blueprints ended up at Lynn, Thomson said that he needed to study the Tesla motor designs to make sure that his were different.

The intrigue must have triggered a variety of emotions in Steinmetz. He had already lived a clandestine life in Germany; by editing a radical socialist newspaper under a pseudonym during the so-called Reign of Terror, he had learned to use secret passwords at radical meetings and write with invisible ink, as when he carried love notes between his leader, the charismatic revoutionary Heinrich Lux, who had been jailed for his activities, and Lux's girlfriend. Although Steinmetz never renounced his affiliation with the socialist movement, he supported a rather unscrupulous capitalistic corporate structure that was motivated not only by the all-consuming profit motive but also by its ability to subvert the law to achieve its ends. This new situation thereby only served to heighten his contradictory nature.

His affiliation with the Machiavellian policies of GE induced Steinmetz to abandon his ideals. His opus on AC, *Theory and Calculations of Alternating Current Phenomena,* coauthored with Ernst Julius Berg, a colleague educated at the Royal Polytechnikum in Stockholm, and first published in 1897, just three years after Tesla's own compendium, omitted any reference to Tesla at all. (By the turn of the century, Berg's name on the cover, like Lux's love notes, disappeared.)

At the time, Tesla's book *The Inventions, Researches and Writings of Nikola Tesla,* edited by T. C. Martin, was a veritable bible for all engineers in the field. It included chapters on alternating-current motors, the rotating magnetic field, synchronizing motors, rotating field transformers, polyphase systems, single-phase motors, etc. That it does not appear in the bibliography of Steinmetz's work is astounding.

In the foreword to Steinmetz's second text, *Theoretical Elements of Electrical Engineering,* written in 1902, the author tries to explain why he omitted reference to the inventor of the AC polyphase system. "Of later years," Steinmetz wrote, "the electrical literature has been haunted by so many theories, for instance of the induction motor, which are incorrect."[21] This was a natural opening that might have catapulted Steinmetz into a discussion which would set the record straight, but he chose a pusillanimous path instead. This decision not only aided in obfuscating the truth as to the origin of the invention; it also bolstered his own image in the corporate community.

As these texts on AC would serve as important templates for subsequent writers, it was quite common in the later years for engineers to obtain degrees, study AC, and even write textbooks on the topic themselves and never come across Tesla's name.

Clearly, it was to GE's benefit to pretend that Tesla never existed, and it was to Westinghouse's benefit to pretend that the Lauffen-Frankfurt transmission had never occurred. The next generation of engineers, and those that followed, never realized that obfuscation had taken place; that is one of the main reasons why Tesla's name practically vanished.

Perhaps the most blatant case of misrepresentation occurred a generation later, when Michael Pupin published his Pulitzer Prize–winning autobiography *From Immigrant to Inventor*. Pupin was able to write long passages on the history of AC and ignore Tesla almost completely. Tesla's name appears only once, in passing, in the 396-page book.[22]

In this work Pupin described "four historical events, very important in the annals of electrical science," that is, the Lauffen-Frankfurt transmission, the harnessing of Niagara Falls, the formation of GE, and the lighting of the Chicago World's Fair by AC. Mentioning the Westinghouse concern only once as a company that was interested in AC, Pupin concluded, "If the Thomson-Houston Company had contributed nothing else than Elihu Thomson to... [GE], it would have contributed more than enough.... [Thus] the senseless opposition to the alternating current system... vanished quickly.[23]

In the preface Pupin had the audacity to write that "the main object of [my] narrative [is]... to describe the rise of idealism in American science, and particularly in physical sciences and the related industries.... [As] witness to this gradual development,... [this] testimony has competence and weight." Considering that Pupin is generally remembered fondly by the engineering world, it is my opinion that he failed to live up to the standards to which he aspired.

These attempts to alter the past turned the stomach of a number of key players, most notably C. E. L. Brown, of Oerlikon Works in Switzerland, and one of his top engineers, B. A. Behrend. A staunch man with a granite profile and hound-dog eyes, Brown, who, with Dobrowolsky, had been the first engineer to transmit electrical power over long distances with Tesla's AC invention, had learned of Tesla's work from British engineer Gisbert Kapp, who published Tesla's 1888 talk in his magazine *Industries*. Kapp, who authored one of the most "brilliant" textbooks on induction motors, wrote Tesla on June 9, 1888, to request the use of his paper for the magazine.[24]

Based on Tesla's treatise and Kapp's refinements, Brown was able to construct "[*before*] Westinghouse... probably... the first successful motor... in 1890."[25] Brown's succinct response, conspicuously placed in *Electrical World*, was directed specifically to Carl Hering, one of the first writers to imply that the invention was Dobrowolsky's. "The three-phase current as applied at Frankfort," Brown wrote, "is due to the labors of Mr. Tesla, and will be found clearly specified in his patents."[26]

Hering's first response was to continue the artifice. "I do not think,"

Hering said, "[that] Mr. Brown does proper justice to the real inventor of this modification of the Ferraris-Tesla system, namely, Dobrowolsky."[27] But Tesla demanded a more clear-cut communiqué. After a discussion with W. J. Johnston, who would later allow Hering to take over the editorship of *Electrical World,* Tesla was able to obtain the following response: "We desire to state right here," Johnston said, "that the *Electrical World* has over and over put itself on record as upholding Mr. Tesla's priority."[28] The magazine was also able to extract from Hering the following: "Dobrowolsky, though he may have been an independent inventor, admits that Tesla's work is prior to his."[29]

Although Hering was loath to admit Tesla's priority, at the same time he placed his finger on an important point: Tesla himself had not demonstrated physically that his system could be used for long-distance transmission. Certainly Westinghouse at that time was not aware of the vast benefits of the system. Had it not been for the success at Lauffen-Frankfurt, Tesla's apparatus might have evolved differently in America. Hering did not have access to various details of the Westinghouse motors, and that was because the work was not in the public domain. Great amounts of money were spent to keep the work private. Had a Lauffen-Frankfurt type of transmission occurred in America without Westinghouse's permission, it would have clearly been a case of patent piracy. Tesla issued patents in most of the industrialized countries, and it appears likely that Brown and Oerlikon licensed Tesla's patents and paid him for the privilege of using them.

Coincidentally, Gisbert Kapp's treatise, which was initially published in two installments in December 1890 in the *Electrician* in London, appears also to have been used extensively by Charles Steinmetz in 1891 and 1892, while he was constructing AC motors at a machine shop in New York before he was hired by Thomson, according to B. A. Behrend, author of one of the first definitive works on the AC motor. An émigré from Switzerland, Behrend came to work for the New England Granite Company, a division of GE, in 1896. Particularly upset by the tactics of such writers as Steinmetz in using other people's work and leaving their names out of the bibliography, Behrend would later become one of Tesla's most important allies. In the foreword of his book, Behrend stated: "The tendency to write books without references is due largely to the desire to avoid the looking-up of other writers' papers. The reader is not benefited by such treatment, as he may frequently prefer the original to the treatment of the author whose book he is reading. Besides, a knowledge of the literature of our profession is essential to an understanding of the art and to an honest interpretation of the part played therein by our fellow workers."[30]

Writing to Oliver Heaviside specifically about such authors as Steinmetz, Behrend quoted Huxley: "*Magna est veritas et praevalebit!*" translating and modifying the quote as follows, "Truth is great, certainly, but considering her greatness, it is curious what a long time she is apt to take about prevailing." The body of his book began with this sentence: "The Induction Motor, or Rotary Field Motor, was invented by Mr. Nikola Tesla, in 1888." Tesla's picture also appeared as the frontispiece.

Throughout his life, Behrend sought to set the record straight as to who the real author of the AC polyphase system was. When Westinghouse sued New England Granite for patent infringements, Behrend was placed in "an embarrassing and disagreeable" position; the high command, which stemmed from Wall Street, wanted him to testify against Tesla.

On May 3, 1901, Behrend wrote back to the attorney Arthur Stem, "My dear sir... You will see that I am now, even more than I have been before, of the opinion that it is not possible for us to bring forth arguments that could go to show the invalidity of the Tesla Patents in suit.... I cannot undertake this duty."[31]

10

THE ROYAL SOCIETY (1892)

The lecture given by Mr. Tesla...will live long in the imagination of every person...that heard him, opening as it did, to many of them, for the first time, apparently limitless possibilities in the applications and control of electricity. Seldom has there been such a gathering of all the foremost electrical authorities of the day, on the tiptoe of expectation.

ELECTRICAL REVIEW[1]

The rapid progress in the field of electromagnetic radiation, opened up by the findings of Sir William Crookes, Sir Oliver Lodge, and especially Hertz, induced in Tesla a mania to complete as many patents as he could. Summoning his prodigious powers of self-denial, depriving himself of sleep, and exerting the full potential of his will, Tesla unfurled his creations as fast as he could. It was at this time that the grand vision arose of wireless transmission of electrical power, and he simply abhorred the thought that someone else should invent it before he could. Thus, he began to build ever more powerful coils while at the same time continuing his numerous experiments in high-intensity lighting, ozone production, converting AC to DC, and wireless communication.

In February 1891, Tesla applied for the first of three portentous patents for the conversion and distribution of electrical energy.[2] This invention, which was finalized after his return from Europe, was the mechanical oscillator, a completely unique, multipurpose device. Unlike the Hertz spark-gap apparatus, which produced slow, rhythmic discharges, the Tesla oscillator supplied a smooth, continuous current which could not only generate hundreds of thousands, or even millions, of volts but could also be tuned to specific frequencies. Over his lifetime, Tesla said, "I developed not less than fifty types of these transformers...each complete in every detail."[3]

The device was, in essence, a small engine, with almost no moving parts. The "work-performing piston was not connected with anything else but was perfectly free to vibrate at an enormous rate. In this machine," Tesla proclaimed, "I succeeded in doing away with all packings, valves and lubrication [although the utilization of oil was intrinsic to its design.].... By combining this engine with a dynamo... I produced a highly efficient generator... [which propagated] an unvarying rate of oscillation."[4] Since the current was so "absolutely steady and uniform... one could keep the time of day with the machine."[5] In fact, the inventor also used the oscillator as a clock.

In June 1891, Tesla came upon an article by Prof. J. J. Thomson. This British scientist, whose work would lead him to a Nobel Prize as the discoverer of the electron, was in the process of directing electrical beams from cathode-ray tubes so as to study the structure of electromagnetic energy. These investigations prompted a vigorous exchange in the electrical journals between these two men[6] and inspired Tesla to "return with renewed zeal to my own experiments. Soon my efforts were centered upon producing in a small space the most intense inductive action."[7] Tesla would describe these exciting results to Thomson in person, six months later, during the lectures he gave in London.

That same year, Tesla took out two more patents on AC motors which he owed Westinghouse; he also took out a patent on an electrical meter and a condenser, and two on incandescent lighting.

On January 8, 1892, T. C. Martin, Josh Wetzler, and George Sheep sent Tesla an invitation "to dine, and spend an evening... before your journey to Europe."[8] Tesla's glass blower, David Hiergesell, provided all of the tubes necessary for the trip. Sailing on the sixteenth, Tesla arrived in London on the twenty-sixth. Sir William Preece provided a horse and carriage for the young inventor and invited Tesla to stay at his house.[9] Tesla's plan was to speak before the Institution of Electrical Engineers a week later "and leave immediately for Paris" to lecture before the Société Française des Electriciens.

It must have been gratifying that Preece took an interest, as he was part of the old guard. Twenty-two years Tesla's senior and one of the patriarchs of the British scientific community, Preece was an amiable gentleman, with a full rich beard, high forehead, wire-rimmed glasses, and an air of self-assurance. As head of the government's Postal Telegraph Office, Preece had worked with telegraphy as far back as 1860 and had brought Bell's telephone, along with Bell himself, to the British Isles in the mid-1870s. He had also been associated with Edison since 1877, having coined the term "Edison effect" after visiting the wizard in 1884 to study his work with vacuum lamps and a peculiar "effect" whereby electronic particles flowed through space from the negative pole to the positive.

Using this device as a voltage regulator Preece returned to England to show his colleagues, especialy Ambrose Fleming.[10]

After a few days of enjoyable company and a tour of London, Tesla relaxed, and on Wednesday, February 3, his discourse, entitled *Experiments with Alternate Currents of High Potential and High Frequency,* was presented.

"For a full two hours, Mr. Tesla kept his audience spellbound. Before such colleagues as J. J. Thomson, Oliver Heaviside, Silvanus P. Thompson, Joseph Swan, Sir John Ambrose Fleming, Sir James Dewar, Sir William Preece, Sir Oliver Lodge, Sir William Crookes, and Lord Kelvin, Tesla proclaimed the driving force of his motivation: 'Is there, I ask, can there be, a more interesting study than that of alternating current?'...We observe how th[is] energy...tak[es] the many forms of heat, light, mechanical energy, and...even chemical affinity....All of these observations fascinate us....Each day we go to our work...in the hope that someone, no matter who, may find a solution [to] one of the pending problems,—and each succeeding day we return to our task with renewed ardor; and even if we are unsuccessful, our work has not been in vain, for in these strivings...we have found hours of untold pleasure, and we have directed our energies to the benefit of mankind."[11]

"Any feature of merit which this work may contain," Tesla humbly stated, "was derived from the work of a number of scientists who are present today, not a few who can lay better claim than myself." Looking about the room and, with a gleam in his eye, Tesla continued.[12] "One at least I must mention....It is a name associated with the most beautiful invention ever made: it is Crookes!...I believe that the origin [of my advances]...was that fascinating little book [on radiant energy] which I read many years ago."[13]

Firing up his great coil, amid erupting thunderbolts, Tesla spoke as if a sorcerer; he announced that with his knowledge he had the ability to make animate that which was inert. "With wonder and delight...[we note] the effects of strange forces which we bring into play, which allow us to transform, to transmit and direct energy at will....We see the mass of iron and wires behave as though...endowed with life."[14]

Lamps suddenly burst forth in a variety of "magnificent colors of phosphorescent light." Tesla touched a wire, and sparks ejaculated from its end; he created sheets of luminescence, directed electrical "streams upon small surfaces," lit wireless tubes simply by picking them up, and "erased" them by "holding a wire from a distant terminal" [i.e., grounding the effect] or by grasping the tube with both hands, thereby "render[ing] dark" the area in between and pulling his hands apart in a steady stroke. And just as easily, he would rotate the tube in the "direction of axis of the coil" and reignite the glow.[15]

His theories on the relationship of wavelength to the structure and manufacture of light and his displays of wireless fluorescent tubes prompted one viewer to postulate that the future mode of lighting a dwelling might occur by actually "rendering the whole mass of the air in the room softly and beautifully phosphorescent."[16]

Tesla unveiled the first true radio tube in this second month of 1892, in the presence of all of the key forefathers of the invention of the wireless. In order to obtain the most perfect vacuum possible, the adept had extracted the air from a bulb that was contained inside another vacuum tube. Within this inner chamber, Tesla generated a beam of light "devoid of any inertia." By producing extremely high frequencies, he created an electric "brush" that was so sensitive that it responded even to the "stiffening of the muscles in a person's arm!" This brush tended to "circle away" from an approaching person, but always in a clockwise direction. Noting that the ray was extremely "susceptible to magnetic influences," Tesla speculated that its direction of rotation was probably affected by the geomagnetic torque of the earth. He further expected that this brush would rotate counterclockwise in the Southern Hemisphere. Only a magnet could get the stream of light to reverse its direction of rotation. "I am firmly convinced," Tesla stated, "that such a brush, when we learn how to produce it properly, may be the means of transmitting intelligence to a distance without wires."[17]

"Of all these phenomena," Tesla began, in the next phase, "the most fascinating for an audience are certainly those which are noted in an electrostatic field acting through a considerable distance. By properly constructing a coil," he continued, "I have found that I could excite vacuum tubes no matter where they were held in the room."[18]

Referring to the work of J. J. Thomson and J. A. Fleming on the creation of a luminous thread within a vacuum tube, Tesla went on to discuss different methods of exciting vacuum tubes by altering the wavelength or the length of the tube.

Setting up a fan as an analog and discussing the research of Preece, Hertz, and Lodge on the radiation of electromagnetic energy into the earth and space, Tesla then displayed "no wire" motors: "It is not necessary to have even a single connection between the motor and generator," he announced, "except, perhaps, through the ground...[or] through the rarefied air....There is no doubt that with enormous potentials...luminous discharges might be passed through many miles of rarefied air, and that, by thus directing the energy of many hundreds of horse power, motors or lamps might be operated at considerable distances from stationary sources."[19]

Based on research conducted the year before, which had been prompted by the work of J. J. Thomson in propagating streams of electrical energy, Tesla expanded upon his high-intensity button lamp, a device that could dematerialize or "vaporize" matter. This arrangement, as we shall see, is precisely the configuration required to create laser beams. Most likely, Tesla displayed actual laser beams at this time. However, neither he nor the other scientists present at the time recognized the unique importance of the directed ray, as it was part of a combination of other lighting effects which resulted in the disintegration of the material that was being bombarded.

There are two types of standard lasers which correspond to Tesla's work: (1) a ruby laser, which reflects energy back to its source, which in turn stimulates more atoms into emitting special radiation, and (2) a gas laser, which consists of a tube filled with helium and neon. High voltage is applied across two electrodes near the ends of the tube, causing a discharge to take place. In both instances, the excited atoms are contained in an enclosure and then reflected into one specific direction. They differ from ordinary flashlights not only because they emit a uniform wavelength of light but also because there is a pausing (metastable) state before the light is emitted.[20]

Tesla worked with lamps constructed in exactly these ways. The first he called a button lamp; the second, an exhausted or phosphorescent tube. Their prime function was as efficient illumination devices. Their secondary functions were as laboratory apparatus for a variety of experiments. In one tube filled with "rarefied gas...once the glass fibre is heated, the discharge breaks through its entire length instantaneously."[21] Another bulb "was painted on one side with a phosphorescent powder or mixture and threw a dazzling light, far beyond that yielded by any ordinary phosphorescence."[22]

"A common experiment [of mine]...was to pass through a coil energy at a rate of several thousand horsepower, put a piece of thick tinfoil on a stick, and approach it to that coil. The tinfoil would...not only melt, but...it would be evaporated and the whole process took place in so small an interval of time that it was like a cannon shot....That was a striking experiment."[23]

Tesla also constructed a type of button lamp which could disintegrate any material, including zirconia and diamonds, the hardest substances known to exist. The lamp was, in essence, a globe coated inside with a reflective material (like the Leyden jar) and a "button" of any substance, most often carbon, which was highly polished and attached to a source of power. Once electrified, the button would radiate energy which would bounce off the interior of the globe and back onto itself, thereby intensifying a "bombardment" effect. In this way the button would be "vaporized."[24]

Tesla next described precisely the invention of the ruby laser, over five decades before its reappearance in the middle of the twentieth century. The description is quite explicit:

> In an exhausted bulb we can concentrate any amount of energy upon a minute button...[of] zirconia...[which] glowed with a most intense light, and the stream of particles projected out...was of a vivid white....Magnificent light effects were noted, of which it would be difficult to give an adequate idea....To illustrate the effect observed with a ruby drop...at first one may see a narrow funnel of white light projected against the top of the globe where it produces an irregularly outlined phosphorescent patch....In this manner, an intensely phosphorescent, *sharply defined line* [emphasis added] corresponding to the outline of the drop [fused ruby] is produced, which spreads slowly over the globe as the drop gets larger....A more perfect result used in some of these bulbs [involves]...the construction of a zinc sheet, performing the double office of intensifier and reflector.[25]

The inventor's talk ended with the speculation that with improvements in the construction of long-distance cables, per his suggestions, telephony across the Atlantic would soon be possible. It is significant to note that at this moment he did not yet envision wireless transmission of voice, but rather wireless transmission of intelligence (i.e., Morse code), light, and power. His discussions with Preece, however, on the existençe of earth currents was beginning to take hold, and shortly afterward, Tesla began to conceptualize the idea of transmitting voice and even pictures by means of wireless.

"It has been my chief aim in presenting these results to point out phenomena or features of novelty," Tesla concluded, "and to advance ideas which I am hopeful will serve as starting points of new departures. It has been my chief desire this evening to entertain you with some novel experiments. Your applause, so frequently and generously accorded, has told me that I have succeeded."[26]

At the end of the lecture "Mr. Tesla tantalizingly informed his listeners that he had shown them but one-third of what he was prepared to do, and the whole audience...remained in their seats, unwilling to disperse, insisting upon more, and Mr. Tesla had to deliver a supplementary lecture....It may be stated, as Mr. Tesla mentioned but which hardly seems to be realized, that practically the whole of the experiments shown were new, and had never been shown before, and were not merely a repetition of those given in...America."[27]

Having seen the inventor handle such enormous voltages "so unconcernedly," many of the attendees mumbled surprise among themselves and gathered the courage to inquire how Tesla "dared to take the current through his body."

"It was the result of a long debate in my mind," Tesla replied, "but though calculation and reason, I concluded that such currents ought not to be dangerous to life any more than the vibrations of light are dangerous.... Consider a thin diaphragm in a water-pipe with to and fro piston strikes of considerable amplitude, the diaphragm will be ruptured at once," the inventor explained by analogy. "With reduced strokes of the same total energy, the diaphragm will be less liable to rupture, until, with a vibratory impulse of many thousands per second, no actual current flows, and the diaphragm is in no danger of rupture. So with the vibratory current." In other words, Tesla had increased the frequency, or alterations per second, but reduced the amplitude or power greatly. The wizard thereupon fired up the coil once again, sending tens of thousands of volts through (or around) his body and illuminated two fluorescent tubes which he held dramatically in each hand. "As you can, see," Tesla added, "I am very much alive."

"That we can see," one member responded, "but is there no pain?"

"A spark, or course, passes through my hands, and may puncture the skin, and sometimes I receive an occasional burn, but that is all; and even this can be avoided if I hold a conductor of suitable size in my hand and then take hold of the current."

"In spite of your reasons," another concluded with a shake of his head, "your speculation resembles to me the feelings that a man must have before plunging off the Brooklyn Bridge."[28]

In listening to Tesla's statement that he had only shown part of what he had prepared, the perspicacious Professor Dewar, inventor of the Dewar flask, or everyday hot or cold thermos, took the inventor at face value and realized that there was more information to impart. The wizard had simply run out of time. As a member of the board of the Royal Institution, also situated in London, Dewar knew that there were many dignitaries who missed the grand event, especially Lord Rayleigh, so he set himself the task of persuading Tesla to present an encore the following evening.

After the talk Dewar escorted Tesla on a tour of the Royal Institution, where he displayed the work of his predecessors, especially Michael Faraday's apparatus. "Why not stay for one more performance?" Dewar inquired.

"I must go to Paris," Tesla insisted, keeping foremost in his mind his desire to limit the time of his visits at each stop so that he could return to the States as quickly as possible.

"How often do you think you will have the chance to visit the laboratories of such men as Crookes or Kelvin?" Dewar asked in his Scottish brogue. At the same time, he invited Tesla to visit his own lab, where he was creating extremely low temperatures that approached absolute zero and conducting pioneer studies of electromagnetic effects in such environments as liquid oxygen.[29] "You've already lived in Paris. Now see London!"

"I was a man of firm resolve," Tesla admitted later, "but succumbed easily to the forceful arguments of the great Scotchman. He pushed me into a chair and poured out half a glass of wonderful brown fluid which sparkled in all sorts of iridescent colors and tasted like nectar."

"Now," Dewar declared with a twinkle in his eye and a grin that brought one of reciprocation on the face of his captive, "you are sitting in Faraday's chair and you are enjoying whiskey he used to drink."

"In both aspects," Tesla recalled, "it was an enviable experience. The next evening I gave a demonstration before the Royal Institution."[30]

At the culmination of the lecture, much of which, again, was new material not presented the previous evening (but integrated into the above discussion), Tesla presented Lord Kelvin with one of his Tesla coils,[31] and Lord Rayleigh took over the lectern for the conclusion. Tesla recalled, "He said that I possessed a particular gift of discovery and that I should concentrate upon one big idea."[32]

Coming from this "ideal man of science," one who had worked out mathematical equations concerning the wavelength of light and who had also calculated the atomic weights of many of the elements, this suggestion made a great impression. A new sense of destiny swirled through Tesla as he began to realize that he would have to figure out a way to surpass his earlier discoveries in AC.

The next day, Tesla received an invitation from Ambrose Fleming to visit his lab at University College on the weekend. Fleming had been successful in setting up "oscillatory discharges with a Spottiswoode Coil as the primary and Leyden jars as the secondary," and he wanted to show Tesla his results.[33] Having been a consultant to Edison in connection with the lighting industry, Fleming would four years hence work with Marconi in the development of the wireless and a few years after that, come to invent the rectifier, a device for converting the incoming electromagnetic waves of AC into DC upon entering the receiving apparatus.[34] Having attended both lectures, Fleming "congratulated [Tesla] heartily on your grand success. After th[is] no one can doubt your qualifications as a magician of the first order." The English aristocrat concluded by dubbing Tesla a member of the new fictitious "Order of the Flaming Sword."

Tesla had sparked the imagination of his British colleagues, and rapidly a number of them began to replicate his work and make their own advances. At Sir William Crookes's lab, Tesla constructed a coil as a gift and taught Crookes how to build Tesla coils on his own, but Crookes complained: "The phosphorescence through my body when I hold one terminal is decidedly inferior to that given with the little one [that you made for me]."[35]

As was his custom, Tesla toiled incessantly until the eclectic Crookes forced him to take a break, and at night, after dinner, the two scientists sat back and prognosticated. Topics ranged from discussions of the ramifications of their own research and potential future of the field to religion, Tesla's homeland, and metaphysics.

Twirling an elongated waxed mustache that fanned out like the tail feathers of a bird of paradise, the bearded mentor revealed that he had experimented in wireless communication before even Hertz began his investigations in 1889. Crookes discussed the possibility that electrical waves would be able to penetrate solid objects, such as walls, and he argued against Kelvin's suggestion that the life force and electricity were at some level identical. "Nevertheless, electricity has an important influence upon vital phenomena, and is in turn set in action by the living being, animal or vegetable." Here Crookes was referring to various species, such as electric eels, iridescent sea slugs, and lightning bugs. Further speculation caused the two men to discuss the possibility that electricity could be utilized to purify water and treat "sewage and industrial waste."

"Perhaps," Crookes suggested, "proper frequencies could be generated to electrify gardens so as to stimulate growth and make crops unappealing to destructive insects."

Expanding on the work of Rayleigh, Crookes discussed with Tesla the possibility of setting up millions of separate wavelengths so as to ensure secrecy in communication between two wireless operators. They also reviewed the work of Helmholtz on the structure of the physical eye, noting that receptors on the retina are "sensitive to one set of wave-lengths [i.e., visible light], and silent to others." In the same way, a receiving device for accepting electromagnetic signals might also be so constructed to receive certain transmissions and not others.

"Another point at which the practical electrician should aim," Crookes said in response to one of Tesla's more dauntless speculations, "is in the control of weather." Such goals as the elimination of fog or the ghastly "perennial drizzle" that plagued the island and creation of great amounts of rain scheduled for specific days were also discussed.[36]

And if this were not enough, Crookes also introduced Tesla to a vigorous discussion of his experiments in mental telepathy, spiritualism, and even human levitation. As a member of the Society of Psychical Research and later president, Crookes was in good company. Other scientists who would rise to the helm of the psychic society included Oliver Lodge, J. J. Thomson, and Lord Rayleigh.[37] Crookes straightforwardly presented a plethora of convincing evidence, including drawings by receivers that matched those created by senders, photographs from seances of ectoplasmic materializations generated by the clairvoyant Florence Cook, and eyewitness accounts of levitation by himself and his wife.[38]

Those statements were enough to raise the eyebrows of anyone, and they served to rattle Tesla's worldview. As a staunch materialist, up to that time Tesla had absolutely no belief in any aspect of the field of psychic research, including relatively tame occurrences, such as thought transference. But with Crookes's documentation and the support of other members of the cognoscenti, especially Lodge, and with Tesla already exhausted from the strain of his severe schedule, the Serb's mind began to spin. He would drop off in the middle of conversations and subsequently frightened his host. The reality that he had constructed and the world of superstition he thought he had left behind when he emigrated from the Old World swarmed through his brain like a hive of bumblebees and shattered mightily his worldview.

The pressure Tesla was under caused Crookes to offer some friendly advice in a letter. "I hope you will get away to the mountains of your native land as soon as you can. You are suffering from over work, and if you do not take care of yourself you will break down. Don't answer this letter or see any one but take the first train." Ending the letter on a waggish note, Crookes added, "I am thinking of [going] myself, but I am only thinking of going as far as Hastings."[39] Tesla wanted to take his advice, but he had to address the Paris society first.

Tesla crossed the English Channel during the second week of February and booked a room at the Hotel de la Paix. At his upcoming lecture "before a joint conference of the Société de Physique and the Société International des Electriciens," which was held on February 19,[40] the inventor sought out the well-known French physician Dr. d'Arsenoval, a pioneer in the field of diathermy. Tesla said later:

> When...Dr. d'Arsenoval declared that he had made the same discovery [concerning the physical effects caused by sending extremely high frequency through the body], a heated controversy relative to priority was started. The French, eager to honor their countryman, made him a member of the Academy, ignoring entirely my earlier publication. Resolved to take steps for vindicating my claim, I...met [with] Dr. d'Arsenoval. His personal charm disarmed me completely and I abandoned my intention, content to rest on the record. It shows that my disclosure antedated his and also that he used my apparatus in his demonstrations. The final judgement is left to posterity.
>
> Since the beginning, the growth of the new art [of electrotherapy]...and industry has been phenomenal, some manufacturers turning out daily hundreds of sets. Many millions are now in use throughout the world. The currents furnished by them have proved an ideal tonic for the human nerve system.

> They promote heart action and digestion, induce healthful sleep, rid the skin of destructive exudations and cure colds and fever by the warmth they create. They vivify atrophied or paralyzed parts of the body, allay all kinds off suffering and save annually thousands of lives. Leaders in the profession have assured me that I have done more for humanity by this medical treatment than all my other discoveries and inventions.[41]

(More recently, a number of researchers, particularly Dr. Robert O. Becker of Syracuse University, have utilized electrical currents to help heal bones that have difficulty knitting. Having studied regeneration capabilities of such reptiles as salamanders whose tails he amputated, Becker has discovered that these animals generate a particular electrical frequency which serves somehow as a field for promoting the total regrowth of the missing appendage. By "artificially duplicating" the signal, Becker reports, "we have been able to produce partial limb regeneration in [mammals such as] rats by a similar technique, and some clinical applications are under study in human beings at this time."[42])

The Paris lecture ignited "the French papers... [with Tesla's] brilliant experiments. No man in our age has achieved such a universal scientific reputation in a single stride as this gifted young electrical engineer," the *Electrical Review* reported.[43]

Tesla met with a number of dignitaries while in Paris, including Prince Albert of Belgium, who was interested in supplying his country with a more economical means of distributing electric power; Monsieur Luka, of the Helios Company of Cologne, with whom Tesla sold his AC motor patents for use in Germany;[44] and André Blondel, an important theoretician in advanced theories with alternating currents.

Forty years later, Blondel recalled "with immense interest and admiration" the Paris conference and congratulated Tesla on the elegant simplicity with which he advanced his concepts in alternating current well beyond the work of his French colleague Deprez and his Italian neighbor Ferraris.[45]

Shortly after the lecture and in a state of "oblivion" associated by "my peculiar sleeping spells, which had been caused by prolonged exertion of the brain," Tesla received a dispatch at the hotel informing him that his mother was dying. "I remembered how I made the long journey home without an hour of rest."[46] Greeted in Gospić by his three sisters, all of whom were married to Serbian priests, and by his Uncle Petar, the regional bishop, Tesla was in a terrible state. Entering the bedroom, he found his mother in "agony."

During this time, Tesla suffered from a peculiar malady similar to amnesia; where he claimed to have lost all memory of his earlier life. He also said that he slowly regained this information before his return to the States. One aspect of this episode which is noteworthy is that it occurred over a long period of time, beginning at the close of 1891, and culminating with his mother's death in April 1892. Tesla said that although he could not remember historical occurrences, he had no trouble thinking about the details of his research including "passages of text" from his writings "and complex mathematical formulae."[47] Simultaneously, Tesla also experienced a psychic event which "momentarily impressed me as supernatural. I had become completely exhausted by pain and long vigilance, and one night was carried to a building about two blocks from our home. As I lay helpless there, I thought that if my mother died while I was away from her bedside she would surely give me a sign."[48]

Having been influenced by "my...friend Sir William Crookes, when spiritualism was discussed," Tesla lay in anticipation. "During the whole night every fiber in my brain was strained in expectancy, but nothing happened until early in the morning. [Awakening in] a swoon, [I] saw a cloud carrying angelic figures of marvelous beauty, one of whom gazed upon me lovingly and gradually assumed the features of my mother. The appearance slowly floated across the room and vanished, and I was awakened by an indescribably sweet song of many voices. In that instant [or] certitude, [I knew] that my mother had just died. And that was true."[49]

Surprised and perhaps even frightened by the clairvoyant vision, Tesla wrote to Crookes for advice. For months, and maybe even for years after, the inventor "sought...the external cause of this strange manifestation." Note how Tesla assumes a priori that the cause came from "outside" as opposed to "inside," that is, from the unconscious. Although he readily accepted the concept of wireless communication, in no way was Tesla capable of allowing for the possibility that the human brain could also act as a receiver of mental vibrations. The idea of telepathy, or spiritualism, for that matter, was a true threat to the paradigm he was operating from, and so Tesla manufactured a physical mechanism as the cause of his noetic experience:

"To my great relief," Tesla wrote, "I succeeded after many months of fruitless effort" in solving the conundrum. The vision of the angels rising up, Tesla attributed to the memory of an ethereal painting of the same subject which he had gazed on prior to the experience, and the sound of the serenading voices, he linked to a nearby church choir that was singing for an Easter mass.[50] Whether or not Tesla's mother died on a Sunday morning is not known. But what is clear is that this analysis greatly relieved the tension Tesla was under, for it supported, once again, a materialistic viewpoint.

A question which remains, however, is whether or not Tesla's excesses in work-related endeavors is enough to explain his onset of amnesia? One theoretician speculated that the enormous voltages that Tesla passed through his body may have contributed to the problem.[51] From a psychoanalytic point of view, one could speculate that Tesla was repressing, that is, unconsciously, but purposefully forgetting events that he did not want to remember. Possible unwanted memories included the way he felt as a child after his exalted brother died and the recent eradication of the royalty clause with the Westinghouse Corporation.

After the death of his mother, Tesla stayed on in Gospić for six weeks to recuperate. On the positive side, it enabled the lone son to reconfirm emotional ties to his family; it also provided Tesla with probably the only extended vacation he would ever take.

He traveled to Plaski to visit his sister Marica, to Varazdin to see his uncle Pajo, and to Zagreb to lecture at the university. He sojourned to Budapest to confer with Ganz & Company, as they were in the midst of constructing a substantial 1,000-horsepower alternator. He also met with a delegation of Serbian scientists who accompanied him down to Belgrade, where an audience was arranged with the king. Young Alexander I conferred upon Tesla a special title of Grand Officer of the Order of St. Sava, and the official plaque was shipped to him a few months later after his return to the States.[52] Tesla also visited the great Serbian poet Jovan Zmaj Jovanovich,[53] and he attended an assembly where he was honored by the mayor.

In front of a welcoming committee, Zmaj read his poem "Pozdrav Nikoli Tesli," and then Tesla took the podium. "There is something in me which is only perhaps illusory," Tesla began, "[It is] like that which often comes to young, enthusiastic persons; but if I were to be sufficiently fortunate to bring about at least some of my ideas it would be for the benefit of all humanity." Referring back to Zmaj's poem, Tesla concluded with a message that would deeply touch the hearts of his people. "If these hopes become one day a reality, my greatest joy would spring from the fact that this work would be the work of a Serb."[54]

On his return leg, Tesla made a special trip through Prussia to see the eminent patriarch Hermann Ludwig von Helmholtz, in Berlin, and his most famous student, Heinrich Hertz, in Bonn. A bearded, youthful man with soft features, a high forehead, and elongated face, Hertz had gained the world's attention by performing the first significant experiments in wireless, many of which Tesla replicated or expanded upon.

In attempts to clarify the findings of James Clerk Maxwell on the nature of electromagnetic phenomena and its relationship to light and the structure of the ether, in 1886, Hertz constructed "flat double-wound spiral coils" which he used in induction experiments in attempts to measure the propagation of electromagnetic waves. Hertz, like Tesla shortly after him, displayed resonance effects between primary and secondary circuits and "established the existence of standing waves with their characteristic nodes and troughs in a long straight wire." He was also able to measure the wavelength of the waves in the wire.[55] Hertz, however, differed markedly with Tesla in terms of his interpretation of the meaning of Maxwell's equations and his subsequent conceptualization of the structure of the ether.

Deriving his clarifications more from theory than from actual experimentation, Hertz had created an elegant mathematical interpretation of Maxwell's equations, but at the expense of some aspects of Maxwell's theory, most notably vector (a quantity that has magnitude and direction) and scalar (a quantity that has magnitude, but no direction, such as a point or field) potentials. In duplicating Hertz's work, Tesla postulated that these components should not have been eliminated.[56] What he tried to tell Hertz, and what he wrote a few months later, was that electromagnetic waves might "more appropriately [be] called electric sound-waves or sound-waves of electrified air."[57]

"When Dr. Heinrich Hertz undertook his experiments from 1887 to 1889," Tesla told an interviewer, "his object was to demonstrate a theory postulating a medium filling all space called the ether, which was structureless, of inconceivable tenuity...and yet possessed of [great] rigidity. He obtained certain results and the whole world acclaimed them as an experimental verification of that cherished theory, but in reality what he observed tended to prove just its fallacy.

"I had maintained for many years before that such a medium as supposed could not exist, and that we must rather accept the view that all space is filled with a gaseous substance. On repeating the Hertz experiments with much improved and very powerful apparatus, I satisfied myself that what he had observed was nothing else but effects of longitudinal waves in a gaseous medium, that is to say, *waves propagated by alternating compression and expansion* [emphasis added]. He had observed waves in much of the nature of sound waves in the air," not transverse electromagnetic waves, as generally supposed.[58]

Tesla tried to open a dialogue by noting that his experiments tended to contradict the polished mathematical results Hertz had achieved, but Hertz rebuked him. "He seemed disappointed to such a degree," Tesla recalled, "that I regretted my trip and parted from him sorrowfully."[59]

Having replicated Hertz's experiments, Tesla tried to show the German professor that his own oscillator could produce a much more efficient frequency for transmitting wireless impulses. Tesla already had his eye on the idea of transmitting power through the ambient medium, and the Hertzian paradigm virtually disallowed this possibility. But egos clashed, as one *Weltanschauung* threatened the other, and Hertz would never come to realize that his device was obsolete. Perhaps this was to

Hertz's advantage, for even to this day wireless frequencies are referred to as Hertzian waves, when, in fact, they are really Tesla's, as they are produced by high-frequency continuous-wave oscillators, not by the primitive Hertzian interrupted spark-gap apparatus.[60]

During the voyage home, Tesla walked the deck of the ship and pondered an incident that occurred to him during a hike he had taken in the mountains during the trip. Having witnessed an oncoming thunderstorm, he had noted that the rain was delayed until a flash of lightning was perceived. This "observation" confirmed Tesla's speculations, with Martin and Crookes, that weather control was possible because it was the production of large amounts of electricity, in Tesla's eyes, that caused the downpour.

Revealing a megalomaniacal streak, Tesla recalled his thoughts that day in the Alps:

> Here was a stupendous possibility of achievement. If we could produce electric effects of the required quality, this whole planet and the conditions of existence on it could be transformed. The sun raises the water of the oceans and the winds drive it to distant regions where it remains in a state of most delicate balance. If it were in our power to upset it when and wherever desired, this mighty life-sustaining stream could be at will controlled. We could irrigate arid deserts, create lakes and rivers and provide motive power in unlimited amounts.... It seemed a hopeless undertaking, but I made up my mind to try it, and immediately upon my return to the United States in the summer of 1892, work was begun... for the successful transmission of energy without wires.[61]

11

FATHER OF THE WIRELESS (1893)

The day when we shall know exactly what "electricity" is, will chronicle an event probably greater, more important than any other recorded in the history of the human race. The time will come when the comfort, the very existence, perhaps, of man will depend upon that wonderful agent.

NIKOLA TESLA[1]

Tesla disembarked from the *August Victoria* in the last week of August 1892.[2] The trauma associated with the death of his mother was alleged to have caused a shock of hair on his right temporal lobe to temporarily turn white.[3] Whether this occurred cannot be determined; however, what is clear from studying photographs taken before and after the excursion is that a qualitative alteration in his appearance took place, the virginal look of adolescence supplanted by the cocksure demeanor of manhood.

After three years at the Astor House, Tesla moved on to the Hotel Gerlach. Set up on the "European plan" by Charles A. Gerlach, its manager, the Gerlach was equipped with "elevators, electric lights and sumptuous dining rooms." The establishment was family oriented and fireproof.[4]

Located on Twenty-seventh Street, between Broadway and Sixth Avenue, the Gerlach was just a few blocks from the new, magnificent Madison Square Garden, a modern galleria with shops, theaters, restaurants, a thirty-story tower, and a coliseum with a seating capacity of seventeen thousand. The Garden, which was still under construction, was financed by banker J. Pierpont Morgan, who was funding Edison at the time; and it was designed and managed by Stanford White, the flamboyant architect of the prestigious firm of McKim, Mead & White, who would later become an important associate of Tesla's.

Having unpacked his bags and cache of missives at the new hotel, on he went to South Fifth Avenue to his lab, which he had been away from for so long. In long strides, the inventor weaved his way through "Washington Square, [to] the heart of that picturesque neighborhood known as the French quarter. [The streets were] teeming with cheap restaurants, wine shops and weather-beaten tenements," establishments Tesla would never frequent himself. To his surprise, he noticed shop owners waving, whispering among themselves; some even displayed awe. Having been elected to the Royal Society of Great Britain, now he had become a *célébrite internationale* and the neighborhood had been awaiting his return. He came upon what one reporter described as the "uninviting...huge yellowish brick building of some half-dozen stories"[5] which housed his lab. Eagerly, the "murky interior" was entered as Tesla traversed the stairs, taking two at a time. He climbed past the oily, smelly lower floors, which were devoted to a pipe-cutting factory, even managed a smile for the owners of the dry-cleaning service on the third floor, and then entered his secluded haven on the fourth.

The inventor had brought a number of books which he had purchased abroad, and he placed them in his library before proceeding into the machine room, where he spent some time removing dust and cobwebs. Tesla's prime concern was to exploit his advances in fluorescent lighting and wireless transmission of power. Over the next few weeks he hired several workers and a secretary and began by dictating an article on experiments he had conducted with Hertzian frequencies and their relationship to the surrounding medium.[6] He refined his oscillators and designed an experiment whereby one of the terminals of a sizable transmitter was attached to one of the city's water mains, and he recorded electrical vibrations at different positions around town. "By varying the frequency," he said, "I was able to watch for evidence of resonance effects at various distances....I think that beyond doubt it is possible to operate electrical devices in a city through the ground or pipe system by resonance from an electrical oscillator located at a central point."[7] Using vacuum tubes and other tuned circuits as detectors, Tesla began to study the principles of harmonics and standing waves, noting that his instruments would respond at certain points along the pipes but not at other positions.

There was also mail to answer and equipment to order. In September correspondence was begun with Mr. Fodor, a German scientist who, with Tesla's help, translated his world-famous discourses into German.[8] Shortly thereafter, Thomas Edison sent an inscribed photograph "To Tesla from Edison."[9] Tesla also conferred with Professor R. H. Thurston, a physics teacher from Cornell who had expertise in thermodynamics.[10]

At the end of the month George Westinghouse stopped by with Albert Schmid to welcome the inventor home and discuss the fate of the Tesla AC system.[11] In May of that year, Westinghouse had won the bid to furnish the power for the upcoming Columbian Exposition, which was going to be held in Chicago, and he reportedly had taken a million dollar loss in order to secure the contract. But even at this juncture he was still not convinced that the Tesla system would prove to be more useful than compressed air and hydraulic power for long-distance transmissions.[12] Although Tesla had great respect for the descendent of Russian noblemen, he still had difficulty hiding his dissatisfaction. Schmid was relieved that the burden of convincing Westinghouse had shifted.

"My conviction, Mr. Westinghouse, is that a motor without brushes and commutator is the only form which is capable of permanent success. To consider other plans I consider a mere waste of time and money."[13]

Westinghouse asked for Tesla's help, particularly in aiding Schmid, Scott, and Lamme, and Tesla agreed.

Having been assured once again that the Tesla system was all that it promised to be and more, Westinghouse returned to Pittsburgh with a new sense of purpose. "In the early part of 1893," Lamme wrote, "much entirely new and novel apparatus was built for our Chicago World's Fair Exhibit."[14] Tesla would commute to and from Pittsburgh during this hectic time to guide the workers on the construction of the large dynamos, or Lamme, Schmid, or Scott would stop by in New York for advice. They were also helping Tesla construct his own exhibit, which would appear under the Westinghouse banner. Scott was in charge of resurrecting Tesla's ingenious spinning egg, a device which not only aptly displayed the principles of the rotating magnetic field but also paid homage to Christopher Columbus, the explorer whose accomplishments were being honored on this 400th anniversary of his transatlantic journey. Hence the title of the fair: the Columbian Exposition. The fair was slated to open in May, and this gave them only a few months to complete what was truly a Herculean task.

Westinghouse may have won the right to light the fair, but Edison would not allow him a license to produce *his* lightbulb. Fortunately for Westinghouse, he did have a viable patent on a Sawyer-Man "stopper lamp," which had a rubber bottom where the filament was attached in place of the Edison all-glass evacuated construction. Although less efficient, the Sawyer-Man lamp worked. With less than six months left until opening day, he had to produce 250,000 of these inferior bulbs. Coupled with the costs of legal disputes, the company was involved in a great risk venture. However, the prize, if all went successfully, would be the right to harness Niagara Falls. Potential revenues from such a contract would be immense.

Tesla arranged for Mr. Luka of the Helios Company of Cologne, to come to Pittsburgh to discuss supplying the German concern with AC equipment for their contract in Germany. "He has been sent here to gather information about railway, steam and other motors," Tesla told Westinghouse. "I believe they would be ready to make a small cash payment and pay a moderate royalty, and I have done what I could to facilitate an understanding."[15] Tesla had also secured other European connections, and soon revenues from abroad began to roll in.

Nevertheless, there remained a good deal of animosity toward Tesla by some other members of the Westinghouse organization, partially because Tesla was paid so handsomely for an invention that they considered had also been conceived by Shallenberger and partially because they simply did not like the pompous foreigner. There were also great financial costs incurred in dismantling the hundreds of profitable Gaulard-Gibbs power stations which were dotted across the nation.

In November 1892, Grover Cleveland, former hangman and sheriff of Buffalo, running on an antilabor ticket, was re-elected president of the United States. Cleveland's second inauguration inflamed many segments of the population and no doubt helped trigger the Panic of 1893.

The calamity began in 1892 with the financial collapse of four major railroads. Then banks failed, and tens of thousands of people became unemployed;[16] and the Westinghouse Company was just beginning a decade-long course of incurring enormous debt. Westinghouse realized that he had to back Tesla unconditionally as the sole inventor of the AC polyphase system. Had there been any ambiguity in the matter, competitors could seize an advantage by obscuring the origins of the invention, and thus they would be able to produce Teslaic technology without royalty payments to Westinghouse.

On January 16, 1893, Westinghouse came out with an announcement touting the Tesla multiphase, or polyphase, system which was circulated to the electrical magazines and major competitors. Having "secured exclusive right to manufacture and sell apparatus covered by [Tesla's] patents" the Westinghouse company promised to use such apparatus to economically harness the many waterfalls which were wasting so much energy.

Now that the problems in Pittsburgh were somewhat alleviated, Tesla could devote more time to his upcoming lectures, which were going to be held at the Franklin Institute in Philadelphia at the end of February and again, the following week, in March at the annual meeting of the National Electric Light Association in St. Louis. He was met in Philadelphia by Prof. Edwin Houston, formerly the partner of Houston's former student, Elihu Thomson.

Tesla began his lecture in Philadelphia with a discussion of the human eye, "nature's masterpiece.... It is the great gateway through which all knowledge enters the mind.... It is often said, the very soul shows itself in the eye."[17]

The study of the eye suggested a number of different and distinct lines of inquiry. For instance, it enabled Tesla to envision the precursor to television, with its numerous transfiguring pixels corresponding to the light-sensitive receiving cells of the retina. In another vein, in conjunction with instruments such as microscopes and telescopes, the eye also opened up new vistas for scientific inquiry. Alluding to the concept of the plurality of worlds, Tesla would say, "It was an organ of a higher order."[18]

"It is conceivable," Tesla continued, "that in some other world, in some other beings, the eye is replaced by a different organ, equally or more perfect, but these beings cannot be men."[19]

Obtaining information from all corners of the universe, at the same time, the eye interacted with that elusive realm called the mind. Furthermore, this organ was also a perfect analog of Tesla's Aristotlean worldview, as the eye had to be triggered from an external source in order to function.[20]

If we go back to one of Tesla's earlier experiments with the "brush phenomena," that is, the creation of a brush or stream of light generated within an insulated vacuum bulb that responded to the faintest electromagnetic reverberations, we see that to Tesla this precursor to the radio tube was actually based on the principles inherent in the construction of the human eye. The brush, we remember, not only reacted to magnetic influences but also to the approach of a person and to the torque of the earth, just as the eye also reacts to faint impulses from near or far. It is "the only organ capable of being affected directly by the vibrations of the ether."[21]

The ether was a nineteenth-century theoretical construct of an all-pervasive medium between the planets and stars. In 1881, Michelson and Morely unsuccessfully tried to measure the ether in their famous experiment with light beams and mirrors. The ramifications of their findings did not become evident until after the turn of the century, a full decade after Tesla's lecture, when Einstein used the Michelson-Morley experiment to suggest that, by its nature, "the ether cannot be detected,"[22] and further, that it was unnecessary for explaining how light could travel through space.

Physics professor Edwin Gora, of Providence College, whose mentors included Arnold Sommerfeld and Werner Heisenberg, stated that the ether could not be detected with nineteenth-century techniques and that Einstein replaced the old ether with a new non-Euclidean space-time construct. This new more abstract ether had such unusual properties as allowing space to curve around gravitational bodies.

Completely disagreeing with Einstein, and never abandoning the concept of the all-pervasive either, Tesla said that space cannot be curved because "something cannot act upon nothing." Light, according to Tesla, bent around stars and planets because they were attracted by a force field.[23]

Gora agreed that the two concepts of curved space and force field may actually be different viable ways of describing the same thing.

Returning to the 1893 lecture, for Tesla, the relationship of electrical phenomena to the structure of the ether appeared to be an important key to understanding how it could be transmitted without wires in an efficient manner.

The problem of the transmission of electromagnetic energy through space was discussed in all three of his lectures on high-frequency phenomena. One question he considered was whether the ether was motionless or in motion. When vibrations were transmitted through it, it appeared to act like a still lake, but at other times, the ether acted like "a fluid to the motion of bodies through it." Referring to the investigations of Kelvin, Tesla concluded that the ether must be in motion. "But regardless of this, there is nothing which would enable us to conclude with certainty that, while a fluid is not capable of transmitting transverse vibrations of a few hundred or thousand per second, it might not be capable of transmitting such vibrations when they range into hundreds of million millions per second."[24]

Tesla would later claim spectacular results in wireless transmission never duplicated by any other researcher; he states that his system was not bound to the inverse-square laws, and it appears that his success, if indeed it was a success(!), was based on the premise that above certain frequencies the ether revealed novel and heretofore unknown features. Perhaps threshold values were involved.

Tesla continued his discussion on the structure of the ether and its relationship to electromagnetic phenomena by making two observations: (1) "that energy [could be transmitted] by independent carriers" and (2) that atomic and subatomic particles whirled around each other like little solar systems.[25] These two concepts, which were tied to the mystery of the structure of the ether, predated similiar ideas proposed by quantum physicists Ernest Rutherford, Niels Bohr, and Albert Einstein by at least a decade.

In Rutherford's case, he is often credited as the first physicist to view the atom as structured somewhat like a solar system. It is evident, however, that Rutherford referred to Tesla's high-frequency lectures in 1895, when he constructed high-frequency AC equipment for conducting long-distance wireless experiments.[26]

Tesla stated that he could create electromagnetic oscillations that displayed transverse and also longitudinal wave characteristics. The first (transverse) case corresponds to the concept of the ether as a medium for propagating wavelike impulses; and the second (longitudinal) case corresponds to what today is known as a quantum of energy analogous to the

A drawing depicting Nikola Tesla displaying wireless experiments at the Chicago World's Fair of 1893.

way sound waves travel through air. Tesla maintained, against all opposition, even to this day, that his electromagnetic frequencies traveled in longitudinal, bulletlike impulses, and thus they carried much more energy than can be ascribed to Hertzian transverse waves. In fact, as alluded to before, Hertz wanted to eliminate the idea of mass from the Maxwellian electromagnetic equations.

Tesla's idea of longitudinal waves in the ether appear to be a direct outcropping of the research undertaken by Ernst Mach, who was still at Prague at the time. Mach's radical views on the relationship between consciousness, space and time, and the nature of gravity were beginning to alter greatly the thinking of a number of key individuals. His idea, which came to be known as "Mach's Principle," hypothesized that all things in the universe were interrelated, for example, the mass of the earth, according to this theory, was dependent on a supergravitational force from *all* stars in the universe. Nothing was separate. This view, which Mach realized corresponded to Buddhist thinking, paralleled closely views espoused by Tesla. Although the following quote was written almost a quarter of a century later, its link to Tesla's 1893 lecture is clear: "There is no thing endowed with life—from man, who is enslaving the elements, to the nimblest creature—in all this world that does not sway in its turn. Whenever action is born from force, though it be infinitesimal, the cosmic balance is upset and universal motion results."[27]

This idea was extended and interlinked between living organisms and inert matter by Tesla. All are "susceptible to stimulus from the outside. There is no gap between, no break of continuity, no special and distinguishing vital agent. The same law governs all matter, all the universe is alive."[28] The source of power which runs the universe is that found within "the sun's heat and light. Wherever they are there is life." As these processes were electrical in nature, to Tesla, the secret of electricity held the secret of life.

Looking at the world around him, Tesla realized that it was a finite place and that the natural resources which gave humans the fuel to produce electricity would eventually run out. "What will man do when the forests disappear, when the coal deposits are exhausted?" he asked his Philadelphia audience. "Only one thing, according to our present knowledge, will remain; that is to transmit power at great distances. Man will go to the waterfalls, [and] to the tides," Tesla speculated, because these, unlike coal and oil reserves, are replenishable.[29]

Having set up the premise that it could be possible to derive inexhaustible amounts of energy with properly constructed equipment, that is, "to attach our engines to the wheelwork of the universe," Tesla described, for the first time ever, his invention of wireless transmission. Cloaking his true goals in more palatable language, he announced, "I...firmly believe that it is practicable to disturb by means of powerful machines the electrostatic conditions of the earth and thus transmit intelligible signals and perhaps power." Taking into consideration the speed of electrical impulses, with this new technology, "all...ideas of distance must...vanish," as humans will be instantaneously interconnected. "First, we must know what capacity the earth is, and what charge it contains." Tesla also speculated that the earth was "probably a charged body insulated in space and" and thus had a "low capacity." The upper strata, much like the vacuum created in his Geissler tubes, would probably be an excellent medium for transmitting impulses.[30] We see here the precursor to the discovery by Heaviside and Kennelly of the ionosphere.

Tesla had already thrust large amounts of electrical energy into the earth to try to measure its period of frequency, but he had yet to come up with a figure that appeared accurate. Nevertheless, he knew the size of the earth and the speed of light and thus was already at this time formulating optimum wavelengths for transmitting impulses through the planet.

During his talk, Tesla demonstrated *impedance phenomena* by turning on and off a lightbulb by placing it at various positions along an electrified metal bar. Based somewhat on the work of Hertz, this experiment demonstrated the concepts of wavelength and standing waves. He constructed circuits with two or three bulbs independently connected in a row and placed metal bars at various points along the way, thereby illuminating or extinguishing one or another of these bulbs by impeding or not impeding the electrical flow. He also displayed electric lamps illuminated with only a single wire and therefore was able to establish that the wire itself could be replaced by connecting the lamp directly to the earth, which also was a conductor, as no return circuit (as found in the Edison bulbs) was necessary. As before, Tesla also displayed lamps illuminated with no connections whatsoever.

With pure resonance, Tesla suggested, wires become unnecessary, since impulses can be "jumped" from sending device to receiver. Naturally, the receiving instruments would have to be tuned to the frequency of the transmitter. "If ever we can ascertain at what period the earth's charge, when disturbed [or] oscillates with respect to an oppositely electrified system or known circuit, we shall know a fact possibly of the greatest importance to the welfare of the human race."[31]

Tesla proceeded to present a diagram which depicted how to set up the aerials, receivers, transmitter, and ground connection. The son of one of his assistants described the apparatus:

> In the transmitter group on one side of the stage was a 5-kva high-voltage pole-type oil filled distribution transformer connected to a condenser bank of Leyden jars, a spark gap, a coil and a wire running up to the ceiling. In the receiving group at the other side of the stage was an identical wire hanging from the ceiling, a duplicate condenser bank of Leyden jars and a coil—but instead of the spark gap, there was a Geissler tube that would light up...when voltage was applied. When the switch was closed, the transformer grunted and groaned, the Leyden jars showed corona sizzling around their foil edges, the spark gap crackled with a noisy spark discharge, and an invisible electromagnetic field radiated energy into space from the transmitter antenna wire [to the receiver antenna wire].[32]

Tesla elaborated: "When the electric oscillation is set up," he said, "there will be a movement of electricity in and out of [the transmitter], and alternating currents will pass through the earth.... In this manner neighboring points on the earth's surface within a certain radius will be disturbed." Although Tesla's main goal was to *transmit power*, he also noted that "theoretically,...it [w]ould not require a great amount of energy to produce a disturbance perceptible at great distance, or even all over the surface of the globe."[33]

In Tesla's autobiography, written a quarter of a century later, the inventor informs the reader that there was such opposition to his discussion of wireless telegraphy at that time that "only a small part of what I had intended to say was embodied [in the speech].... This little salvage from the wreck has earned me the title 'Father of the Wireless.'"[34] Tesla stated that it was Joseph Wetzler who told him to deemphasize his work in wireless in this lecture. Wetzler probably edited out a number of key passages which, in the long run, could have helped Tesla establish more easily his priorities in the field. Nevertheless, the entire Philadelphia speech runs a hundred typeset pages and covers numerous other topics as well. What is important to realize is that for the first time ever, a major inventor announced bold possibilities in the field of wireless communication; simultaneously, he explained in step-by-step fashion all of the major components that would be needed for success.

The question of who invented the radio is complex, for there was no single developer. Experiments in wireless can be traced back to Joseph Henry, who, in 1842, transmitted electrical energy across a thirty-foot room between magnetized needles and sensitive Leyden jars, and to Samuel Morse, who sent messages in 1847 by means of induction across an eighty-foot-wide canal by using something called "current leakage."[35]

The first individual to transmit messages over long distances using aerials (in the form of kites) and a ground connection was Mahlon Loomis. A dentist and experimentalist who also used electricity to stimulate growth in plants, Loomis not only received a patent on the device in 1872 but also successfully introduced the "Loomis Aerial Telegraphy Bill" before the U.S. Congress. Loomis made such an impact that $50,000 was appropriated to help him in his pursuit. In 1886, Loomis sent wireless messages fourteen miles between two mountains in Virginia, and a few years later, he also sent messages between ships two miles apart in Chesapeake Bay. There is little doubt that Tesla was aware of Loomis. For one thing, his patent was registered, and Tesla always made it a practice to study the work of his precursors. Also, it should be noted that some of the wording from Loomis's patent applications and published writings sound eerily like the wording in some of Tesla's discourses. For instance, Loomis discusses the passing of "electrical vibrations or waves around the world," and principles of harmonics and resonance, and he also refers to harnessing "the wheelwork of nature," a favored term of Tesla's.[36]

In 1875, Thomas Edison, while working with Charles Batchelor, noticed an unusual sparking effect emanating from the core of an electromagnet which leaped to noncharged bodies several feet away. By using an electroscope, he was *unable* to distinguish a charge.[37] In actuality, he had created a high frequency that could not be detected by his equipment. "By charging a gas main, Edison was able to obtain sparks from the fixtures in his house several blocks away.... Edison thought that since energy can take various forms, and it was possible to change electricity to magnetism, magnetism might be transformed into something else."[38] Edison therefore announced to the scientific community that he had discovered a new "unknown force." Possibly, Tesla's ideas of connecting an oscillator to the water mains of a city may have been influenced by this research.

In the early 1880s, William Preece, electrical engineer for the British Post Office, began directing experiments in wireless communication by means of an inductive apparatus. He was probably also the first inventor to realize that the *earth itself was an integral component in the successful implementation of any wireless system.* After isolating the role of the earth as either a primary or secondary circuit, Preece utilized telephone receivers as detecting devices and concluded that "on ordinary working telegraph lines the disturbance reached a distance of 3,000 feet, while effects were detected on parallel lines of a telegraph 10 to 40 miles apart in some sections of the country." Preece's work of detecting earth currents, which was duplicated by Western Union engineers in the United States, significantly influenced the theories expounded by Tesla.[39]

Preece had displayed a long-standing interest in wireless communications. He had visited Edison in the mid-1880s, just after Tesla emigrated to America, to witness firsthand Edison's latest invention, which he called the "grasshopper telegraph," a device for jumping messages from dispatch stations to moving trains. By means of induction or resonance, a metal strip attached to a telephone receiver on a moving railroad car would send or receive messages from a similar strip strung parallel to the track at the station. Although the invention never evolved beyond this primitive early stage, the patent would later have important legal significance in priority battles over the invention of the wireless.

Thus, Edison is clearly one of the fathers of wireless transmission, as are Henry, Morse, Loomis, and Preece. Concerning the history of radio tubes, Edison also had an important invention, discussed above, of a dual-filament lightbulb which displayed a flow of current between them, Preece having named it the "Edison effect." J. J. Thomson used it to help in his discovery of the electron, and Tesla combined information from this device with Crookes's work on radiation effects inside evacuated glass tubes to invent his "brush phenomena," which was the first such vacuum tube explicitly created for wireless transmission of intelligence.

Other precursors to Tesla included Heinrich Hertz, Oliver Lodge, and Edouard Branly. A French professor of physics, Branly, perhaps influenced by the knowledge of the Edison effect, noticed that the gap of Hertz's tuned circuits could be replaced by a glass-enclosed tube which contained finely scattered metallic particles. When current passed through the tube by means of wireless induction, the particles aligned themselves along the path of the gap and closed the circuit. A light tapping on the tube opened the circuit once again until transmission occurred. Lodge perfected Branly's 1890 discovery of particle cohesion and labeled it the "coherer."[10]

These scientists were not thinking about "wireless telegraphy" at the time of their initial work. They were explorers in a new field of electromagnetic induction, and it was not until 1894, by Lodge's own calculation, that he thought in terms of utilizing the equipment as a means of conveying information.[11] On the other hand, we remember that Crookes, writing in 1892, at the very time he was meeting with Tesla in England, noted that he had experimented in the wireless transmission of Morse code from one end of a house to another at about the same time as Hertz and Lodge were experimenting in the late 1880s, but he never publicized his work or furthered the invention beyond this casual experiment.

Tesla realized, as did Hertz, that the Hertzian frequencies through space were not conductive for long-range communication, but unlike Hertz, Tesla sought a way around this limiting factor. Therefore, he devised not only the means of securing more powerful transmitters but also "concatenated tuned circuits," which were, in essence, sensitive radio tubes for receiving information.[12] During this speech in Philadelphia, Tesla also introduced the concept of using *both* an aerial and ground connection and a single wire as a return for the operation of "all kinds of devices." This system of wireless transmission was outlined in detail in highly visible articles which appeared in 1891, during his first public demonstrations of wireless Geissler tubes at Columbia College, in 1892 in Europe, and were explicitly delineated in 1893. It would be another full year before a high school boy by the name of Guglielmo Marconi would begin his first tinkerings in the field.

12

ELECTRIC SORCERER (1893)

It was known that Mr. Tesla, who enjoys a high reputation as an electrician, had been experimenting upon a practically new electric light, but it was not known outside his laboratory that he had achieved such wonderful results or come so near [to] revolutionizing the theory of light. Other electrical explorers, especially Dr. Hertz and Dr. Lodge, had evolved the theory that the phenomena of light were related to the electro-magnetic vibrations of ether or air, but it remained for Mr. Tesla to demonstrate this fact and make the knowledge practically available.

NEW YORK RECORDER[1]

Tesla left Philadelphia by rail at the end of February for the National Electric Light Association convention in St. Louis. Accompanied by T. C. Martin, who was covering both lectures for *Electrical Engineer*, they discussed the creation of a textbook based on the inventor's collected writings. The first half would be devoted to the full range of inventions associated with the AC polyphase system, with chapters on motor design, single phase and polyphase circuits, armatures, and transformers; the second half would contain Tesla's three discourses on high-frequency phenomena, which he had presented in New York, London, and Philadelphia. With an introduction written by Martin and a few miscellaneous articles at the end, the treatise would run almost five hundred pages. Josh Wetzler would be second editor. Publication date was set for the end of the year.

Martin had achieved a great coup, solidifying a deal with what many people were saying was "the greatest living electrician."[2] *The Inventions, Researches and Writings of Nikola Tesla* would be a landmark text, becoming a virtual bible for the numerous electricians who read it.

On February 28, Tesla arrived in St. Louis, invited to speak by James

I. Ayer, general manager of the local Municipal Electric Light & Power Company. The arrival of the inventor created a jolt of excitement. His speech was billed as a replication of the London Lecture. "Over four thousand copies of the journal containing [a] biographical sketch were sold upon the streets...something unprecedented in the history of electrical journalism."[3] A procession of eighty electrical utility wagons and meter-men carts rolled down Main Street as thousands of people clamored to get tickets for the performance.[4]

At the opening ceremony, both Tesla and Ayer were inducted into the association as honorary members;[5] afterward, Ayer introduced Tesla to one of his engineers, H. P. Broughton, a recent graduate of Cornell University, who was assigned as an assistant for the duration of the convention.[6]

The chamber allocated proved to be too small, so the event was moved to the Grand Music Entertainment Hall, a large auditorium with a seating capacity of over four thousand. Tickets were hawked on the steps for anywhere from three to five dollars. Yet even this theater proved to be inadequate, for the house was "crowded to suffocation."[7]

Mr. Ayer introduced the inventor to the audience "with a sort of reverence as one who has an almost magic power over the vast hidden secrets of nature" and presented Tesla with "a magnificent floral shield, wrought in white carnations and red Beauty roses."

Peering out at the sea of faces, Tesla realized that it would be wise to restrict his lecture to the more "spectacular" experiments. With Broughton, he displayed his invention of wireless transmission of electrical energy by lighting up wireless receiving tubes by throwing a switch from the opposite side of the room.

"Of all the many marvelous things we observe," he said, "to me it seems, a vacuum tube excited by an electric impulse from a distant source, bursting forth out of the darkness and illuminating the room with its beautiful light, is as lovely a phenomenon as can greet our eyes."[8]

"By way of amusement," Tesla created sheets of electricity between two condenser plates. He illuminated lightbulbs with and without filaments, ignited phosphorescent globes that "threw a most dazzling light far beyond that yielded by any ordinary phosphorescence," and created strobe effects by whirling tubes that "look[ed] like white spokes of a wheel of glowing moonbeams."

Tesla then approached his most powerful coil.

Noticing Prof. George Forbes in the audience, the engineer from Glasgow who had so highly recommended his AC system to the Niagara Power Commission, the inventor bowed in respect. Acknowledging his appreciation, Tesla predicted that ere long great surges of electrical energy based on this work would soon usher from the great waterfall.

> These next series of experiments are shown with some reluctance; yet forced thereto by the desire to gratify those who had shown such interest and formed so large an audience, I see I have little choice....Charg[ing] my body with electricity from an apparatus which I have devised, I can make the electricity vibrate at the rate of a million times a second. The molecules of the air are then violently agitated, so violently that they become luminous, and streams of light then come out from the hand. In the same manner I am able to take in the hand a bulb of glass filled with certain substances and make them spring into light.[9] I had the pleasure of performing these very experiments privately before Lord Rayleigh, and I will always remember this distinguished scientist trembling in eagerness and excitement when he witnessed them. The appreciation I have received from such a distinguished scientist as he has repaid me fully for the pains to which I have worked to achieve these ends.[10]

Tesla thereupon turned to the coil and announced that due to the enormous potentials about to be generated, he had constructed the machine with hard rubber insulation, "as even dry wood is by far too poor" for protection.[11]

> I now approach the free terminal with a metallic object held in my hand, this simply to avoid burns. The sparks cease when the metal...touches the wire. My arm is now traversed by a powerful electric current, vibrating at about a million times a second. All around me the electrostatic force is felt, and the air molecules and particles of dust flying about are acted upon and are hammering violently against my body. So great is this agitation of the particles, that when the lights are turned out you may see streams of feeble light appear on some parts of my body. When such a streamer breaks out, it produces a sensation like the pricking of a needle. Were the potentials sufficiently high and the frequency low, the skin would probably be ruptured under the tremendous strain, and the blood would rush out with great force in the form of a fine spray.[12]

Extending his fingers like a preening peacock, the electric sorcerer issued flames of lightning as if he were Thor himself. "The streamers offer no particular inconvenience," he assured the audience, "except that in the ends of the finger tips a burning sensation is felt."

With his coil still charged and flames still shooting from his head, a number of other effects were displayed, including the running of motors by way of energy through his body and the lighting of a variety of colorful tubes which the inventor waved about as if they were phosphorescent rapiers. His finale included the stretching of a series of long cotton strings across the stage which he zapped with a pencil-line corona of violet that etched a dazzling glow along its length, illuminating the stage in an eerie iridescence. Shouts of "Bravo" accompanied the thunderous applause as "the lecturer bowed again and again" in response.

Shimmering from aftereffects and still "emitting ethereal flames and fine halos of splintered light"[13] after the lecture, Tesla was called to the lobby so that he could be greeted by many members of the admiring public. "So great was the desire...to see Mr. Tesla closer...[that] several hundreds of the leading citizens seized the opportunity and Mr. Tesla's hand in a very vigorous manner."[14]

Upon his return to New York, Tesla completed his citizenship requirements. No longer would his patent applications refer to him as "a subject of the Emperor of Austria," but rather as "a citizen of the United States." This was a proud moment. He would keep these papers safely stored away in a vault in his room for the rest of his life.

Now a full "American," Tesla decided once and for all to take on the mighty icon Thomas Edison. Setting up one of his most provocative experiments, the upstart attacked head-on Edison's carbon filament lamp. Taking two identical bulbs, one filled with air and the other containing a vacuum, and attaching to it "a current vibrating about one million times a second," Tesla demonstrated that the lamp filled with ordinary air did not glow but the one with the vacuum glowed brightly. "This showed the great importance of the rarefied gas in the heating of the conductor"; and furthermore, the new light was cool to the touch. Brazenly, Tesla could conclude: "In incandescent lighting, a high resistance filament [Edison's invention] does not at all constitute the really essential element of illumination."[15]

Taking the same two bulbs, Tesla then lowered the frequency, converted it to DC, and showed that the filament in the nonexhausted bulb now began to glow, though not as intensely as the other. He concluded that with DC the filament was the essential component; with high-frequency AC, the atmosphere around the filament, and thus the vacuum, was paramount. The higher the frequency, the more efficient the illuminant. In fact, he noted, if DC were abandoned altogether and extremely high frequencies used, there would be no need for a filament at all!

Edison became perturbed not only because he had sent men to the Orient, Central America, and the Amazon in order to find the perfect filament but also because one or two of them died in the quest.[16] That pompous Serb was not only saying that AC was intrinsic to the development of any practical lighting system; he was also proclaiming that Edison's most famous work would eventually be for naught.

At a cottage by Edison's rock-crushing factory in Ogden, New Jersey, a local newspaper reporter asked the inventor about the "possibility of producing light from electricity without heat."[17] Edison murmured that this feat had been achieved, but the reporter pressed him. "I specifically mean with reference to discoveries in the line taken up by Nikola Tesla."

Edison "rose from his easy chair," spat another wad of tobacco off the porch, and barked, "It is a problem surrounded by difficulties."

"And what about Tesla?"

"He has made no new discovery," the chief wizard began, "but has shown considerable ingenuity in increasing vibrations. He gets his results from the induction coil and the Geissler tube."

"Do you think it will replace your light bulb?"

"Light without heat may be obtained some day, but I do not care to prophesy that it will be a pleasant light. Its biggest trouble is its quality. It is a ghastly color, more like the light of lightning bugs than anything else."

Certainly, in this regard, Edison was entirely correct, as even today the lightbulb generates a more pleasant glow than fluorescent tubes. However, as anyone who has changed a lightbulb knows, Edison's invention is constructed so as to break down within a few months, whereas Teslaic fluorescent lights can last for years, even decades. Furthermore, according to Tesla's view, the incandescent lamp created a harsh, bright light from a small central source. It was Tesla's plan to create a pleasing light from a very large, spherical surface.

Now, for the first time, Nikola Tesla was sought after by the popular press. Quoted as being "modest" by many interviewers and sometimes self-effacing, initially Tesla avoided publicity. But his sensational accomplishments had made him a folk hero, and it was now up to the journalists to find out why.

The *New York Herald* was "the first" newspaper to capitalize on the rising star. In a long feature article adorned with an etching drawn from an outdated photograph, the *Herald* described the inventor's demeanor and his heritage and discussed the full range of his future plans.

"Mr. Tesla is such a hard worker that he has little time for social pleasures, if indeed he has any taste in that direction. He is a bachelor, tall, very spare of build, has dark, deep-set eyes, jet black hair and an expression that suggests at once the deep thinker. Though polite and even friendly to newspaper men, he has no desire to exploit himself in print."

Tesla described his parents and schooling, his invention of the rotating magnetic field, and also his new system of lighting, which he promised would be "a more practical illuminant than we have at present." He discussed the possibility of sending vast amounts of electricity along wires from Niagara Falls and also the concept of wireless transmission of intelligence and power through the earth and air "to any distance."

"From present experimental evidence it can be quite safely concluded that an attempt to transmit intelligible sounds through the earth from here, for instance, to the European continent without any cable will succeed beyond a doubt....Such a result...which I have advocated for two years...if achieved, would, of course, be of incalculable value to the world and greatly advance the progress of the human race."

The *Herald* ended with a question about how it felt to send hundreds of thousands of volts through the body, and Tesla replied, "If you are prepared [for the shock] the effect on the nerves is not nearly as great. Initially, you feel a burn, but beyond that the feeling is scarcely noticeable. I have received currents as high as 300,000 volts, an amount which, if received in any other way, would instantly kill."[18]

Although the travel and the lectures were digressions, Tesla accepted the engagements to solidify his place in the history of electrical science. For the cloistered conceptualist, naturally, the accolades and interaction with colleagues were also inducements. However, it was only now, as he was being sought after by reporters and the adoring public, that he began to realize, in a conscious sense, his latent desire for recognition. Every hour, every moment, that was not spent working on inventions was time away from his purpose. Even the intervals spent eating and sleeping delayed progress.

Although both his parents had passed away at relatively young ages, there were other kin whose life span had traversed more than ten decades. Perhaps in part as a defense, Tesla would exclaim that he expected to attain centenarian status. He had come to see the human body in its essence as a machine, one that could be efficiently regulated by the stern application of willpower, and so Tesla exerted his will to reduce his sleeping to a minimum and his eating to the bare necessities. Although over six feet tall, he kept his weight to a scant and unvarying 142 pounds.[19] The strain was beginning to show, but the Serb was on a quest; his goal was nothing short of saving the whole of humanity through the application of his fertile brain.

Having transmitted energy by means of wireless from one end of a room to another, the inventor's next task was to devise a way to extend this principle to generate larger amounts of power and to create separate, noninterfering channels. Tesla began to experiment with more and more powerful oscillators, ones that would not only radiate high-frequency AC but also ones capable of engendering physical pulsations.

"The first gratifying result was obtained in the spring...when I reached tensions of about 1,000,000 volts with my conical coil. That was not much in the light of the present art," Tesla wrote a quarter of a century later, "but it was then considered a feat."[20]

According to calculations performed by Professor John Tyndall, the Edison electric lightbulb had an efficiency of about 5 percent, meaning that 95 percent of the electricity produced went into the production of heat or was simply lost in transit. The gas flame, which was still by far the most common form of artificial luminescence, had an efficiency of "less than one percent." As Tesla told Martin, if "we were dealing with a corrupt government, such wretched waste would not be tolerated." This squander was "on a par with the wanton destruction of whole forests for the sake of a few sticks of lumber."

"The energy," as Martin wrote, "is more or less frittered away, just as in July the load in an iceman's cart crumbles and melts... along the street." Whereas "armies of inventors had flung themselves on the difficulties involved in these barbaric losses occurring at every stage," Tesla had designed an entirely original series of inventions which either converted electricity to rotary action or practically the reverse, that is, converted steam power into electrical energy.[21]

In the first instance, Tesla constructed a high-frequency oscillator which was immersed in a vat of oil. By modulating the frequency of an AC, he caused the oil to flow at varying rates. Just like a waterwheel turning a turbine, this oil caused a blade to rotate. By increasing the frequency, the flow would change, and the rotation of the turbine would be increased.

In the second instance, that is, in the case of the steam generator, Tesla combined into one invention both an engine and a dynamo and created a device that was as much as *one-fortieth* the size of comparable traditional constructions. In the old-fashioned steam engine, the reciprocating back-and-forth piston action had to be converted to a rotary effect by way of a crankshaft and flywheel. This device, in turn, was connected to a turbine, which thereby produced electricity. In the Tesla generator, a steam driven piston attached to a condenser was made to bob up and down within a magnetic field, cutting the lines of force, thereby producing a current. This procedure reduced markedly the loss involved in converting mechanical action to electrical power. There was no flywheel or crankshaft. Martin eloquently wrote: "We note immediately the absence of all the governing appliances of the ordinary engine. They are non-existent. The steam chest is the engine, bared to the skin like a prize-fighter with every ounce counting.... Denuded in this way of superfluous weight, and driven at high pressure, the engine must have an economy far beyond the common. With an absence of friction due to the automatic cushioning of the light working parts, it is also practically indestructible. Furthermore, whereas the ordinary steam piston, "weighing, perhaps as much as a thousand pounds," could only change its direction, say, ten times per second, the Tesla oscillator could oscillate at "one hundred strokes a second." Tesla hoped not only to reduce the complexity of existing equipment but also to create a current that would "maintain a vibration with perfect constancy." He took out patents on both of these devices in the summer.[22]

With these various oscillators, Tesla could provide a plethora of effects. Electrically, he could generate precise frequencies that could be used to transmit information or electrical energy. When the oscillator was pulsating at the frequency of light, he could manifest luminescence as well. And mechanically he could create pulsations through metal bars, or pipes, and test for harmonic frequencies and standing waves. By studying impedance phenomena, he could transmit electromagnetic energy through such a conduit and cause a light to illuminate at certain positions along its length but not at others, and if he increased the vibrations and struck a harmonic frequency, he could get the iron rod to begin to vibrate with such intensity that it would crack itself in two. This effect was equivalent to that produced by Joshua's horns at Jericho or to the danger that could be initiated by soldiers marching in lockstep across a suspension bridge. If their beat struck the resonant frequency, the bridge could sway violently and possibly collapse. Thus, soldiers are trained to break stride while walking across bridges to avoid this amazing type of catastrophe.

The World's Columbian Exposition

The Scene by Night

> Inadequate as words have been found to convey a realizing idea of the beauty and grandeur of the spectacle which the Exposition offers by day, they are infinitely less capable of affording the slightest conception of the dazzling spectacle which greets the eye of the visitor by night....
>
> Indescribable by language are the electric fountains. One of them, called "The Great Geyser," rises to a height of 150 feet, above a band of "Little Geysers." [By adding rotating colored lights, the effects created] are so bewildering no eye can find the loveliest, their vagaries of motion so entrancing no heart can keep its steady beating.
>
> *W. E. Cameron*[23]

On May 1, the Chicago World's Fair, or Columbian Exposition, opened. This auspicious occasion came at a paradoxical time for the country, for on the one hand, the United States had announced that it was the new leader in the creation and deployment of new technology; on the other hand, it was embroiled in the Panic of 1893. Simultaneously, and for one exceptional time, the world at large was experiencing universal peace.

The Columbian Exposition covered almost seven hundred acres, had sixty thousand exhibitors and cost $25 million. With 28 million attendees, the Chicago Fair boasted a $2.25 million profit. Whereas the Paris Exposition of 1889 had the stupendous 984-foot tall Eiffel Tower, the Chicago Fair boasted the Ferris wheel. Revolving on the largest one-piece axle ever forged, the wheel stood 264 feet high and had a seating capacity of over two thousand.[24]

Every day, hundreds of thousands of visitors from every corner of the globe streamed into Chicago's "White City." The chief architect, Chicagoan Daniel Hudson Burnham, along with the other planners, based their design on a city of waterways, much like Venice, with a "Court of Honor" centrally placed to house the major "palaces." With wooden façades made to resemble marble, the buildings rivaled even the great stone achievements of the ancient Romans and Greeks. The Manufacturers & Liberal Arts Pavilion, more than twice the size of the others, was "by far, the largest building in the world." Nearly a third of a mile long and well over two football fields wide, the structure covered 30.5 acres and had a seating capacity of seventy-five thousand.

Abutting Lake Michigan, at the far end above the Court of Honor, with columns extended like the arms of a mighty conductor, was the magnificent Arch and Peristyle. Designed by head sculptor August Saint-Gaudens, this statue depicted a general, representing Christopher Columbus, thunderously entering the fair with his mighty team of horses. Beneath this quadriga was the great, gilded sixty-five-foot Statue of the Republic, which rose from the main canal like a benevolent behemoth as it blessed not only the Manufacturer's Pavilion but also palaces for agriculture (designed by Stanford White), machinery, administration, and electricity.

And that was just the Court of Honor.

Running perpendicular to the "great quadrangle" for nearly a mile was yet another opulently lined moat containing most of the other large-scale pavilions and displays described in one text as "a cacophonous confusion of advanced cultures, converted heathens, peculiar tongues, and queer importations."[25] With a fleet of fifty gondolas driven by brand-new electrical engines, visitors could motor to any pavilion they wanted.

The Electricity Pavilion, adorned with a dozen elegant minarets, four of which rose 169 feet above the hall, was over two football fields in length and nearly half that measure in width. Covering three and one-half acres, this "spacious and stately" structure "befit[ted] the seat of the most novel and brilliant exhibit of the Columbian Exposition."

The ex-governor of the state of Illinois, William E. Cameron, described the palace as housing the "magical achievements of Mr. Edison and his brother wizards."[26] Containing the names of the forefathers of the science above a formidable statue of Benjamin Franklin launching his kite at the main vestibule, Electricity Hall presented a fascinating potpourri of advanced technologies.

Foremost among the exhibitors were the megalithic corporations, such as Westinghouse and GE, from America, and the more modest sized AEG, from Germany. Whereas AEG reproduced some of the AC equipment Brown and Dobrowolsky used in their "epoch-making" 108-mile Lauffen-to-Frankfurt transmission, GE presented its own AC system. Westinghouse, having won the bid to light the fair with the only bona fide patents, was in an odd situation. Legally, it should have been able to block the competitors from advertising pirated apparatus, but pragmaticaly, considering time limitations and other factors, such a tactical action was out of the question. In fact, in some ways, they owed AEG gratitude for pointing the way. Their course was to make it clear that there was only one inventor. And so they erected a forty-five-foot high monument in the center aisle of Electricity Hall which proclaimed the truth to the world. The testimonial read in big bold letters: Westinghouse Electric & Manufacturing Co. Tesla Polyphase System. And with this method, from an annex of Machinery Hall, Westinghouse illuminated the entire World's Fair. Having manufactured a quarter of a million Sawyer-Mann stopper lamps for the occasion, Westinghouse generated three times more electrical energy than was then being utilized in the entire city of Chicago.[27]

Not to be eclipsed, GE constructed a scintillating eighty-two-foot high Tower of Light in the exact center of Electricity Hall. It comprised eighteen thousand lamps running the full length of the pedestal, capped by a single enormous Edison lightbulb ignited at the apex.

Other features within Electricity Hall included, on the second floor, electrical gadgets to cure all ills, such as charged belts, electrical hairbrushes, and body invigorators, and on the main floor, displays by the most prominent inventors of the day. Elihu Thomson, for instance, unleashed a high-frequency coil that created sparks five feet long; Alexander Graham Bell unveiled a telephone that could transmit voice via light beams; and Elisha Gray presented his teleautography machine, a precursor of today's fax machine. For a few cents, it would reproduce a person's signature electronically at a distance.

Tom Edison presented his own cornucopia of gadgets, including the multiplex telegraph, the fantastic talking machine known as the phonograph, and his kinetescope, which for the first time in a public forum displayed "the varying labile movements" of a human being in motion.

Tesla's exhibit, which occupied part of the Westinghouse space, featured a number of his early AC devices, including motors, armatures, and generators, phosphorescent signs of noted electricians, such as Helmholtz, Faraday, Maxwell, Henry, and Franklin, and a sign for his favorite Serbian poet Jovan Zmaj Jovanovich. Tesla also displayed vacuum tubes illuminated by means of wireless transmission, his rotating egg of Columbus, sheets of crackling light created by high-frequency discharges between two insulated plates, and other neon signs reading Westinghouse, and Welcome Electricians. These last two displays "produc[ed] the effect of a modified lightning discharge...accompanied by a similar deafening noise. This was probably one of the most novel attractions in a sensational way seen in the building, as the noise could be heard anywhere within Electricity Building and the flash of the miniature lightning was very brilliant and startling."[28]

Tesla traveled to Chicago in August not only to visit the fair and present a week of demonstrations but also to attend the International Electrical Congress that was held there for that month. "At Electricity Hall, Professor Tesla announces he will send a current of 100,000 volts through his own body without injury to life, an experiment which seems all the more wonderful when we recall the fact that the currents made use of for executing murderers at Sing Sing, N.Y., have never exceeded 2,000 volts. Mr. Tesla also shows a number of other interesting experiments, some of which are so marvelous as to be almost beyond description."[29]

On August 25, Nikola Tesla spoke before "one thousand electrical engineers" at the International Electrical Congress in Agriculture Hall. In attendance was a "galaxy of notables," including Galileo Ferraris, Sir William Preece, Silvanus Thompson, Elihu Thomson, and honorary chairman Hermann Ludwig von Helmholtz, whom Tesla also took on a tour of his personal exhibit.

"People crowded about the doors and clamored for admittance....The great majority of those who came, came with the expectation of seeing Tesla pass a current of 250,000 volts through his body....Ten dollars was offered for a single seat, and offered in vain. Only members of the Electrical Congress, with their wives, were admitted, and not even they unless they were provided with credentials." Before the lecture, a Chicago reporter inquired of Professors William Preece and Silvanus Thompson as to the use of the various equipment scattered about the hall, but they "gazed in wonder and confessed they could not guess....They [simply] lumped off the whole lot under the generic term of 'Tesla's animals.'

"Presently, white-haired Elisha Gray was seen escorting a tall, gaunt young man towards the platform. The young man smiled with pleasure but modestly kept his eyes on the floor. His cheeks were hollow, his black eyes sunken...but sparking with animation;...Intense and continuous application of his work has sapped his energy until his friends say he has almost reached the point of dissolution. A gentleman who dined with him a week ago says he could scarcely make himself heard across the table, he was so worn out. He has glossy black hair parted in the middle, a mustache, heavy under his aquiline nose, but fading to a suggestion at the sides of his mouth...his ears are large and stand out from his head. He wore a neat four-button cutaway suit of brownish gray." Gray said to a rousing applause: "I give you the Wizard of Physics, Nikola Tesla."

"I have with great reluctance accepted these compliments, because I had no right to interrupt the flow of speech of our chairman," Tesla began with characteristic humor. Appearing somewhat like a resurrected cadaver, Tesla spoke to allay the fears of all those concerned for his frail health. "A number of scientific men urged [a group of electricians] to deliver a lecture. A great many promised that they would come, [but] when the programme was sifted down I was the only healthy man left...and so I managed to take some of my apparatus...and give you a brief outline of some of my work."[30]

Tesla proceeded to display his new steam generators and mechanical oscillators, some of which were so compact "that one could readily carry them in the crown of one's hat." He told the audience that his goals were multifaceted. Such a device could, among other things, be used to run motors with perfect synchrony, or electric clocks. He had also produced a continuous-wave radio transmitter, although no one at the time understood the complete ramifications of the equipment. Nevertheless, when the resonant frequency was reached, wireless lights would again illuminate, and in that way intelligence was wirelessly transmitted.

One of Tesla's more unusual exhibits, which was similar to his egg of Columbus, was another ring which displayed not only the principles of the rotating magnetic field but also his theory of planetary motion.

> In this experiment one large and several small brass balls were usually employed. When the field was energized all the balls would be set spinning, the large ones remaining in the center while the small ones revolved around it, like moons about a planet, gradually receding until they reached the outer guard and raced along the same.
>
> But the demonstration which most impres[sed] the audiences was the simultaneous operation of numerous balls, pivoted discs and other devices placed in all sorts of positions and *at considerable distances from the rotating field.* When the currents were turned on and the whole animated with motion, it presented an unforgettable spectacle. Mr. Tesla had many vacuum bulbs in which small light metal discs were pivotally arranged on jewels and these would spin anywhere in the hall when the iron ring was energized.[31]

Tesla returned to New York exhausted but exhilarated.

13

THE FILIPOVS (1894)

Mr. Tesla has been held a visionary, deceived by the flash of casual shooting stars; but the growing conviction of his professional brethren is that because he saw farther he saw first the low lights flickering on tangible new continents of science. The perceptive and imaginative qualities of the mind are not often equally marked in the same man of genius.

T. C. MARTIN[1]

Fame had arrived, and at all levels. In engineering circles Tesla was known as "one of the most remarkable discoverers of the age,"[2] to the magazines as (heaven forbid!) "the New Edison,"[3] to the newspapers as "Our Foremost Electrician,"[4] to the masses as a wondrous wizard from a strange land, and to the financiers as bankable.

Having recognized the virtuoso from the outset, T. C. Martin, as head trumpeter, was now helping choreograph Tesla's vault into the public arena. Commerford, as he was known to his friends, had generated a propitious relationship to all concerned when he met with Robert Underwood Johnson, associate editor of the *Century*, to propose a feature article on the thriving electronic savant.

"Take a seat," Johnson offered as he scanned through toppling stacks of manuscripts scattered about the office to find a chair that could be easily cleared. A brother of a congressman and a poet, Johnson had, in 1889, been honored by the city when he was chosen to contribute an original sonnet for the unveiling of the Washington Square arch which he read before President Harrison and other dignitaries. A temporary structure designed by Standford White commemorating one hundred years since the inauguration of George Washington, this arch predated by a few years the permanent marble Arch also designed by White. Johnson was also knowledgeable about inventors, having visited Edison's laboratory in the early

1880s as a reporter for *Scribner's Monthly,* the precursor of the *Century.* Tesla, no doubt, had already piqued his interest.

Impressed with the eloquent portrait Martin had painted of Tesla, Johnson invited his colleague over for dinner. "Why not bring the wizard along. Perhaps there is more than one article here for the offing."

On a first-name basis with such personalities as mayoral candidate Theodore Roosevelt and writer Mark Twain, two frequent visitors to his Union Square office, Johnson and his ebullient wife, Katharine, were the very essence of the phrase "gracious hosts." Through their dinner parties at their home at 327 Lexington Avenue, a visitor could dine with any number of luminaries, such as sculptor August Saint-Gaudens; actress Eleonora Duse; poet and editor in chief of the *Century,* Richard Watson Gilder; naturalist John Muir; activist for children's rights Mary Mapes Dodge; conductor of the Boston Symphony Monsieur Gericke; composer and pianist Ignace Paderewski; thespian Joseph Jefferson; or writer extraordinaire Rudyard Kipling. Johnson had just returned with his wife from their second trip to Europe, where they had, by chance, run into Twain in Venice. The world was a cozy and romantic place for the Johnsons, in part because they found it that way and in part because they molded it that way.[5]

Martin arrived with a "pallid, drawn and haggard" Tesla at the Johnson home in the middle of the Christmas season. They were greeted by Katharine and Robert Johnson and their two children, Agnes, who was about sixteen, and Owen, who was two or three years younger.

The inventor's frail condition, which one reporter described as having "reached the limits of human endurance,"[6] took the Johnsons, especially Katharine, by surprise. A facile conversationalist, Katharine was a striking woman of Irish descent, poised, with shoulders swept back, her head held erect. Although her hair was now beginning to streak gray, she still exuded an air of youth. It was her eyes in particular that visitors were drawn to. Slightly coquettish, they displayed a daring sense of play beneath a wistful stare.

Triggering a constellation of emotions ranging from maternal instincts to beguilement, Katharine became enthralled, some would say mesmerized, by the Serbian superman who had entered her world; and he had taken to her.

Tesla talked about his European tour, particularly his meeting with Sir William Crookes, and the Johnsons invited him for Christmas.

"You are overworked, Mr. Tesla, and you should take a holiday," Katharine declared. "Perhaps a good Christmas meal will keep you going through the first months of the new year."

"I get all the nourishment I require from my laboratory," Tesla retorted. "I know I am completely worn out, and yet I cannot stop my work. These experiments of mine are so important, so beautiful, so fascinating, that I can hardly tear myself away from them to eat, and when I try to sleep I think about them constantly. I expect I shall go on until I break down altogether. Come, let us go there for our dessert." A carriage was called, and in short order the Johnsons were lured to the "magician's den."[7]

"Be prepared for a surprise or two," Tesla said as he, in the words of a reporter who experienced a similar episode, "ushered [them] into a room some twenty five feet square, lighted on one side by two broad windows, partially covered by heavy black curtains. The laboratory was literally filled with curious mechanical appliances of every description.... Snakelike cables ran along the walls, ceiling, and floor. In the center was [an electric dynamo which sat upon] a large circular table covered with thick strips of black woolen cloth. Two large brownish globes, eighteen inches in diameter, [were sus]pended from [the] ceiling by cords. Composed of brass, coated [and insulated with] wax, [these globes] served the purpose of spreading the electrostatic field...."

Tesla shut the doors and drew the curtains "until every chink or crevice for the admission of light was concealed and the laboratory bathed in absolutely impenetrable gloom. [As we] awaited developments, exquisitely beautiful luminous signs and devices of mystic origin began to flash about. Sometimes they seemed iridescent, the entire room filled with electric vibrations, [as] tubes [and] bulbs [which we held] became luminous.... What impressed [us] most of all, perhaps, was the simple but cheerful fact that [we] remained unscathed, while electrical bombardments were taking place on every side."[8]

A few days later, in honor of the celebration of the Serbian Christmas, on January 6, Katharine sent Tesla a bouquet.

"I have to thank Mrs. Johnson for the magnificent flowers," Tesla wrote Robert. "I have never... received flowers [before], and they produced upon me a curious effect."[9]

Tesla would return to the Johnson home on a regular basis, for dinner or after an evening on his own for a late nightcap; and the inventor reciprocated by taking the Johnsons out on the town. They would be attending the gala performance of Dvořák's *New World Symphony*.

"Upon receipt of your first note," Tesla wrote Robert, "I immediately secured the best seats I could for Saturday. Nothing better than the 15th row! Very sorry, we shall have to use telescopes. But I think the better for Mrs. Johnson's vivid imagination. Dinner at Delmonico's."[10]

Intrigued with Tesla's fascinating heritage, Robert became interested in Serbian poetry, and so Tesla began to translate for him. They obtained permission from Zmaj Jovanovich to feature some of his poems for inclusion in the *Century*, and also in a book Johnson entitled *Songs of Liberty*.

Their favorite was undoubtedly the ballad about a warrior from a Montenegrin battle which took place in 1874.

Luka Filipov

One more hero to be part
Of the Servians' glory!
Lute to lute and heart to heart
Tell the homely story:
Let the Moslem hide for shame,
Trembling like the falcon's game,
Thinking on the falcon's name—
Luka Filipov.

The verse goes on to describe a fierce battle, in which Luka captures a pasha and marches him back to the prince. However, on the way, Luka becomes wounded in an ambush, so his soldiers decide to retaliate by killing the arrested Turk.

We'd have fired, but Luka's hand
Rose in protestation,
While his pistol's mute command
needed no translation:
For the Turk retraced his track,
Knelt, and took upon his back
(As a peddler lifts his pack)
Luka Filipov!

How we cheered him as he passed
Through the line, a-swinging
Gun and pistol—bleeding fast—
Grim—but loudly singing. . . .
[But] as couriers came to say
That our friends had won the day,
Who should up and faint away?
Luka Filipov![11]

One can easily envision Robert's and Katharine's eyes aglow as Tesla spontaneously translated Zmai's ode in their living room one special evening and how Robert worked to hone it for publication. From this moment on, brother Robert would be known as "Dear Luka," and Katharine would be addressed as "Mrs. Filipov."

To Katharine, Tesla was more than a man; he was an icon of historical importance, a trophy to display before her lady friends, and even more than this, he was a tangible symbol of intangible longings. In her essence, Katharine was a frustrated artist whose ever-present yearnings pulled her from moments of exaltation to the edge of despair. She must have been a difficult person to live with, but like Tesla, she had the capability to spark that realm that makes life all the more worth living. Selfish, egocentric, and histrionic at times, Katharine also had a dominant streak that not only pulled people to her but manipulated them to her needs. Even the elusive hermit became a fly in her web.

"Dear Mr. Tesla," she wrote in January 1894, 'We have had a hospital here since Thursday. Robert and Agnes the invalids. Robert is better but not out, and we want you to come...this evening and brighten us up....As a great favor come...to us immediately."[12]

This beckoning to her auxiliary mate became a recurrent theme that replayed itself again and again through the next many years. For instance, in 1896 she wrote, "Dear Mr. Tesla, I shall expect to see you tomorrow evening";[13] in 1897, "Come soon";[14] and in 1898, "Will you come to see me tomorrow evening and will you try to come a little early....I want very much to see you and will be really disappointed if you do not think my request worthy [of] your consideration."[15]

But Katharine was not alone in her infatuation; T. C. Martin had discovered him, but Robert would soon become Tesla's closest confidant. A triad of intimacy quickly transpired as Commerford Martin's hold on the inventor began to loosen. And just as much as the Johnsons would come to cherish Tesla, he would cherish them. Here was a loving family the isolate could bind to.

Tesla's health constantly was an issue not only for their own concerns but also, for the good of the species. Martin, whose new article in the *Century,* would certainly serve to advance his own career as well as Tesla's, was just as solicitous, and he discussed the precariousness of the situation with Katharine.

"I do not believe...that he will give up work at any very early date," Martin wrote. "Talking of California with him in a casual way elicited the fact that he had a couple of invitations to lecture there so that I don't want to jam his head into that lion's mouth. I believe he is going to take more care of himself and you may have done us all a great deal of service by your timely words. Yet in spite of that," Martin continued, "I fear he will go on in the delusion that woman is generically a Delilah who would shear him of his locks. If you can manage it, I believe it would be a good scheme to have that Doctor get hold of him....My prescription is a weekly lecture from Mrs. RUJ."[16] This passage, by a Tesla colleague, is a rare glimpse at one of the enduring mysteries about Tesla, namely his supposed celibacy and mysterious sexual inclinations.

By coincidence, only a fortnight later, Heinrich Hertz was dead. He was only thirty-six. Martin wrote to Tesla, "For God's sake, let it be a warning to you. All Europe mourns for such an untimely taking off."[17]

But Tesla would take no heed, and his custom of driving himself to the limit would be a style of living that he would continue for many years. The future of the race and his role in it was clear to him; nothing could stand in his way.

"The time will come," Tesla told Katharine, "when crossing the ocean by steamer you will be able to have a daily newspaper on board with the important news of the world, and when by means of a pocket instrument and a wire stuck in the ground, you can communicate from any distance with friends at home through an instrument similarly attuned."[18] His gaze and seeming powers of divination took her breath away. Penetrating his inner sanctum only added to her captivation.

At the outset, Tesla's unparalleled achievements, explosion into international stardom, and promise for the future generated an express desire for all to cash in, to sell his multitudinous array of inventions so that they could ascend together to the next social rung. Their goal was to become millionaires. Tesla's "cold light" appeared to be the first ticket.

As Martin's erudite biographical essay had received words of praise even from competitors, Johnson suggested that they plan a second piece, one that featured the lab. Johnson would get some of his more famous friends to come, and photos would be taken. This would be an exclusive, the first ever with the new revolutionary cold light. Johnson writes:

> We were frequently invited to witness his experiments, which included...the production of electrical vibrations of an intensity not before achieved. Lightning-like flashes of the length of fifteen feet were an every-day occurrence, and his tubes of electric light were used to make photographs of many of his friends as souvenir of their visits. He was the first person to make use of phosphorescent light for photographic purposes—not a small item of invention in itself. I was one of a group consisting of Mark Twain, Joseph Jefferson, Marion Crawford and others who had the unique experience of being thus photographed.[19]

Naturally, the Twain pictures would become the centerpiece. S. L. Clemens, as he signed his letters to Tesla, came to the laboratory on March 4, 1894, and again on April 26.[20] Twain wrote to postpone the initial meeting for one day, and Tesla and Johnson exchanged notes about it. Curiously, Twain's private log for this period makes no reference to the occasion. He had dined with Stanford White at his Madison Square Garden tower in January and, the following month, had received a notice that $160,000 in royalties were coming his way because of the Paige typesetting machine which he had backed, but that was all he jotted down, even though Twain had been aware of Tesla from the very first moment that the inventor had gone public with his creation of the AC polyphase system.[21]

Back in November 1888, Twain had written: "I have just seen the drawings & description of an electrical machine lately patented by a Mr. Tesla, & sold to the Westinghouse Company, which will revolutionize the whole electric business of the world. It is the most valuable patent since the telephone."[22]

Twain had run into Tesla on occasion, at the Players' Club or Delmonico's, or at the artist Robert Reid's studio. One night, in Twain's words, "the world-wide illustrious electrician" had joined the Reid party. Jokes and stories were swapped, and songs were sung, particularly Kipling's "On the Road to Mandalay."[23] Tesla related the story of how Twain's books saved his life when he was a boy of twelve struck down with a case of malaria, and this tale served to endear Tesla to Twain to the point of bringing tears to the writer's eyes.[24]

Interested in inventions and their exploitation, Twain asked Tesla at the lab if it would be all right if he could sell high-frequency electrotherapy machines to rich widows in Europe upon his next sojourn; the inventor naturally agreed. Tesla, in turn, showed the great writer yet another creation which he claimed would help these widows digest their meals.

This contraption, he explained, "consists of a platform supported on elastic cushions that are made to oscillate by means of compressed air. One day, I stepped on the platform and the vibrations imparted were transmitted to my body.... Evidently, these isochronous rapid oscillations stimulated powerfully the peristaltic movements which propel the food-stuffs through the alimentary channels."

"You mean, it'll make me regular?" Twain inquired.

"Precisely, and without the use of elixirs, specific remedies or internal applications whatever."

Without further ado, Twain stepped aboard as Tesla tried to stop his assistants from chuckling. As Twain had been so enthusiastic, Tesla neglected to inform him that peristaltic action is induced almost immediately.

"Suddenly, Twain felt an unspeakable and pressing necessity which had to be promptly satisfied," Tesla told the Johnsons the next day to their tearful glee, for he had to jump off the platform and find his way swiftly to the lavatory.

"I think I'll start with the electrotherapy machines," Twain said upon his return. "Wouldn't want the widows to get too healthy all at one shot."[25]

The photos of Twain and the others were processed nearly one full year before they would appear in print. Tesla was elated and commented to Johnson that the one of Joseph Jefferson was "simply immense. I mean the one showing him alone in the darkness. I think it is a [work] of art."[26] Katharine suggested that they all celebrate at Delmonico's and that he then join them for a summer holiday at the Hamptons. Tesla wrote back: "I fear that if I depart very often from my simple habits I shall come to grief." However, realizing that he would soon be missing "the pleasure of your company" he reluctantly agreed to go to dinner, but not to the holiday, and therefore signed off, "In the anticipation of the joy [of dining with friends] and subsequent sorrows, I remain, Yours sincerely, N. Tesla."[27]

Concerning the photographs, which were the first ever taken with phosphorescent lights, Tesla, who had his eye on potential investors, became impatient for the publicity; but Martin and Johnson were aghast.

"I think that we ought to have a little talk about giving to the daily newspapers a hint that Mr. Tesla has succeeded in taking photos by phosphorescence," Martin warned. "*It will leak* out some hour and then someone...with the customary arrogance [will place it] in the papers....[We need] to get our priorities established. I think R. U. Johnson feels the same way."[28]

This became the beginning of a series of disagreements between the inventor and his editor. *The Inventions, Researches and Writings of Nikola Tesla* was now in press, and Martin and Tesla were making money from the sale of the book; however, Tesla kept wanting to give copies away freely. He sent the text to each of his uncles and his three sisters in Bosnia/Croatia and also mailed his article on Zmai to his uncle Pajo and his sister Marica.[29]

Martin had to tread carefully, for although he was unhappy with Tesla's lack of concern for the financial side of the situation, he also in no way wanted to alienate himself from the wellspring. Martin wrote, "Your request [for more free copies] is just too hard. It seems to me that the Pittsburgh boys, if they love you, ought to be willing to blow a little money of their own on the book. But you are the best judge of relations to them." Martin promised to send Tesla a dozen copies at reduced prices. "Perhaps," Martin requested, "you would like to make us a bid on the whole edition," closing as follows: "When you write me, make it autographic as often as you can. People are beginning to deplete my stock."[30]

From Tesla's point of view, it was his book, and Martin should simply do as he asked. This would become a sore point, especially because Martin would come to lend Tesla money based on profits that were due him as editor and Tesla would never repay him.[31] Martin would overlook it, for now.

After the first photographs arrived, including the discreetly stylized engraving of Tesla based on his most recent portrait, Martin requested a sneak preview. "I will lock [them] up or put [them] in a safe deposit vault, if you wish, until the hour of publication," promised Martin. "But I want to get one of the *first* as an historical souvenir." At the same time, Martin informed Johnson that the University of Nebraska had offered Tesla an honorary doctorate in celebration of their twenty-fifth anniversary. "I have urged him to accept. I want you and Mrs. Johnson to bring your influence on him also. Her spell is now a potent one, I fancy, with him, so far as any woman's can be, next [to] his sisters."[32]

It is unlikely, however, that Tesla thought much about getting a doctorate from the obscure University of Nebraska. To a person of his background and education, the offer was virtually meaningless. Johnson thought it more appropriate for a prestigious institution, such as Columbia College, to grant such an honor. Tesla had just received the Elliott Cresson Gold Medal Award from the Franklin Institute for "his earnest and indefatigable work as a pioneer in this field, and on account of the great value to science of his researches."[33] However, this was not the same as a doctorate, and so Johnson wrote to Hewey Fairfield Osborn, one of the dignitaries at Columbia, urging that they make the offer instead.

Johnson wrote:

> There would be a particular appropriateness in Columbia giving him a degree, since his first lecture was, if I mistake not, delivered at the College and since New York City is the scene of his most important discoveries.... I think it may truly be said that there are few men occupying this unique position... in both the theoretical and practical phases of scientific work.... As to his general culture, I may say that he... is widely read in the best literature of Italy, Germany and France as well as much of the Slavic countries to say nothing of Greek and Latin. He is particularly fond of poetry and is always quoting Leopardi... or Goethe or the Hungarians or Russians. I know of few men of such diversity of general culture or such accuracy of knowledge.

Johnson ended the letter with a character reference. His personality, he said, "is one of distinguished sweetness, sincerity, modesty, refinement, generosity and force."[34]

Since Professor Osborn knew Tesla well and had been witness to the Columbia lecture, he concurred with Johnson and thereupon spoke to Seth Low, the president of the university. "Poulton tells me," Osborn said, "that Tesla was covered with honors while in England and France. We certainly must not allow any other university to anticipate us in honoring a man who lives under our very eyes."

"Isn't he a countryman of Pupin's?" Low inquired.

"Yes, of course. It was at the urging of Professor Pupin and Professor Crocker that Tesla spoke here to begin with."

"Didn't they have a row between them?" the president responded cautiously.

"I have learned confidentially that there has sprung up some slight difference between them, but I'm sure it will probably be healed. In any event, there seems little doubt that Tesla is the leading electrician in the country."[35]

Within a few weeks, the inventor was given an honorary doctorate from Columbia, and shortly thereafter he received a similar honor from Yale.

Having received professional recognition among his peers, Tesla had arrived. Feature articles were appearing in prestigious periodicals, and he was hobnobbing with the most esteemed literary and social figures of the day.

14

NIAGARA POWER (1894)

How extraordinary was my life an incident may illustrate.... [As a youth] I was fascinated by a description of Niagara Falls I had perused, and pictured in my imagination a big wheel run by the Falls. I told my uncle that I would go to America and carry out this scheme. Thirty years later I saw my ideas carried out at Niagara and marveled at the unfathomable mystery of the mind.

NIKOLA TESLA[1]

Conquest of Niagara was by no means assured. It hinged on very specific factors. The first serious scheme to harness the mighty cataract, the Evershed plan, was proposed in 1886. Thomas Evershed, a civil engineer, who had worked on the Erie Canal, had conceived the idea of the creation of a complex network of canals and tunnels running adjacent to the Falls, whereby two hundred waterwheels and accompanying industrial mills would be placed. He had probably ruminated over the scheme for twoscore years, as Evershed had worked as a surveyor at Niagara in his youth as far back as the 1840s. Although the idea was attractive, it was costly and dangerous to implement, as most of the nine miles of excavation required for the canals and wheel pits had to be done by blasting through stone; estimates ran as high as $10 million. Thus, the officials of the Cataract Construction Company sought counsel from notable engineers and inventors.[2]

In 1889, Edison submitted a plan which boldly asserted that DC could be transmitted to Buffalo, approximately twenty miles away. As appreciable amounts of DC electricity had never traversed a distance in excess of one or two miles, this suggestion appeared highly optimistic, and it was doubted by most other engineers, particularly Sprague and Kennelly, two of Edison's coworkers. Westinghouse was also dubious of the feasibility of transmitting electrical energy, and he suggested the implementation of a sophisticated system of cables and compressed air tubes to transfer the power to Buffalo.[3] Thus, for these reasons, plans for harnesing the falls centered for the most part on constructing an industrial complex adjacent to the falls.

The long-distance transmission of electrical *power* was simply beyond imagining. It must be remembered that what Edison generated was meager amounts of electricity capable of illuminating lightbulbs, and then only in close proximity to the source of the power. Since his DC apparatus still made use of a commutator, he was incapable of transmitting appreciable amounts of power, although he could run a few motors if they were close to the generator. That is why the 1891 Lauffen-Frankfurt transmission was so startling. Brown and Dobrowolsky had not only surpassed, by a factor of about one hundred, Edison's long-distance record; they had also transmitted significant amounts of power, a spectacular achievement with no comparable precedent.

Brown and Dobrowolsky had had predecessors. Two years earlier, Sebastian Ziani de Ferranti, the son of an Italian musician living in Liverpool, England, had been the first to use Tesla's apparatus at a plant at Depford. Ferranti, a brilliant engineer with talents purported to have rivaled Edison's, had already made important modifications on the Gaulard-Gibbs AC apparatus for Siemens Brothers and also Ganz & Company at their London branch. His bold idea was to create a central station along the Thames so that electricity could be pumped to numerous substations around town.[4] In 1889, from Depford, Ferranti transmitted an unprecedented 11,000 volts to four substations of six to seven miles away, where 10,000 horsepower alternators were driven.[5] This was a magnificent accomplishment, but it appears doubtful that many people understood that it was Tesla's system[6] or even that true success had been achieved. In no way did it gain the publicity of the Lauffen-Frankfurt enterprise, nor did it spark the idea of spreading the energy of Niagara Falls beyond its surrounding area.

Nor had Westinghouse's initial achievements in electrical power transmission demonstrated the capabilities of the Tesla system. He had succeeded at Telluride, Colorado, with Stillwell, Shallenberger, and Scott, in transmitting 60,000 volts of AC for a distance of four miles to run a 100 horsepower Tesla motor, and he had illuminated the Chicago World's Fair of 1893. Both of these were great triumphs, but neither demonstrated that electrical power could be transmitted over long distances. In short, without the Lauffen-Frankfurt success, there would have been no proof that AC was capable of traversing the twenty miles from Niagara to Buffalo, let alone from Niagara to New York City, which was over three hundred miles away. That is why the financial backers of the Niagara project sent Edward Dean Adams, president of the Cataract Construction Company, to Europe to confer with Brown and Dobrowolsky, and that is why Dobrowolsky intimated that the invention was his. There was no physical proof to the contrary, as clearly he and Brown had been the first and only engineers to realize such a feat.

Adams, from the firm Winslow, Lanier & Company, out of Boston, was a slight, gentle-looking man with large round eyes, small head, the face of a teenager, and a mammoth handlebar mustache. He began his relationship with longtime J. P. Morgan associate Charles Lanier in 1881. Working his way up to full partner, Adams was placed on the board of directors of a number of major railroads, including Henry Villard's Northern Pacific, and also the Ontario and Western Railroad, with lines stemming from Buffalo to New York City.[7] He also sat on the board of directors of the Edison Electric Light Company, as its second largest stockholder.

In 1889, along with Villard, who was attempting to combine all the major electric companies into one large corporation, Adams tried to curtail the highly expensive lightbulb patent dispute between Edison and Westinghouse by getting them to confer with one another, but, of course, Edison wanted no part of such an arrangement.[8]

As president of the Cataract Construction Company, Adams sold his shares in the Edison concern so that he could be impartial in his investigations, and in 1890 he "established the International Niagara Commission, with headquarters in London.... His intention was to consult with leading European scientists and engineers and to examine the most advanced hydraulic-power [compressed air] techniques, a branch of engineering in which Switzerland excelled."[9]

In 1890, Adams traveled to Europe with Dr. Coleman Sellers, another executive of the Cataract Construction Company, where they conferred with engineers in France, Switzerland, and England. In London they visited Ferranti at his electrical station at Depford; they also met with Professor Rowland, who had traveled from Johns Hopkins University, and Gisbert Kapp, electrical engineering editor and author of the classic text *Electrical Transmission of Energy*. Rowland argued in favor of AC, and Kapp recommended C. E. L. Brown as the most prominent engineer to undertake the project. He was located at the Maschinenfabrin Oerlikon Works in Switzerland. Adams wired J. Pierpont Morgan, who was in Paris, with the suggestion that he return to Switzerland to meet with Brown. Morgan concurred.[10]

Before leaving England, Adams met with Sir William Thomson (Lord Kelvin), whom he placed in charge of the International Niagara Commission, and a contest was created awarding cash prizes totaling $20,000 for the best plans submitted for harnessing the falls. L. B. Stillwell, who was in London with H. H. Byllesby, at the Westinghouse branch, wired to Pittsburgh to request permission to compete by giving Adams a plan based on the Tesla system; but Westinghouse turned down the idea because he did not want to give away $100,000 worth of advice for such a paltry sum.

Of the twenty proposals submitted, most involved compressed air and hydraulic equipment. "Of the six electrical plans, four used direct current...[one] proposed single phase [AC], but 'details were not fully described.' The remaining plan by Prof. George Forbes advocated polyphase installation."[11] Forbes, who was a professor from Glasgow and who was later hired as consulting engineer to the Niagara Power Company, wrote to the commission: "It will be somewhat startling to many, as I confess it was at first to myself, to find as the result of a thorough and impartial examination of the problem that the only practical solution lies in the adoption of alternating current generators and motors....The only [workable one] is the Tesla motor manufactured by the Westinghouse Electric Company and which I have myself put through various tests at their works at Pittsburgh."[12]

Although initially rejected by the commission, the Forbes report caught the eye of Adams. Nevertheless, Adams went to Switzerland to confer with C. E. L. Brown, who declined the offers to head up the Niagara project.

Morgan's emissary was Francis Lynde Stetson, a lawyer who was also part of the Cataract Construction Company. He was sent to Switzerland and London to review the prevailing technology for their company, GE, but it was starting to become obvious that the major patents were all owned by Westinghouse. In Tivoli, where there were waterfalls 334 feet high, Ganz & Company of Budapest, another Westinghouse-linked operation, was constructing a hydroelectric plant to transmit electricity to Rome, which was eighteen miles away, and in Portland, Oregon, at Willamette Falls, Westinghouse was also transmitting thousands of volts of AC over distances of twelve miles. Although Kelvin sided with Edison in insisting that DC was superior, Adams now knew that Westinghouse held all the trump cards.[13]

In America, however, and from the point of view of GE, the outcome was not all that obvious. The Panic of 1893 had taken its toll, and Charles Coffin, CEO of GE, was forced to "ruthlessly" lay off a large number of their workers and cut the pay of many others. Not only had production of electrical equipment fallen dramatically, in-fighting between the Edison and Thomson camps peaked.[14] Although Thomson and Steinmetz now realized that AC was vastly superior to DC, they were unable to guarantee to Coffin that they could devise equipment superior to Tesla's. Desperate to compete, it appears that a memorandum was sent by Thomson to E. G. Waters, general manager of the GE plant in Pittsburgh, for the purpose of recruiting an informant who worked for Westinghouse.[15]

Noticing that blueprints were missing, Westinghouse accused GE of industrial espionage, and charges were brought against Thomson's Lynn plant, where a sheriff, acting under court order, found the missing documents. The GE officials claimed that their interest was in seeing whether Westinghouse was pirating their protected lightbulbs, and the jury split six to six on the decision. Westinghouse suspected that a janitor was the culprit, but the man was never prosecuted.[16]

Simultaneously, Steinmetz and Thomson were submitting patent applications for an AC motor which used a "teaser current" instead of a full-fledged polyphase one,[17] but it was obvious to the patent office that the apparatus was based on the Tesla system, and their patent application was denied. This did not stop Thomson from insisting that he was the real inventor of the AC system, and by 1894, with Thomson's great expertise, he devised an induction motor that in some respects was superior to the one produced by Westinghouse.[18] Ironically, even to this day, biographies of Elihu Thomson often accuse Tesla of the piracy instead of the reverse![19] Although GE was brought to trial for stealing blueprints, they audaciously continued for the next several years, through Waters, to pay spies to obtain information from the Westinghouse plant.

Nevertheless, Westinghouse's success at Telluride and at the Chicago World's Fair eliminated any remaining doubts as to who would be awarded the Niagara contract. In the first months of 1893, Forbes, Rowland, and Sellers visited Pittsburgh to test their equipment, and in May of that year the deal was signed with Westinghouse.

As J. Pierpont Morgan was the major force behind GE, it is interesting to speculate about why he allowed Westinghouse to gain the bid. First of all, when the actual contract was signed, because the operation was so enormous (and because of Morgan's ties to GE), a large portion of the work was also given to GE. Westinghouse constructed the "generators, switch-gear and auxiliary equipment in the powerhouse, [and] GE was awarded contracts for the transformers, the transmission line to Buffalo, and the equipment for the substation there."[20] Thus, although Westinghouse got the larger share, GE was by no means cut out and in fact ended up with a licensing arrangement which gave them their first legal foothold on the fundamental patents held by the other company.

Morgan had close ties to August Belmont, who was one of Westinghouse's financial backers, and it is possible that this connection had something to do with the arrangement. He acquiesced in part because of his respect for the commission Adams had set up and also because of the advice he received from his lawyer William B. Rankine, who lived in Buffalo and who had devoted his life to the enterprise, and his close associate, Francis Lynde Stetson, who told Morgan of Tesla's "daring promise [as far back as 1890] to place 100,000 hp on a wire and send it 450 miles in one direction to New York City, the metropolis of the East, and 500 miles in the other direction to Chicago, the metropolis of the West, [to] serve the purpose of these great urban communities."[21]

In 1894, Tesla hit his stride. The Martin article in the *Century* opened the floodgates, and an army of reporters from newspapers and magazines descended upon him. That year would find features on Tesla in such prestigious periodicals as *New Science Review, Outlook* and *Cassiers; McClure's* and *Review of Reviews* boldly announced that Tesla was the founder of the discoveries which lay behind "the largest electrical enterprise in the world,"[22] and the *New York Times* profiled him in a four-column spread complete with a large stylized portrait and an in-depth account of his philosophies and newest creations.[23] The following year, the *Times* wrote, "To Tesla belongs the undisputed honor of being the man whose work made this Niagara enterprise possible.... There could be no better evidence of the practical qualities of his inventive genius."[24]

Since the contracts with Westinghouse and GE had been signed, Adams no longer had to pretend to be unbiased; now he was free to seek business ventures. He came down to New York to visit Tesla at his laboratory, and there Adams was introduced to the inventor's new mechanical and electrical oscillators and to a revolutionary new AC system of lighting that was vastly superior to the existing one. Adams offered Tesla $100,000 for a controlling interest in "fourteen U.S. patents, many foreign patents," and any future inventions which Tesla might conceive. The inventor accepted the proposal.[25] In February 1895, the creation of the Nikola Tesla Company was formally announced, with the directors being Tesla, Alfred S. Brown of New York, Charles Coaney of New Jersey, William B. Rankine of Buffalo, Edward D. Adams, and Adams's son Ernest of Boston.[26]

Tesla had reached the inner sanctum of the corporate world. He now had on his board two of the most important members of the Niagara project, and he was sitting on at least a half-dozen entirely new inventions, each of which had the potential for creating completely new industries. His mechanical oscillators looked as if they were going to replace the steam engine; his electrical oscillators were intrinsic to his system of fluorescent lighting, remote control, and his now secret work in wireless transmission; and there were other ideas as well, such as his work in artificial intelligence, ozone production, "cheap refrigeration, and cheap manufacture of liquid air, the manufacture of fertilizers and nitric acid from the air."[27] However, all of these inventions were still, to a great extent, in their developmental stages, and Tesla's strong suit was never in manufacturing. He gained his backers ostensibly because of his track record in AC and because of the promise held by his cold light and various oscillators. But his real interest, his passion, remained in the wireless transmission of power, and most of his time was spent in that direction.

15

EFFULGENT GLORY (1894)

Dear Mr. Tesla,

Early in 1894, I told our mutual friend... T. C. Martin, that your book "The Inventions, Researches and Writings of Nikola Tesla," would still be considered a classic a hundred years hence. I have not changed my opinion.... The application of the principles therein first enunciated have definitely changed for the better, life on this planet.

D. McFarlan Moore[1]

In February 1894, Tesla went with T. C. Martin to the laboratory of Mr. Bettini, an inventor who had advanced Edison's work in the perfection of the phonograph. "He wants to display his marvelous collection of song cylinders—Madam Melba, deReszke, Salvini and Bernhardt," Martin wrote. "He wishes the honor of catching your voice as well."[2]

During this time, Martin continued to forward reviews from *Book News, Physics Review,* and *London Electrician.* "Praise from Sir Herbert [Spencer] is praise indeed," the editor wrote. "Thanks for continuing the Russian translation; the German edition is selling as well."[3]

On the home front, Tesla had formed a new company with two of the coordinators of the Niagara project, and he was now turning his attention to designing a wireless system that would be able to transmit light, information, and electrical power. Thus, the inventor worked to perfect his mechanical and electrical oscillators and vacuum lamps. The mechanical oscillator could efficiently convert steam to electrical power, the electrical oscillator could produce the frequencies necessary for power transmission, and the wireless fluorescent bulbs could be illuminated when the correct wavelengths were reached capable of engendering light. According to Tesla's theory, if the ether could be made to oscillate at 500 trillion times per second, pure light would be created. Below that figure, heat was produced.[4]

T. C. Martin continued to play a role somewhat like a manager. In the spring he set up a sitting for the inventor before the sculptor Mr. Wolff, and to Tesla's dismay, he also arranged an interview with S. S. McClure, the well-known magazine editor. Tesla had too many events crammed into his calendar, but T. C. persisted.

"I cannot very well call off McClure now," Martin wrote, "after your little dinner which has simply made him more eager than ever for the article." Ending the missive with customary praise, Martin wrote, "He knows now personally, what he knew before only by hearsay, viz that you are a great man and a nice fellow. I excuse your blushes."[5]

It would take nearly a year for the *Century* to complete its own article and organize the photographing of the celebrated individuals who came to 33-35 South Fifth Avenue, but finally, in April 1895, Martin's follow-up piece appeared. Even today, a century later, it remains a spectacular testament to the wizard's fabled laboratory, for not only did it display many fascinating inventions and prognostications; it also presented dazzling multiple-exposure photographs of such stars of the day as actor Joseph Jefferson, poet F. Marion Crawford, and sorcerer Nikola Tesla. "Martin's article should be read very carefully to be appreciated," Tesla wrote Johnson.[6]

With thirty-five homes illuminated by gas for every one lit by electricity, the market was ripe for a more efficient source. However, Tesla's fluorescent lamps were still in an experimental stage, as evidenced by the ten-minute time-lapse photograph of Mark Twain taken for the Martin article.[7] The story for Tesla's steam dynamo was analogous. According to Martin's figures, nine-tenths of all dynamos in the country were steam driven. The rest gained their power from either compressed air, waterwheels, or gasoline. Since the Tesla oscillator was forty times more efficient than the prevailing steam-driven dynamo, it seemed that the marketing of this product would be highly profitable. But again, the perfection of the device was not completed and therefore not yet marketable. Nevertheless, it was revolutionary, and Tesla contacted Babcock & Wilcox to begin manufacturing prototypes.

Before a large number of physicians and engineers of the American Electro-Therapeutic Association at his laboratory in October 1894, Tesla used the oscillator to generate electricity for over fifty incandescent lamps and vacuum tubes. Also, "arc lights were shown in operation." Naturally, the oscillator also generated the extremely high frequencies necessary for engendering beneficial electrotherapeutic effects.[8]

Refining his theory on wireless communication, the inventor realized that electrical energy could be transmitted in two distinctly different ways, one as radiation through the air, and the other as conduction through the ground. Today this difference corresponds to FM and AM radio. (As an experiment to verify that FM travels mostly through the air and AM mostly through, or along, the ground, drive your car into an underpass and turn on the radio. When you switch on AM [amplitude modulation], you will probably hear static, whereas FM [frequency modulation] comes in loud and clear.) "When I showed my experiments to an audience," Tesla said, "it would happen that when I wanted to operate single circuits with some devices more than one circuit would respond, and people would call my attention to this....I would then have to say that the circuits were not carefully tuned."[9] Thus the inventor turned his attention to the problem of creating separate channels.

By studying Herbert Spencer's theories on resultant forces and nerve conduction,[10] Tesla began to realize that he could design vacuum tubes which would respond only when a precise combination of two or more circuits were being triggered. One of the lamps held by Twain was illuminated in this manner, its dual circuit triggered by a corresponding dual circuit created by two cables laid around the room. Like the combination of a safety lock, this invention not only explains the mechanism behind tuning into different stations on the radio, it also explains the principle behind telephone and TV scramblers. In this way, Tesla attained "the exclusiveness and non-interferability of impulses.[11]

Transmitting messages from an outer circuit to an inner circuit twenty feet away was one thing; sending information over long distances was another matter entirely. Tesla had realized for a number of years that the earth carried a charge, and therefore he decided to *utilize the planet* itself as a carrier of electrical energy. If this were so and individualized channels could be created, transmission lines appeared to be superfluous. He therefore began to construct many different-sized coils to connect to the ground (usually via the water-main system) in order to ascertain the terrestrial period of frequency.

With an oscillator constructed for this purpose, Martin wrote, "if he has not yet actually determined the earth's [precise] electrical charge, or 'capacity,' he has obtained striking effects which conclusively demonstrate that he has succeeded in disturbing it....[When his oscillations] are in harmony with the individual vibrations of the [earth], an intense vibration or surging will be obtained."[12] Martin ended the article with the suggestion that with this device not only could information and power be transmitted, but also the weather might be modified. "Perchance, we shall 'call up' Mars in this way some day, the electrical charge of both planets being utilized as signals," Martin concluded, quoting Tesla's wish at the finale of the exposé.[13]

Classifying Tesla was becoming more and more difficult. Martin portrayed him in a variety of ways—as a practical inventor, a wizard, and an iconoclast; critics portrayed him as "an impractical...visionary enthusiast."[14] "His inventions already show how brilliantly capable he is," one newspaper reported, but his "propositions...seem like a madman's dream of empire."[15]

"One is naturally disappointed that nothing practical has as yet proceeded from the magnificent experimental investigations with which Tesla has dazzled the world," Lt. F. Jarvis Patten wrote in *Electrical World*.[16] But his solid accomplishments at Niagara could not be ignored. The article concluded:

> Before the general public, he stands as a phenomenal inventor from the Eastern world, from whom is expected little less than if he carried Aladdin's lamp in his hand, which, of course, is wrong, and an injustice both to the public and to Mr. Tesla....He has doubtless much importance in store for us, but the difference should never be lost sight of between the search for nature's truths in the lab of the physicist and the reduction of the results attained, however great their promise, to a form suitable for commercial use....If some of the dreams which Tesla and others are cherishing today ever come to realization, the magnitude of the ensuing change in the material life of the world simply defies the imagination.[17]

In Manhattan, Tesla soon caught the eye of Joseph Pulitzer, the German-Hungarian immigrant and owner of the *New York World*. Pulitzer sent his newest reporter, Arthur Brisbane, on the interview. Already one of the most widely read columnists in the nation, Brisbane had recently jumped from Charles Dana's *New York Sun*, but at age thirty, Brisbane would soon also leave Pulitzer to become head shock reporter for William Randolf Hearst's up-and-coming yellow *Journal*. He met Tesla in July 1894.

While dining nightly at Delmonico's restaurant, the reporter was introduced by Mr. Delmonico to "a very handsome young man with a pointed black beard." Brisbane wrote that "Mr. Delmonico lowers his voice when he speaks of Mr. Tesla, as Boston cab drivers used to lower their voicess in speaking of John L. Sullivan."

"That Mr. Tesla can do anything," Delmonico contended. "We managed to make him play pool one night. He had never played, but he had watched us for a little while. He was very indignant when he found that we meant to give him fifteen points. But it didn't matter much, for he beat us all and got all the money." Delmonico noted that they were only playing for quarters, but "it wasn't the money we cared about, but the way he studies out pool in his head, and then beat us, after we had practiced for years. [It] surprised us."[18] Considering that Tesla was very proficient at the sport and had slyly kept that information confidential makes the story all the more interesting. Tesla could be deceptive, particularly in pecuniary matters.

Although the article appeared somewhat disjointed and redundant, it nevertheless aptly described Tesla's habits and appearance at that time. Tesla would dine frequently at the popular establishment, always choosing a table by the window and usually reading a newspaper. Brisbane described Tesla's "eyes [as being] set very far back in his head. They are rather light. I asked him how [that] could [be, as he was] a Slav. He told me that his eyes were once much darker, but that using his mind a great deal had made them many shades lighter." This confirmed a theory on brain usage and eye color that Brisbane had heard about.

Brisbane went on to write that Tesla "stoops—most men do when they have no peacock blood in them. He lives inside of himself." And yet the article was adorned with the most fantastic full-body engraving of Tesla standing quite unstooped, proud as a peacock, illuminated like a virtual Christmas tree.

This full-page spread remains one of the most spectacular portraits of the inventor ever concocted, and it caused numerous people to ask Tesla how it felt to pump such huge amounts of electricity throughout his body. "I admit that I was somewhat alarmed, when I began these experiments," Tesla admitted, "but after I understood the principles, I could proceed in an unalarmed manner."[19]

When pressed for details a number of years later at the Harvard Club, the inventor responded:

> [When] the body of a person [is] subjected to the rapidly alternating pressure of an electrical oscillator of two and one half million volts [this] presents a sight marvelous and unforgettable. One sees the experimenter standing on a big sheet of fierce, blinding flame, his whole body enveloped in a mass of phosphorescent streamers, like the tentacles of an octopus. Bundles of light stick out from his spine. As he stretches out the arms, roaring tongues of fire leap from his fingertips as myriads of minute projectiles are shot off from him with such velocities as to pass through the adjoining walls. He is in turn being violently bombarded by the surrounding air and dust.[20]

Tesla was walking a difficult line and revealing contradictory natures: reclusive versus natural showman. The interest of the press made it difficult to protect against piracy. He wanted to disclose, for public and historical purposes, his overall goal, but he had to keep vital details confidential. One reporter who had spent a day with "this kindly wizard of Washington Square" revealed that Tesla "confided to me that he was engaged on several secret experiments of most abundant promise, but their nature cannot be hinted at here. However, I have Mr. Tesla's permission to say that some day he proposes to transmit vibrations through the earth [so] that it will be possible to send a message from an ocean steamer to a city, however distant, without the use of any wire."[21] Even his laboratory workers were insulated from the details of his secret experiments.[22]

After calculating the size of the planet and the hypothetical wavelength of solar rays, Tesla constructed a transmitter with several circuits capable of engendering their electrical charge into the earth. One end of the instrument was attached to the ground via the radiator or water main, and the other end was connected to a cable which Tesla brought to the roof of the building as an aerial connection. With Mr. Diaz Buitrago, his draughtsman, in charge of the transmitter, the inventor would take his receiving instruments as far as five miles away. His first experiments were set up on the roof of the Hotel Gerlach, approximately thirty blocks from the lab. With balloons filled with hot air, helium, or hydrogen to hold the elevated terminal (aerial) high above the buildings and a cable attached to the water main of the hotel, Tesla set up his receivers and verified that, indeed, electrical energy could be secured from his transmitter.[23]

Had Tesla taken a reporter with him on any one of these excursions, it is quite possible that the generally accepted history of the invention of wireless transmission would be completely different, because in all of Tesla's years, he never demonstrated before a viewing body any long-distance wireless effects.

Still three years away from expressing details of his wireless work in patent applications, Tesla had actually hidden some of his plans in patents already secured on his mechanical and electrical oscillators drawn up in 1891 and 1893.[24] This would present a problem for Michael Pupin, who was also experimenting with resonance effects and the transmission of simultaneous messages. However, Pupin's goal involved the improvement of the prevailing telephone and telegraph lines. He was not attempting to send messages without them.

Pupin realized that by equally spacing pulsations of AC, the rapidity and number of transmissions could be increased greatly. Unfortunately, his February 1894 patent prospectus was easily covered in Tesla's existing inventions and high-frequency lectures.[25]

Tesla had announced in London, in February 1892, that "if the wave length of the impulses is much smaller than the length of the wire, then corresponding short waves...would [greatly] reduce the capacity [making it] possible to send over the wire high frequency currents at enormous distances. [Furthermore], the character of the vibrations would not be greatly affected." The creation of a "screen" to "cut the wire into smaller sections" would make it possible to transmit many telephonic messages over the transatlantic cable.[26]

These patents and published announcements in no way deterred Pupin. He felt that he had discovered something, and he began a long campaign against the U.S. Patent Office in attempts to bulldoze his way into a legal foothold. The prize, if successful, would be enormous, as he would have exclusive rights on a way to successfully transmit at the same time large numbers of noninterfering long-distance telephonic and telegraphic conversations over the same wires. Pupin's nemesis was John Seymour, commissioner of patents.

The Columbia University professor's first strategy was to write up a patent application. He submitted it on February 10, 1894, claiming, "I certainly consider myself the first to make a practical application of this principle to multiple telegraphy."[27]

Seymour's reply a few months later was that Pupin's "claims...are rejected on the arrangement of apparatus shown...by patents by Thomson and Rice...[and by] Tesla's article Experiments in Alternating Currents." Seymour also cited the exact page and figure number, concluding that Pupin had merely "multiplied Mr. Tesla's electric light circuits," which in no way was a new invention.[28]

Hiring a lawyer to help him put together a legalistic-looking circular, Pupin argued that he had indeed been the inventor of the "Art of Distributing Electrical Energy by Alternating Currents." The typeset brief, which resembled an official court document, read in part as follows: "Tesla produced luminous effects, and did not consider multiplex signalling....He does not disclose several exciting circuits acting on the main line, with means for tuning each exciting circuit independently....The applicant was the first to apply the principle upon which the claims are based and did more than merely multiply the Tesla electric circuit."[29]

Seymour wrote back: "Claims 1, 2 and 3 are again rejected on Tesla. It is well known in the art that several periodicities may be simultaneously impressed on the same line....The examiner can see no more in these claims than a multiplication of Tesla's circuit in a manner well understood in the art."[30]

Pupin was adamant. He was convinced that he had been the first to invent the obvious and, furthermore, that he had a total right to make use of Tesla's oscillators, as they were now being generally accepted as the optimum-frequency generators for long-distance electrical transmissions. By studying each rejection notice by the U.S. Patent Office, Pupin kept refining the language of his patent in order to try and come upon a way that would secure a legal foothold for him. Moreover, Pupin continued to convince himself that indeed this invention was his.

Altering history in his mind and erasing Tesla in his classroom, Pupin would continue his battle to secure this highly lucrative patent on the means of transmitting many simultaneous messages over long distances. His battle would continue for another six years, until after John Seymour retired.

The year 1894 was a banner one. In July the spectacular portrait of Tesla appeared in the *World.* He had major coverage in *Electrical World,* the *New York Times* and *Review of Reviews,* and his AC polyphase system was going to be utilized at Niagara Falls. Tesla had formed a partnership with financiers from Wall Street, he had invited the historical giants of his age to his workplace, and he had made marked improvements in experiments in wireless communication.

The year closed with an invitation by Tesla to the Johnsons to come to his shop. "Dear Luka," Tesla wrote on December 21, "You have not forgotten the visit to my laboratory tomorrow I hope. Dvořák will be there and a number of other celebrities in America's elite."[31]

Anton Dvořák, fifteen years Tesla's senior, had immigrated from his native Czechloslovakia in 1892, to be appointed director of the National Conservatory of Music. Forever homesick, Dvořák stayed in the United States for only three years, but during that time he composed some of his most famous works, particularly the *New World Symphony.* After the performance, Dvořák visited the wizard's lab. Christmas and New Year's Eve with the Johnsons would round out one truly remarkable year.

16

FIRE AT THE LAB (1895)

The destruction of Nikola Tesla's workshop, with its wonderful contents, is something more than a private calamity. It is a misfortune to the whole world. It is not any degree an exaggeration to say that the men living at this time who are more important to the human race than this young gentleman can be counted on the fingers of one hand; perhaps on the thumb of one hand.

CHARLES DANA[1]

It was "one fine Sunday afternoon in 1894," as Tesla was strolling up Fifth Avenue with twenty-five-year-old D. McFarlan Moore, a colleague of great promise in the field of fluorescent lighting, when the Serbian savant "deliberately stopped" and pensively proclaimed, "Moore, after we have signalled from any point to any point on the earth, the next step will be signalling to other planets."[2]

Before the inventor could undertake such an enormous task, he first had to perfect wireless effects over long distances on the earth. One of his plans was to send messages from his laboratory to receiving equipment which he was going to place on a Hudson River steamboat.[3] Unfortunately, on March 13, 1895, Tesla's laboratory burned to the ground. The "whole floor collapsed and equipment dropped to [the] second floor."[4]

For one fleeting moment, the civilized world was in shock, for the demolition of the maestro's atelier was a tragedy of incalculable proportions. Fortunately, Tesla was not injured, for he was asleep in his hotel at the time. "Two tottering brick walls and the yawning jaws of a somber cavity aswim with black water and oil were all that could be seen [that fateful] morning...of a laboratory which to all who had visited it was one of the most interesting spots on earth."[5]

This "coming great man," *Current Literature* noted, "[who] lives his life, as in a dream, forgetful of the lapse of time, and living only for the future...[was about to] revolutionize the cost and economies of electric lighting, and place it within the reach of the humblest and poorest of people....To have all of...his innumerable marvels...swept away at one stroke is a calamity to the whole world as well as to himself."[6] Perhaps in part to help cheer Tesla up and "in honor of the Serbian-American pioneer of electric communication, the Postal-Telegraph Union of Serbia caused a sensation by connecting a simultaneous concert in Belgrade and Nis by telephone so that both audiences could hear it."[7]

"The Tesla laboratory was, in a sense, a private museum," T. C. Martin wrote. "The owner kept in it many souvenirs of bygone toil and experiment." After describing in detail its contents, Martin concluded: "Perhaps the most painful loss of all is the destruction of Mr. Tesla's notes and papers. His memory is all right, and flashes on any experiment of the past with the revealing power of a search-light, but the time it will take for the inventor to recreate his ongoing investigations will also cost other experimenters years of sweat and pain....[Nevertheless,] while the ashes of his hopes lay hot...Tesla was at work again with clenched determination."[8]

The strain was enormous, and Tesla steeled himself to fight off depression. One paper reported that he suffered a "physical collapse."[9]

To help keep his spirits up, Martin met with the inventor at a local cafe to give him more free copies of their book; he may have also given him money.[10] "Ere, it please your majesty," the editor said with a bow, "your experiments have been repeated in Berlin under your name with the Emperor's brother, Prince Henry, assisting. If you did not get a daily bolster from me you would have a relapse into dullness as you do when you miss your daily dose of electricity." The duo sat down to review Martin's article on the burning of the lab so that a more accurate description of its lost contents could be enumerated.[11]

Westinghouse was still in the midst of battling William Stanley of the William Stanley Company and Elihu Thomson of GE in patent litigation, as each was continuing to produce AC induction motors illegally while at the same time implying that the invention was of their own design. Together, these two concerns outsold Westinghouse by 10,000 kilowatts for the period 1893–97.[12] According to some, their motors and generators were more efficient. Stanley, who continued to brazenly promulgate the sale of his polyphase system in advertisements in the electrical journals, had by this time increased his operation from fifteen men to a few hundred, and GE was more than twice that size.

Embroiled in the numerous patent disputes, the Westinghouse Company decided to take out a full-page advertisement which proclaimed:

Westinghouse Electric & Manufacturing Company
sole owners of the
Tesla Polyphase System

The display continued: "The novelty of Mr. Tesla's inventions was recognized by Prof. Elihu Thomson, who said in discussing the Tesla inventions before the American Institute of Electrical Engineers in 1888: 'I have certainly been very much interested in the description given by Mr. Tesla of his new and admirable little motor. I have, as probably you may be aware, worked in somewhat similar directions and towards similar ends. The trials which I have made have been by the use of a single alternating current circuit, not a double alternating circuit.'"[13]

Although GE intimated that the Tesla motor was dangerous because it gave off too many sparks, there is no evidence that the fire in Tesla's lab was caused by his equipment. It had started in the floor below, in a dry-cleaning establishment. Some investigators intimated that a careless night watchman may have been responsible, perhaps by smoking near oily rags.[14]

Estimates of the uninsured loss ranged as high as a million dollars, but the actual damage was probably closer to $250,000. O'Neill suggests that Adams came to the rescue by advancing $40,000 in return for a piece of the company.[15] However, Adams was already a partner, and thus, he, Tesla, and the other partners all suffered losses. Still, there is some evidence that Adams did provide further assistance at this time.[16] Royalties from Europe and continuing modest annual payments from the Westinghouse Company helped offset the misfortune, but clearly Tesla now needed to raise additional revenue in order to open up a new place.

In particular, Tesla received a number of letters from friendly members of the Westinghouse Company. Ernest Heinreich, an engineer and author, wrote, "I hasten to offer you my sincere regret," Not knowing the situation, he continued, "I trust that you were well insured and will be able very soon to find another suitable location to carry on your work."[17]

Tesla was uninsured, but he had generated too much momentum to slow his progress. Within a few days, he was out scouting new locations. In the interim, he turned to the one lab where he knew vital equipment would be on hand. For the next few weeks, Nikola Tesla rolled up his sleeves at Tom Edison's workshop at Llewellyn Park, New Jersey, a laboratory that "shuts out everybody who has not been given a pass by Mr. Edison himself, or one of his assistants."[18] Simultaneously, he contacted Albert Schmid for more equipment, scoffing, as was his nature, at any expense which might be incurred. "I shall rely, as to the price, entirely on the fairness of the Westinghouse Co.," Tesla said, concluding: "I believe that there are gentleman in that company who believe in a hereafter."[19]

Vice president and general manager Samuel Bannister shipped, as a gift, some early Tesla models which had been saved from the World's Fair and wrote of his regret concerning "your misfortune.... I am glad to know that you took off your coat to work to get everything back into shape as soon as possible."[20] However, this was little consolation, as the Westinghouse Company would begin to bill Tesla for the cost of machinery lost in the fire which was on loan; they also charged him for new equipment.

In April, partly in response to Brisbane's announcement that Tesla was "greater even than Edison,"[21] the rivalry between the two men, at least in the press, intensified. "Who Is King, Edison or Tesla?" the *Troy Press* of New York inquired.[22] Joseph Jefferson, speaking in Boston, left no doubt about his position. "Edison has been deposed," the thespian proclaimed, "and Tesla has been coronated [new potentate]."[23]

The "Twin Wizards of Electricity" met in May in Philadelphia, along with Alexander Graham Bell, at the National Electrical Exposition. For the first time ever, Tesla's AC was transmitted the grand distance of five hundred miles. Tesla was disappointed that no appreciable amount of power was transported along the existing telephone lines for fear, by the underwriters, that damage or a fire could result. Nevertheless, the experiment was a complete success and dwarfed the achievements of the old hundred-mile Lauffen-Frankfurt record.

"The most amazing thing at this exposition," Edison remarked, "is the demonstration of the ability to deliver here an electric current generated at Niagara Falls. To my mind it solves one of the most important questions associated with electrical development." Bell concurred, stating, "This long distance transmission of electric power was the most important discovery of electric science that had been made for many years."

"[Bell] with Edison, looking into the future, realized that by means of this discovery cities and towns remote from the places of electrical generation would be able to obtain the services of this agent...with great economy...[and with] a practical convenience far superior than is now possible."

Tesla, "who solved the problem," stated, "I am now convinced beyond any question that it is possible to transmit electricity...by water power...to commercial advantage over a distance of 500 miles at half the cost of generation by steam [or coal].... I am willing to stake my reputation and my life upon this declaration."[24]

No record exists of what was said between Tesla and Edison on this occasion, but it seems likely that each was privately amused by the rivalry played out in the press, that Tesla thanked Edison for the temporary use of his laboratory, and that Edison expressed his condolences for the loss of Tesla's workplace.

GE was now losing the propaganda campaign on all fronts. Even Edison was admitting Tesla's accomplishment. Rumors began to circulate that an agreement with Westinghouse was reached on the pooling of patents, but a total solution was a number of years away, partly because GE had so many divisions that were pirating the apparatus, and partly because Westinghouse saw no immediate advantage in making a deal.[25] After one significant loss in the courts, the headquarters of GE was forced to abandon its position; but they would continue to bargain because their financiers controlled the Niagara enterprise. Moreover, the size of that venture prevented Westinghouse from tackling it alone. T. C. Martin, however, naively thought that a complete reconciliation and vindication for Tesla was at hand. "This would [now] mean the acknowledgement of supremacy of your polyphase patents," he wrote. "I suppose each will license the other."[26]

It does not appear that Tesla enlightened him on the finality of the deal he had signed, because one full year later, the editor discussed the subject again. "I cannot tell you how pleased I am about the news of the recognition of your patents by G.E. Company, and I hasten to congratulate you.... [This] ought to fill your pockets with your own money," Martin concluded.[27]

At least from a historical perspective, vindication was at hand, for now it was accepted even by the opposition that Tesla's work alone was making possible the "yoking into service of old Niagara herself";[28] but the inventor would never reap any financial gains other than those he had already settled for.

"This discovery forms the basis," *Review of Reviews* announced, "of the Niagara Company's attempt to utilize that enormous power which for centuries has been running to waste, and thus to turn machinery in towns and cities so far away as Buffalo, 20 miles distant, and perhaps New York and Chicago." And then the magazine dropped the bombshell. "And it underlies the hardly less bold venture of the Westinghouse and Baldwin companies to drive a through railway express by electricity. It is not too much to say that the Tesla motor is behind all the large attempts at power transmission by electricity which are being made throughout the country, not only in the fields of manufacture and transportation, but also in mining, irrigation and farming."[29]

August, 7, 1895

My dear Mr. Westinghouse,

I learn from the journals of your friendly agreement with the Baldwin Locomotive works....The news of your consolidation has been an agreeable surprise. Such a splendid union of means and abilities cannot fail to be of interest to both parties concerned.

Yours very truly,
N. Tesla

Tesla was in need of new capital. The Westinghouse Corporation was billing him for lost equipment and for new machines ordered, and yet they had just gained two gargantuan contracts in two entirely separate fields, both based on his creation. Surely the corporation would overlook a few thousand dollars owed, especially when it was Tesla's personality that played a key role in gathering such principals of the Niagara Company as Edward Dean Adams, John Jacob Astor, and William Birch Rankine into the fold.

It also occurred to the inventor that use of his induction motor in an entirely novel field such as railway transportation was, in a sense, outside the original intent of their contract. Should not he also benefit from such a fortuitous turn of events? Tesla would continue, somewhat naively, to bring new potential clients to Pittsburgh, almost as if he were their private ambassador, but he would never receive additional compensation for the service.

17

MARTIAN FEVER (1895–96)

If there are intelligent inhabitants of Mars or any other planet, it seems to me that we can do something to attract their attention. . . . I have had this scheme under consideration for five or six years.

NIKOLA TESLA[1]

John Jacob Astor III graduated from Harvard University at the age of twenty-two in 1888. He was one of the wealthiest men on the planet, with assets in the neighborhood of $100 million. By comparison, J. Pierpont Morgan's wealth was perhaps $30 million. As a youth, Astor had been an inventor, having patented a bicycle brake and a pneumatic walkway which won a prize at the 1893 Chicago World's Fair. Other inventions included a storage battery, an internal-combustion engine, and a flying machine.[2]

During his college years, when he was known by the unfortunate nickname of "Jack Ass," Astor, who now sported long, tapered sideburns and a waxed mustache, had undertaken courses with the inimitable astronomy professor William Pickering. One of Astor's pet projects was to find a way to create rain by "pumping warm, moist air from the earth's surface into the upper atomosphere," but the U.S. Patent Office had turned him down.[3] Thus, when Pickering mentioned that the seasons were due to the inclination of the earth's axis off of the ecliptic, Astor became intrigued. If the earth were not tilted away from the sun, Pickering suggested, it would probably have one uniform, moderate climate even at the extreme north and south latitudes.

As part of the curriculum, Astor was introduced to the Harvard Observatory. There, along with such up and comers as Perceival Lowell, brother of the president of the university, Astor could peer through the great telescope and view such wonders as the craters on the moon, the satellites of Jupiter, and Saturn's spectacular rings.

In April 1890, Professor Pickering made headlines when he photographed what he said was a snowstorm on the planet Mars. He calculated that the area covered was almost equal to that of the United States.[4] Two years later, during a celebrated trip to Harvard's observatory in Arequipa, Peru, the bushily bearded professor announced another major discovery: "lakes in great numbers on Mars. The canals," Pickering proclaimed, "have dark as well as bright regions. We also observed clouds, and the melting of snows, and this confirmed Herschel's hypothesis that there was vegetation around the regions of water."[5]

The idea of attempting to signal "Marsians," as they were then called, was a familiar ambition of the day, and Astor, like Tesla, was caught in the fancy of it. In 1894, Perceival Lowell announced in *Nature* his description of the canals of Mars. At the same time, Astor, just thirty years old, completed a science fiction novel about space travel. Entitled *A Journey in Other Worlds,* his book offered a futuristic vision of what takes place one century into the future. A few months after publication, in February 1895, the financier presented a copy to the great inventor.

Although Tesla did not appear to be particularly impressed by the work, the inventor promised Astor to keep it "as an interesting and pleasant memento of our acquaintance."[6]

Adorned with ethereal outer-space illustrations by Dan Beard, Astor's tale begins in the year 2000, with a meeting at Delmonico's restaurant of the Terrestrial Axis Straightening Company, whose task it is to create fair weather throughout the planet.

Astor envisions for "the close of the 20th Century" a picture telephone, an airplane with the ability to fly to Europe in one day, an electric automobile, hidden phonographs by the police to record conversations of criminals, color photography, a rain-making device, the idea of colonizing the solar system, and the understanding that the earth would appear like a crescent moon when seen from outer space.

Perhaps Astor's most impressive prediction is the path that his "spaceship" *Callisto* takes on its journey to Jupiter. Astor hypothesizes that just as magnetism has a repelling force, gravity should as well. This energy, which he called apergy, is simply the opposite of gravity. By harnessing apergy, the astronauts in the story aim their ship first toward the sun, then "change their course to something like a tangent to the earth, and [receive] their final right direction [back out toward Jupiter] in swinging near the moon...to bring apergy into play."[7] Exactly a century after this book appeared, NASA actually did send a spacecraft, named the *Galileo,* on a voyage along a remarkably similar trajectory, using Venus instead of the moon as the pivot for the swing-back out toward Jupiter. Whereas this modern trip will take a few years, Astor's weary travelers cover the distance in a matter of days. Jupiter is abundant with life. Flowers greet them by "sing[ing] with the volume of a cathedral organ." The red spot, they find out, was caused by a forest changing color due to a cold snap.[8] Armed, the astronauts are able to hunt down animals resembling mastodons which they kill for food. Fortunately, they also have the wherewithall to hop back onto the *Callisto* so that they can return to Earth.

Fueled by a competitive spirit, the newspapers and magazines continued to promulgate the idea that Mars was inhabited by beings possibly more intelligent than we. As Tesla made headlines in the New York dailies and electrical journals for his bold prediction that he would "signal the stars" and Astor made the bookstores with his space traveling "romance of the future," other luminaries were also capitalizing on the extraterrestrial fervor.

In 1895, George Lathrop, son-in-law of novelist Nathaniel Hawthorne, had earthlings battle warriors from the Red Planet on the pages of the *New York Journal.* Their weapons were disintegrating death rays invented by the Wizard of Menlo Park, Thomas Edison. The following year, George duMaurier, grandfather of Daphne, wrote the novel *The Martian,* in which he describes telepathic winged beings "that descend from no monkey" but are able to adorn marble statues and irrigate the entire planet.[9] And the year after, H. G. Wells gained notoriety with his serialized *Person's* magazine horror story *War of the Worlds,* in which ghastly octupus-like Martians storm Earth in their egg-shaped spaceships and take over.

Although fictional, these stories were based on prognostications put forward by supposedly sober scientists. The major culprit was the French astronomer and psychic researcher Camille Flammarion. In *Stories of Infinity: Lumen—History of a Comet in Infinity,* published in 1873, Flammarion interviewed "Lumen," a sagacious returning comet, on such topics as the speed of light, time travel, and life on other planets. Lumen: "Ah, if you knew the organisms that vibrate on Jupiter or Uranus.... you would know that some living beings can understand without eyes, ears or smell; that there are other faculties of an unascertainable number in nature essentially different than yours."[10]

This idea, called the plurality of worlds hypothesis, is an antediluvian concept which, through the ages, has counted numerous scientists among its adherents. Early astronomers such as Kepler, Newton, Laplace, and Herschel took this position, along with such modern astrophysicists as Carl Sagan.

Human beings, grasping the immensity of the cosmos, know that life is not necessarily unique to Earth. Roman and Greek mythology, which concerned the lives and responsibilities of specific deities and included a god for each of the known planets, probably served as a psychological template for astronomers' speculations and corresponding religious beliefs.

Carl Jung has linked such mythological thinking to the belief in UFOs, the search for meaning, and the search for God. Identifying God with the unknowable, the unconscious, and the wisdom revealed in dream interpretation, Jung says that the myth arises through attempts by the conscious to understand the unconscious.[11] Thus, the mysteries of outer space are connected with those of inner space. Primordial instincts, archetypes, would therefore be the mechanism evolved through attempts to explain celestial natural phenomena. Over time, these were transformed into the myths of our forefathers.

This belief in ancient sky gods and extraterrestrial existence stems from a common motif, that is, that humans cannot be the highest beings in the cosmos and, furthermore, that there is a supreme creator. Because the idea strikes a deep chord, through the centuries numerous scientists, artists, and authors have been seized by this notion.

In 1835, Richard Adams Locke of the *New York Sun* created a series of front-page articles on astronomer Sir John Herschel, discoverer of Uranus and alleged discoverer of advanced life-forms on the moon. Locke's hoax, which spread around the world before it was exposed, was made possible by the fact that Herschel was in South Africa at the time and therefore out of contact with the press. Herschel's supposed discoveries of unicorn-like animals and winged humanoids were made via a marvelous (and fictitious) telescope that was 150 feet long and could magnify the heavens forty-two thousand times.[12] Thirty years later, Jules Verne journeyed his readers to the moon, but by the late 1870s the destination had shifted to Mars.

The first attempt to create a map of Mars and delineate the features seen on it can be traced back to Bernard de Fontana and Christian Huygens in the mid-1600s. More detailed drawings were sketched by Herschel in 1830 and by numerous other scientists in the 1860s and 1870s, such as Camille Flammarion,[13] and in the 1880s by Giovanni Schiaparelli, who named these channels *canali*.

In two scientific treatises, *The Plurality of Worlds* and *Mars and Its Inhabitants*, Flammarion stated his belief that Mars not only housed life but also intelligent beings. Dwarfed in stature, positioned next to his fifteen-foot-tall telescope, the bearded French astronomer described in detail the mountains, valleys, craters, lakes, and oceans of Mars in *North American Review* in 1896. "It is obvious," Flammarion concluded, "the world of Mars is…vigorously alive." Perhaps unconsciously influenced by the 1835 *New York Sun* hoax, or du Maurier's story, Flammarion suggested that due to the lightness of the atmosphere, "the inhabitants…may have received the privilege of flight.…May they not rather be like dragon-flies fluttering in the air above the lakes and the canals?"[14]

Whereas Flammarion only set out to describe the Martians, Tesla actually made plans to contact them. The most influential American proponent, however, was undoubtedly the erudite Perceival Lowell, descendant of the famous Massachusetts Lowell family. Influenced by

Flammarion, Lowell would capture the front page of the newspapers on many occasions with "Mars inhabited" headlines. He would also come to author a number of scholarly accounts covered in such prestigious journals as *Nature* and *Scientific American*,[15] all culminating in his weighty text *The Canals of Mars*, disseminated by the distinguished Macmillan Publishing Company.[16]

Unlike the warlike humans of Earth, Martians dwelled in a coordinated world, Lowell speculated. They had outgrown their savage instincts and "consciously practice peace." These Martians were "sagacious builders" who conserved their precious water and learned to live in a civilized global society.

Mars was an older and thus more experienced planet. Its people had lived through the technological revolution aeons ago, so they had learned to harvest and cultivate their planet from a global perspective.

With humanity at the dawn of a new technological society, it was comforting to think that we would not have to face this rapidly advancing condition alone. As one of a community of intelligent planets, we had neighbors to whom we could turn for guidance.

By the late 1890s, Lowell had completed construction of his own gargantuan telescope at Flagstaff, Arizona, where it is still, today, one of the finest in the world. There he would report each new discovery, including the cataloging of galaxies, which at the time were called "island universes."

It is hard to overestimate Lowell's impact on contemporary thinking. For instance, the vegetation hypothesis was echoed by Wernher von Braun, Willy Ley, and P. Bonestell, who cowrote in their 1956 text *The Exploration of Mars:* "And this is the picture of Mars at mid-century: a small planet which 3/4ths is cold desert, with the rest covered with a sort of plant life [most likely lichen]....Mars is not the dead planet...but neither can it be inhabited by the kind of intelligent beings that many people dreamed of in 1900."[17]

A society's beliefs determines its reality. But society is made up of individuals, and in the case of the idea that Mars was inhabited, these individuals often embellished their supposed objective scientific findings. Supported vigorously by the press, the most important proponents of the "life on Mars" scenario were the astronomers, but the position was also championed by the inventors.

Elihu Thomson, a longtime stargazer and friend of Professor Pickering's, was so enthusiastic that he often took his telescope to his factories so that workers could see the Martian canal system with their own eyes.[18] Other eminent scientists included Lord Kelvin, who, upon his arrival in America in September 1897, announced to the press his idea to flash a signal at night from the glittering metropolis of New York City to Mars to let them know we are here. No doubt he discussed this plan with Tesla when he visited his laboratory during the journey.[19] Edison, too, was caught up in esoteric causes, but his wish was to invent a telephone-like device to contact departed spirits rather than living Martians.

"The possibility of beckoning Martians was the extreme application of [my] principle of propagation of electric waves," Mr. Tesla told the interviewers in 1896 in the article "Is Tesla to Signal the Stars?" "The same principle may be employed with good effects for the transmission of news to all parts of the earth.... Every city on the globe could be on an immense circuit. [Thus] a message sent from New York would be in England, Africa and Australia in an instant. What a grand thing it would be."[20]

18

HIGH SOCIETY (1894–97)

Nikola Tesla is one of the great geniuses and most remarkable men who have ever had anything to do with electricity.... It is as much of an honor to propose him for membership as his membership would be an addition to the club.

LETTER TO THE PLAYERS' CLUB BY STANFORD WHITE[1]

Commuting to New Jersey was only a temporary solution, and within a few weeks, Tesla returned to New York, where he secured a laboratory just below Greenwich Village, near Chinatown, at 46 and 48 Houston Street. Nervous and perpetually on the brink of exhaustion, Tesla began to experiment with the healing properties of his oscillators as reports began to filter in from around the country about its remarkable curative properties. These high-frequency "vitality boosters" would generate a "universal healing agent" that, when applied, would enable the body to "throw off all diseases," said Dr. F. Finch Strong. "Effects obtained [included]...increase of strength, appetite and weight, induction of natural sleep, and elimination."[2] Other doctors reported the ability to cure tuberculosis.

"Tesla believes that electricity is the greatest of all doctors and says that when his laboratory was burned, nothing but regular daily applications of electricity kept him from sinking into a state of melancholia."[3] "My high frequencies," he would say, "produce an anti-germicidal action." Thus, as part of his routine for a day at the lab, the inventor would disrobe, step upon his apparatus, and turn on the juice. A prickly corona would envelop his body and restore it to a more pristine condition.

Electricity had become the new panacea; it could cure the sick, eradicate the criminals, and even eliminate recalcitrant undergrowth that "interfere[d] with the running of trains.... Weeding has always been considered very hard work, but with the aid of the electrical weeder, a man has only to touch a button and the weeds vanish."[4]

In May 1895, Tesla and the Johnsons attended the dedication ceremony for the new arch, situated as the gateway, looking south, to Greenwich Village at Washington Square Park and looking north to the beginning of Fifth Avenue. Designed by Stanford White, "the dashing man with the red moustache," the lofty edifice stood taller and wider than any comparable one erected by the mighty ancient Romans or Greeks. Johnson had recited one of his poems in honor of the original wooden one which had been built in 1889.

This was but one more connection between Tesla and the celebrated Stanford White, whose many other monuments and buildings were rapidly reshaping the city into a regal testament to the great and vibrant gay era. Tesla would often run into White at the offices of the *Century*, where the artist would be commissioned to illustrate their covers, and at meetings involving the Niagara Falls enterprise at Delmonico's, the Waldorf, and the theater or the Madison Square Garden roof restaurant. Having designed the Players' Club in Manhattan in 1887, the Tennis Club in Newport, churches, and numerous mansions, White was also an interior decorator. It was said that he knew the color of the boudoir of every woman of note in the city. A trendsetter and sensualist, White became one of the leading choreographers of the ambience of the percolating metropolis. Good friends with the inventor, White often talked with Tesla about their shared vision of the future.

Tesla had met White in 1891 when the piano virtuoso Ignace Paderewski played the Garden for five breathtaking performances. Edward Dean Adams was in the midst of courting both men, Tesla for his inventions and White for his architectural prowess, and he wanted to get the two together to discuss the best way to place Westinghouse's behemoth electric generators in the upcoming powerhouse.

"My dear Mr. Adams," White wrote, "I duly received the information of the Cataract Construction Company and will pitch in [as soon as possible]. . . . With McKim in Chicago and Mead in Canada I am here alone in a sort of maelstrom of work. . . . If it were not for the Roof Garden and the ballet girls to cheer me up, I should have been dead long ago."[5] By the end of the year, White had sent Adams designs for the proposed buildings. Adams reciprocated by sending White a magnificent text on precious stones and a "stunning gift" (most likely a ruby or emerald).[6]

In 1893, Tesla and White crossed paths again, as both were cardinal participants in the Chicago World's Fair. The following year, White, then forty, urged the inventor to become a member of the Players' Club. "Will you not let me put you up for membership?" he inquired. "It is an inexpensive club and the character of men I think you would like, and I know it would give me the greatest pleasure to meet you there now and then."[7] Tesla requested that Johnson be included for membership, and White agreed.

In the heart of winter, in early 1895, White invited Tesla "for a little supper [for] the artist, Ned Abbey, in my room in the Tower," and Tesla "sharpened his appetite for the occasion."[8] There, in White's sanctuary, where one's mind could spin a thousand tales, the duo gazed out over the entire city. This moment symbolized the pinnacle of social achievement, for only the elite could enter White's chamber and only the imagination of the outsiders could discern what might transpire. A month later, the inventor reciprocated by asking White, his wife, Bessie, and their son, Lawrence, to his den.

> March 2, 1895
>
> My dear Tesla,
>
> I cannot thank you too much for your kindness in showing all your wonderful experiments the other day. They made a deep impression on me, as they did everyone, and I am going to see them again someday, if you will let me.
>
> Sincerely yours,
> Stanford White[9]

A fortnight later, the laboratory was in cinders, but for the electronic savant, it was almost as if he were simply shifting gears. In the spring he received a risqué invitation to White's outrageous "Girl-in-the-Pie Banquet." As the story goes, and there are a number of versions, a dozen scantily clad maidens served a twenty-course meal at the notorious photography studio of Jimmy Breese at 5 West Sixteenth Street, the dinners having been shipped from Sherry's. Attending the clandestine affair were other friends of White's, including artists August Saint-Gaudens and Robert Reid, and the inventor Peter Cooper Hewitt. At the culmination of the feast, with the band playing, the young ladies returned in even more provocative outfits, singing and wheeling out a pie the size of a small automobile. To the tune of "Four and Twenty Blackbirds," the crust burst open with the flutter of a flock of canaries, and out popped a topless young woman. Mum was the word until sketchy details were published in the *World*.[10]

Tesla became privy to the architect's salacious activities and may have partaken himself in discreet entanglements, although it is just as likely that his phobia for germs or monastic inclinations would have inhibited him. White admired Tesla, as each, in his own way, was a sculptor of the New Age. Meeting occasionally for a round of pool at the Players' Club or at a boxing match, perhaps with Twain, at the Garden, Tesla also accompanied

White for sailing jaunts out at Southhampton with a dozen members of the clique.

On one occasion, White asked Tesla to join him for an outing with Mr. William Astor Chamber, an African explorer. As usual, Tesla was busy at work, but after some tactful prodding, he relented. "I am so delighted that you have decided to tear yourself away from your laboratory," White said. "I would sooner have you on board than the Emperor of Germany or the Queen of England."[11]

The year 1895 was a peculiar one. The U.S. government was nearing bankruptcy. In the Panic of 1893 bondholders had wished to secure gold instead of paper money, and the mint had made good by depleting its reserves. By January 1895 the United States was within days of being unable to meet its debts. Quietly, President Cleveland had asked August Belmont, a wealthy Jewish businessman (and backer of the Westinghouse Company), to meet with the European Rothschilds to secure replacement gold reserves. The reality of the day, however, included an unfortunate worldwide wave of anti-Semitism. Only the year before, in a famous trial in France, the Jewish captain Alfred Dreyfus had been convicted on a "trumped up charge of treason." The Rothschilds were Jewish. How would it look to have Jewish financiers bail out an entire nation? It was for this reason, according to Morgan biographer George Wheeler, that J. Pierpont Morgan, an upstanding Episcopalian, was brought into the picture.[12] Morgan, with Belmont's help, was able to secure $60 million in foreign gold reserves, and the country was saved from insolvency. The incident also marked the anointment of Morgan as King of Wall Street.

In October, a twenty-two-year-old, well-mannered stenographer named George Scherff walked into Tesla's laboratory and applied for a job.[13] The inventor reviewed the secretary's credentials and hired him. Although Scherff knew nothing about electrical engineering, Tesla was impressed with his demeanor and intelligence, and within a matter of days the youth was busy at work transcribing papers and taking over the general management of the office.

In the same month, Tesla forwarded a book on Buddhism to Luka, whom he hadn't seen since the end of the summer. Johnson had traveled with his wife to Italy to receive a decoration from King Humbert for his work on securing a law for international copyright and during this period, Tesla had taken some time to attend lectures in Brooklyn on Buddhism by Swami Vivekananda.[14]. "My dear Friend and faithful stranger," Johnson wrote back, "I am touched by your remembrance of me in sending the book.... [I'll] drop into your laboratory some day for old acquaintance's sake."[15]

"Glad to know that you are again in town and established in the beautiful Johnson Mansion," Tesla wrote Mrs. Filipov. "I cannot say as much for my laboratory which is [still in need of] furnishing."[16]

Tesla reported the local gossip, such as how Stanford had difficulty deciding between which of two beautiful sisters to spend an evening with; the gist of swami Vivekananda's lectures on the external nature of God and the transmigration of souls, and of his progress in netting more millionaires. He was meeting with railroad magnate and U.S. senator Chauncey DePew; J. Beavor Webb, a fleet captain, shipbuilder, and Morgan man; Darius Ogden Mills, a stock market manipulator and principal in GE; and John Jacob Astor.

The wealthiest of the crew, except for Astor, was undoubtedly Mills, who had made his fortune in San Francisco during the California gold rush. Owner of the *New York Tribune,* and a palace on Fifth Avenue "opposite St. Patrick's Cathedral...of which a Shah of Persia might have been proud,"[17] Mills had been the second private citizen in history, after J. Pierpont Morgan, to have his abode illuminated by electricity. As Herbert Satterlee tells the story, Mills was so impressed with the Edison invention that he insisted on becoming a partner in the company. "Only if for every share of Edison stock you purchase for yourself, you purchase one for me," Morgan replied, and Mills agreed.[18] Tesla had much to tell his European traveling friends.

At the end of the year, Tesla began to apply more pressure on Edward Dean Adams to influence John Jacob Astor. The Colonel, as he was now called, was funding, of all people, mountebank John Worrell Keely. This was a situation that had to be changed. Keely's motor hadn't motored in twenty years; Tesla's had turned the world. Martin wrote the inventor of his astonishment at Astor's gullibility;[19] Tesla pressed Astor for a commitment.

Attempting, perhaps, to capitalize on the Christmas spirit, Tesla met with Astor and his nautical counselor, J. Beavor Webb, on December 19 and pitched his cause. "I am impressed with your endeavor, Mr. Tesla," Astor commented, "although, as I understand it, your latest inventions are yet to reach the point of being marketed. Nevertheless, I'll speak to Mr. Adams. By all means, let's keep the door open."

Tesla telephoned Adams that afternoon and wrote Astor the following day:

> My dear Mr. Astor,
>
> [Adams] would be only too glad to have you with us. We agreed that we would jointly provide from 500–1000 shares of the Parent Company for yourself and Mr. Webb at the price of $95 a share of a par value of $100 each.
>
> The Parent Company owns my patents...[and rights in foreign and domestic markets, which I believe] will profoundly affect the present state of the mechanical and electrical arts, and

will create a greater revolution in their applications than my ideas on the transmission of power which are at present, generally adopted.[20]

Christmas was drawing near, and with it the renewal of the Serb's link to his adopted American family. The invitation from the Johnsons was wholeheartedly welcomed. "My dear Luka," Tesla wrote, "I am, as you know, very fond of millionaires, but the inducements you offer are so great that I shall set [them] aside... to partake in the splendid lunch which Mme. Filipov will [prepare]....[For] Christmas, I want to be at home—327 Lexington Avenue—with my friends, my dear friends—the Johnsons. If you will prepare a dinner for a half dozen and invite nobody, it will suit me.... We shall talk of bless[ed] peace and be merry until then."[21]

Tesla did his best to overlook the erotic tension emanating from Katharine as she directed the servants, with Agnes, to set the dinner table and Tesla talked shop with Robert and his son, Owen. Katharine could never be part of the bond that existed between Tesla and Robert. Her heart ached for what it could not have and yet, simultaneously, was filled with what it now possessed.

With Robert out of the room, Katharine was too intense. She claimed a telepathic link to the wizard, her breast palpitated when he was near, hormones gushed. On one occasion she edged them to the line. He had no choice but to withdraw.

On the last day of the year, Stanford dropped off a note. He wanted Tesla to hire a promising lad, the son of his friend Charley Barney, a banker with ties to Whitney and Vanderbilt. "My dear Mr. White," Tesla wrote back, "I heartily agree that the young fellow who has two awfully pretty sisters, ought to be helped by all means. Unfortunately," Tesla continued, he still had the responsibility of "carrying three superfluous [work]men" who were not really working because of the delay caused by the fire.[22]

As the relationship with the Johnsons became more intimate, there may have been rivalries between them over which one had the greater access to "Him." Tesla wrote after the new year: "My dear Luka, I am glad to know you shall love me, but I am much disappointed to learn that the boil has bothered you so much. I doubt, however, that you are a hero, because heroes do not go to bed on account of a boil."[23]

Although Katharine had seen Tesla twice in December, it only served to ignite her passion even more. Torn between loving a professorial and delightful gentleman, who could count among his friends Mark Twain, John Muir, Rudyard Kipling, and Teddy Roosevelt, even if he did suffer from boils, and an exotic internationally known virtuoso whose singular talents promised to transform an entire world, Katharine wanted "to feel herself *en rapport*" with the wizard so that she could discuss their psychic link:

February 12, 1896

Dear Mr. Tesla,

I have had such a wonderful experience the past three years. So much of it is already [gone?] that I sometimes fear it will all pass away with me and you of all persons ought to know something of it for you could not fail to have a scientific interest in it. I call it thought transference for want of a better word. Perhaps it is not at all that. I have often wished and meant to speak to you of this, but when I am with you I never say the things I had intended to say. I seem to be only capable of one thing. Do come tomorrow.

Sincerely yours,
Katharine Johnson

Stanford may have been able to leave his wife on Long Island while he courted young starlets at his bachelor pad at Gramercy Park or his private loft atop the Garden Tower, but "dear Mr. Tesla" was cut from different cloth. He would often dine with women and tantalize with his eyes, but that would have to be the extent of a relationship.

Apparently Tesla undertook a self-imposed vow of chastity, having been influenced in part by Swami Vivekananda, who preached chastity as the path to self-transformation and enlightenment.

Tesla met the Swami on February 13, 1896, at a dinner with Sarah Bernhardt after one of her performances in the play *Iziel.* As with the rest of the world, Tesla had first heard of the Swami during the summer of 1893 when the "Hindoo" gained overnight prominence after speaking at the Congress of World Religions, which had been held at the Chicago World's Fair. As Tesla had been in Chicago within a month of the talk, it is conceivable that he met or saw the Swami speak at that time.

Vivekananda told "the great electrician" about "Vendantic Prâna [life force] and Akâsa [ether], which according to [Tesla], are the only theories modern science can entertain."

Having studied Madam Blavatsky's theosophical teachings, Tesla was already versed in the idea of Akâsa and the Akâshic Records, which are, in essence, the records of all historical events existing in some vibratory state in this ether.

"The Brahmâ, or Universal Mind," the Swami continued, "produces Akâsa and Prâna."

Tesla agreed with the essential premise of this Buddhist view, replying that the theory could be "proved mathematically by demonstrating that force and matter are reducible to potential energy," and then the inventor invited Swami Vivekananda, some of his devotees, and Sarah

Bernhardt to his laboratory for the following week to demonstrate through experiments this principle.

After Tesla showed the swami some of his "creations," the swami advised that pure creation, in the sense that "something" was born from "nothing" was not possible. To Swami Vivekananda, creation was a process of combining existing elements into a new synthesis. This idea of the eternal nature of existence with no beginning and no ending was appealing to Tesla, and he later referred to this and related concepts in some of his writings. Today, this theory in cosmology refers to the steady-state theory of eternal creation, which would oppose the more generally accepted big-bang theory, which hypothesizes a particular date for the beginning of time. The reason that the big-bang theory is the more generally accepted one is because the universe is expanding. Working backward, it appears logical that all matter in the universe was together at one time in one location. Current estimates place the big bang at about 15 billion years ago.[24]

How Astor could be taken in by Keely, was a mystery to Tesla, for the financier declined at this time to partake in the venture. He had taken a month to consider the proposal:

> January 18, 1896
>
> Dear Mr. Tesla,
>
> Your letter offering me some of your oscillator stock received....95 seems rather a high price; for though the inventions covered by the stock will doubtless bring about great changes, they may not pay for some time as yet, and, of course, there are always a good many risks.
>
> Wishing the oscillator as much success as I could if financially interested, and hoping soon to be able to use one myself,
>
> [I remain] Yours sincerely,
> John Jacob Astor[25]

Although a rejection, the letter was not a complete denial. It was going to take one or two more go-arounds to land this big fish.

The oscillators, for Tesla, of course, were never ends in and of themselves. Tesla's goal was to send energy into the earth and use *it* as a conduit to transmit messages and power. Details of the plan, however, were such a tightly held secret that even his workers were not completely confided in. Somewhat surreptitiously, Tesla took a train to Colorado Springs in late February 1896 to look over a prospective site for a new laboratory and also to conduct the kinds of wireless experiments he had wanted to undertake before his laboratory burned to the ground. Tesla instructed a colleague, perhaps a local engineering professor, to transmit a musical song on an autoharp through Pikes Peak to his receiving equipment, which included another autoharp, attuned to the first, four miles away, on the other side of the mountain.

The experiment was a success; the song "Ben Bolt," played on one side of the mountain, was picked up by means of a resonant earth frequency on the other. Tesla, however, completely confounded the details of the instrumentation involved. By implying to the press that the energy utilized derived from the earth and not from one of his oscillators, Tesla also succeeded in generating hyperbolic headlines as well.

Based on this false premise, page 1 of the March 8, 1896, Sunday magazine section of the *World* announced not only Tesla's historic wireless achievement but also the supposed experimental verification that the earth was imbued with "free energy" of essentially unlimited amounts. By tapping this reservoir, the future was clear: "Electricity would be as free as air.... The end has come to telegraph, telephone companies...and other monopolies...with a crash."

19

SHADOWGRAPHS (1896)

The rising claims of the inventors revives an incident in connection with the discovery of the Roentgen ray.... Oliver Lodge [announced] apparatus by which he saw through a man. A few days later Mr. Edison [proclaimed] that he had apparatus with which he had seen through two men. Within a week, Mr. Tesla produced rays of such penetrating power that they went clear through three men. When this was shown to Mr. Edison, the great man, who hasn't a spark of jealousy in his nature, smiled and said, "Well, let's stop it at three. What do you say? I think three men will do as well and prove as much as a regiment."

NEW YORK MAIL & EXPRESS[1]

A few days before the New Year, the scientific world was shaken by the remarkable discovery by Wilhelm Roentgen of a queer, unknown energy that he called X rays, which emanated from his Lenard and Crookes tubes. Michael Pupin wrote, "No other discovery within my lifetime had ever aroused the interest of the world as did the discovery of the X-rays. Every physicist dropped his own problems and rushed headlong into the research." Astonishingly, Pupin added, "To the best of my knowledge I was at that time the only physicist here who had had any laboratory experience with vacuum-tube research.... I obtained the first X-ray photograph in America on January 2, 1896, two weeks after the discovery was announced in Germany."[2] As Pupin could so neatly do, he foreswore any mention of his compatriot. To Pupin, Tesla was a nonperson.

Roentgen gained world recognition virtually overnight with his announcement that he had discovered a new energy emanating from cathode-ray tubes that could illuminate light-sensitive chemicals at the far end of a room, penetrate solid objects, and photograph the internal organs and bones of living beings. As Pupin noted, scientists from all over the world dropped their current projects to join in this exciting new venture. Tesla himself wrote no fewer than nine articles on the topic in a two-year period. Although Tesla may have noticed these rays and their effects on photographic paper years earlier,[3] he did not pursue the investigations and left no doubt that the discoverer of what he liked to call "shadowgraphs" was Wilhelm Roentgen.

Tesla had taken the word "shadowgraph" from Søren Kierkegaard, who described them in his essay "Either/Or." To the existential philosopher they were sketches "that derive from the darker side of life...[but] are not directly visible....The [shadowgraph] does not become perceptible until I see through the external....Not until I look through it, do I discover that inner picture too delicately drawn to be outwardly visible, woven as it is of the tenderest moods of the soul."[4]

In Europe meager X rays were being produced by static machines and Ruhmkorff induction coils; Tesla suggested, instead, the use of a high-frequency disruptive coil attached to a special bulb with two electrodes, a cathode inside the vacuum, for generating the "cathode streams," and an anode placed as far away as possible outside the bulb to limit the reduction of the potential. With this apparatus, "effective pressures of about 4,000,000 volts were achieved."[5] At first, the bulb will get hot and glow with a purplish hue, then the electrode will disintegrate, and the bulb will cool. Use of a fan helps. "From [this] point on...the bulb is in a very good condition for producing the Roentgen shadows." When the electrode is too hot, it is probably because the vacuum is not high enough.[6]

Generating such high voltages, this work not only was set up to measure the quality of the energy emanating from the bulb and to test its ability to pierce living and nonliving objects or be reflected; it also laid the foundation for Tesla's later experiments with particle-beam weapons.

Here, in 1896, Tesla discussed the idea promulgated by the quantum physicists a few years later that the energy had both particle-like and wavelike properties. Having set up a target to shoot the streams at, Tesla wrote: "The effects on the sensitive plate are due to projected particles or else to vibrations [of extremely high frequencies]."[7] The inventor further speculated that "the streams are formed of matter in some primary or elementary condition...Similiar streams must be emitted by the sun and probably by other sources of radiant energy."[8] Tesla also appeared to have come close to the idea of breaking up the electron into subatomic particles. "The projected lumps of matter act as inelastic bodies, similarly to ever so many small lead bullets....These lumps are shattered into fragments so small as to make them lose entirely some physical properties possessed before the impact....[Might it not be possible] that in the Roentgen phenomena we may witness a transformation of ordinary matter into ether?[9] [Or] we may be confronted with a dissolution of matter into some unknown primary form, the Akâsa of the old Vedas."[10]

The inventor as physicist then proceeded to take X rays of small animals, such as birds and rabbits, as well as his workers, and of his own skull, ribs, limbs, and vertebrae. As some shadowgraphs took as long as an hour to obtain, Tesla noticed that he would sometimes fall asleep while he was being bombarded by the machine.

Week after week, Tesla would crank out yet another article on his "Latest Results." On March 18, 1896, he announced in *Electrical Review* that he had produced shadowgraphs of humans at distances of forty feet and affected photosensitive paper at a distance of sixty feet from the source of the rays. The inventor also tested different metals to see which ones reflected the energy in the best way. Adorned with a lavish X ray of the bones of the wizard's own rib cage, the article conveyed an eerie impression.[11]

"I told some friends," Tesla wrote, "that it might be possible to observe by the aid of [a]...screen objects [and skeletons] passing through a street....I mention this odd idea only as an illustration of how these scientific developments may even affect our morals and customs. Perhaps we shall shortly get used to this state of things."

For Tesla, Roentgen rays were a gateway to a world invisible and ripe for new possibilities. "Roentgen gave us a [wonderful] gun to fire...projecting missiles of a thousandfold greater penetrative power than that of a cannon ball, and carrying them probably to distances of many miles.... These missiles are so small that we may fire them through our tissues for days, weeks, and years, apparently without hurtful consequence."

Throughout the year, the inventor suffered from "the grippe." Although his illness made the papers, nobody seemed to link it to his excessive experimentation with the mysterious energy. In fact, concerning the risk to one's health, Tesla wrote: "No experimenter need be deterred from...investigation of Roentgen rays for fear of poisonous or generally deleterious action, for it seems reasonable to conclude that it would take centuries to accumulate enough of such matter to interfere seriously with the process of life of a person."[12] We now know, of course, that this view is wrong as long-term exposure to X rays can be very dangerous to one's health.

Tesla did, however, refer to pain in the center of his forehead when experimenting with the rays and to "the hurtful action on the skin, inflammation and blistering," but this he attributed to the production of ozone, which in small quantities was "a most beneficial disinfectant." Nevertheless, there was a severe accident in the lab "to a dear and zealous assistant...without a protective screen present. The worker suffered severe blistering and raw flesh exposed," the inventor undertaking "the bitter duty of recording the accident" in order to lessen the danger for others."[13]

Edison was also making headlines with his work with Roentgen rays, especially when he noted that the streams caused blind people to experience sensations in their eyes. "The X-rays succeeded in eliciting from the blind the ejaculation, "I see; yes, I see a light!"[14]

Edison, whose fluoroscope was already being used for lighting the eye during eye surgery, saw the possibility that eyesight, in some way, might be restored with the use of X rays.[15] Tesla doubted it, and so the press took up the charge and created a new round of headlines placing the two pioneers once again against each other. "The humorless dark Hungarian [had the] unpleasant duty to say, 'Is it not cruel to raise such hopes when there is so little ground for it.... What possible good can result?'"[16]

Time proved Edison wrong, as X rays have not been used to "stimulate the retina" in such a way as to restore sight, but the two wizards did perform a number of successful miracles with the strange energy when each used the instruments to locate bullets lodged in the bones of various patients. Fortunately, the Kentucky School of Medicine helped bring the battle between the two to a close when they "combin[ed] the devices of Tesla and Edison" to extract bird shot from the wounded foot of a voter who had received the injury in a fight at an election poll. After developing the X ray, which took only ninety seconds to make, "every bone was distinctly shown, and the shot, about thirty in number, were plainly located."[17]

To celebrate the triumph and quell any purported hostilities, T. C. Martin was able to coax Tesla into joining Edison and a number of other electricians for a day of fishing on a topsail schooner off Sandy Hook. The event was sponsored by the Safety Insulated Wire and Cable Company. Although a storm erupted, accompanied by dark clouds and lightning, the "bold fishermen were undismayed.... In stately grandness... as happy and well satisfied a party as ever rode the waves of the Atlantic's billows.... Toward nightfall, the [ship] turned her prow homeward.... Nicola Tesla [caught] a flounder of large dimensions... [and] Edison caught a shocking big fluke."[18]

20

FALLS SPEECH (1897)

Nikola Tesla said much in a notable speech at a banquet to celebrate the conveyance of power from Niagara to Buffalo. Not [just] a plodding workman, he is a dreamer of wise dreams, a poet, and a humanitarian, working with new tools for the benefit of all. He is a man who wonders at the folly of men who invent guns when they might invent tools. His spirit is naturally hopeful.... He looks not so much at the world as at the universe. He finds power in the waterfall, and at the same time looks forward to a time when we may, perhaps, tap the unseen forces of the planets and use the cosmic energy that swings the stars in their courses. He looks to a time when power shall be so cheap, so universal, that all labor shall be done by tireless machines and every man's life be thus so much more worth living.

CHARLES BARNARD[1]

In July 1896, Tesla journeyed to Niagara Falls for his first survey of this great enterprise. He traveled with George Westinghouse, Edward Dean Adams, William Rankine, and Comdr. George Melville of the U.S. Navy. Also present was Thomas Ely, supervisor of motive power for the Pennsylvania Railroad. Tesla was important to all five for almost as many reasons.

A reporter for the *Niagara Gazette* greeted them upon their arrival. "Tesla is an idealist," the journalist wrote, "fully six feet tall, very dark of complexion, nervous and wirey. Impressionable maidens would fall in love with him at first sight, but he has no time to think of impressionable maidens. In fact, he has given as his opinion that inventors should never marry. Day and night he is working away at some deep problems that fascinate him, and anyone that talks with him for only a few minutes will get the impression that science is his only mistress and that he cares more for her than for money and fame."

Rankine predicted that Buffalo would receive electrical power by November, and Westinghouse predicted that costs would be cheaper than steam. "You could say it will cost one half what steam power cost[s]," Rankine added in support.

"Mr. Tesla, what is your opinion of the effect of this development of power on Buffalo and Niagara Falls?"

"The effect will be that both cities will stretch out their arms until they meet."[2]

Tesla looked up at the roaring cataract overcome with emotion as he and the others donned their rain gear before entering the mighty wonder. He had grown up just fifty miles from the magnificent maze of cascading flumes known as Plitvice Lakes, but those were Lilliputian compared to this thundering colossus. Pride overcame the inventor as he trailed behind for a few moments to think, as he so often did, of his mountain homeland. It had been four years since he had seen his family, fifteen years since his first successful construction of a turbine that could be driven by water-power, and nearly thirty-five years since he had told his uncle of his dream of one day harnessing Niagara Falls. Humbled by this awesome manifestation of nature, he sat for a moment to reflect as he watched his cohorts disappear along the catwalks into a mist of rainbows.

"Let's go, Mr. Tesla," Adams called out, having waited as patiently as he could, for the next stop on the itinerary was the Edward Dean Adams Hydro-Electric Power Station, the first of two that would be built in his name. Designed by Stanford White, the edifice housed nearly a dozen gargantuan Tesla turbines, capable of generating collectively over 35,000 kilowatts. The men appeared like dwarfs sauntering amid a lustrous gadgetry assembled as if by giants—one long row of towering, kettle-shaped engines. From this chamber, an efficient, nonpolluting, never-ending source of electrical energy was about to be generated capable of driving the factories and illuminating the streets and homes of nearly one-fourth of the entire continent. The echoes of their steps faded as they stood for a moment in silence in the chapel of the dawning New Age.

Upon his return to New York, Tesla found a letter from Sir William Preece.[3] A young man, half British, on his mother's side, and half Italian, had stopped by Preece's office with a wireless Morse-code apparatus based on the work of Heinrich Hertz. Guglielmo Marconi, just twenty-two years old, had brought a notebook which reviewed the literature in the field (most likely the writings of Hertz, Lodge, and Tesla). Marconi had chosen wisely, as Preece was head of the British Post Office and had experimented himself in testing induction effects through the ground from telegraph lines.[4]

"After the experiments with the classical Hertz devices under the auspices of the Imperial Post Office in England," Tesla reported many years later, "Preece wrote me a letter conveying the information that the tests had been abandoned as of no value, but he believed good results [would be possible by my system]. In reply, I offered to prepare two sets for trial and asked him to give me the technical particulars necessary to the design. Just then, Marconi came out with the emphatic assertion that he had tried out my apparatus and that it did not work. Evidently he succeeded in his purpose, for nothing was done in regard to my proposal."[5] Tesla's first patent specifically for wireless transmission was filed a year later, on September 2, 1897 (no. 650,353).

The following month, in August 1896, Tesla received a histrionic plea from Katharine, who was vacationing with her family at a cottage in Bar Harbor, Maine. Wanting desperately for Tesla to join them, she could only allude to her wish.

> August 6, 1896
>
> Dear Mr. Tesla,
>
> I am so troubled about you. I hear you are ill.... Leave work for a while. I am haunted by the fear that you may succumb to the heat.... Find a cool climate. Do not stay in New York. That would mean the laboratory every day....
>
> You are making a mistake my dear friend almost a fatal one. You think you do not need change and rest. You are so tired you do not know what you need. If somebody would only pick you up and carry you bodily. I hardly know what to expect to gain by writing you. My words have no effect, forgotten as soon as read perhaps.
>
> But I must speak and I will. You do not send me a line? How delighted I should be if it bore an unfamiliar postmark.
>
> Sincerely yours,
> Katharine Johnson[6]

Robert, having some perspective on Katharine's sense of drama, also wrote to invite him up. "But I know it isn't safe for you to get more than three miles away from Delmonico's. The rumor is that you have melted in your laboratory."[7]

Perhaps Katharine was right, for Tesla was unavailable to the Johnsons even upon their return. Tesla was also ignoring letters from his sisters from Croatia, particularly Marica, who, much like Katharine, asked him why he would not respond. Roentgen rays had been left behind many months ago, but he was still gaunt from illness and overwork. Now he was in a wireless race against newcomers like Marconi. Fearing that his invention would be pirated, Tesla's lab became a more mysterious place.

> November 7, 1896
>
> Dear Mr. Tesla,
>
> It may seem presumptuous [for] a stranger to address you,

> but Mrs. Johnson, (my wife) whom you may remember having met, cannot refrain from uniting with me in congratulating you on the success of the Buffalo experiment.... If this seems taking too great a liberty with one whom we know so slightly, I trust you will attribute it to our interest in the progress of humanity.
>
> Respectively yours,
> Robert Underwood Johnson[8]

Tesla's holiday spirit prevailed, and he joined his beloved Johnsons for Christmas dinner, apologizing for being so distant by bringing Mrs. Filipov an exquisite bouquet of flowers.

The celebration of the inauguration of the Niagara power station was held at the Ellicott Club in Buffalo in the midst of winter's most dangerous month. Fortunately, the weather was permitting, and 350 of the nation's most prominent businessmen made the January trek. Hosted by Morgan's advance man, Francis Lynde Stetson, a law partner of Grover Cleveland's, the list of attendees included a veritable who's who of commerce. Curiously missing from the event, although invited, were such notables as John Jacob Astor, J. Pierpont Morgan, and Thomas Alva Edison.

"Mr. Stetson spoke of the pall of smoke hanging over Buffalo and said that the day should come when power would come from Niagara and not from smoke and steam.... The introduction of Nikola Tesla, the greatest electrician on earth, produced a monstrous ovation. The guests sprang to their feet and wildly waved napkins and cheered for the famous scientist. It was three or four minutes before quiet prevailed."[9]

A constellation of psychological peculiarities accompanied the wizard's lecture. He began in a self-deprecating manner: "I have scarcely had courage enough to address an audience on a few unavoidable occasions.... Even now as I speak... the fugitive conceptions will vanish, and I shall experience certain well known sensations of abandonment, chill and silence. I can see already your disappointed countenances and can read in them the painful regret of the mistake of your choice."[10]

Why did Tesla "poison the well" with this dreadful opening? A deep sense of inferiority appears evident and yet Tesla was also completely aware that this dinner was in his honor and was therefore the pinnacle of his life to date—and through him an apotheosis for the whole of humanity. Why didn't he simply congratulate himself or accept praise well deserved? We see here the first tangible manifestation of an overpowering feeling of inferiority, a clear-cut self-destructive element in his nature. The dark legacy of a deep-seated repression flooded through his veins, like a hydra about to annihilate.

Nevertheless, it was *his* inventions that would change an entire world. It was the name Nikola Tesla which appeared a dozen times on the patent plaque of his new system. It was Nikola Tesla who was praised with "wild enthusiasm" by the corporate and engineering intellegentsia. And it was Nikola Tesla who changed, in precise and measurable ways, the very direction humanity was taking. This was a moment of anointment; through his specific action, the evolution of the race and the texture of an entire planet would be permanently changed in a positive way.

Yet at this moment of the fulfillment of his greatest wish, a deep neurotic constellation was also triggered. From the psychoanalytic perspective, Tesla could now repay his family for the death of his brother by symbolically bringing the brother back to life—and, on the larger scale, give the world a new life, his AC polyphase system. But the shadow had its hold, and he was simply unable to accept the happiness of the moment without throwing a monkey wrench into it. His speech went on: "These remarks, gentleman, are not made with the selfish desire of winning your kindness and indulgence of my shortcomings, but with the honest intention of offering you an apology for your disappointment.... But I am hopeful that in my formless and incomplete statements... there may be something of interest... benefiting this unique occasion."[11]

Tesla's unconscious plan, the heart and soul of his neurosis, was to completely undermine himself by downgrading the Niagara endeavor. It is possible that Stetson had read the speech on the train ride up to Buffalo and foresaw the tragic consequences, as it appears that he waited for a propitious moment to cut the tail end off.

Now that Tesla had arrived, he began to see himself as more than a mere inventor. He was a creator, not of great paintings or of great musical compositions but of great technologies. Niagara Falls was but a steppingstone to the larger plan. His speech went on to pay homage to the "philanthropic spirit" of the businessman, and the great contribution of the scientist. Tesla also hailed such individuals as arc lighting designer Charles Brush, vacuum tube inventor Philip Lenard, and railroad engine designer Frank Sprague as well as Wilhelm Roentgen, Lord Rayleigh, Elihu Thomson, Thomas Edison, and George Westinghouse. "All of these men and many more are untiringly at work investigating new regions and opening up unsuspected and promising fields."

> Among all these many departments of research, there is one which is of the greatest significance for the comfort and existence, of mankind, and that is the electrical transmission of power.... We have many a monument of past ages exemplif[ying] the greatness of nations, the power of men, the love of art and religious devotion. But that monument at Niagara has something of its own, worthy of our scientific age, a true monument of enlightenment and of peace. It signifies the

> subjugation of natural forces to the service of man, the discontinuance of barbarous methods, the relieving of millions from want and suffering.... Power is our mainstay, the primary source of our many-sided energies.[12]

Stetson saw his moment and returned to the stage to whisper in Tesla's ear. "I am just informed," Tesla suddenly announced, "that in three minutes we have to leave.... What can I say? (Cries of 'No.').... I can congratulate the courageous pioneers who have embarked in this enterprise and carried it to success. Buffalonians, I would say friends, let me congratulate you on the wonderful expanse of possibilities opened and let me wish that in no time distant your city will be a worthy neighbor of the great cataract which is one of the great wonders of nature."[13] There was a train to catch. The rest of the speech would be published in the electrical journals.

It was a fortuitous break. For here we see a positive statement concerning the stupendous achievement at Niagara, and we also see the seeds of a new vision Tesla was planting for the world. He was not a mechanic but an artist. Monetary gain was not an end; in fact, the providing of cheap power for the masses was a goal. Businessmen were not greedy capitalists but noble philanthropists. This was a utopian dream which perhaps one day might come true. And as we shall see, it was also a justification, maybe even a rationalization, for some of the audacious ways Tesla chose to spend the "contributions" of the financiers who came to support his Promethean campaign.

Playing the "Master Game" Tesla was gambling for all the stakes. His goal was no less than the ability to transform himself into a deity. This, as O'Neill writes, was his "superman complex."

> We shall not satisfy ourselves simply with improving [present day methods], we have a greater task to fulfill to evolve means for obtaining energy from stores which are forever inexhaustible, to perfect methods which do not imply consumption and waste of any material whatever.... [I have] examined for a long time the possibilities of operat[ing] engines on any point of the earth by the energy of the medium [and] am glad to say that I have devised means which has given me fresh hope that I shall see the fulfillment of one of my fondest dreams; namely, the transmission of power from station to station without the employment of any connecting wire.[14]

Tesla had audaciously proclaimed in the written and published part of his speech that this great enterprise which they were about to usher in (and the one we still use a century later) was already obsolete! He had a better plan. There was no need for the millions of telephone poles that were to be erected, no need for the megatons of copper that would have to go into an endless array of interconnected power lines, no need for the enormous production of rubber for insulation or for the tens of thousands of acres required to support the system, and no need for the workmen who were soon to be hired to maintain the equipment, for all of this, the transmission of electrical power, light, and information, could be achieved without wires. No wonder Stetson cut Tesla short.

This speech was the pivotal moment in Tesla's career. He set out full force to achieve this end. Nothing but death would stop him in his attempts to realize the dream.

21

LUMINARIES (1896–98)

A memorable occasion was the first meeting of Tesla and Paderewski. Two more intellectual or lovable men I have never known. They were most congenial and became friends at once. On comparing notes, they discovered that they had both been in Strasburg years before at the same time [1882], Tesla as an electrical assistant on a small salary, and Paderewski as a student in music, and they laughed heartily at the change of their conditions since that time of storm and stress.

ROBERT UNDERWOOD JOHNSON[1]

Tesla traveled back to Manhattan with T. C. Martin, Francis Lynde Stetson, Darius Ogden Mills, and John Hays Hammond's brother, Richard Hammond, who was considering placing Tesla turbines at a dam in California. A Roentgen symposium was being coordinated by Martin and R. U. Johnson under the auspices of the *Century,* and Tesla would help in the preparations, but disagreements with Tom Edison, Elihu Thomson, and Michael Pupin had peaked, and Tesla declined an invitation to a dinner of all the participants. "I can not explain, but it is really impossible for me to join the company,"[2] he said.

Edison was about to join forces with Marconi. Thomson was still pirating Tesla's induction motor; Pupin, the Tesla oscillators. Stetson, as Morgan's "attorney general," could smooth over the bumps in GE's attempts to lay the groundwork for a pooling of patents with Westinghouse—GE would get the AC polyphase system in exchange for the Vanderpoel trolley patents—but he could never erase the growing vendetta against Tesla by a number of key men in his industry.

This was a particularly difficult time for Martin and so a gap between Martin and Tesla began to form.

At the end of January, Tesla sponsored another lavish feast followed by a tour of his lab. He invited John Jacob Astor and his stunning wife, Ava

Willing, and Mr. and Mrs. Stanford White. Due to the late hour, Lady Astor was "dreadfully disappointed" to pass on the electrical pyrotechnics.[3] However, White and company were able to take advantage of the entire affair.

"My dear Tesla," wrote White, "I cannot tell you how impressed I was the other night at your laboratory, and how delighted I was to be there." Signing the letter "Affectionately yours," White also congratulated Tesla on his "address at Buffalo...[which was] so full of beautiful thoughts."[4] This was White's second visit to the lab, and his feelings for Tesla continued to grow.

March 28, 1896

Dear Luka,

I happen to be free this evening. If you have visitors (ordinary mortals) I will not come. If you have Paderewski, Roentgen or Mrs. Anthony I *will* come.

Yours sincerely,
TGI[5]

"Tesla Great Inventor" had met Paderewski for the first time at a dinner at the Johnson "salon" in April of 1896. After an invitation from Robert, Tesla wrote back, "Hope that Miss—I mean Mr. Paderewski will come."[6] He was referring to Paderewski's trademark, a lush, unbridled mane which flared above his head as he played the piano during his concerts.

Referring to this first meeting, Johnson wrote, "Like Tesla, he [Paderewski] has a marvelous mind, which is a store-house of knowledge on all sorts of topics." Johnson, who had written a poem about the virtuoso, likened Paderewski's music to "choiring angels in Paradise."[7]

Latter-day prime minister of Poland, Paderewski was the highest-paid performer of the decade. Richard Watson Gilder, editor in chief at the *Century*, frequently invited, in Paderewski's words, "all the great visiting artists, musicians, distinguished writers, sculptors, painters and politicans who came to America....A connoisseur of art and of life, [Gilder] appreciated and recognized at once the unusual in everything—in people as well."[8] Through Gilder, Johnson met many luminaries, and through Johnson, Tesla followed.

At this same time, in the spring of 1896, Tesla and the Johnsons were busy reading Rudyard Kipling's recently published *Jungle Book*. "The stories of Kipling's are charm itself," Tesla wrote Mrs. Filipov. "I think Rikki Tikki Tavi is the best."[9] Two days later, a party for Kipling was organized at the Johnson home. "Sorry I can not come to dinner," Tesla wrote, "but will show up as soon as possible afterwards."[10]

Kipling, who was thirty at the time, had purchased a house from his brother-in-law in Brattleboro, Vermont, where he had written *The Jungle Book,* and now he had come to New York to help promote it.

Much like Twain, Kipling was known as a "globe-trotter," having visited such places as Ceylon, India, New Zealand, and Australia. After the party, and his stay in New York, Kipling traveled to England and then on to South Africa. On the boat trip over he ran into John Hays Hammond, and when he reached Cape Town, he talked Cecil Rhodes into giving him a tour of the front lines of the Boer War. Upon his return to New York, in early 1899, he spent time with Tesla before another formal dinner in his honor was prepared by the Johnsons. The following letter, although written three years later, reflects the kind of friendship the two men shared:

> Dear Mrs. Filipov,
>
> What is the matter with inkspiller Kipling? He actually dared to invite me to dine in an obscure hotel where I would be sure to get hair and cockroaches in the soup?[11]

On the very day of the party, Kipling rocked the world, for he was taken ill with typhoid and nearly died. Katharine would spend the next few months helping to take care of the ailing novelist.[12] Kipling survived the ordeal, but tragically, his daughter Josephine perished. He was in such a weakened condition that his wife was forced to hide the young girl's death from him until he was well enough to be able to handle the news. As the newspapers carried daily reports of Kipling's condition on their front pages, the world mourned the young girl's loss, just as they rejoiced with his recovery. Having experienced family tragedy in his own life, Tesla hoped that the death would not impair Kipling's ability to write. "I am delighted Kipling has recovered," Tesla told the Johnsons. "I just hope there will be no evil consequences except a sorrow though hard to bear." Philosophically, Tesla added, "It will probably give his works more dignity and depth."[13]

> *Do not thus drift with the mob. I invite you to join me in a month's worship with Nature in the high temples of the great Sierra Crown beyond our holy Yosemite. It will cost you nothing save the time and very little of that for you will be mostly in eternity.*
>
> JOHN MUIR[14]

Another frequent visitor to the Johnson household was the conservationist and poet John Muir, who, a few years earlier, had taken Johnson on a tour of Yosemite, what Muir called, "the great creations of the Almighty." On his trips to New York, Muir typically dressed in a seasoned three-piece suit complete with a gold watch dangling from a high vest pocket; but his demeanor was "mountain man" all the way. Sixty years old, still in his prime, his hair was gray and long and his face adorned with a beard that extended to his navel like scraggy sagebrush. His eyes may have sparked in different directions because of an early industrial accident, but his gaze proclaimed the aura of an enlightened one.

Having dined with Muir at the Johnsons' during the holiday season, and having invited the naturalist back to his lab, Tesla later told Katharine, that he appreciated Muir's contribution to society. "I am always grateful to [Muir]," he wrote, "for his magnificent description of Yosemite Valley which I have read through in one breath."[15]

A founder of the Sierra Club with Johnson, Muir was, in a sense, Tesla's doppelgänger. Muir was a somewhat unkempt naturalist who spent most of his time in the forest, but part of him longed for civilization, and Tesla, ever the fastidious and smartly dressed metropolitan, spent most of his life in the city but longed for the mountains. A former inventor himself, having fashioned an award-winning bed that toppled the sleeper out onto the floor in the morning, Muir was not opposed, in principle, to human progress, only to callous obliteration of the precious wonders of the planet. Since Tesla's inventions sought to utilize renewable energy and minimize destruction of natural resources, in that sense, both Tesla and Muir shared a common end.

Muir's writings remain his most powerful statement to the spiritual path. The importance of his friendship with Tesla helps support the modern-concepts conservation and ecological thinking. Placing it all in perspective, Muir wrote, "Who wouldn't be a mountaineer! Up here all the world's prizes seem nothing."[16]

22

SORCERER'S APPRENTICE (1896–97)

Tesla had his rooms in the laboratory on Houston Street darkened [and] a current was turned on invisibly. As the group ga[z]ed at this the apartment was filled with a terrific lightning display, with the snapping, crackling sound, displacing the reverberation of heaven's artillery, and all remarked the weird and awing effect of the exhibition....

His face lighting up and his spare figure vibrant with pride, [Tesla announced], "I am producing an electrical disturbance of intense magnitude [which] by means of certain simple instruments can be felt and appreciated at any point of the globe without the aid or intervention of wires of any [kind] at all."

NEW YORK JOURNAL[1]

One of Tesla's most ardent admirers was Yale student Lee De Forest, who had studied the inventor's collected works the previous semester. Tesla's writings are "the greatest exciters to zealous work and study," De Forest wrote in his diary. "His New York laboratory [is] a fabulous domain into which all ambitious young electrical students aspire...to enter and there remain....How I pray that I may equal or excel him, that all this belief in my genius is not idle and conceit."

In May 1896, De Forest was able to arrange a visit to the East Houston Street den. "Now is the critical point in the curve of my life," the youth scribbled on his pad as he sat on the train waiting for his stop, "for I am about to seek work with Nikola Tesla."

The expert greeted the novice and gave him a tour, but was unable to take him on as an apprentice. Noticing the dejected countenance on the budding inventor's face Tesla told De Forest: "I see a great future for you as you are endowed with a keen mind. You will not need this job to succeed." Tesla wished him well and suggested De Forest contact him again.

De Forest would apply once more in the spring of 1898, and probably again in 1900 or 1901, but for various reasons he was never hired. In retrospect, Tesla's decision to reject the talented engineer was unfortunate, for De Forest would soon rise to a premier position as a pioneer in the field of wireless communication. He had a commercial mind and became a tenacious rival to Tesla's other young competitor, Guglielmo Marconi.[2]

Throughout 1896, Tesla applied for and received eight patents on his wireless system. These were mostly different types of oscillators for generating electromagnetic currents of high frequency and high potential. His first application specifically in the field of radio communication was in 1897; his second, remote control, in 1898. Earlier patents on oscillators dating back to 1891 and 1893 also covered this work, though in veiled form. Over the next five years, the inventor's arsenal grew to thirty-three fundamental patents, covering all essential areas of "transmitting electrical energy through the natural medium."[3]

As part of his overall scheme, Tesla also began working on perfecting a system of telephotography. His interest could be traced to 1893 and the Chicago World's Fair, where Elisha Gray's teleautographic machine was displayed. But over the summer of 1896 competition peaked, especially when Edison announced his plans to market an "autographic telegraph." "I'm getting it ready for you newspaper fellows," Edison said, "and when I get working, all you will have to do is hand your copy to the operator say in New York, for example, the cover will be shut down and presto! the wires will transmit it letter for letter to the machine at the other end in Buffalo. The wires will transmit 20 square inches of copy a minute and will carry sketches and pictures as well."[4]

In attempts to outdo Edison, Tesla told the *New York Herald* of his own advances. Under the imaginative assumptions that images from the retina could be captured and transmitted, Tesla included this esoteric idea along with a more realistic plan to transmit text and pictures over the phone lines and by wireless.[5]

Even today, the facsimile machine holds a special place in our imagination, for something typed in an office in New York can be transmitted instantaneously by satellite or over the phone lines to a receiver in San Francisco, Moscow, or Tokyo. One can only imagine the sense of disbelief facing Tesla as he tried to convince readers that pictures could be sent by means of wireless from one city to another. Even primitive Morse-coded messages had yet to really be effectively transmitted.

Nevertheless, Marconi was closing in rapidly on successfully demonstrating his wireless apparatus as Tesla scoured the libraries to study the history of telephotography.

Tesla traced early work in inventions that led to the development of the facsimile machine and television to English physicist Alexander Bain, who, in 1842, first transmitted pictures by using a grid of electrical wires imbedded in wax held firmly beneath a sheet of chemically treated paper. All of these wires could be channeled into one cable and strung to a receiving station where an identical grid was constructed. If the receiver traced out the letter *A*, for instance, with an electric stylus, the particular wires comprising the design of the letter would be electrified, and these, in turn, would treat that area of the paper on the receiving end to spell out the matching character. Autographs and drawings could also be transmitted in the same manner. As the process evolved, pictures could be broken down into a finite number of picture elements so that they could also be dispatched. In the 1860s, this grid was replaced by a single wire through the implementation of spinning disks and "perfect synchronism between transmitter and receiver," Tesla's synchronous AC motor helping augment the procedure in the late 1880s and early 1890s.[6]

With the development of the photographic process, the first wireless pictures were sent in 1898 by Küster and G. Williams, "but the arrangements involved the employment of Hertz waves and were impractable." In 1892, Tesla recalled many years later, "the attention of the scientific world was directed to a wonderfully sensitive receiver, consisting of an electron stream maintained in a delicately balanced condition in a vacuum bulb, by means of which it was proposed to use photography in the transmission of telegraphic and telephonic messages through the Atlantic cables, and later also by wireless."

In 1904, Dr. Arthur Korn, an electrical engineer from the University of Munich, gained the attention of the scientific community when he successfully transmitted wire photos from Munich to Nuremburg. According to Korn, who is often credited as the inventor of the television tube, the apparatus utilized "Tesla currents."[7] Tesla notes that once Korn introduced "a sensitive place [and] a selenium cell to vary the intensity of the sending current," the process which is called the television advanced a giant step. Korn's "tube is excited by a high-frequency current supplied from a Tesla transformer and may be flashed up many thousand times per second," thereby attaining the moving television image.[8]

Tesla dates Korn's first experiments to 1903. A May 1899 article states that Tesla was working on a "visual telegraphy" system with the light-sensitive element selenium, thus predating Korn's work by four years. It is probable that Tesla was replicating the experiments of Küster and Williams on precursors to the video camera, although Korn, not Tesla, perfected it.[9]

In essence, the modern television works in a way analogous to Bain's first 1842 construction. An electronic tube, similar to Tesla's brush vacuum tube, is moved across an entire TV screen in an instant. During its pass over the area where the image is formed, it releases synchronized impulses for each pixel. The position of the beam and its precise sequence of firing are directed by impulses stemming from the broadcast station. Each pass over the entire screen creates a separate image that when played in sequence portrays natural moving pictures.

Tesla followed the advances in the field of telephotography as they occurred and experimented himself. His first task was to figure out the optimal way to transmit the energy.

Having tested the Hertz spark-gap apparatus, Tesla found that this device, which Marconi was utilizing, was subject to static interference and involved arbitrary *pulsed* (damped) frequencies that were weak. They transmitted themselves transversely through the air and did not take into account longitudinal properties which were augmented when making use of extremely high potentials and a ground connection. Working out calculations that took into account the speed of light and the size of the earth, Tesla designed carefully constructed *continuous* (undamped) electromagnetic waves that were in harmony with those of the planet.

By 1897, Tesla amassed all of the essential patents for generating, modulating, storing, transmitting, and receiving wireless impulses. In a letter to his lawyer, Parker W. Page, Tesla wrote, "I forward herewith M. Marconi's patent which was just allowed.... I notice that the signals have been described as being due to Hertzian waves, which is not the case. In other words, the patent describes something entirely different than what actually takes place.... How far does this affect the validity of the patent?"[10] Clearly, Tesla already suspected that Marconi was utilizing his equipment.

In Tesla's first patent specifically for wireless transmission, no. 649,621, filed on September 2, 1897, he discusses the need for a "terminal...preferably of large surface...maintained by such means as a balloon at an elevation suitable for the purposes of transmission...[and another] terminal of the secondary...connected to [the] earth.... At the receiving station a transformer of similar construction is employed." The specification goes on to describe how the wavelengths can be constructed and modified so as to tune the circuits and take into account the natural properties of electromagnetic energy.

Having taken into account the size and capacity of the earth, Tesla had calculated that with a coil fifty miles in length oscillating at 925 times per second, a resonant relationship to the frequency of light was realized. Since he had worked with exhausted tubes, Tesla knew that electricity traveled more easily through them than through air. Thus, he reasoned, if his transmission towers were placed on high elevations above obstructions and further raised by the use of balloons, the upper atmosphere (or ionosphere) itself would serve as a means of transmission. Large power

stations located near waterfalls would supply the energy necessary for the wireless transmission to this upper stratum.[11]

In a separate way, the earth itself would also serve as a medium. The following passage from the same 1897 patent application criticized Marconi's use of the more primitive Hertzian apparatus: "It is to be noted that the phenomenon here involved in the transmission of electrical energy is one of true conduction and is not to be confounded with the phenomena of electrical radiation which have heretofore been observed and which from the very nature and mode of propagation would render practically impossible the transmission of any appreciable amount of energy to such distance as are of practical importance."[12]

Marconi, who was working with Lloyds of London in ship-to-shore experiments, was using a more trial and error method. In July 1896, in experiments with Preece, the Italian had successfully transmitted messages through walls and over distances of seven or eight miles. In December he applied for a patent, which Preece felt was "very strong,"[13] although he knew the youngster had been anticipated by Lodge and Tesla. The patent was not original, and it did not put forth any new principles; nevertheless, Marconi was definitely succeeding in the real world, while Tesla advanced in his laboratory in refinements of apparatus and in the theoretical realm. The differences in sophistication of knowledge on the subject are aptly described by one of Marconi's associates, Mr. Vyvyan. "We knew nothing then about the effect of the length of a wave transmitted governing the distance over which communication could be affected," the colleague stated. "We did not even have the means or instruments for measuring wave-length, in fact we did not know accurately what wave-length we were using."[14]

Preece's initial work in his study of earth currents and induction effects generating from normal telegraphic lines in the 1880s and 1890s led him to realize the strength of Tesla's system. Marconi, at that stage, had no understanding of the role of the earth in conveying electrical energy. He was utilizing the principle of "radiation" through the air according to Hertz's apparatus. Without understanding why, he did make use of an aerial and ground connection, but this setup had already been published widely in 1893 by Tesla. Other principles were taken from Oliver Lodge, who was in a patent dispute with Guglielmo Marconi. William Preece was well aware that they had been anticipated, but he could also see Marconi advance markedly, while his precursors basically stood still.

After Marconi rejected Preece's suggestion that they request the use of Tesla's apparatus, the British nobleman was placed in a conflictual situation. In August 1897 he mailed off a "terse" dispatch. "I regret to say that I must stop all experiments and all action until I learn the conditions that are to determine the relations between your company and the [British] Government Departments who have encouraged and helped you so much."[15] But the die was cast, and Preece became helpless to stop what he knew was a complex form of piracy. He became ill and retreated to Egypt, where he stayed a year.

Marconi was also being aided by H. M. Hozier, director of Lloyds, who, according to one account, "succeeded in sending reasonably clear messages...[through Hozier's apparatus]...in one place, at least where Marconi himself had failed."[16] Lloyds also contacted Tesla "to rig up a wireless set, ship-shore in 1896 to report the international yacht race, [but] Tesla refused the offer, claiming that any public demonstration of his system on less than a world-wide basis would be confused with the amateurish effort being made by other experimenters."[17]

Instead, Tesla performed a long-distance clandestine experiment which he told no one about, not even his workers. Sometime in 1896 or early 1897, the inventor turned on his generator to "produce continuous trains of oscillations" and took a cab to the Hudson River. There he caught a boat and ferried up to West Point with a battery-operated machine "suitable for transportation." "I did this two or three times," he told the courts in 1915. "[But] there were no signals actually given. I simply got the note, but that was for me just the same." In other words, having brought a receiving instrument with him, the inventor simply tuned it to the point where it began to respond to the oscillations emanating from his laboratory back at East Houston Street. "That is, I think, a distance of about thirty miles," Tesla said.[18]

Tesla also considered harnessing wind power, the tides, solar and geothermal energy, and also energy released during the process of electrolysis. If water was separated into oxygen and hydrogen, these explosive substances could theoretically be used to generate the heat to create steam. Working along varying lines of research, Tesla also patented ozone-production machines and devised a scheme whereby nitrogen from the air would be electrically separated out and blended with conveyor belts of soil to create a fertilizer machine.

"All the agriculturist needs," Tesla suggested, "is...[to] shovel...a quantity of loose earth, treated by a secret chemical preparation in liquid form...into the cylinder. An electric current is passed through the confined atmosphere, the oxygen and hydrogen are...expelled, and the nitrogen which remains is thus absorbed into the loose earth. There is thus produced as strong fertiliser for a nominal price [right] at [the farmer's] home."[19]

On April 6, 1897, Tesla spoke again to the public at the New York Academy of Sciences. Over four thousand people attended.[20] With large photographs of dozens of radio tubes Tesla had designed gracing the walls, Tesla set out to explain his advances in Roentgen rays. Certainly the audience was interested in seeing a remarkable device that revealed the human skeleton of a living person, but unquestionably most of them came to behold the sorcerer hurl his thunderbolts.[21]

Tesla's world telegraphy system had finally come into clear focus. His plan was to disturb the electrical capacity of the earth with gigantic Tesla oscillators and thereby use these earth currents themselves as carrier waves for his transmitter. In 1897 he explained precisely how his world telegraphy system would operate:

> Suppose the whole earth to be like a hollow rubber ball filled with water, and at one place I have a tube attached...with a plunger....If I press upon the plunger the water in the tube will be driven into the rubber ball, and as the water is practically incompressible, every part of the surface of the ball will be expanded. If I withdraw the plunger, the water follows it and every part of the ball will contract. Now, if I pierce the surface of the ball several times and set tubes and plungers at each place, the plungers in these will vibrate up and down in answer to every movement which I may produce in the plunger of the first tube.

There is a peculiar addition to this paragraph: "If I were to produce an explosion in the centre of the body of water in the ball, this would set up a series of vibrations in the whole body. If I could then set the plunger in one of the tubes to vibrating in consonance with the vibrations of the water, in a little while and with the use of a very little energy, I could burst the whole thing asunder."

The water corresponds to "terrestrial currents" (which are today known as telluric currents), and the plungers refer to his transmitters and receivers. "The inventor thinks it possible that his machine when perfected may be set up, one in each great centre of civilization, to flash the news of the day's or hour's history immediately to all other cities of the world; and stepping for a sentence out of the realms of the workaday world, he offers a prophecy that any communication we may have with other stars will certainly be by such a method."[22]

This article, which appeared in *Scribners*, also discussed Marconi's successful wireless transmission of eight miles in Europe. We see in these passages stemming from 1896 and 1897, that Tesla had already conceived of a total plan for his world telegraphy system and that it utilized a variety of wireless modes, one being through the upper air strata, another by means of mechanical resonance, which he called telegeodynamics, and a third, and his most important, by riding terrestrial currents. His next plan was to measure precisely the frequency of the planet and construct transmitters in harmonic relations to it. Nodal points could then be mapped out from, say, a wireless transmitter placed at Niagara Falls to precise positions for receiving towers on different continents.

For all intents and purposes, no one, except for Marconi, had demonstrated that wireless messages could be transmitted more than a few hundred feet. And success only involved the modest goal of sending Morse coded messages. Marconi's next plan, which would capture the world's attention, was to radiate the impulses across the English Channel. Certainly Tesla had demonstrated all of the principles found in the modern radio years earlier, but his public demonstrations were held within lecture halls. He had established that he could illuminate lamps from a transmitter placed on the roof of his laboratory on Houston Street to his hotel twenty-six blocks away, but these experiments were held in secret and were never publicized.[23] The 1895 fire also thwarted his efforts to display long-distance effects.

And when Lloyds of London contacted him, to the dismay of his secretary, George Scherff, Tesla rejected the offer to demonstrate his system's capabilities.

Tesla, however, was not content with merely setting up a world broadcasting system that, from a conceptual perspective, is superior to our prevailing technology in that it would transmit *power* as well as information; he also suggested that he could contact the stars, create rain in the deserts, or cause wide scale havoc. Tesla became the quintessential mad scientist. Through his inventions, the world could be dominated according to whim.

The contradictions within the inventor began to pull him in opposite directions. In letters to the Johnsons he calls himself Tesla Great Inventor and implies that, like Paderewski and other luminaries, he was not a mere mortal, but during his Niagara Falls speech he denigrated himself. Tesla was somewhat wealthy during this period, but the requirements of his operations were beyond his means. And yet Tesla shunned billing engineers when they sought his aid and voided a royalty contract with Westinghouse, though it was now worth a fortune.

In June 1897 it was reported that Westinghouse had paid $216,000 for his patents.[24] As Tesla and his partners, Brown and Peck, were receiving yearly checks of $15,000, with an initial down payment of probably $70,000,[25] this works out to about a quarter of a million dollars for a ten-year period. In a letter to Astor, Tesla places the figure at $500,000,[26] but in either case, it was still millions of dollars less than its actual value.

By this time, Westinghouse and GE had formalized their "entente cordiale." This meant that a second gigantic corporation with numerous subsiaries would be benefiting from Tesla's invention but the inventor would not receive a dime for it. The electric subway trains would also be implementing the Tesla motors and system, and again the inventor would get no compensation.

Tesla's new plans would require enormous expenditures. Westinghouse was making it clear that his company would not be a source of funds beyond their former signed agreement (although it is possible that Tesla did receive additional revenues for other inventions, such as for his oscillators). At the end of the year, the inventor wrote his friend Earnest Heinreich, an engineer at the Westinghouse Corporation who also was a novelist. "My dear Heinreich," Tesla wrote, "It is true that I have not been quite well of late, but can assure you that I am physically and mentally all right at present. I have however, still a little ailment which may be best designated as financial anemia, from which you are yourself suffering, if I am not mistaken. I wish you would remember me to all the boys at about Xmas time for it is just possible that some of them might feel disposed to send me a token."[27]

There were other pressures as well. Tesla was having difficulties with his financial backer Edward Dean Adams, who was opposing his efforts to promote his wireless enterprise; there was this continuing competition with Marconi; and there were the painful echoes of his Serbian past, from Kosovo and his lost youth. His parents and brother were dead, and he was away from his family not just in distance but also in spirit. In yet another of so many letters sent, sisters Marica and Angelina pleaded for a response. Tesla had sent funds on many occasions and a copy of his Martin text, but they wanted more. "Remember what your name is and where you come from," Marica wrote, ending her letter with the customary "I am kissing you in spirit."[28] Part of him ached from the pain, and just as he was coasting high and ahead of the pack on the crest of the wave.

Maybe it was the influence of the Boer War or the unrest brewing in Cuba, but Tesla's destructive streak began to emerge. His previous inventions were already reshaping human events; his newest creation could interlink every remote hamlet or tear the world apart. He decided to experiment.

With George Scherff present, Tesla placed one of his mechanical oscillators on the center support beam in the basement of the Houston Street building where his laboratory was located and adjusted the frequency to the point where the beam began to hum. "While he was attending to something else for a few moments, it attained such a crescendo of rhythm that it started to shake the building, then it began shaking the earth nearabout [and other buildings with support beams in resonant frequencies]....The Fire Department responded to an alarm frantically turned in; four tons of machinery flew across the basement and the only thing which saved the building from utter collapse was the quick action of Dr. Tesla in seizing a hammer and destroying his machine."

"The device could be a Frankenstein's monster," Tesla confided many years later. "If not watched, no substance can withstand the steadily applied rhythm when its resonance point is reached. Skyscrapers could easily be destroyed with the steady building up of resonance from the timed strokes of a five-pound hammer.[29]

In another rendition of the story, told at another time, Tesla claimed that he had taken his alarm clock–sized oscillator to a building site "in the Wall Street district." Finding one under construction, about "ten stories high of steel framework..." he clamped the vibrator to one of the beams and fussed with the adjustment until he got it.

"In a few minutes I could feel the beam trembling," Tesla told a reporter. "Gradually, the trembling increased in intensity and extended throughout the whole great mass of steel. Finally, the structure began to creak and weave, and the steel-workers came to the ground panic-stricken, believing there had been an earthquake. Rumors spread that the building was about to fall, and the police reserves were called out. Before anything serious happened, I took off the vibrator, put it in my pocket and went away. But if I had kept on ten minutes more, I could have laid that building flat in the street. And, with the same vibrator, I could drop the Brooklyn Bridge into the East River in less than an hour."

Tesla told the reporter that he could split the earth in the same way, putting an end to mankind.

"The vibrations of the earth," he said, "have a periodicity of approximately one hour and forty-nine minutes. That is to say, if I strike the earth this instant, a wave of contraction goes through it that will come back in one hour and forty-nine minutes in the form of expansion. As a matter of fact, the earth, like everything else, is in a constant state of vibration. It is constantly contracting and expanding.

"Now suppose that at the precise moment when it begins to contract,

I explode a ton of dynamite. That accelerates the contraction, and in one hour and forty-nine minutes, there comes an equally accelerated wave of expansion. When the wave of expansion ebbs, suppose I explode another ton...and suppose this performance be repeated time after time. Is there any doubt as to what would happen? There is no doubt in my mind. The earth would be split in two. For the first time in man's history, he has the knowledge with which he may interfere with cosmic processes."

Tesla calculated that this procedure might take more than a year to succeed, "but in a few weeks," Tesla said, "I could set the earth's crust into such a state of vibration that it would rise and fall hundreds of feet, throwing rivers out of their beds, wrecking buildings, and practically destroying civilization. The principle cannot fail."[30]

23

VRIL POWER (1898)

We entered an immense hall, lighted by... [a] lustre... but diffusing a fragrant odor. The floor was in large tesselated blocks of precious metals, and partly covered with a sort of matlike carpeting. A strain of low music, above and around, undulated as if from invisible instruments....

In a simpler garb than that of my guide, [a figure] was standing motionless near the threshold. My guide touched it twice with his staff, and it put itself into a rapid and gliding movement, skimming noiselessly over the floor. Gazing on it, I then saw that it was no living form, but a mechanical automaton.... Several [other] automata... stood dumb and motionless by the walls.

THE COMING RACE, EDWARD BULWER-LYTTON[1]

One of Tesla's major inventions in terms of ingenuity, originality, and complexity of design was a remote-controlled robotic boat which he called the telautomaton. This device was unveiled at the Electrical Exposition held at Madison Square Garden during the height of the Spanish-American War in May 1898, but earlier precursors could be traced to wireless motors which he displayed before the Institute of Electrical Engineers in 1892.

This single invention not only established all of the essential principles of what came to be known a few years later as the radio; it also lay as the basis of such other creations as the wireless telephone, garage-door opener, the car radio, the facsimile machine, television, the cable-TV scrambler, and remote-controlled robotics. The precise nature of the invention, virtually its patent application, was published in most of the technical journals at the time of its inauguration.[2]

The telautomaton paralleled precisely a model developed by British novelist Edward Bulwer-Lytton in 1871, although Tesla insisted in a missive to Johnson, written two years after the invention's inauguration, that he had not been inspired by this science-fiction tale.[3]

As Bulwer-Lytton was perhaps the most popular author next to Charles Dickens at that time, it is unlikely that Tesla was unaware of this story when he conceived of the invention. In *The Coming Race,* Bulwer-Lytton describes a concept which he called "vril power." This was an energy transmitted from the eye and body of the fictional advanced species which was used to animate automatons.[4] In essence, Tesla built a working model that substituted electricity for the novelist's "vril." The story begins when the protagonist falls into a hole in the earth and comes upon an advanced civilization: "In all service, whether in or out of doors, they [the people of Vril-ya] make great use of automaton figures, which are so ingenious, and so pliant to the operations of vril, that they actually seem gifted with reason. It was scarcely possible to distinguish the figures I beheld, apparently guiding...the rapid movements of vast engines, from human forms endowed with thought."[5] As we shall see, key aspects of Bulwer-Lytton's story correlate quite closely with positions espoused by Tesla.

The electrical exhibition was organized by Stanford White, who worked with Tesla to fashion a rainbow room of neon lights at the entrance, and it was presided over by Chauncey Depew, another Tesla friend, who was also one of the principals of the New York Central Railroad and a U.S. senator from New York. It had been hoped that President McKinley would illuminate the exposition by means of telegraph lines from Washington, but something went awry, so Vice President Garret Hobart opened the proceedings instead. Representing the Marconi company was Tom Edison's son, Tom Junior, who obtained the position through T. C. Martin. This liaison marked the beginning of a partnership between Marconi and Edison, as the Menlo Park wizard had wireless patents which the Italian wanted to own in order to boost his legal position on priority of discovery. The event also portended the upcoming break in the friendship between Tesla and Martin.

Animosities between Spain and the United States had run high for a number of years. Ever since 1895, when the Spaniards took repressive measures against rebelling Cubans, many Americans began to champion the cause of Cuban annexation.

The sinking of the battleship *Maine* in Havana harbor in February of 1898 eliminated any doubts, and war was officially declared two months later. Tesla had been meeting with John Jacob Astor throughout this period in his continuing attempts to woo the financier as Astor spelled out more clearly his position on relevant issues.[6] While his wife played mah-jongg at home, the colonel jaunted along the deck of his mighty ship, the *Nourmahal,* which he had armed with four machine guns in order to protect against potential pirates. Labeled as insipid and henpecked by gossip columnists, Astor sought his freedom on the high seas.

Perhaps it was during an outing on Astor's yacht that the inventor conceived of the idea of fashioning the teleautomaton in the form of a torpedo. "Come to Cuba with me where you can demonstrate your work upon the insufferable scoundrels," Astor suggested.

Tesla may have been tempted, but in the midst of a whirlwind of invention, he graciously declined as he had been called "for a higher duty."[7]

Tesla finalized construction of his remote-controlled boat and considered how to make amends as Astor conferred with President McKinley in Washington and then hastened to the front lines. The colonel had donated $75,000 to the U.S. Army to equip an artillery division for use in the Philippines and lent the *Nourmahal* to the navy for use in battle. The tall ship, nearly a hundred yards in length, was equipped with a corps of military seamen. Able to feed sixty-five at one sitting, the steam-driven three-masted schooner made a formidable warship. With his honorary rank stepped up to inspector general, Colonel Astor sailed his battalion down to Cuba, where he could "watch Teddy Roosevelt in the Battle of San Juan Hill through a pair of field glasses."[8]

Beating the Spanish with modern instruments of destruction became the overriding theme of the exposition. Tesla would have far and away the most sophisticated construction, but he chose to portray it by deceptively emphasizing mysterious features: "In demonstrating my invention before audiences, the visitors were requested to ask any question, however involved, and the automaton would answer them by signs. This was considered magic at the time, but was extremely simple, for it was myself who gave the replies by means of the device."[9]

The boat, approximately four feet in length and three feet high, was placed in a large tank in the center of a private auditorium, set up for special viewing for key investors like J. O. Ashton, George Westinghouse, J. Pierpont Morgan, and Cornelius Vanderbilt.[10] By means of a variety of transmitters and frequencies, the inventor could start, stop, propel, steer, and operate other features, such as putting lights on or off. Tesla was also planning on constructing a prototype submersible, perhaps to compete in the mock battles that were staged between models of the American ships and the Spanish fleet, but it was never built.

Due to the lack of access the press had to this exclusive invention the newspapers featured Marconi's wireless detonation system instead. By means of a bomb planted onboard the enemy frigate and a simple button placed in the hands of Tom Junior, "Spanish" ships were blown to smithereens. Marconi, however, had not solved the problem of tuning a frequency, and so, on one occasion, Edison's son accidentally blew up a desk in a back room that had housed other bombs. Fortunately, no one was injured.[11]

It appears that the public appreciated the dramatic Marconi contraption, which appealed to baser instincts, as compared to Tesla's masterwork, which was sixteen years ahead of its time operationally and at least a century ahead of its time conceptually, that is, as envisioned in final form. Only the scientific journals explained with any clarity the complexity of the device.[12]

Tesla's coy portrayal kindled a blitzkrieg of epithets from the press. The following fantastic prognostication particularly upset them:

Torpedo Boat Without a Crew

> My submarine boat, loaded with its torpedoes, can start out from a protected bay or be dropped over a ship side, make its devious way along the surface, through dangerous channels of mine beds...watching for its prey, then dart upon it at a favorite moment...discharge its deadly weapon and return to the hand that sent it....I am aware that this sounds almost incredible and I have refrained from making this invention public until I had worked out practically every detail.[13]

By allowing the following editorial to appear in his journal *Electrical Engineer,* T. C. Martin was, in a backhanded way, another to lead the assault.

Mr. Tesla and the Czar

> Mr. Tesla fools himself, if he fools anybody, when he launches into the dazzling theories and speculations associated with his name....Just of late Mr. Tesla has been given publicity to some of his newest work....We should be glad personally to see him finish up some of the many other things that have occupied his energies these ten years past.

The editorial then went on to criticize Tesla's oscillator and his method of "delivering large quantities of current...without wires, say from Niagara Falls to Paris [which has also yet to happen]....Mr. Marconi has already telegraphed from balloon to balloon without wires...over twenty miles, thus proving in advance the tenability of Mr. Tesla's proposition.[14]

Discredit of the wireless torpedo followed. This was in reaction to Tesla's suggestion that the ultimate weapons could be "devil automata." Caught up in war fever, Tesla emphasized nefarious implications of his work: Automatons would fight while humans would live. He wrote, "The continuous development in this direction must ultimately make war a mere context of machines without men and without loss of life—a condition [which will lead]...in my opinion...to permanent peace."[15]

This position was refuted by a number of individuals, the most eloquent by Frenchman M. Huart:

The Genius of Destruction

> Like all inventors of destructive machines, [Tesla] claims that his [devil automata] will make the governments which are inclined to create international conflagrations hesitate. On this account Nikola Tesla claims a right to be called a benefactor of humanity. The genius of destruction would seem to have, then, two aims. It creates evil but mostly good. Through its help the abolition of wars may no longer be a utopia of generous dreamers. A blessed era will open up to the people, whose quarrels will be settled in view of the terror of the cataclysms promised by science. What contradictions of conception is the human mind subject to?[16]

Coincidentally, this view was espoused by Mark Twain, who wrote to Tesla from Europe wanting to sell the patents to cabinet ministers in Austria, Germany, and England, by Bulwer-Lytton, and by Czar Nicholas of Russia, whom Tesla himself was negotiating with.[17] (And that was how "Nicholas" Tesla became associated with the czar.) In the modern era, Edward Teller, one of the inventors of the hydrogen bomb, and, more recently, President Ronald Reagan in his 1980s Star Wars speeches, have also expounded on this position. But Tesla (much like Einstein) came to regret his initial view of how the agents of Armageddon could lead humans to peace.

The brazen essay, which had appeared in Martin's journal, continued as an introduction to Tesla's "thoughtful" paper on electrotherapeutics and then concluded with the following convoluted backhand compliment:

> It is not our desire to pose as apologists or publicists for Mr. Tesla. He needs no assistance of that kind; and so long as he commands freely whole pages of the Sunday papers, for which Mr. Wanamaker pays gladly his thousands of dollars, the scientific journals have little to do with the matter. All we wish to say is that it is not fair to condemn, as many do, Mr. Tesla as a visionary and impractical. No man has finished his work till he is dead, and even then there are long, long centuries in which his ideas can prove themselves true. The visionaries are thus often in the end the most sordid of realists—something Mr. Tesla will never be.[18]

As Martin had been, in a sense, Tesla's advance man, his decision to allow this critique in his journal became a tacit sanction for other writers to unfurl their condemnation. For instance, another scathing review appeared in both *The Scientific American* and the more popular *Public Opinion.*

The article appeared on the same page as the obituary of mountebank inventor John Worrell Keely.

Was Keely a Charlatan?

> In the death of J. W. Keely of Keely Motor fame... the world has been robbed of one of its most unique and fascinating characters....[Keely] was always going to startle the world but never did. It is sincerely to be hoped that Keely's alleged secrets have died with him.

Science and Sensationalism

> ...That the author of the multiphase system of transmission should, at this late date, be flooding the press with rhetorical bombast that recalls the wildest days of the Keely Motor mania is inconsistent and inexplicable to the last degree....The facts of Mr. Tesla's invention are few and simple as the fancies which have been woven around it are many and extravagant. The principles of the invention are not new, nor was Tesla the original discoverer.[19]

This implication that Tesla was not the author of his system of wireless communication echoed previous charges that he was not the genuine inventor of the AC polyphase system. This was what particularly angered him, as it was essential that his work be original. "I wish I could lay upon the fellow all the forked lightning in my laboratory," Tesla told the Johnsons at dinner at their home.[20]

"Perhaps it would be more effective if an outside person came to your defense," Robert suggested.

"My dear Luka, I know that you are a noble fellow and devoted friend and I appreciate your indignation at these uncalled for attacks, but I beg you not to get involved under any condition as you would offend me. Let my 'friends' do their worst, I like it better so. Let them spring on scientific societies worthless schemes, oppose a cause which is deserving, throw sand into the eyes of those who might see. They will reap their reward in time."[21]

"Then how can we redress such an outrageous individual?"

"Let us have a profound contempt for the creature," Tesla concluded.

"I don't see Commerford in this same category," Katharine offered, trying to set the stage for a reconciliation.

"I know you and Luka want me to forgive your friend Martin for his disparaging editorial. It was well done, but not so painstakingly as many others before. He renders me more and more valuable services."

"At least talk to him," Katharine pleaded.

Tesla grabbed his hat, coat, and gloves and waved his hand. "Sorry for him. That is all," he said as he departed.[22]

Tesla counterattacked with a spirited response to *Electrical Engineer* which they were forced to publish:

> On more than one occasion you have offended me, but in my qualities both as Christian and philosopher I have always forgiven you and only pitied you for your errors. This time, though, your offense is graver than the previous ones, for you have dared to cast a shadow on my honor....Being a bearer of high honors from a number of American universities, it is my duty, in view of this slur, to exact from you a complete and humble apology....On this condition I will again forgive you, but I would advise you to limit yourself in your future attacks to statements for which you are not liable to be punished by law.[23]

Tesla, of course, was angry with the general tone of their editorial; but what particularly upset him was the implication that he had abandoned his mechanical and electrical oscillators and his cold lamps without filaments (fluorescent lights). He was in the midst of negotiating a large business transaction with a number of investors, particularly Astor. In no way did he want there to be any implication that these endeavors were being abandoned.

Martin's rebuttal was published right after Tesla's letter:

His Friends to Mr. Tesla

> One foremost electrical inventor [unnamed—probably Elihu Thomson]...has been kind enough to say that the *Electrical Engineer* made Mr. Tesla.

This statement was disputed by Martin as "a person's actions make the person"; however, the journal (i.e., Martin) did cite the fact that in the past it published Tesla's articles and book of his inventions and lectures and, furthermore, that it was their editor who has "striven with all the ability possessed to explain Mr. Tesla's ideas." This is completely true. For an eight-year period, 1890–98, *Electrical Engineer* published 167 articles by or about Tesla, 40 more than *Electrical Review* and 70 more than *Electrical World*.[24] Moreover, it was clearly Martin who choreographed the reclusive inventor's entrée into the American electrical arena.

As Tesla often promised more than he delivered, the journal, as a "true friend," felt an obligation to urge him to complete "a long trial of beautiful but unfinished inventions." They (i.e., Martin) also took keen exception to the fantastic statements about Tesla's remote-controlled flying machine that could change its direction in flight, "explode *at will* and...never make a miss." Martin continued: "Our past admiration of Mr. Tesla's real, tangible work is on record, and stands; but we draw the line at such things as these. We are sorry Mr. Tesla feels so keenly, but we cannot help it."[25]

Given that this assault stemmed from one of Tesla's closest allies, it bears careful consideration. However, from a historical perspective, we should also consider hidden agendas. For instance, in 1894, Tesla was freely distributing his collected works and not paying for additional copies.

"I made some money out of my Tesla book," Martin confessed to Elihu Thomson many years later, "[but it was] promptly borrowed from me by the titular component, so that two years of work went for nothing."[26] The following year, in 1895, Tesla's laboratory burned to the ground, and Martin wrote an admirable tribute.[27] Perhaps that is why he did not insist Tesla repay him.

Martin had been placed in an awkward position, for he was also a good friend of Tom Edison's, a powerful Tesla rival; and as a reporter he had to be objective in covering advances from other rivals, such as Marconi. Tesla's irritating habit of living beyond his means and seeing projects completed before they actually materialized was forever a source of frustration to his longtime protector. And history has so proved, in many ways, that Martin was right. Tesla's oscillators were never a commercial success; his wireless system of distributing light, information, and power (in its total form) was never realized; and for reasons difficult to understand, Tesla's fluorescent lights were never marketed.

On the other hand, Tesla was extremely prolific. He did build working models of all of his inventions. For instance, the telautomaton was a fully functioning prototype. And it takes many years for one's endeavors to come to fruition. Tesla had already proved himself in a variety of ways. That all of his projects never materialized is understandable given the great scope of his efforts.

Tesla's telautomaton remains one of the single most important technological triumphs of the modern age. In its final form, it was conceived as a new mechanical species capable of thinking as humans do, capable of carrying out complex assignments and even capable of reproduction. The invention also comprised all of the essential features of wireless transmission and selective tuning. Here was a true work of genius.

Surprisingly, Tesla was an adherent of a stimulus-response model for explaining human behavior and consciousness rather than a proponent of a model espousing a creative unconscious. The rivalry between Ernst Mach and Carl Stumpf, Tesla's philosophy teacher, discussed earlier, and the work of Descartes on the self-propelled automata correlate to a number of key positions Tesla took which influenced directly the development of his telautomaton. The mind, according to this proposition, was nothing more than a simple compilation of cause-and-effect sensations. What we call ideas were secondary impressions derived from these primary sensations.

Paradoxically, although Tesla's achievement was highly original and although he touted himself as the "creator of new principles," in no way did the inventor think that he had ever produced a new idea that did not stem from something external, for example, a mechanism present in nature or deriving from the work of others. A reader of great philosophers, Tesla fully understood what the adoption of his telautomaton would mean to the world. He saw clearly the implication in the "coming race." Machines would not only replace laborers, they would think for themselves. Tesla's genius, therefore, was not only in the appreciation of the advanced thinking of others but also and more important, in implementing on a practical basis their abstract ideas. Whereas others sought to change the world with thoughts, Tesla manifested on the physical plane real working models. He certainly was the father of remote-controlled electronic "beings," but he would have been the last person to claim that he was the father of the idea.

In a famous article printed in the *Century* in 1900, Tesla explained the entire conceptualization behind his telautomaton: "I have, by every thought and every act of mine, demonstrated and do so daily, to my absolute satisfaction that I am an automaton endowed with a power of movement, which merely responds to external stimuli beating upon my sense organs, and thinks and acts accordingly. I remember only one or two cases in all my life which I was unable to locate the first impression which prompted a movement, or a thought, or even a dream."[28]

Tesla neglects to mention that one of these two instances was the revelation he had in Professor Poeschl's class: He had seen that the commutator could be eliminated in the DC machines. In other words, Tesla's most successful invention, the AC polyphase system, was initiated from intuitive insight. Nevertheless, Tesla stubbornly clung to the "tabula rasa" premise. No inspiration, according to this pundit, began from within; self-directed responses were initiated only after external stimuli were received. This is a complex idea, as it seems from the first paragraph below that Tesla believes the reverse.

How Cosmic Forces Shape Our Destinies

> Every living being is an engine geared to the wheelwork of the universe....There is no constellation or nebula, no sun or planet...that does not exercise some control over its destiny—not in the vague and delusive sense of astrology, but in the rigid and positive meaning of physical science.
>
> More than this can be said. There is no thing endowed with life—from man who is enslaving the elements, to the humblest creature—in all the world that does not sway in its turn.[29]

At the time he conceived these ideas, that is, in the early 1890s, Tesla was studying Herbert Spencer and also Buddhist writings. He even gave his friend Johnson a copy of a book on Buddhism to read. Nevertheless, the influence of Mach's principle and Newton's laws concerning such correlates as the angular momentum of the earth, sun, and galaxy also figured into his cosmological paradigm. "The Buddhist expresses it one way, the Christian in another, but both say the same: We are all one.... Science, too, recognizes this connectedness of separate individuals, though not quite in the same sense that it admits that the suns, planets and moons of a constellation are one body, and there can be no doubt that it will be experimentally confirmed in times to come."[30]

Having studied will psychology (occult psychological principles) as a youth, Tesla was a firm believer in self-determination and the incredible power of the will. Somehow, however, he reconciled this internal procedure, which the philosopher George Gurdjieff links to a direct expression of the soul, to his external-behavioristic paradigm. For Tesla, the spark of life is not only biological but also present in the structure of matter: "Even matter called inorganic, believed to be dead, responds to irritants and gives unmistakable evidence of a living principle within."[31]

Such things as metals respond to stimuli (e.g., magnets). Tesla refuses to separate the motive forces involved in electromagnetic effects from reactions of "living" matter. This in essence was Bulwer-Lytton's "vril power." The energy that runs the universe directs life. "Thus, everything that exists, organic or inorganic, animated or inert, is susceptible to stimulus from the outside. There is no gap between, no break in continuity, no special and distinguishing vital agent. The momentous question of Spencer, What is it that causes inorganic matter to run into organic forms? has been answered. It is the sun's heat and light. Wherever they are there is life."[32]

As Tesla himself was a "self-propelled automaton entirely under the control of external influences," he could use the model of himself to build his telautomaton. What we call memory, Tesla further stated, "is but increased responsiveness to repeated stimuli." Creative thinking and also dreaming would be derived from secondary reverberations of these initial external stimuli.

> Long ago I conceived the idea of constructing an automaton which would mechanically represent me, and which would respond, as I do myself, but of course, in a much more primitive manner to external influences. Such an automaton evidently had to have motive power, organs for locomotion, directive organs and one or more sensitive organs so adapted as to be excited by external stimuli....
>
> Whether the automaton be of flesh and bone, or of wood and steel, it mattered little, provided it could provide all the duties required of it like an intelligent being.[33]

To Tesla, his remote-controlled boat was not simply a machine, it was a new technological creation endowed with the ability to *think.* In Tesla's view, it was also, in a sense, the first nonbiological life-form on the planet. As a prototype, this first new life-form was "embodied," in Tesla words, with a "borrowed mind," his own! "[It will be able] to follow a course laid out or...obey commands given far in advance, it will be capable of distinguishing between what it ought and what it ought not to do...and of recording impressions which will definitely affect its subsequent actions."[34]

Very few individuals could comprehend the magnitude of the creation in 1898, and so they lashed out at Tesla instead.

In November 1898, the examiner in chief of patents came to witness a demonstration of Tesla's telautomaton before granting a patent, so "unbelievable" was the claim. "I remember that when later I called on an official in Washington, with a view of offering the invention to the Government," Tesla wrote, "he burst out in laughter....Nobody thought then that there was the faintest prospect of perfecting such a device."[35]

24

Waldorf-Astoria (1898)

November 29, 1897

My dear Luka,

I have concluded that so important a literary event as the appearance of your splendid book of poems should be firstly commemorated by a dinner at the Waldorf as Mrs. Filipov suggested in her peculiarly delicate way.... As this does not seem possible because of your immense popularity, I would like you to name another evening very soon, however, for my money may run out.

Yours sincerely,
N. Tesla[1]

The Waldorf-Astoria was the tallest hotel in the world, a center for banquets, concerts, and conventions in the city, and the permanent or temporary residence of the wealthiest and most eminent citizens of the day. Residing there became a goal to which Tesla aspired; it would be one he would achieve before the end of the year, and one he would maintain for the next two decades. Built in two parts, the original Waldorf was completed by William Waldorf Astor in 1893; the Astoria, by his cousin, John Jacob Astor, in late 1897. At first, Jack was reluctant to tear down his mother's home to erect a hotel, but after the Waldorf grossed $4.5 million in its first year, he changed his mind. Its opening "marked the beginning of a new concept in living," extolling the essence of exclusiveness, cordiality, pomposity, and elegant grandiosity to the masses.[2]

The manager, George C. Boldt, was a Prussian immigrant from the island of Rügen, situated near Denmark in the Baltic Sea. "Mild mannered, dignified and unassuming," Boldt resembled "a typical German professor with his close-cropped beard which he kept fastidiously trimmed...and his pince-nez glasses on a black silk cord." Described also as "a martinet, and a man of mercurial moods," Boldt was a socialite, in certain ways, of the most superficial kind. "I would rather see Mrs. Stuyvesant Fish enjoying a cup of tea in an all but empty Palm room," Boldt declared, "than a dozen lesser-known guests there feasting."[3]

The manager also adored mechanical contrivances, sprinkling the hotel with such modern conveniences as pneumatic tubes, electric bulb carriage calls, flashing control panels on the elevators, and "his network of hushed but authoritative buzzers."[4] A few years later, the Waldorf would be the first hotel with a radio tower. No doubt, Tesla, himself an elitist, was attractive to Boldt. With his position of inventor extraordinaire well established, it is likely that the inventor was accepted in a distinct class above the manager. Once he moved in, Tesla may even have avoided paying rent in lieu of his connection with Astor, or he may have negotiated a favorable deal.

With over nine hundred on staff, the acclaimed "Oscar of the Waldorf" as chef, and Boldt's capable wife overseeing the decor, there was no finer establishment. A regal fragrance wafted from every corner of the hotel, with exquisite porcelain, exotic flowers, and expensive furniture decorating the halls, dining rooms, and suites. This was the alley where the peacocks came to strut. At over six feet two inches and dressed in suede high-tops, tails, cane, top hat, and ever-present white gloves, Nikola Tesla was one of the proudest and most renowned.

The Spanish-American War dragged on through most of 1898 as Tesla continued to try to exploit his telautomaton for use as a naval weapon. He had offered his wireless transmitters to aid in the organizing of ship and troop movements but was turned down by the secretary of the navy for fear, as Tesla reported a year later, "that I might cause a calamity, as sparks are apt to fly anywhere in the neighborhood of such apparatus when it is at work." Tesla tried to guarantee that he had overcome "these defects and limitations," but it was to no avail.[5] Public demonstrations and photographs of lightning bolts spewing from his ingenuity worked to hinder any assurances he might give. Instead, during the conflict, the navy utilized hot-air balloons connected to ships by telegraph lines instead. Being up in one would "make a man's hair turn white," since the balloon was an easy target, but soldiers had to "obey orders, and that was all there was to it."[6]

Tesla contacted shipbuilder Mr. Nixon, designer of the *Oregon* (the Villard ocean liner he had fixed while working for Tom Edison in 1884), and also submarine builder John P. Holland.[7] Two years later, Holland would sell the navy its first submersible; it would weigh a commanding seventy-four tons and become a quintessential fighting machine,[8] but in 1898 he was still having difficulty negotiating a deal. "The Navy Department was obliged to decline...[Holland's offer] to go into Santiago Harbor and destroy the Spanish warships...as it smacked of privateering and was in violation of international law."[9] And Tesla also invited military personnel to his laboratory, particularly U.S. Navy Rear Admiral Francis J. Higginson, chairman of the Light House Board, to discuss the use of his wireless transmitters. But dealing with the government was anything but easy.[10]

During this period, in June 1898, Richmond Pearson Hobson rocketed to stardom, capturing the hearts of America through his heroic efforts in the war. A few months later, Hobson would become the key attraction of the Tesla-Johnson social net, and a decade after that, his fame so well established, he would become a presidential candidate.[11]

On June 4, the *New York Times* reported that a bold American fighting frigate, the *Merrimac,* had "made a dash" into Santiago Harbor under "a lively cannonade of fire" in attempts to attack the waiting Spanish armada. The ship was sunk, and "an officer, an engineer and six seamen were taken prisoners." The *Times* concluded, "Everybody is astounded at the audacity of the American vessel."[12]

The following day, it was revealed that enemy fire had not sunk the ship at all. Rather, it had been deliberately scuttled by Lieutenant Hobson for the purpose of locking in the harbor the entire Spanish fleet. "This splendid stroke" effectively removed the feared Admiral Cervera from the war. "In a day, in an hour, the potent, all-pervading force of electricity...flashed his fame over the round world."[13] As Hobson was imprisoned in the dungeon of Morro Castle, he continued to make headlines as the world waited for the war to end and his hoped-for release.[14]

As ever, Katharine continued to spread her Irish charms, inviting the Serbian mystic to dinner so that she could be "hypnotized" by his presence.[15] The very day of the sinking of the *Merrimac,* the inventor received the following provocative missive:

> June 6, 1898
>
> Dear Mr. Tesla,
>
> I want very much to see you [tomorrow evening], and will be really disappointed if you do not think my request worthy of your consideration. Robert is giving himself a birthday party and he is going to have some of your Servian songs sung.
>
> You must save this evening for us. After this date I am going away to Washington for a visit, so if anybody cares to see Mrs. Filipov?
>
> When you come tomorrow evening, we'll talk about the hand which is before me now but which is doomed to seclusion....I cannot stand it. It is too strong, too virile—when I enter the room without thinking it makes me start—*It is the only thing in it.* But it is not satisfactory for it does not give the proper

idea of your hand which is large and free. Like yourself, this seems sticky, short, I know what causes this, it is the shadows. You must try again and make your hand be as large and grand as it is.

Faithfully yours,
Katharine Johnson[16]

In January, Tesla had published a full-page photograph of his hand for *Electrical Review* in his attempts to dramatically display the improvements he had made in the efficiency of his vacuum lamps.[17] Every line on the palm was visible, so effective was the illuminant (although, as Katharine points out, the shape is marred by an underlying shadow). The famous palmist Cheiro was the rage of the day, having published analyses of the hands of such celebrities as Tom Edison, Sarah Bernhardt, and Theosophist Annie Besant. Mark Twain commented: "Cheiro has exposed my character to me with humiliating accuracy. I ought not to confess this accuracy, still I am moved to do it."[18]

It seems likely that Tesla's clever vanity had come upon a way to exploit his fluorescent tubes and at the same time reveal in a veiled way the magnanimity of his being by exposing his hand to the world. Referring to a professional palmist, Tesla's hand supposedly reveals "a flirtatious streak and hypersensitivity" seen in the Girdle of Venus; from the head line "incessant worries stemming from the past," a close association with his mother (because it is tied to the heart line), "an irrational streak and blind spot in his thinking" (due to chaining, an undulating course, and foreshortening of the line); and all of this is counterbalanced by a "remarkable fate line, which, rising like a firm oak tree, reveals stability, vision, creative aspiration, stubbornness and an ability to withstand great stress and turmoil.... The fate line is the strongest line in the hand."[19]

During these months playful banter peaked between the inventor and the elusive Madame Filipov; it seems she tried to get Tesla married. In February, Katharine wrote, "Another charming lady is to be here who does not believe that you are my friend, that I even know you. I wish to convince her that you are on my list and you shall sit beside her. Come and shed the radiance of your happy countenance upon us all especially the Johnsons."[20]

In March she demands that Tesla join them for lunch to "bring solace to your friends" and a few days later invites him over again. "A very charming girl is to be here who wants very much to meet Mr. Tesla. A real one, I assure you."[21] Tesla decides to have everyone for dinner first and writes, "I shall send my private equipage... to dine... at the Waldorf and I am getting up the appetite for the occasion."[22]

Tesla's celibacy has always been a question mark. It seems probable that he and Katharine had engaged in a "sticky" liaison a few years earlier, but at this stage, due in part to Katharine's arrangements, Tesla had become active with other women, and Katharine had enjoyed his pleasures vicariously. Three ladies who interested him were Mrs. Winslow, Miss Amatia Kussner, and Miss Marguerite Merrington. The first, alas, was married; to the second, Tesla wanted to display his inventions at his laboratory; "There is another reason why she should come but that is difficult to say.... Well I do not want to say anything disparaging of a lady."[23]

For yet another "Johnson blowout" Tesla wrote, concerning the third lady, "I'd sooner be glad to rely on your choice and only remember to suggest Miss Merrington—if she would come. I know I would be her victim—before dinner, but after that I think I could hold my own for she does not drink claret."[24] And a few months later:

March 9, 1899

My dear Mrs. Johnson,

I shall be glad to have any of your friends come but we must have one lady for each gentleman—else you must dine without me.... Agnes must come by all means and—wouldn't you invite Miss Merrington? She is such a wonderfully clever woman. I would say wise had she but married. Really, I would like to have her with us.

Sincerely,
Nicholas I of Houston St.[25]

Born in England, Marguerite Merrington was raised in a convent in Buffalo before she studied the piano and became a teacher at her alma mater, Normal College. Resigning to move to New York City to follow her passion of "dramatic author," Miss Merrington first made headlines in 1891 with her well-received romantic play *Letterblair.* This success was followed two years later when she won a $500 prize from the National Conservatory of Music for her libretto *Daphne,* which was reviewed by, among others, Antonin Dvořák. "Tall, graceful and charming," the regal Miss Merrington was a "frequent dinner guest at the Johnson home." An integral part of the Gramercy Park set, Miss Merrington accompanied the young Owen Johnson to Mark Twain's birthday party in 1905 and maintained her creative spark throughout her life, writing a book on General Custer and his wife in 1950, the year before she died, unmarried, at the age of ninety-one.[26]

In August 1898, John Jacob Astor returned from the battlegrounds, but it was not until December that Tesla met with Astor at his home. While Jack was considered by many to be "cold-hearted, humorless, weak-minded and almost completely absent of personality,"[27] his wife, Ava, was seen as the most beautiful woman in America. Tesla was particularly taken by Lady

Astor's loveliness, and it seems that she was enthralled by the inventor's experiments. The three dined together on occasion at Delmonico's or the Waldorf, and when Tesla arrived at the Astor residence, he often brought along a bouquet of flowers. But although "Ava…sparkle[d] at every lighted candle and Jack…followed her about like a bedraggled and slightly bad tempered spaniel," all was not well with the marriage; where Astor was able to leave his exquisite wife for months on end, touring the high seas for adventure and the noble cause, she retreated by entertaining a zealous interest in bridge.[28]

Thus, even though Ava was on his side, the inventor was uncertain of his standing. "My dear Astor," Tesla began, "I would like to explain why I could not go down to Cuba with you."

"I understand," Astor replied. "During the gunfire, perhaps that is when it dawned on me, I realized that your life was too precious to risk on such a trip. I see, however, from recent reports that you have been attacked after all, but it has been by reporters instead."

"I'm glad," Tesla quipped, "that I am living in a place in which, though they can roast me in the papers, they cannot burn me at the stake."[29]

Tesla thereupon called a meeting with Astor and two of his cronies, Mr. Clarence McKay and Mr. Darius Ogden Mills, so that he could display his continuing progress with his oscillators and fluorescent lights as well as show them patent applications, articles which had appeared in the technical journals, and reports on tests performed by the Royal Society in London and the Roentgen Society in Germany. "Let me read you the following dispatch from Sir William Crookes," Tesla said. "Congratulations. The performance of your machine is marvelous." And Tesla presented another report, which hailed his oscillator as "one of the most significant of the age."

"You will see how many enterprises can be built up on that novel principle, Colonel. It is for a reason that I am often and violently attacked, because my inventions threaten a number of established industries. My telautomaton, for instance, opens up a new art which will sooner or later render large guns entirely useless, and will make impossible the building of large battleships, and will, as I have stated in my patent long before the Czar's manifesto, compel the nations to come to an understanding for the maintenance of peace."[30]

"You are taking too many leaps for me," Astor said, causing the others to reconsider as well. "Let us stick to oscillators and cold lights. Let me see some success in the marketplace with these two enterprises, before you go off saving the world with an invention of an entirely different order, and then I will commit more than my good wishes. Stop in again when you have a sound proposal or call me on the telephone."

Tesla waited until the new year and then hit the colonel with a direct assault. "My dear Astor," Tesla said, "It has always been my firm belief that you take a genuine, friendly interest in myself personally as well as in my labors.... Now I ask you frankly, when I have a friend like J.J.A., a prince among wealthy men, a patriot ready to risk his life for his country, a man who means every word he says—who puts such a value on my labors and who offers repeatedly to back me up—have I not a foundation for believing that he would stand by me when, after several years of hard work I have finally brought to commercial perfection some important inventions which, even at the most conservative estimate, must be valued at several million dollars."

Informing Astor that George Westinghouse had given him $500,000 for the AC polyphase system and that Edward Dean Adams had invested $100,000 to become a partner in his later endeavors when he had "14 [new] U.S. and as many foreign patents," Tesla remarked that there was a "powerful clique" which still now opposed him. "And it is chiefly for this reason that I want a few friends, like yourself, to give me at this moment their valuable financial and moral support."

Having "placed faith" in Astor's words, Tesla reveals that he had sold off securities to buy back control of his company, although "Mr. Adams still has a minority interest." Having stated that his laboratory in the past has "paid $1500 for every $100 invested, on the average," the inventor proclaims, "I am fully confident that the property which I have now in my hands will pay much better than this."

"I now produce a light superior by far to that of the incandescent lamp with one third of the expenditure of energy, and as my lamps will last forever, the cost of maintenance will be minute. The cost of copper, which in the old system, is a most important item, is in mine reduced to a mere trifle, for I can run on a wire sufficient for one incandescent lamp more than 1000 of my own lamps, giving fully 5000 times as much light. Let me ask you, Colonel, how much is this alone worth when you consider that there are hundreds of millions of dollars invested to-day in electric light in the various chief countries in which I have patented my inventions in this field?

"Sooner or later," Tesla continues, "my system will be purchased either by the Whitney Syndicate, G.E. or Westinghouse, for otherwise they will be driven out of the market."

The inventor closed: "Then consider my oscillators and my system of transmitting power without wires, my method of directing the movement of bodies at a distance by wireless telegraphy, the manufactures of fertilizers and nitric acid from the air, the production of ozone...and many other important lines of manufacture as, for instance, cheap refrigeration and cheap manufacture of liquid air, etc.—and you will see that, putting a fair estimate on all, I cannot offer to sell any considerable amount of my property for less than $1000 a share. I am perfectly sure that I will be able to command that price as soon as some of my inventions are on the market."

Telling Astor that he had contracts pending with "the Creusot Works in France, the Helios Company in Germany, Ganz and Company in Austria and other firms," Tesla requested an investment of $100,000. "If you do not take that much interest you will put me at a great disadvantage." Should Astor come in, other Astor associates, such as Mr. McKay and Darius Ogden Mills, "would do the same." If, Tesla wrote, "after six months you should have any reason to be dissatisfied, it will be my first duty to satisfy you."[31]

Astor stressed interest in seeing Tesla exploit his fluorescent lights, and the inventor agreed. On January 10, 1899, papers were signed whereby Astor gave Tesla $100,000 for five hundred shares of the Tesla Electric Company; in return, Astor was elected director of the board.[32] At the same time, Tesla moved into the Waldorf-Astoria. Tesla also received $10,000 from the dry-goods manufacturer Simpson and Crawford,[33] and he may also have received funds from Mr. Mills or Mr. McKay. The old Tesla Company, with William Rankine and Edward D. Adams, was, for all intents and purposes, dissolved, along with his relationship with Alfred Brown and Charles Peck, although all of these individuals may have in one way or another been involved in the new enterprise.

The first letter on Waldorf stationery written to the Johnsons is dated November 3, 1898. It is one of the rare letters in which Tesla refers to Mrs. Johnson as "My dear Kate." Tesla was about to hook a big fish, perhaps the wealthiest fish on the planet, and his sense of self-importance escalated accordingly. In haughty fanfare, the Serbian aristocrat separated his exalted ilk from "other social tribolites... plebians, drummers, grocerymen, [and] Jews."[34] Anti-Semitic references are rare in the Tesla correspondence, but no doubt he was anti-Semitic, at least in the social sense, that is, as a common reflection of the times. Anti-Semitism against ethnic ghetto dwellers, such as those newly arrived in Williamsburg, Brooklyn, and New York's Lower East Side, was common among the upper classes, even though the Rothschilds, August Belmont, Jacob Schiff, and Bernard Baruch were well respected and known to be Jewish. It is, however, clear evidence of one of Tesla's prejudices. The letter also refers to Tesla's wish to meet Lieutenant Hobson, whose Cuban exploits had been featured in the *Century*. "The interest in [Hobson] was at a fever heat [at our offices]," Johnson recalled, "[and] the estimates of the sale of his [proposed] book ran into the hundred thousands." Unfortunately, shortly after his account appeared, a scandalous event occurred which was egged on by the "sensational press," and the idea for the book "fell flat."[35] The incident had to do with Hobson's handsome appearance and the inability of women to refrain from kissing him when he came into their midst.

"I would have cut off my right arm rather than to offend one of [those ladies]," Hobson declared, concluding, "The kissing episodes, what little there was to them, were entirely beyond my control, and my conscience is clear."[36]

The dashing lieutenant was well liked by Tesla and the Johnsons, and his entrée into their circle added a wonderful spark to their lives. He would join a group that included at this time the Gilders, Miss Kussner, Miss Merrington, Mrs. Winslow, Mrs. Robinson, Mrs. Dodge, Rudyard Kipling, and John Muir. Playful jealousies became evident as Tesla and the Johnsons vied for the war hero's attention, Tesla daring even to tender to Hobson one lady with whom the inventor may have had intimate relations.

"Remember, Luka," Tesla teased, "Hobson does not belong to the Johnsons exclusively. I shall avenge myself on Mme. Filipov by introducing him to Mme. Kussner and somebody will be forgotten."[37]

Tesla spent many "delightful" hours with Hobson, inviting him to the laboratory, to dinner, and out on the town. "He is a fine fellow," the inventor concluded.[38] Their friendship would last.

Hobson, twenty-eight, a southerner, was a striking presence in his uniform, with deep-set, penetrating eyes, hair swept back, a firm, prominent chin, and a handlebar mustache. A graduate of the U.S. Naval Academy in 1889, he had spent three years in Paris studying at a maritime college and had worked for the Office of Naval Intelligence.

Gifted with a keen mind, Hobson had worked for the secretary of the navy during the China-Japan war. His family, on both sides, included lawyers, judges, a governor, and a general.[39] He was a ready-made hero, and he was a social catch of the first order.

All the pieces of Tesla's puzzle were now in place. He had obtained fundamental patents on wireless communication and remote control, he had calculated the type of energy he needed in order to disturb the electical conditions of the planet, he had obtained a sizable sum of working capital from one of the wealthiest men in the world, he had begun serious negotiations with the U.S. Navy, and, as a social triumph, he had moved into the Waldorf-Astoria. The budding entrepreneur settled upon a plan for marketing his oscillators and cold lamps—well, they could wait for now—and then Tesla took the next bold step. He would test his wireless theories on a grand scale.

The laboratory at Houston Street was simply too small and vulnerable to fires and potential spies. With few people aware, Tesla had scouted the country to seek out potential sites for his new "Experimental Station." George Scherff, his capable secretary, tried to get Tesla to reconsider, to stay in New York and do something tangible, something that would pay an immediate return, but he was talking to a deaf ear. Destiny was urging Tesla westward.

25

COLORADO SPRINGS (1899)

Nikola Tesla, the Servian scientist, whose electrical discoveries are not of one nation, but the pride of the world, has taken up his abode in Colorado Springs.... On East Pike's Peak avenue, with limitless plains stretching to the eastward, and a panorama of mighty mountains sweeping away north and south, to the west—Tesla has caused to be constructed a [wireless] station for scientific research.

DESIRE STANTON, COLORADO SPRINGS, 1899[1]

Having been invited to Colorado Springs to build his laboratory by Westinghouse patent attorney Leonard E. Curtis, a longtime adviser and friend through the difficult years of the "battle of the currents," Tesla shipped his equipment in early spring, 1899. Before he left New York and as the coup de grâce to his relationship with T. C. Martin, Tesla met with competing editor Charles W. Price of *Electrical Review* and professional photographer Dickenson Alley to choreograph a spectacular piece on the wizard's laboratory. Complete with a rich description of his experiments and a sensational series of photographs, the article ran in the March 29, 1899, issue. Starting with a full-length portrait of the inventor grasping a basketball-sized wireless vacuum lamp glowing resplendently, the essay went on to describe the evolution of other inventions, such as his high-tension transformer, which resulted in Tesla's flat, spiral transmitting coil. This eight-foot transmitter, "easily recognizable by its spider web appearance" was the first which efficiently enabled the inventor to generate two individualized vibrations, or tuned circuits, simultaneously and also produce many millions of volts.[2] Other prints depicted the flamboyant engineer transmitting high currents through his body to illuminate a variety of vacuum tubes, such as one which he whipped around his head in a multiple exposure. With one hand seeming to pluck a refulgent rod out of the midst of a spiral galaxy of blurred light and the other grasping a sparking, circular high-tension coil, "the operator's body...[was] charged to a [great] potential."[3]

Tesla arrived in Colorado Springs on May 18, 1899, after stopping in Chicago to display his telautomaton to the Commercial Club, a local electrical society. Situated at the very gateway to the Rocky Mountains, at the edge of a plain that stretched for hundreds of miles, the Colorado site would prove an excellent choice not only for monitoring wireless energy generated from his transmitter but also for studying a common phenomenon in the region, the electrical storms.

Stepping from the station after his weeklong journey, the inventor was met by Curtis and a few dignitaries. A horse and carriage took him to his hotel, the Alta Vista, where he stayed, in room 207.[4] Much like his childhood home in Smiljan, the town was situated at the foothills of a mighty chain of mountains. The Rockies arose so suddenly that they looked almost as if they were still being formed. A view on a clear day stretched virtually to Wyoming to the north and New Mexico to the south, and it was a common sight to witness lightning storms in the distance while standing in sunshine.

Hoping to be the "Little London" of the West, the people of "the Springs" welcomed the great inventor by honoring him with a banquet, sponsored by Curtis, at the El Paso Club. Well known throughout the region because his AC power transmission system had been adopted at lead, silver, and gold mines in such camps as Telluride and Cripple Creek, Tesla was happily met by society people, town officials, and the governor.[5] A few days later, another notable also came to town, Admiral Schley, recently back from his victory in Santiago Bay. The entire town celebrated the hero's visit.[6] No doubt Tesla had easy access to the admiral, and they probably discussed the potential use of his telautomaton as a weapon to help abolish war.

As part of Tesla's arrangement, with Curtis's guarantee, the El Paso Electric Company provided Tesla with free electrical power to support the great quest. Local contractor Joseph Dozier was introduced to discuss construction of the lab. Off they drove down Pike's Peak Avenue to what is now the corner of Coyote Street, near Prospect Lake, to look the site over. Dozier, it appears, had a mystical bent, so discussions drifted to talk of life on other planets and unusual ways to search for gold in the nearby hills.[7]

The inventor had traveled west for a variety of reasons, particularly his wish to experimentally verify that he could transmit light, information, and power to vast distances by means of wireless. "I wanted to be free of the disturbing influences in the city which make it very difficult to tune circuits," the inventor added.[8] Tesla had embarked on a massive plan that presupposed an understanding of a technology which even challenges today's comprehension of power-distribution systems. Details of the work would remain secret, Tesla not even revealing his intent to erect the station until almost the day he departed.[9]

From the very first day of his arrival, he announced rather optimistic plans, telling town columnist Mrs. Gilbert McClurg, wife of the secretary of the chamber of commerce, who wrote under the nom de plume Desire Stanton,[10] "With [my] wireless telegraph oscillator, [I]...could talk to the inhabitants of the planet Mars...if they know enough to take a message...and will talk to the people of the earth, at any distance away, without the assistance of wires."[11]

There was a popular notion and even astronomical speculation that other planets were inhabited and that Mars could be contacted. Growing large fields with flora planted in different-colored symbols or creating giant reflecting surfaces to flicker signals were two other proposals espoused by writers in the technical journals that received serious consideration. With Percival Lowell's magnificent telescope only a few hundred miles away, in Flagstaff, Arizona, reports of Martian activities were a commonplace topic of conversation.

With Tesla's arrival, in Desire's words, "the day of 'vril power' is not far distant."

"I would light whole cities and give to mere machines all the motions of intelligence," said Tesla. "But my first plan is to simply collect experimental data, mount instruments and record experiments at different atmospheric levels."

"Tesla's plan for cabling across the Atlantic is to erect two terminal stations, one in London and one in New York, with the oscillators placed at the top of high towers, communicating thence with great disks suspended in captive balloons floating 5,000 feet above the earth to catch the strata of rarefied air through which electrical waves travel most easily. A message could be flashed instantly by these lightning rays from the oscillator to the disk in the balloon, and across the thousands of miles of intervening space to the second disk....Mr. Tesla says he is ready to put his wireless system into operation as soon as the practical details can be arranged."[12]

His scheme was multifaceted. He could utilize the ionosphere to act as a conduit or a reflector of the electrical waves;[13] he could use the intrinsic electrical impulses of the earth itself, that is, its geomagnetic pulse as a carrier wave, or he could transmit energy in the more conventional wireless way "with one single tuned circuit on the transmitting and receiving end,"[14] such as he had demonstrated in his public lectures in London, Paris, New York, Philadelphia, and St. Louis, that is, by using a resonant tuned circuit, comprising a transmitter and a receiver, an aerial and ground connection, or he could use conventional copper lines.[15]

With Colorado Springs six thousand feet above sea level, one of his first experiments involved the transmission of very high frequencies up long wires to terminals situated two miles in the sky.[16] Helium-filled balloons more than ten feet long were ordered from a "balloon farm" in Germany run by Professor Meyers, and thousands of feet of wire and cable were shipped from the Houston Street lab.[17]

Other equipment included batteries, receiving and measuring instruments, switches, transformers, vacuum pumps, and dozens of types of electronic tubes specially prepared by a Manhattan glassblower. Also, his huge oscillators and gargantuan Tesla coils were shipped, although his novel, flat, spiral transmitter, which appears so prominently in the 1898 photographs of his New York laboratory, was not sent. Louis Uhlman, one of his key engineers, was placed in charge of equipment at Houston Street, with George Scherff business manager and liaison.

Tesla's wireless experimental station was a large barnlike structure approximately sixty feet wide and eighteen feet high, with a 200-foot-tall bulbous-topped aerial whose height could be adjusted to differing lengths. Situated on East Pike's Peak Avenue, within walking distance of the center of town, the building was connected by transmission lines to the El Paso Electrical Station only "a few miles away," From their circuit, Tesla was able "to draw, according to necessity, 100 horse-power and more."[18] Guarded by a sign which read Great Danger, Keep Out, the lab housed a high-frequency transformer and a Tesla coil with a diameter of forty-five feet.[19] All of his experiments would be carefully recorded in his private Colorado notebook (which was discovered in the 1950s among his papers at the Belgrade Museum). Theories, experiments, occasional personal observations, and highly technical mathematical equations filled the pages.[20]

Tesla stated that his main reason for coming to Colorado "was to produce a [resonant transformer] which would be capable of disturbing the electrical condition [of part], if not the entire globe... thus enabling me to transmit intelligence to great distances without wire."[21] The plan was actually quite simple; Tesla assumed that the earth had a resonant frequency and therefore could be measured and utilized as a gigantic carrier wave to distribute electrical power. Since the entire earth was in a harmonic relationship with his equipment, Tesla claimed that there would be "no diminution in the intensity of the transmitted impulses. It is even possible to make the actions increase with distance from the plant according to an exact mathematical law."[22]

By regulating the height of the aerial, different wavelengths could be created which could be measured in terms of their harmonic relationship to the natural electrical properties of the earth.

A few weeks later, Tesla summoned Fritz Lowenstein. Just twenty-five years old and a recent German immigrant, Lowenstein had only been working for Tesla for a month. Thus, there was a question about his ability to be reticent about the work. Nevertheless, he was, as Tesla said, "a man possessed of the highest technical training"[23] and probably the most formally educated of his crew.

In June, Tesla experimented with a wireless telephone, although it is not known whether he actually transmitted spoken words.[24] To Astor, he wrote, the following year, "There is nothing novel in telephoning without wires to a distance of five or six miles, since this has been done often before.... In this connection, I have obtained two patents."[25]

Whether or not Astor was completely aware of Tesla's plan to leave New York to set up an experimental station in Colorado Springs is unknown. Astor had been in Europe during much of the planning stages of the venture, returning to New York on June 14. Certainly Tesla notified Astor of his plans, but this author suspects that the financier did not find out until after their business arrangement was solidified. As Astor was expecting progress on the oscillators and cold light, he probably had mixed feelings when he contacted Scherff to check on Tesla's progress.[26]

Professor Meyers's balloons had finally arrived. "They should only be about two-thirds inflated," Meyers warned, "as otherwise they may burst when they attain some height."[27] "He has also included some kites," Scherff wrote with the shipment. The balloons were launched in July, "but they [are] too heavy and do not work well."[28] The scheme, although plausible, was cumbersome, as energy would have had to be transmitted up a long wire, which weighed down the balloon (or kite) in a somewhat haphazard fashion, and then another balloon placed at some distance would be needed to receive the transmission and then send it down long wires to instruments on the ground. For numerous reasons this line of investigation was abandoned.

Next on the agenda was the measurement of the electrical properties of the earth, monitoring electrical storms, and creating electromagnetic oscillations that would be in a harmonic relationship to terrestrial currents.

Uppermost in Tesla's mind was the problem of individualization of messages and protected privacy. Therefore, most of his experiments involved the combination of two or more frequencies and the construction of receiving instruments tuned to these specific arrangements. "The chief feature of a practical wireless telegraph system," Tesla told Lowenstein, "is the secrecy, immunity and selectivity of the oscillating and receiving apparatus." The inventor thereby set about to create a variety of mechanisms to produce multiple wavelengths. Whereas Marconi and the others were using inefficient Hertzian "pulsed oscillations at very high frequencies," Tesla worked with "continuous [undamped] oscillations in the low HF [high-frequency] range."[29]

"Do you understand what we are now trying to achieve?"

"Yes, Mr. Tesla," Lowenstein replied, "I understand that in this invention the elements of the receiving apparatus respond to the elements of the transmitter, and that only the co-action of the responding of all these elements of the receiver apparatus make the registering apparatus register."

"Excellent."[30]

On June 16, Tesla set out to create an efficient ground connection. His workers were instructed to dig a hole twelve feet deep near a water main, and a copper plate twenty feet square was buried there. "Water was kept constantly flowing upon the ground to moisten it and improve the connection," but the dry earth and the problem of rock formations interfered with the creation of a completely efficient setup. Nevertheless, "purposely unsensitive receiving instruments placed 200 feet from the shop responded when connected to the ground. The action of the device was strong even though it was concluded that the earth resistance was still too great."[31]

For the next series of experiments, Lowenstein was in charge of the transmitter, and Tesla attended to his numerous receivers. Lowenstein recalled, "I handled myself the big transmitter sending two vibrations through the ground by two separate secondary circuits.... Mr. Tesla would then...go outside of the building leaving me instructions for continuously switching the oscillator on and off in certain intervals.... I don't know how far he went, but by the time he came back again...[in] the afternoon, you may easily build an idea how far Mr. Tesla could have gone at the time I was standing at the switch."[32] In 1916, Tesla stated that he occasionally conducted experiments as far as ten miles from the station.

On the eve of July 4, one of the most stupendous electrical storms ever recorded in the region rocked Pikes Peak. "Observations made last night. They were not to be easily forgotten for more than one reason. First of all a magnificent sight was afforded by the extraordinary display of lightning, no less than 10,000–12,000 discharges being witnessed inside of two hours.... Some...were of a wonderful brilliancy and showed often 10 or twice as many branches."[33]

While tracking the storm with his sensitive receiving apparatus, Tesla noticed that even though the storm had passed out of sight, the instruments "began to play periodically." This was experimental verification of "stationary waves," periodic electronic vibrations impressed upon the earth itself. Also troughs and nodal points were detected. "It is now certain that they can be produced with an oscillator," Tesla wrote in his notebook, and then added in brackets, "[This is of immense importance.]"[34]

Tesla wrote to his secretary the same day: "Dear Mr. Scherff, I have received messages from the clouds 100 miles away." And two days later: "We have just about finished all [the] details; my work is really to begin in earnest right now."[35]

26

CONTACT (1899)

My dear Luka,

Everybody is after me since I was favored by the "Martians."... My friend J. Collier...has persuaded me to make a short statement regarding the subject of interplanetary communication.

Yours sincerely,
Nikola Tesla[1]

The Colorado notebook is virtually a daily record of Tesla's work at the time. Nowhere in the notes can there be found a distinct passage of the pivotal moment when he received unidentified impulses that he came to attribute to extraterrestrials; however, he does refer, on December 8, to this event, writing to friend and columnist Julian Hawthorne: "The art of transmitting electrical energy through the natural media...will...perhaps make it possible for [man] to produce...wonderful changes and transformation on the surface of our globe as are, to all evidence, now being wrought by intelligent beings on a neighboring planet."[2]

And just a few weeks later, during the holiday season, while still in Colorado, Tesla, in a Christmas message to the local Red Cross Society "when it asked me to indicate one of the great possible achievements of the next hundred years,"[3] wrote: "I have observed electrical actions, which have appeared inexplicable. Faint and uncertain though they were, they have given me a deep conviction and foreknowledge, that ere long all human beings on this globe, as one, will turn [their] eyes to the firmament above, with the feelings of love and reverence, thrilled by the glad news: 'Brethren! We have a message from another world, unknown and remote. It reads: one...two...three....'"[4]

Throughout July, Tesla was carefully monitoring the electrical activity of the earth, verifying that it had a specific geomagnetic pulse and harmonics off of that pulse. On the twenty-eighth, he worked on increasing the sensitivity of his receivers by "magnifying the effects of feeble disturbances." The inventor had tuned his equipment so carefully that "in one instance the devices recorded effects of lightning discharges fully 500 miles away, judging from the periodical action of the discharges as the storm moved away."[5] Thus, he reasoned, he did not have to test transmitted oscillations by installing a receiver fifty, two hundred, or five hundred miles away, as he was already proving that this could be accomplished simply by monitoring these distant electrical storms. This was one way Tesla rationalized his decision not to conduct long-distance experiments; he had verification that they would work.[6] Three days later, on August 1, the inventor departed from writing out his growing file of complex equations to compose a discourse of four thousand words on the atmosphere and the climate. In these passages, he describes the "baffling power of the moonlight" for taking night photographs, the "amazing brilliancy of the stars," magnificent sunsets and shooting stars, the peculiar ability of voices to travel several miles from the center of town to his laboratory, the "curious phenomena of the rapid formation and disappearance of cloud formations," and the numerous unusual shapes that appeared therein.

"The days were clear with just enough clouds in the sky to break the monotony of the blue," he wrote. "No wonder... people in feeble health are getting on here so well.... I soon learned there were thousands of consumptives in the place... and concluded that while this climate is certainly in a wonderful degree healthful and invigorating, only two kinds of people should come here: Those *who have* the consumption and those *who want to get it*...." Placing himself back into the mood of scientific observations, he ended the essay with the following line: "But the most interesting of all are the electrical observations which will be described presently."[7]

It appears likely that this sudden burst of poetic reverie could be attributed to the mystical moment he had encountered three nights earlier while monitoring his equipment alone at the lab.

This event, which was to alter his destiny in many ways, as we have seen, did not come out of the blue. Tesla had been planning for nearly a decade to make contact.

Talking With the Planets

Nikola Tesla

The idea of communicating with other worlds... has been regarded as a poet's dream forever unrealizable....

> [Having] perfected the apparatus...for the observation of feeble effects [from] approaching thunderstorms...so far from my laboratory in the Colorado mountains, I could feel the pulse of the globe, as it were, noting every electrical charge that occurred in a radius of eleven hundred miles.
>
> I can never forget the first sensations I experienced when it dawned upon me that I had observed something possibly of incalculable consequences to mankind. I felt as though I were present at the birth of a new knowledge or the revelation of a great truth....There was present something mysterious, not to say supernatural, but at the time the idea of those disturbances being intelligently controlled signals did not yet present itself to me....
>
> It was sometime afterward when the thought dashed upon my mind that th[ose] disturbances might be due to intelligent control. Although I could not decipher their meaning...the feeling is constantly growing on me that I have been the first to hear the greeting of one planet to another.[8]

As the inventor admits, the night he received the signals he did not attribute them to extraterrestrials. Most likely he first thought that they were periodic oscillations stemming from the thunderstorms that he was monitoring. A few days later, it began to dawn upon him that the metronome nature of the beats did not correlate with the supposition that they were linked to lightning discharges. The article also speculated that they may have stemmed from Venus or Mars. Two decades later, in 1921, Tesla wrote:

> Others may scoff at this suggestion...[of] communicat[ing] with one of our heavenly neighbors, as Mars...or treat it as a practical joke, but I have been in deep earnest about it ever since I made my first observations in Colorado Springs....
>
> At the time, there existed no wireless plant other than mine that could produce a disturbance perceptible in a radius of more than a few miles. Furthermore, the conditions under which I operated were ideal, and I was well trained for the work. The character of the disturbances recorded precluded the possibility of their being of terrestrial origin, and I also eliminated the influence of the sun, moon and Venus. As I then announced, the signals consisted in a regular repetition of numbers, and subsequent study convinced me that they must have emanated from Mars, the planet having just then been close to the earth.[9]

Note slight alterations from the original article and letter to the Red

Cross. In the 1901 article Tesla does not single out Mars as the only possible source of the impulses. Venus or other planets are also mentioned. In the 1921 article he says that Venus had been ruled out. Clearly this had to occur over two years after the event, that is, some time *after* the 1901 article which still included Venus as a possibility. "After mature thought and study," Tesla himself dates his "positive conclusion that they must [have] emanate[d] from Mars" to the year 1907.[10] In the Red Cross letter and the 1901 article Tesla is very specific in mentioning three beats. In the 1921 article the number of beats is obscured. Julian Hawthorne, who had written to Tesla while he was in Colorado Springs and who met with Tesla in New York upon his return, also refers to "three fairy taps." Tesla also alters the facts with regard to competing wireless operators. While Tesla was in Colorado, he received a number of letters from George Scherff concerning competitors. For example on August 1, 1899:

> Dear Mr. Tesla,
>
> Mr. Clark, the experimenter in wireless telegraphy called this morning seeking a powerful oscillator or information on how to build one.[11]

In August and September 1899, Scherff continued to inform Tesla about Clark, who could send messages three miles, thereby obtaining employment by a New York newspaper to report yacht races. Other wireless operators at this time included Professor D'Azar in Rome, Professor Marble in Connecticut, and Dr. Riccia in France.[12] And, of course, there was Guglielmo Marconi, who captured the imagination of the media during the America's Cup races that autumn. Although Scherff wrote, "The *New York Times* continues to boom Marconi,"[13] Tesla confidently replied, on September 22, 1899, to Scherff, "Do not worry about me. I am about a century ahead of the other fellows."[14]

Tesla, as one of numerous adherents to the group-fantasy belief that Mars was inhabited, assumed that the impulses stemmed from there. In 1899 it was frankly inconceivable to him that he could have intercepted a competitor's message. However, the fact of the matter is that Marconi was transmitting messages hundreds of miles across Europe and the English Channel during the summer of 1899 and was using as a signal the Morse-code letter *S* (dot-dot-dot), which precisely corresponds to the three beats Tesla said he intercepted while he was in Colorado.[15]

On July 28, the very date it has been hypothesized that Tesla received the signals, Marconi was with the British Admiralty and the French Navy in the English Channel, demonstrating his wireless apparatus between ships in mock battle maneuvers over distances of thirty miles, fifty-five miles, and eighty-six miles. "On 28 July, Marconi had inspected [the ship] *Alexandra*'s equipment in preparation for hostilities."[16] Most likely he transmitted the letter *S* at that time to see if it was picked up by the other warcraft. If Tesla was monitoring his equipment at twelve midnight, it would have been about 8:00 A.M. in England, so the times correlate as well.

At first, Tesla must have sincerely believed that the source of the impulses were extraterrestrial, for he boldly stated as much in a series of published articles.[17] A few years later, the awful truth dawned upon him. Worse, Tesla may have intercepted Marconi's impulses and made a fool of himself by claiming they had derived from superior intelligences. Marconi for Tesla was anathema. In 1921, ironically, while Marconi was making headlines by trying himself to intercept messages from nearby planets, Tesla wrote: "I was naturally very much interested in [recent] reports that these supposed planetary signals were nothing else than interfering undertones of wireless transmitters. These disturbances I observed for the first time from 1906–1907 occurred rarely, but subsequently they increased in frequency. Every transmitter emits undertones, and these give by interference long beats, the wavelength being anything from 50 miles to 300 or 400 miles."[18]

This statement supports the hypothesis that the 1899 impulses also stemmed from some competitor. Furthermore Tesla suggests the actual mechanism for his encounter: an undertone effect; and it appears that he unfortunately also provided, through Marconi's piracy, the very oscillators used to transmit the signals! The transmitter on the high seas in England, therefore, was attuned to the receiving equipment in Colorado. Coincidentally, this realization in 1906–1907 occurred, as we shall see, during a time of great emotional stress. Rather than face the truth, the mystical Serb clung to a supernatural explanation.

The most ardent proponent of the outer-space scenario was undoubtedly the journalist Julian Hawthorne. The son of Nathaniel Hawthorne, Julian authored a series of elaborate treatises on Tesla's philosophy, laboratory work, experiments in interplanetary communication, and place in history. Perhaps because he had engaged in a spectacular duel of articles with his brother-in-law George Lathrop, who wrote science-fiction tales about Tom Edison battling invaders from Mars in Arthur Brisbane's *New York Journal*,[19] Hawthorne took up Tesla's extraterrestrial cause.

The competition between Edison and Tesla would never abate, and it continued even into the realm of science fiction. Like many creative individuals, Edison had an interest in the occult. With Charles Batchelor, he had studied telepathy,[20] and he had worked with spiritualists on a "telephone" to communicate with departed souls. Edison was interested in space travel and interplanetary communication. Lathrop, who had married Julian's sister, had worked with Edison since the mid-1890s in cowriting a number of articles that became the precursors to latter-day fantastic tales.

In Lathrop's story "Edison's Conquest of Mars," when the Red Planet warriors invaded the earth, the Wizard of Menlo Park "invented a disintegrating ray...and it was 'Edison to the Rescue of the Universe.'"[21] The son of Nathaniel Hawthorne would not be outdone.

And How Will Tesla Reply to Those Signals From Mars?

Julian Hawthorne

> The other day, there happened to Mr. Tesla the most momentous experience that has ever visited a human being of this earth—three fairy taps, one after the other, at a fixed interval travelling with the speed of light were received by Tesla in Colorado from some Tesla on the planet Mars!
>
> No thoughtful man can have much doubt then, that little as we are aware of it, we must for ages have been subjected to the direct inspection of the men of Mars and of the older planets. They visit us and look us over year after year; and report at home: "They're not ready yet!" But at length a Tesla is born, and the starry men are on the watch for developments. Possibly they guide his development; who can tell?[22]

Perhaps more than any other writer, Hawthorne elevated Tesla to the order of an interplanetary Adonis whose mystical destiny upon the earth was to give its inhabitants electrical power, instrumentation, and enlightenment. Note the overblown yet elegant description Hawthorne pens in meeting with the sorcerer: "Ever and anon there appears a man who is both scientist and poet [who] walks with feet on the ground but with head among the stars. Men of this mark are rare. Pythagoras was one; Newton must have had a touch of the inspirational; in our own times Tesla is the man....He was born in Herzogovina, of Greek stock, one of the oldest families there. I believe he is a prince at home."[23]

Hawthorne interviewed Tesla in the same article in which the inventor reiterated his extraterrestrial hypothesis and his technological vision of the future, creating a world in which cheap energy would be available for all and humanity could begin to take its rightful place in the evolutionary hierarchy. And although Hawthorne tried to introduce some doubt as to the definite reality of the extraterrestrial encounter, a quarter of the way through the five-thousand-word treatise, the writer softened the potential criticism by rationalizing that "the hopes Tesla holds out embody things that ought to be true; that would immensely enlarge and beautify the world if they were true...." Concluding with a rhetorical question, Hawthorne wrote, "And what about conversing with Mars?...Tesla will do what he was sent here to do."

Backed by the opposition, factions of the press also lashed out vigorously. One severe critic, under the byline of a mysterious Mr. X, cautioned "intelligent readers": "Mr. Tesla obviously wants to figure in the newspapers. Everyone would be greatly interested if it were true that signals are being sent from Mars. Unfortunately, he has not adduced a scrap of evidence to prove it.... His speculations on science are so reckless as to lose an interest. His philosophizing is so ignorant as to be worthless."[24]

While in Colorado, Tesla negotiated with officials of the U.S. Navy and Light House Board, with nine letters passing between them from the spring of 1899 and, upon his return to New York, through the autumn of 1900.[25] On May 11, Rear Adm. Francis J. Higginson of the U.S. Navy wrote Tesla a letter which was forwarded to Colorado:

> Dear Sir:
>
> I would like to ask you if you can not arrange to establish a system of wireless telegraphy upon the Light-Vessel No. 66, Nantucket Shoals, Mass., which lies off about 60 miles south of Nantucket Island.[26]

Higginson stated explicitly, "The Light-House Board [has] no money...[so funding] will have to be paid from some outside source."

Tesla sent his "humble apologies for a tardy reply," because of a "severe cold," and then ended the note with this seemingly innocuous line: "[It] is also my sincere hope that I am not standing in the way of some other expert more deserving and better able to fulfill the task than myself."[27]

The statement strikes this investigator as odd. Why would Tesla write that he was potentially "standing in the way of some other expert more deserving" when he knew that this was a completely false statement. No other expert was more deserving or more knowledgeable than he. Moreover, he knew that it was very likely that other experts were pirating his work, so why would he encourage more of that activity? This was clearly a self-deprecating and self-destructive element. Be that as it may, the response from Commander Perry, Higginson's associate, was just as peculiar:

> Office of the Light House Board
> Washington, D.C. 16 August '99.
>
> Mr. N. Tesla
> Experimental Station,
> Colorado Springs, Col.
> Sir:
> In acknowledging the receipt of your letter of 11 Aug. '99....from certain expressions used in it the Board fears there may be some misunderstanding, so in order to prevent you from going to any trouble of expense, the board desires to say that it

has taken no action as yet toward providing any apparatus for using wireless telegraphy, as no appreciation is available for the purpose....

When it does take up the question of installing apparatus for communicating with light-vessels, your great name and fame in such matters will insure earnest consideration of yourself.

Respectfully yours,
T. Perry, Commander, U.S.N.

Apparently, Commander Perry read into Tesla's letter information that wasn't there. The inventor did not discuss reimbursement but took the opportunity in another letter to chastise Perry for his miserly response, writing on August 20, 1899, from his "Experimental Station" in Colorado Springs, to the Light House Board, Washington, D.C.:

Gentlemen,

....On this occasion permit me to avail myself of my acquired and precious prerogative as a citizen of the United States and to express my deep astonishment that, in a country of such vast wealth, and leading in enlightenment, so important a body as the Light House Board, instead of being provided with unlimited resources, should be trivially hampered and placed in...such an awkward position.

Very respectfully yours,
N. Tesla[28]

Although clever, Tesla's response was short-sighted, as the benefits that would accrue from installing wireless apparatus on this boat would clearly outweigh any short-term loss. The equipment would have been far more advanced than Marconi's, the press and public would see, without doubt, Tesla's superiority, and other branches of the navy and armed forces would have created contracts with the inventor. Furthermore, it would have been the first public demonstration by Tesla of long-distance wireless telegraphy. Unfortunately, throughout Tesla's long career, he never demonstrated this capability to anyone other than himself.

Nevertheless, Tesla's letter did not destroy his chances with the Lighthouse Board. On September 14, Commander Perry responded by offering Tesla the contract because the navy "preferred to award home talent" over Marconi.[29]

Here was the opportunity of a lifetime. Certainly it was understandable that Tesla needed to stay in Colorado through the autumn. He would return to New York the first week of January 1900. The Colorado experiment was a costly endeavor, and the inventor had now built up momentum, pushing toward the grand conclusion, that is, his wish to send impulses around the globe. Perry wanted action "quickly." But asking him to delay ninety days would not have been an unreasonable request. However, from the emotional point of view, Perry had said the wrong thing. He had mentioned the name of Signor Marconi.

> Gentlemen,
>
> ...Much as I value your advances I am compelled to say, in justice to myself, that I would never accept a preference on any ground, the merit of my own work excepted, particularly not in this case, as I would be competing against some of those who are following in my path and as any pecuniary advantage which I might derive by availing myself of the privilege, is a matter of the most absolute indifference to me.
>
> But since you have reasons for preference, permit me to state...that a few years ago I laid down certain novel principles on "wireless telegraphy" which I have been since perfecting.

Tesla went on to describe the seven features of his system: (1) an oscillator; (2) a ground and elevated circuit; (3) a transmitter; (4) a resonant receiver; (5) a transformer "that scientific men have honored me by identifying it with my name" (Tesla coil); (6) a powerful conduction coil; (7) a transformer in the receiving apparatus. Having "carefully perused all the reports of the more successful experimenters as they appeared," Tesla discovered that "they are all using, with religious care, these devices and principles, without the slightest departure, even in minor details...." He ended the letter by offering his services once again but requesting that the navy purchase an even dozen transmitters, with the caution that

> in the end one is apt to be accused of making outrageous prices. It is more than probable that my apparatus will cost more than that offered by others as I look to every detail myself.
>
> With many thanks for your good intentions I remain,
>
> Very respectfully yours,
> N. Tesla[30]

The navy never responded to this letter. One year later, on October 4, 1900, Tesla wrote Admiral Higginson. Four days later, the admiral responded: "It will...be necessary before asking Congress for money to carry on this work to have further estimates of cost."

Tesla's style of writing to the navy was particularly irritating and filled with contradictions. He claims to be "absolutely indifferent" to gaining a "pecuniary advantage," and yet he tells Commander Perry that the cost might appear to be "outrageous." At the dawning stage of a completely new industry, instead of building one or two prototypes to display before the government, Tesla insists on a sizable order. In an early letter he states that he does not want to stand in the way of any competitor; in another he claims he did not know there were any other competitors. In one passage, he accuses all his competitors of piracy (which was probably true), and yet in another he wishes them "hearty success." His position was incongruous to say the least, and it served to scuttle his own cause. This would turn out to be one of the most significant blunders of his career.

27

THOR'S EMISSARY (1899)

Th[e] problem was rendered extremely difficult, owing to the immense dimensions of the planet... But by gradual and continuous improvements of a generator of electrical oscillations... I finally succeeded in reaching rates of delivery of electrical energy actually surpassing those of lightning discharges.... By use of such a generator of stationary waves and receiving apparatus properly placed and adjusted in any other locality, however remote, it is practicable to transmit intelligible signals, or to control or actuate at will any one apparatus for many other important and valuable purposes.

NIKOLA TESLA[1]

Tesla went to Colorado in part for reasons of secrecy. His all-important transmitting oscillators and general design had already been pirated, and he would shortly be involved in a variety of priority battles. Looking at the Colorado project from the technical point of view, the inventor was in a virgin field and needed to experiment in order to determine a workable plan for distributing light, information, and power by means of wireless. The measuring of standing waveforms from the electrical storms throughout July confirmed what he had suspected, namely, that the earth had a resonant frequency and could therefore be used as a carrier wave to transmit his signals.

Letters between Scherff and Tesla continued on an almost daily basis throughout the summer. In August, Tesla received an "invitation to attend the banquet honoring the birthday of the Emperor Francis Joseph."[2] Correspondence also came in from Austria, India, Australia, and Scandinavia. "The [last] one," Scherff wrote, "is a proposition to become agent or manufacturer of your new light for Sweden, Norway and Denmark."[3] To the numerous business inquiries, Tesla wrote, "[Tell them] that I am on a scientific expedition, and will return in a few weeks."[4] There was also correspondence with William Rankine, E. D. Adams, Mr. Coaney (a stockholder), and Alfred Brown.

Bills were forwarded, and the inventor, in turn, would periodically mail off funds to cover these expenses. Wages for the crew ran about ninety dollars a pay period. The New York laboratory, in turn, prepared new equipment to be shipped west, as Scherff continued to send details about their construction.

> September 6, 1899
>
> Dear Mr. Scherff,
>
> Can you write about something more interesting than the pump. There are many things happening in a great city.... [Try] and make your correspondence more interesting... [such as in sending] press clippings.
>
> Sincerely,
> N. Tesla[5]

As was his custom, the inventor lived in the future, writing to Scherff in late August that he expected to return to New York in a few weeks. It would be more like four months. At the same time, Lowenstein requested permission to take leave, as he wanted to return to Germany for some family matter. Tesla feared that he might be an industrial spy, but he was only going home to get married. Koleman Czito was called on to replace him.

"Czito has just arrived," Tesla wrote, "and I [am] glad to see a familiar face again. He looks a little too fat for the work I expect of him."[6] He would come just in time to take part in some of the most spectacular electrical experiments ever performed. A hearty and trustworthy companion, Czito would stay at Tesla's call until he was an old man. By that time, he had trained his son Julius to take over. Julius would eventually come to aid Tesla in some of his more clandestine earth-lunar experiments as well as in day-to-day responsibilities.[7]

Throughout September, Tesla designed a large number of electronic tubes for his glassblower in New York to fashion, and for Scherff to ship, as he continued to document his work with a local photographer. Electrical energy generated exceeded 3 million volts. Tesla reported, "I drew 1-inch sparks between my body and an iron pipe buried in the ground about 100 feet from the laboratory."[8]

On the twenty-ninth, time-lapse pictures were mailed to John Jacob Astor, sugar refinery king, H. D. Havemeyer, his wife and their daughter, Mrs. E. F. Winslow, Stanford White, socialite Mary Mapes Dodge, and the Johnsons. He also shipped copies to Lord Kelvin, Sir William Crookes, Sir James Dewar, William Roentgen, Philip Lenard, and Adolph Slaby.[9] "Look them over carefully before delivery," the inventor instructed his liaison, "and do not allow the workers, other than yourself and Mr. Ulman to see them."[10]

Throughout the autumn, the inventor continued to change the height of the ball at the top of the antenna to measure the change in capacity and relationship to generated wavelengths in order to tune the equipment to the earth's frequency and "bring the oscillator into resonance with th[e] circuit."[11] Made out of wood, the ball was coated with metal. He also studied the strange phenomena of fireballs, which, when created by natural means, can appear like tumbleweeds of lightning that can roll down a street and smash into a tree or house. They are rarely seen, though there are documented sightings. Although Tesla himself had not witnessed any natural fireballs, he was able to create smaller ones in his lab. "Sometimes it apeared [*sic*] as if a ball would form above the coil, but this may have been only an optical effect caused by many streamers passing from various points in different directions.... At [other] times, a big cluster of them would form and spatter irregularly in all directions."[12] "He produced them quite by accident and saw them, more than once, explode and shatter his tall mast and also destroy apparatus within his laboratory. The destructive action accompanying the disintegration of a fire ball, he declared, takes place with inconceivable violence."[13]

In one instance, he pushed the experiments too far, and a fire started. Trapped by streamers that could maim or kill, the inventor had to roll to safety to save his life. To Johnson he wrote, "I have had wonderful experiences here, among other things, tamed a wild cat and am nothing but a mass of bleeding scratches. But in the scratches, Luka, there lies a mind. MIND."[14]

A few weeks later, while a photographer was present, Tesla set the roof in flames but was able to extinguish it before much damage was done. "The display was wonderful in spite of this," he wrote in his diary.[15]

Having studied the phenomena, Tesla attributed the generation of fireballs to "the interaction of two frequencies, a stray higher frequency wave imposed on the lower frequency free oscillation of the main circuit." They could also be produced when "stray high frequency charges from random earth currents" interacted with charges from his oscillator.[16]

The following week, he extended the ball to a height of 142 feet and began "propagating waves through the ground."[17]

> Referring to electrical or radio wave action at a distance, I know from experience that if proper precautions are not taken, fires of all kinds and explosions can be produced by wireless transmitters. In my experiments in Colorado, when the plant was powerfully excited, the lightning arresters for *twelve miles around* were bridged with continuous arcs, much stronger and more

> persistent than those which ordinarily took place during an electric storm. I have excited loops (coil aerials) and lighted incandescent lamps at considerable distance from the laboratory without even using more than five or ten per cent of the capacity of the transmitter. When the oscillator was excited to about 4,000,000 volts and an incandescent lamp was held in the hand about *fifty or sixty feet from the laboratory,* [emphasis added] the filament was often broken by the vibration set up, giving some idea of the magnitude of the electromotive forces generated in the space.[18]

Tesla had calculated that the earth pulsated at varying frequencies, especially twelve cycles per second.[19] With his coils wound with lengths of wire in harmonic relationships to the required wavelengths needed to "girdle the globe," he wrote in his diary, the length of the coil was calculated based on the equation:

wavelength/4 = harmonic of total wavelength (or) required length of coil

Taking into account the speed of light, 186,000 miles per second and the circumference of the earth, it was determined that coils would have to be "roughly" a mile in length, or some harmonic of this figure, to be in a resonant terrestrial frequency.[20] Other components included the thickness of the wire itself and horsepower generated. By increasing the frequency between pulsations, the inventor claimed to be able to boost horsepower to a few hundred thousand, although this amount of produced energy would last only a fraction of a second.[21]

Czito arrived for work one day in mid-autumn to find the inventor vigorously watering the ground around the metal plate which he had buried near the lab. "If I could only insulate these wires with liquid oxygen, I could reduce losses another magnitude," the inventor said. "Here, put these on." He gave Czito a pair of rubber-soled shoes as he laced up a pair for himself.

"All the way today, sir?" Czito inquired.

"To the limit, my friend. Now remember," the inventor cautioned, "keep one hand behind your back at all times."

Czito responded with a nod of his head. He had no plans to risk electrocution by creating a circuit through his heart.

"Throw the switch when you see my signal."

"We had better use these, sir." Czito handed his boss two cotton balls and took two for himself, and they plugged them into their ears.

The lanky Serb lumbered from his lab in his high shoes, past the mud, to place testing equipment and cold lamps at various locations in the earth, and positioned himself on a knoll about a mile away, near Prospect

Lake. Even though insulated, sparks jumped from the ground to his feet as he crunched along the path.[22]

The sun was low on the horizon as Colorado Springs began to turn on streetlights and electric lamps in preparation for the night. "Now," Tesla waved as Czito fired up the equipment.

The sound began as a low rumble and built to a "roar [that] was so strong that it could be plainly heard ten miles away." The ground trembled with the noise as the inventor gazed over to a nearby corral to watch a half-dozen horses rear on their hind legs and gallop frantically away. "Butterflies were carried around in... circle[s] as in a [whirlpool] and could not get out, no matter how hard they tried,"[23] as the flume of streamers stormed up the shaft high above the roof of the lab and blustered out from the apex, splitting lightning bolts fully 135 feet in length. *Kaboom! Zip! Zap! Kaboom!* Looking to the sky, the wizard held his wireless torches up in triumph as they flickered in his thunder.

The end came abruptly, the Springs plunged into darkness. He had shorted out the town.

Fortunately "the powerhouse had a second, standby generator which was started up soon after. Tesla was insistent that he be supplied with current from this reserve machine as soon as it was running, but his demand was refused." Forced by El Paso Electric to fix the damaged generator himself, the inventor was back on line in a day or two. "In the future, he was told, he would be supplied with current from a dynamo operated independently from the one supplying the [El Paso Electric] company's regular customers."[24]

By the end of 1899, Tesla was ready to return to New York. He wanted to get home for the holidays, to spend them with the Johnsons, but it would take him a little longer to wind things down. In December, he sent for his photographer, Dickenson Alley, to capture his work in the best possible light. By using multiple exposures, Alley would create what is perhaps Tesla's most famous photograph: that of the inventor sitting calmly reading a book, dwarfed by myriad tongues of explosive lightning. (This picture is a multiple exposure. Tesla, of course, was not sitting there at the time the oscillator was fired up; the electricity would have killed him.)

December 22, 1899

Dear Mr. Tesla,

We will keep your memory green Christmas day.... How lovely it would be if you should suddenly appear in our midst... to spend it with us....

I sometimes wonder if you could make me glad again, just to see you, it is so long since gladness has been in my days. Everything that once was has disappeared. It is as if one had gone to sleep in soft moonlight and had anchored out of place and out of time to find himself in the stone age, himself a stone.

What does it all mean?...

Sometimes I have a little sign of you through Robert by way of the office. I am hoping the New Year may bring you what you most desire and that it may bring to us my dear friend.

Faithfully yours,
Katharine Johnson[25]

28

THE HERO'S RETURN (1900)

Common people must have rest like machinery but the great old Nick—the Busy One—see him go 150 hours without food or drink. Why he can invent with his hands tied behind his back! He can do anything, in short, he is superior to all laws of hygiene and human energy. He is a vegetarian that doesn't know how to vegetate....

ROBERT UNDERWOOD JOHNSON[1]

On January 7, 1900, Tesla left Colorado Springs with every intention of returning. Engaging C. J. Duffner and another watchman to look after the laboratory, the inventor departed with inexplicit promises for future payment. His funds exhausted, he also left without covering outstanding bills he had incurred with the local power company.[2]

The Johnsons were thrilled with the wizard's return, and they celebrated in grand style by dining out. With Gilder's approval, Robert suggested that Tesla compose a discourse on his recent endeavors.

Coincidentally, Marconi was in Manhattan seeking investors and planning on lecturing on his progress in wireless.[3] "When I sent electrical waves from my laboratory in Colorado, around the world," Tesla reported, "Mr. Marconi was experimenting with my apparatus unsuccessfully at sea. Afterward, Mr. Marconi came to America to lecture on this subject, stating that it was *he* who sent those signals around the Globe. I went to hear him, and when he learned that I was present he became sick, postponed the lecture, and up to the present time has not delivered it."[4]

Although fearful of Tesla, Marconi was also desirous of obtaining a greater understanding of the master's equipment. With Michael Pupin as intermediary, Marconi was introduced to Tesla at the New York Science Club.[5] Pupin was in exceptionally high spirits, as John S. Seymour, commissioner of patents had finally retired. After six years of submissions, in his attempts to try and prove that his understanding of resonance and harmonics in the field of AC transmission superseded Tesla's, he had finally won. In December 1899 he applied once again for his patent, "The Art of Reducing Attenuation of Electrical Waves," and the new commissioner, Walter Johnson, sanctioned it.[6] Apparently just one month later, the trio left after dinner to visit Tesla's lab. George Scherff was working late and greeted them at the door.

"I remember [Marconi] when he was coming to me asking me to explain the function of my transformer for transmission of power to great distances," Tesla recalled. Although the inventor obviously had mixed feelings about the meeting, he nevertheless obliged with a discourse on the difference between Hertzian radiations and Tesla currents. "Mr. Marconi said, after all my explanations of the application of my principle, that it is impossible."

"Time will tell, Mr. Marconi," Tesla shot back.[7] Pupin was able to usher Marconi to the door before discussions became more heated.

"I understand completely what you are doing, Mr. Marconi," Pupin began as he walked the young Italian back to his hotel. "I would like very much to act as a consultant in your operation."

"That would be an honor," Marconi said as he discussed with Pupin a way to "persuade Signor Edison to come aboard." Marconi's reason, in particular, was to obtain Edison's grasshopper patent, which described a wireless way for jumping messages from train stations to moving trains and which Edison patented in the 1880s.

Pupin was elated. Not only was he becoming professionally involved in an exciting international wireless enterprise; he had also begun to cash in on his new AC patent. In June, Pupin received a $3,000 advance for selling the rights to John E. Hudson, president of AT&T, and a few months later he negotiated for yearly payments of $15,000 per year, for an amount totaling $200,000 for the invention Commissioner Seymour called "tautological" and "no more…than a multiplication of Tesla's circuit [that utilized principles] well understood in the art."[8] In either case, the patent enabled AT&T to perfect long-distance telephone transmissions and provided Pupin with a handsome income for many years to come. It also vindicated his position that he had understood Tesla's invention better than Tesla did.

Tesla tried again to interest the submarine designer John Holland in telautomatics; he also worked to fashion "dirigible wireless torpedoes" or small airships which could be controlled from the ground. "Everybody who saw them," he revealed a few years later, "was amazed at their performance."[9]

After putting together a prospectus and conferring with his lawyers, the inventor packed his bags for Washington to speak in person with

Admiral Higginson of the Light House Board and Secretary John D. Long of the navy. He planned not only to offer his "devil telautomata" but also a scheme for "establish[ing] wireless telegraphic communication across the Pacific." Met with ridicule and skepticism, the inventor was shuffled into what Mark Twain called "the circumlocution office." "My ideas," Tesla said, "were thrown in the naval waste basket.... Had only a few 'telautomatic' torpedoes been constructed and adopted by our navy, the mere moral influence of this would have been powerfully and most beneficially felt in the present Eastern complication [the Japanese war with Russia]."[10]

Tesla had hoped at least that the U.S. Coast Guard or Navy would come through on a smaller scale by financing the construction of modest-sized transmitters for their lighthouses and ships, but the agencies dodged any serious commitment and continued to hide behind a bureaucratic quagmire involving the need for congressional approval.[11]

"I've circumscribed the globe with electrical impulses," he told Scherff upon his return. "Let them have the Hertzian dabblers. They'll come back around my way soon enough."

"What will you do with Professor Pupin, stealing your work in alternating current?"

"He's involved in sending voice over wires," the inventor replied. "Who can be bothered."

It was at this time that Tesla commissioned an agent in Britain to locate an appropriate place for constructing a receiving station,[12] as he continued to rework blueprints for his transoceanic broadcasting system. Using his English royalties as collateral, he asked George Westinghouse for a loan of a few thousand dollars; he also tried to interest him in the wireless enterprise.[13]

Westinghouse, however, declined to get involved, but he did advance the inventor the requested funds, even though his company had overextended themselves nearly $70 million in their rapid expansion and changeover to the polyphase system. Incessant legal fees due to the never-ending litigation on patent priority battles, mostly with the countless subsidiaries were also a great drain. Swiss emigrant B. A. Behrend, author of one of the early standard textbooks on AC motors, wrote in his treatise that much to the chagrin of New England Granite's (a GE subsidiary) patent attorney, he refused to testify against Tesla, "as such evidence would be against [his] better convictions."[14]

This letter was written in 1901, *a full year after* Judge Townsend's unequivocal ruling vindicated Tesla as the sole author of the AC polyphase system (see chapter 3).[15] Now Westinghouse could finally begin to collect damages and pay back its enormous debt. George Westinghouse sent Tesla a thank you note congratulating himself "for winning the suit" and congratulating Tesla for being "awarded the credit for a great invention." Westinghouse ended the letter as follows "You know I appreciate your sympathetic interest in my affairs."[16]

In the early part of 1900, Tesla filed for three patents related to wireless communication.[17] He made several attempts to contact the elusive Colonel Astor but concentrated most of his efforts on working on an article for the *Century*. Robert had requested that Tesla write an educational piece about telautomatics and wireless communication. The plan was to decorate the essay with photographs of the remote-controlled boat and the inventor's fantastic experiments in Colorado, but Tesla had other ideas. Influenced by Western philosophers Friedrich Nietzche and Arthur Schopenhauer about such ideas as the creation of the Übermensch through activation of the will and renunciation of desire and by Eastern philosophers such as Swami Vivekananda on the link between the soul and Godhead, Prâna (life force) and Akâsha (ether) and its equivalence to the universe, force, and matter,[18] the inventor decided to compose a once-in-a-lifetime apocalyptic treatise on the human condition and technology's role in shaping world history.

Robert pleaded with him "not to write a metaphysical article, but rather an informative one," but Tesla would not listen. Instead, he sent back a twelve-thousand-word discourse which covered such topics as the evolution of the race, artificial intelligence, the possibility of future beings surviving without the necessity of eating food, the role of nitrogen as a fertilizer, telautomatics, alternative energy sources (e.g., terrestrial heat, wind, and the sun), a description of how wireless communication can be achieved, hydrolysis, problems in mining, and the concept of the plurality of worlds.

Robert was now in a bind. Neither he nor Gilder wanted to publish a lengthy, controversial, abstract philosophical essay which might damage the magazine. However, they could not simply cross out sections they were unhappy with, for they were dealing with a man who was born a genius and a friend who had contributed two previous gems that added greatly to the prestige of their publication. How to approach the hypersensitive savant was a difficult problem which Robert did not relish.

> March 6, 1900
>
> Dear Tesla,
>
> I just can't see you misfire this time. Trust me in my knowledge of what the public is eager to have from you.
>
> Keep your philosophy for a philosophical treatise and give us something practical about the experiments themselves.... You're making a task of a simple thing and for all I have said, forgive my clumsy way of saying it because of my love and

respect for you, and because I have had nearly 30 years of judging what the public finds interesting.

Faithfully yours,
(believe me never more faithfully)
RUJ[19]

March 6, 1900

My dear Robert,

I heard you are not feeling well and hope that it is not my article that makes you sick.

Yours sincerely,
N. Tesla[20]

Tesla knew what he was doing. He had decided, once and for all, to put down a significant percentage of the knowledge he had amassed into one treatise, and there was no way he was going to change it. Most likely Robert conferred with Gilder. Clearly, the essay was brilliant and original, and the more they read it, the more they realized its many layers of wisdom. The best tack to take at this point, they reasoned, was to work to clarify the piece by using subheadings, by including all of the startling electrical photos from Colorado, and the telautomaton, and by having Tesla more carefully explain the details of his inventions, and then hope for the best. The published essay began as follows:

The Onward Movement of Man

Of all the endless variety of phenomena which nature presents to our senses, there is none that fills our minds with greater wonder than that inconceivably complex movement which, in its entirety, we designate as human life. Its mysterious origin is veiled in the forever impenetrable mist of the past, its character is rendered incomprehensible by its infinite intricacy, and its destination is hidden in the unfathomable depths of the future.

Inherent in the structure matter, as seen in the growth of crystals, is a life-forming principle. This organized matrix of energy, as Tesla comprehended it, when it reaches a certain stage of complexity, becomes biological life. Now, the next step in the evolution of the planet was to construct machines so that they could think for themselves, and so Tesla created the first prototype, his teleautomaton. Life-forms need not be made out of flesh and blood.

As an environmentalist, Tesla was concerned about personal hygiene, air and water pollution, and the needless waste of natural resources. Through concentration on energy problems, solutions could be achieved. Thus, many of Tesla's inventions were created specifically to maximize efficient use of energy and prove out the principle that a self-directed thinking machine could alter the course of civilization by gaining greater control over the evolution of the planet.

In the middle of the treatise, the inventor explained in vivid detail the mechanism behind his wireless transmitter. Numerous photographs of his experiments at Colorado Springs also enhanced the impact of the message. Thirty-five pages later, he ended the treatise with a discussion of the cognitive hierarchy and the speculation that "intelligent beings on Mars... if there are [any]" most likely utilize a wireless energy-distribution system that interconnects all corners of their planet. Tesla concluded: "The scientific man does not aim at an immediate result. He does not expect that his advanced ideas will be readily taken up. His work is like that of the planter—for the future. His duty is to lay the foundation for those who are to come, and point the way."[21]

When the article appeared in the June issue of the *Century*, it created a sensation. Tesla circulated advance copies to friends, such as Mrs. Douglas Robinson, one of the founders of the Metropolitan Museum of Art,[22] Julian Hawthorne, Stanford White, and John Jacob Astor. In Astor's case, Tesla included his wireless patent applications, forwarding "this matter to your home, instead of your office [for secrecy reasons]....The patents give me an absolute monopoly in the United States not only for power purposes," the inventor continued in another letter to the colonel, "but also for establishing telegraphic communication...no matter how great the distance."[23] Those who were Tesla supporters rallied around him, *Nature* gave it a "favorable response," and the French quickly translated it for their readers,[24] but those who were against him now had a new supply of ammunition for a frontal assault.

The stage was set in March 1900, when Carl Hering was elected president of the AIEE; Professor Pupin was a close second.[25] Hering, who would also become editor in chief of *Electrical World & Engineer*, set a new tone for the electrical community. Just as he had called into question Tesla's priority work on AC a decade before, when he had backed Dobrowolsky, he also challenged Tesla's credibility in the field of wireless. Other opponents included Reginald Fessenden, who was trying to obtain competing patents on tuned circuits, and such traditional rivals as Lewis Stillwell, Charles Steinmetz, Tom Edison, and Elihu Thomson. The first potshots appeared in the *Evening Post*[26] and then in *Popular Science Monthly*.

Tesla had suggested that the sum total of human energy on the planet, which he called M, could be multiplied by its "velocity," V, which was measured by technological and social progress. Just as in physics, the total human force could be calculated as MV^2. If humans go against the laws of religion and hygiene, the total human energy would diminish. In a primitive or agrarian-based society the energy would progress arithmetically. However, if the new generation had a "higher degree of enlightenment," then the "sum total of human energy" would increase geometrically. Tesla was suggesting that with his inventions of the induction motor, AC power transmission, and his remote-controlled robots, human progress would evolve at ever increasing rates.

In a highly visible discourse under the banner title "Science and Fiction," an anonymous writer with the nom de plume "Physicist" vehemently attacked this premise. "Unhappily," this critic wrote, "Mr. Tesla in his enthusiasm to progress...neglects to state which direction is the proper one for the human mass to follow, north, south, east, west, toward the moon or Sirius or to Dante's Satan in the centre of the earth....Of course, the whole notion ...is absurd."

The editorial, which continued for six columns, called into question Tesla's invention of the telautomaton, his belief that fighting machines would replace soldiers on the field—"international bull-fights...or potato-races might do just as well"—his work in wireless, and his support for the plurality of worlds hypothesis. The author suggested that the *Century,* in future issues, should subject these types of articles to a scientific board "for criticism and revision if only for the protection against bogus inventions and nonsensical enterprises." Hurling epithets as if in combat with a mortal enemy, "Physicist" concluded, "The editors [of the *Century*] apparently impute to their readers a desire to be entertained at all costs....They evidently often do not know science from rubbish and apparently seldom make any effort to find out the difference."[27]

The onslaught continued in *Science* and in a follow-up editorial again in *Popular Science Monthly,* this time by a mysterious "Mr. X."

"*Science* (Pseudo) contains an article from XXX. 'Physicist' is not in it," Tesla wrote to Johnson, adding sarcastically, "It is also highly complimentary to the editors of your great magazine."[28] Other daily papers also attacked the inventor's controversial claims.

Tesla, however, maintained a blind eye to this credibility problem and audaciously or foolheartedly followed up this article with the infamous piece "Talking With the Planets" in *Colliers,* which we reviewed in an earlier chapter. Making no secret of his identity, Reginald Fessenden, who was now embroiled in a legal dispute with Tesla, vehemently wrote in Hering's journal that the source of "the so-called Martian signals have long been known...and only the crassest ignorance could attribute any such origin." Having at one time been "a serious obstacle to multiplex systems, [they are now all but] eliminated." Fessenden said the signals were due to "street cars, lightning flashes and the gradual electriciation of the aerial. Furthermore, the different kinds are easily distinguishable. Those ignorant of the subject might mistake them for intelligent signals."[29]

Ever since his return to New York, Tesla made repeated efforts to

Nikola Tesla at the height of his fame in 1894.

(Above) The Chicago World's Fair at night, illuminated by Westinghouse Corporation utilizing the Tesla AC polyphase system.

(Opposite above) Tesla displaying wireless fluorescent tubes before the Royal Society in England, 1892.

(Opposite below) Thomas Edison *(center)* at his Menlo Park invention factory. Seated to Edison's left is Charles Batchelor, key partner and the man who introduced Tesla to Edison, probably in France in 1883.

(Right) Thomas Commerford Martin, editor of the 1893 text *The Inventions, Researches, and Writings of Nikola Tesla,* the only collected works produced during Tesla's lifetime. (MetaScience Foundation)

(Above) The Waldorf-Astoria, where Tesla lived from 1897 to 1920. (MetaScience Foundation)

(Left) Katharine Johnson, who had a long-standing platonic love affair with the inventor. (Little Brown)

(Opposite above) Mark Twain in Tesla's laboratory in 1894. (MetaScience Foundation

(Opposite below) Robert Underwood Johnson, editor of *Century* magazine and one of Tesla's closest friends. (Little Brown)

(Right) Niagara Falls at the turn of the nineteenth century.

(Below right) Edward Dean Adams, one of Tesla's financial backers and the man responsible for the Niagara Falls hydroelectric project. (MetaScience Foundation)

(Below left) C. E. L. Brown, an important Tesla supporter and the first engineer to transmit AC polyphase currents over long distances. (MetaScience Foundation)

(Above) The patent plaque at Niagara Falls, listing Tesla as the primary inventor of the AC polyphase system. (Marc J. Seifer)

(Left) Charles Steinmetz, initially a Tesla supporter and then one of the inventor's most vigorous opponents throughout their lifetimes. (Meta-Science Foundation

Nikola Tesla displaying his invention of the cold wireless lamp without filament. (*Electrical Review*, 1898).

rekindle his friendship with Astor, but the gadabout was proving difficult to corner. Over the summer, the Johnsons tried to woo the inventor to Maine for a vacation, but he was too intent on contacting the multimillionaire.

August 2, 1900

Dear Mr. Tesla,

I have been thinking of you all day and all evening as I do so often.... I sat on a little hillside this afternoon looking over green meadows to the sea beyond...and wishing that I could loan you my eyes that you might have my visions and drink in the beauty of the day.... You are as silent as only you know [how] to be.... Do call us.

Yours faithfully,
Katharine Johnson[30]

August 12, 1900

My dear Mrs. Johnson,

Just a line to tell you that I never can and never shall forget the Filipovs—they have given me too much trouble.

Yours sincerely,
N. Tesla[31]

Unable to relax until he settled things with Astor, Tesla tried again, having forfeited the chance for a needed respite.

August 24, 1900

My dear Colonel Astor,

I still remember when you told me, that if I could only show you great returns on your investment, you would gladly back me up in any undertaking, and I hope not in a selfish, but in a higher interest that your ideas have not changed since.... [Re: oscillators, motors and lighting system] not less than $50,000,000 [can be]...made out of my invention. This may seem to you exaggeration, but I honestly believe that it is an understatement.

Is it possible that you should have something against me? Not hearing from you I cannot otherwise interpret your silence....

Finally, Astor replied, stating that he was "glad to get your letter, and will get back to you."[32] But this was really a decoy procedure, for Astor continued to slip away. Tesla shot off another round of letters outlining his progress with his oscillators, fluorescent light—"the commercial value...if rightly explored, is simply immense"—and wireless enterprise,[33] but again Astor balked.

Astor never directly told Tesla his real feelings. His reluctance in advancing the partnership suggests that he was angry with Tesla for *not* exploiting the oscillators and fluorescent lights in 1899, as he had promised, but had instead run off to Colorado Springs to conduct his wireless folly.

Certainly the onslaught in the papers injured Tesla's reputation, but it is this author's belief that the attack by the press had little if anything to do with Astor's turnabout. Tesla had deceived him. As wealthy as he was, Astor wanted to invest in a sure thing. The oscillators and the fluorescent lights seemed practically ready for market, but rather than perfect these inventions, Tesla took the capital to go off on another venture and returned without a dime. Astor was enraged but too much of a gentleman to even let Tesla know. With Stanford White and Mrs. Douglas Robinson behind him, Tesla explored a fresh lead.

29

THE HOUSE OF MORGAN (1901)

J. P. Morgan towered above all the Wall Street people like Samson over the Philistines.

NIKOLA TESLA[1]

In May 1900, Gentleman Jim Corbett was KO'd by James Jeffries in a championship bout held in Coney Island. Tesla, an avid boxing fan, probably attended. Back at the hotel, a Serbian youth with a familiar name had left a message. It was Anna's son, the one and only girl Tesla had ever fallen in love with. Through the years they had maintained contact, so Tesla had been notified of the boy's arrival. However, he was not prepared for the career that the lad had chosen.

"I want to be a boxer," the boy proclaimed.

Unnerved by this announcement, Tesla conferred with Stanford, who helped set up the youth at a boxing school near the Garden. Every so often Tesla would go down to the gym to follow the boy's progress; finally, it was decided that he was ready to enter the ring. Stanford had done his best to set up a reasonable match, but the youth persisted in seeking out a tougher opponent.

One blow knocked the boy unconscious, and he died shortly thereafter in the hospital. "Tesla grieved for him as though he was his own son."[2]

In the fall of 1900, J. Pierpont Morgan announced the wedding of his daughter Louisa to Herbert Satterlee, Morgan's latter-day biographer. It was a magnificent event with a guest list numbering twenty-four hundred. The Serbian wunderkind felt quite at home at the gala occasion, for many of his friends were there, including John Jacob and Lady Astor, Mrs. Douglas Robinson and her brother Teddy Roosevelt (whom Tesla had met at Mrs. Robinson's Madison Avenue home in March of 1899), William Rankine, Edward Dean Adams, Darius Ogden Mills, Chauncey DePew, and Stanford White. Other guests included Jacob Schiff, Henry Clay Frick, Grover Cleveland, August Belmont, President William McKinley, and Thomas Edison. Morgan was in an exceedingly good mood and personally greeted each guest with a warm handshake.[3] "I read your article in the *Century,* Mr. Tesla, and was very impressed."

Coincident with Tesla's impending liaison with the House of Morgan and recent triumphant return from Colorado Springs, his handwriting and signature began to display a frivolous abundance of ornate embellishments. Although these samples were written in moments of gaiety in letters to the Johnsons,[4] they nevertheless reveal a subconscious, qualitative change in his state of mind as compared to his usual lean, bare-bones handwriting. Graphologists note that "the paper [is] frequently treated as a substitute object.... [Thus] the graphically expansive [writers]... usually are the same who not only dominate the paper, but also their environment, [just as] the graphically timid ones are also timid in other respects."[5]

One could therefore speculate that Tesla was a very visible character at the affair. And just as Tesla adorned his signature, he adorned his body, wearing the latest suits, top hat, cane, and white gloves. He took extreme pride in being the leader of his field and one of the best-dressed men to walk Fifth Avenue. Now tending toward flamboyance, the inventor began to identify more heartily with the opulence and power that surrounded him.

Twenty-eight-year-old Anne Tracy Morgan, Louisa's younger sister, was particularly taken with the dashing inventor, and they began a friendship.

"The Thanksgiving dinner at [Morgan's home that] year was an unusually large and gay affair with its traditional four varieties of pies,"[6] and Tesla was invited to the day-after event, held Friday evening.[7] Anne may have seen this as an opportunity to extend their friendship; they would come to exchange letters for the duration of their lives—but Tesla saw it as a business opportunity. The wizard brought along fascinating multicolored electric bulbs that emanated dancing spiderwebs of lightning, static-electricity devices that made a person's hair stand on end, and other wireless paraphernalia. The inventor exchanged hellos with J. P. Morgan Jr., now in his early thirties, and gave photographs of his work at Colorado Springs to Anne.

After dinner, Morgan met Tesla privately in order to discuss a possible partnership. The tenor of the times is discussed by Herbert Satterlee, a man who knew Tesla personally. Having written a virtual daily log of Morgan's life, Satterlee deliberately deleted any reference to the Morgan-Tesla liaison; but the following paragraph, which coincides precisely with this time, appears to be a justification for the financier's decision to back the wireless venture: "The dying year saw the completion of many combinations of the smaller companies in the steel industry....They were all getting rich. Gates was speculating in Wall Street. Judge Moore began to buy fine horses....Conversely, Reid and the others invested in large country estates....[And Morgan gambled on an oddball inventor.] Dinners at the Waldorf-Astoria and at Sherry's and lavish entertainments were the order of the day, and everywhere there was evidence of rapidly accumulated wealth. They all seemed to think that there was no end to it."[8]

Known among the clique for his collection of mistresses, Morgan extended his passion to amassing an immense hoard of treasures, including ancient coins, precious stones, tapestries, carvings, rare plates, the paintings of the masters, statues, old books, and original manuscripts. Some of his most prized possessions included first drafts of Charles Dickens's novels, a portrait of Nicolaes Ruts by Rembrandt, a number of eleventh-century Byzantine medallions, and a Gutenberg Bible.[9] Hanging in the study was his latest acquisition, *Christopher Columbus* by Sebastiano del Piombo.[10] It was hung next to a painting of the Commodore's three-hundred-foot-long yacht, which Morgan often preferred as a sleeping quarters to his home when it was docked near Wall Street during the boating season. Tesla eyed the del Piombo in great admiration.

"Mrs. Robinson has talked me into donating it to the Metropolitan Museum. Naturally I hate to part with it, but you know how persuasive she can be."

The skittish Tesla had seen Morgan at close range before, but never for long periods in so intimate a way. Plagued from his youth by a series of skin conditions, Morgan's beet-red and deformed proboscis, retouched out of all official photographs, was often swollen and coated with warts. An art dealer, encountering Morgan in a similar circumstance, was quoted as saying:

> I was unprepared for the meeting....I had heard of a disfigurement, but what I saw upset me so thoroughly that for a moment I could not utter a word. If I did not gasp I must have changed colour. Mr. Morgan noticed this, and his small, piercing eyes transfixed me with a malicious stare. I sensed that he noticed my feelings of pity, and for some time that seemed centuries, we stood opposite each other without saying a word. I could not utter a sound, and when at last I managed to open my mouth I could produce only a raucous cough. He grunted.[11]

"I want to know, Mr. Tesla," Morgan began, eyeing one of the inventor's Colorado photographs, "how you survived among all of this lightning."

"I didn't," Tesla said, avoiding a direct stare. "Those are multiple exposures."

"How clever. White tells me you want to build a wireless tower?"

"I have perfected an apparatus which permits the transmission of messages to any distance without wires, making long and expensive cables as a means of conveying intelligence commercially obsolete. This creation also enables the production and manipulation of hundreds of thousands of horsepower, bringing instruments on any point of the globe into action regardless of their distance from the transmitter."

"Instruments?"

"Telegraph keys, phones, clocks, remote photography."

"You have a wireless means for transmitting pictures?" Morgan rebounded, raising his eyebrows.

"There is nothing novel in telephotography. Edison has been working on it since Elisha Gray's device was presented at the 1893 Exposition. My patents simply usurp the need for using wires."

"Don't push my tolerance, Mr. Tesla. Your proposal as far as I understand only deals with telegraphy. I'm a simple man who wants a way to signal incoming steamers during times of fog, to send messages to Europe, maybe get Wall Street prices when I'm in England.[12] Can you do this? Can you send wireless messages such long distances?"

"Indeed I can, Mr. Morgan."

"And the problem of billing? Wouldn't anyone with a receiver have free access to this information? I'm not about to subsidize my competitors, or the public for that matter."

"I can guarantee absolute privacy for all messages. Broad rights have been secured which gives me a monopoly in the States and most of Europe."

"How 'broad' are your costs?"

"Although this work concerns a decade of effort, I know that I am in the presence of a great philanthropist, and therefore do not hesitate to leave the apportionment of my interest and compensation entirely to your generosity."

"Don't flatter me, Mr. Tesla. Let's get down to brass tacks. What will it cost?"

"My plan requires two transmitting towers, one to transmit across the Atlantic and the other across the Pacific. The former would require an expenditure of approximately one hundred thousand dollars, the latter about a quarter of a million."

"Let's talk about one ocean at a time. What would I get for funding the construction of a wireless plant to cross the Atlantic?"

"Its working capacity would equal at least four of the present ocean cables and it would take six to eight months to complete."[13]

"What about Marconi? Stetson says his costs are one seventh of your quote."

"That is so. However, there are key elements missing for his success, elements which can only be found in my patents, in apparatus universally identified with my name and published in writings dating back to 1890 and 1893, when Marconi was still pulling his mother's apron strings."

"He transmitted fourteen hundred words from ship to shore, right here in New York during last year's America's Cup races. I know, I was there. I saw his equipment."

"Mere child's play. He's using equipment designed by others, and also the wrong frequency. The slightest changes in weather will disrupt his messages, and he has no device for creating separate channels. I have tested his Hertzian frequencies at length, Mr. Morgan, and believe there is no commercial viability in them whatsoever."

"Why exactly are they so wrong?"

"For one example, they do not make use of the natural electrical properties of the Earth. The Tesla currents on the other hand are tuned to the frequencies of our planet. These are continuous waves, not pulsed interruptions. In short, my way is best for transmitting substantive information, and insuring total privacy."[14]

"I have a handful of articles with Marconi's pictures all over them which appear to disagree with you. The British Post Office is using the Hertzian method.—Here's a newspaper report I picked up in England from an Admiral who has used Marconi's transmitters for distances exceeding 80 miles: 'Our [ship] movements have been directed with an ease and certainty and carried out with a confidence which, without this wonderful extension of the range of signalling, would have been wholly unattainable. It is a veritable triumph for Signor Marconi.'[15] And I have articles which question whether you have ever sent messages beyond the confines of your laboratory."

"I see that I have taken up enough of your time," the inventor said, looking at his watch. "I thank you very much for your hospitality."

"I'm not saying we can't do business, Mr. Tesla, but I am going to have to think this over."

"Very good."

Upon Tesla's departure, Morgan took out a deck of cards and went through his nightly ritual of playing solitaire. Before him was another file on Tesla's patents, but these were not in wireless: "Mr. Tesla's discoveries do away with the carbon filament...[He] explained that by creating an electrostatic field, [cold vacuum] tubes could be hung anywhere in the room. [They can not burn out, because there is no filament to be destroyed]....The estimated manufacture of incandescent lamps is 50,000 a day...."[16]

December 10, 1900

Dear Mr. Morgan,

Appreciating the immense value of your time...I have withdrawn more or less hastily last Friday preferring to make a few condensed statements at long distance which with a small effort on your part will put you in the possession of knowledge I have gained only after a long and exhaustive study.

This lengthy letter, one of the first of many, began with a quote from Professor Adolph Slaby of Germany, who referred to Tesla as "the father of this telegraphy," and also included quotes from Lord Kelvin and Sir William Crookes on other developments in the field, such as in the construction of his oscillator for generating wireless frequencies. The letter also pointed out Tesla's legal position, having patented all of the fundamentals of the process, in America, Australia, South Africa, and Europe, and noted the specific flaws of Marconi's system (pointed out above). "Apologizing for this digression...I beg you to bear in mind that my patents in this still virgin field, should you take hold of them...will command a position which, for a number of reasons will be legally stronger than that held by those of my own discovery in power transmission by alternating currents."

Tesla ended the letter with a rousing challenge: "Permit me to remind you that had there been only faint-hearted and close-fisted people in the world, nothing great would ever have been done. Rafael could not have created his marvels, Columbus could not have discovered America, the Atlantic cable could not have been laid. You of all should be the man to embark on this enterprise...[which will be] an act of inestimable value to mankind."[17]

THE FIRST BILLION-DOLLAR TRUST

Perhaps the most outrageous character living in the Waldorf was the cigar-chomping robber baron and stock market manipulator, John W. (Beta-Million) Gates, co-owner of the American Steel & Wire Company. During an average day, Gates could win or lose $40,000 in a poker game, and on some occasions, the swing could be ten times that much. Conferring with Henry Clay Frick, another Waldorf occupant and occasional poker companion, Gates helped choreograph one of the biggest deals of the century.

On December 12, a dinner for the steel magnates was held in honor of Charles Schwab at the University Club. Sponsored by Schwab's boss, Andrew Carnegie, the attendees included J. Pierpont Morgan, Edward H. Harriman, August Belmont, Jacob Schiff, John W. Gates, and Carnegie's first manager, Henry Clay Frick.

In an unrehearsed speech after the meal, with Carnegie now absent from the meeting, Schwab expounded, in a clear and forthright manner, on the advantages of creating a giant steel trust.[18] After discussions which lasted until three in the morning, Morgan began to realize the great benefit of Schwab's plan, and within a few months he finalized the merger, placing Schwab at the head of the new $1.4 billion steel trust, the first company ever to be capitalized at over a billion dollars. Carnegie received approximately $226 million, Frick, $60 million, and Rockefeller, for his iron mines, $90 million. Gates, the "gargantuan gambler," playing it like a poker hand, held out as long as possible, until Morgan threatened to erect a wire company without him, and so he, too, walked away with a hefty profit. By March 1901, with the creation of the new corporation, Morgan was able to add steel to his portfolio, which at this time included electrical, shipping, mining and power industries as well as telephone, railroad, and insurance conglomerates.[19] Political cartoons of the day portrayed him as an Atlas with the earth on his back or as a Goliath towering over less powerful individuals, such as the king of England, the German Kaiser, or the president of the United States.

In reaction to "Morganization," anarchy started to become a viable political alternative. Furthermore, although Morgan stood for strength and stability in business, in reality the creation of U.S. Steel was a magnificent gamble. Carnegie knew this and thus stated: "Pierpont is not an ironmaster. He knows nothing about the business of making and selling steel. I managed my trade with him so that I was paid...in bonds, not stocks! He will make a fizzle of the business and default in payments of the interest. I will then foreclose and get my properties back."[20]

Schwab also feared this. Within two years, the crafty conciliator had resigned from the firm to take charge of Bethlehem Steel, a much smaller gem that was sturdy and profitable.

Thus, Morgan was to have his headaches with the steel monopoly partly because of market problems and mostly because of labor disputes, particularly a strike which nearly crippled the company. Probably the main reason why U.S. Steel succeeded was because of the invention of the automobile, which created an enormous new market.

To offset the possibility of the great conglomerate folding and to raise the potential for more revenues, Morgan "enlisted" the famous stock manipulator James Keene to create an artificial interest. Keene bought and sold large blocks of U.S. Steel to bogus investors in order to create the illusion of bullish interest.[21] The sham worked, and within a few weeks the New York Stock Exchange experienced the most active trading days in history. "Big Steel's common stock, which had been offered on the market at 38, rose almost immediately to 55 and Pierpont Morgan became the hero of the financial world and the principal demon of those who feared and hated monopoly."[22]

THE TESLA DEAL

Tesla met with Morgan in the midst of these steel negotiations in attempts to solidify a deal. At the height of Christmas season, he took a cab to Morgan's office at 23 Wall Street and handed the financier some of his specifications.

"Mr. Morgan, my plan, and my patents, which I offer to you, will command a position legally stronger than that held by the owners of Bell's telephone or by the holders of the patents based upon my discovery in power transmission by alternating current."

"Send me the paper work, and I'll look it over."

"Sir, in view of the intense activity in this field, it is desireable that I should be placed without delay in a position enabling us to profit from my advanced knowledge."[23]

Morgan moved over to the window and stared down Wall Street. "If I agree to help build your station to transmit across the Atlantic, I want it understood that I am,"—and then he turned to face the Serb and lowered his voice to barely a whisper—"a silent partner.[24] Do you know what that means, Mr. Tesla?"

"Yes, sir, I do."

"Good. I want to be frank. I do not have a good impression of you.[25] You abound in controversy, you are boastful, and aside from your deal with Westinghouse, you have yet to show a profit on any other creation. On the other hand, I appreciate your talents, so let me put my cards on the table. If we proceed, whatever figure we decide upon shall be firm. I will not be bilked for continuing research funds."[26]

"It is not money I seek, although these inventions, in your strong hands, with your consummate knowledge of business, can be worth an incalculable amount. You know the value of scientific advancement and artistic creation. Your terms are mine."[27]

"That is not good enough. Give me specifics. Give me a figure."

"As I stated at our first meeting, I think a hundred thousand dollars will suffice for the construction of a ninety-foot-high transatlantic transmitter."[28]

"Let's be certain on this. Shall we say one hundred and fifty thousand for the erection of said transmission tower and a fifty-fifty split on the company's stock?"[29] Morgan reached for his checkbook, wrote out a down payment, and handed it to the inventor.

Overawed at his good fortune and humbled in the presence of the king, Tesla could not stop himself. "Let the control be yours, Mr. Morgan. I insist that you take fifty-one percent and I forty-nine."[30]

"You are a strange man indeed. All right, it's a deal. After the papers are drawn up, you may draw on the House of Morgan as the need arises for the full limit."[31]

January 3, 1901

My dear Colonel Astor,

Hearty wishes for the new Century.... Mr. Morgan's generous backing, for which I shall be grateful all my life, secures me my triumphs in wireless telegraphy and telephony, but I am still unable to put my completed inventions [oscillators, fluorescent lights] on the market. I can hardly believe that you, my friend since years, should hesitate to join me in introducing them when I can offer you ten times better returns on your investment than anyone else.

Yours sincerely,
N. Tesla[32]

The Fine Print

March 5, 1901

Dear Mr. Tesla,

I beg to acknowledge receipt of your letter of the first instant together with assignments of an interest in various patents as shown upon the schedule and assignments handed me therewith and to confirm the understanding therein expressed.

Very truly yours,
J. Pierpont Morgan[33]

The octopus was not content with creating a simple partnership in wireless. Unbeknownst to the inventor, Morgan wanted the lighting concern as well and control of Tesla's patents. Boldly, these were added to the agreement. Tesla was now placed in a difficult situation, since Astor was the principal backer in the other enterprise and the inventor had not planned to include the actual patents as part of his collateral. "When I received your formal letter," Tesla wrote three years later, "it specified an interest in fifty-one percent in *patents* on these inventions. That was different though my share was the same. It was a simple sale. The terms were entirely immaterial to me and I said nothing, for fear of offending you. You have repeatedly referred to some stock, and it is just possible, that a mistake was made."[34] Rather than confronting his new benefactor head-on, Tesla acquiesced.

February 18, 1901

Dear Mr. Stedleg [Morgan's intermediary],

....I need scarcely say that I would sign any document

approved by Mr. Morgan, but believe that there exists a misunderstanding in regards to my system of lighting which was not included in the original proposition.

Rather than try to amend the agreement to remove this vital concern, which, in Tesla's words in the same letter, "will create an industrial revolution," the inventor pointed out the great advantages of the lighting venture and included a promotional announcement entitled "Tesla's Artificial Daylight." He concluded by stating that "besides myself, Col. Astor is interested.... [Thus] it will be necessary for me to comply with a formality before making the assignments. I shall attend to this matter at the earliest possible moment."[35]

A month before, perhaps in anticipation of this problem, Tesla had appealed once again to his original benefactor:

January 11, 1901

My dear Colonel Astor,

Since Mr. Adams and his associates are entirely out, I have practically nobody but you and Mr. Morgan with me.... Please let me hear from you.... With me you are not with some wild syndicate, but with a man to whom your name, credit and interest are sacred.[36]

Astor called on the telephone a week later. He told the inventor that he was concerned that Tesla did not have fundamental patents and that other inventors might have priority, particularly on the wireless enterprise.

"Do not be misled by what the papers say, Colonel, I have the controlling rights. Why not come in with Mr. Morgan and myself."[37]

Astor avoided making any definitive statements, and so Tesla apparently simply attached the lighting concern to the wireless deal with the commodore.[38] Now Morgan controlled the fundamentals behind two completely independent *new* industries. Tesla could hardly complain; he had agreed to the proposal. All he had to do was succeed with the capital now provided.

March 5, 1901

Dear Mr. Steele [another Morgan intermediary]:

Now that all dangers of conveying a wrong impression to Mr. Morgan is removed by his kind acceptance of my proposal, I would call to his attention that I consider my fundamental patents on methods and apparatus for the wireless transmission of energy as the most valuable patents of modern times and as to my system of lighting, I am convinced that it constitutes one of the most important advances and is of enormous commercial value.

Yours sincerely,
N. Tesla[39]

On the thirteenth of the month, always a favored day for the superstitious wizard, Tesla paid Westinghouse back a note for $3,045. He was out of debt and on his way.[40]

30

WORLD TELEGRAPHY CENTER (1901)

Dear Mr. Morgan,

How can I begin to thank you in the name of my profession and my own, great generous man! My work will proclaim loudly your name to the world. You will soon see that not only am I capable of appreciating deeply the nobility of your action, but also of making your primarily philanthropic investment worth a hundred times the sum you have put at my disposal in such a magnanimous and princely way!

With many many wishes from all my heart for your happiness and welfare, believe me.

Ever yours most faithfully,
N. Tesla[1]

In March 1900 there was a fire in the East Houston Street building that housed Tesla's laboratory. "The Jews on the lower floor [were] burned out...[and this] frightened me nearly to death," Tesla wrote the Johnsons. "It was a close shave, and if the misfortune had happened, it would have been probably the last of your friend Nikola."[2] Throughout this period the onslaught by the press also continued in an unbridled fashion.

February 25, 1901

My dear Mr. Tesla,

We never forget old friends and defend them against all malicious assailants at all hazards.

Yours truly,
Earnest Heinrich

One of the old guard from the Westinghouse corporation, Heinreich, included a newspaper clipping which he had written. "Anyone who is ignorant," Heinreich wrote, "...does not know that Tesla stands in the front rank of electrical inventors by what he has actually accomplished."[3]

Another who came to Tesla's defense was T. C. Martin, who authored a word of praise in *Science*. "The ship was off its course," Tesla wrote, "but I always had faith in the captain."[4]

December 13, 1900

My dear Tesla,

I am delighted to get your kind favor of Dec. 12. I know of no change whatever in my sentiments towards you these many years from the beginning until now. I shall always be very proud of my modest association with your earlier work....

As ever,
T. C. Martin[5]

Unfortunately, there was also bickering between Tesla and Martin about the previous editorial attacking Tesla in Martin's journal and about Tesla's slow progress on his other inventions. Concerning the vacuum lamps, Martin wrote: "I should be delighted to see you or any other man give us the commercial art." Perturbed, Tesla cut off his correspondence, and so the friendship remained impaired.[6]

Just three months later, in March 1901, Tesla invited an admirer and disciple of Swami Vivekananda, Miss Emma C. Thursby, to his laboratory. "My light will then be permanently installed and you and your friends—Miss Farmer in particular—will be most welcome to see it."[7]

Tesla's New Surprise

Julian Hawthorne

Great preparations for an experiment upon a stupendous scale are being made at the wizard's laboratory on East Houston Street....An unannounced visitor gained entrance today by chance. Tesla was not there. But what the visitor saw chained him to the spot.

A Wonderful Color

From a stout beam [from] the...ceiling hung three dazzling, pulsating clots of purple-violet light. The room glowed with the warmth of a strange, unearthly rich color—a hue that is not listed in the spectrum. Above and below the beams twisted long glass spirals closely coiled—snakes of beating violet flame....

Sudden Darkness

> One of Tesla's workman found the unannounced visitor spellbound. A quick spring to the wall, a concealed button touched, and darkness.
>
> Those who knew say this violet light is the wizard Tesla's new flash signal to the Martians. He will reveal it to the world soon. It is even hinted around the corner of Mulberry St. at Police Headquarters that Tesla has already wig-wagged the red planet and had a response.[8]

Hawthorne was living in Yonkers and often took a train down to Manhattan, dining with Tesla at the beginning of the year.[9] They shared a number of common friends, including Stanford White, whose father, Richard Grant White, had confessed to Hawthorne once "among other sacred confidences of a woman whom he had found and loved in New York." The rumor, in essence, was that he had been essentially a bigamist, maintaining one home for his family and one for his concubine.[10] Perhaps this explained Stanny's penchant for philandering. The son, however, easily surpassed the father, as he was able to maintain five or six retreats, including an estate on Long Island, an apartment in Gramercy Park, the Garden Tower suite, "the Morgue" on West Fifty-fifth Street, which he and his compatriots, like Saint-Gaudens, "used in a pinch," and his "most infamous haunt," at 22 West Twenty-fourth St.[11]

In March, Stanford became smitten by the exotic charms of the sixteen-year-old Floradora siren, Evelyn Nesbit, who had been featured as a Gibson centerfold for *Colliers* and as a nearly bare-breasted Spanish dancer in a popular musical on Broadway. White watched her perform night after night for many weeks before he was able to arrange a rendezvous, which first took place in the heat of the summer at the Twenty-fourth Street studio.[12]

"Stanford had me put in an electric door," the inventor told Hawthorne. "You press a button, and it automatically opens."

White had decorated his bachelor den in shades of red, with velvet curtains on the windows, soft cushions on the floor, and tapestries, statues, and paintings, mostly nudes all around. In his loft, with the room set like a small forest, lit with a bright skylight, could be found a red velvet swing hanging from the ceiling, like one of Chaucer's toys, with green ropes trailing down from the seat, like vines from a tree.[13]

Besides meeting with the mystical son of the renowned Gothic author, Nathaniel Hawthorne, Tesla dined with Stanford White and sometimes Mark Twain at the Players' Club, or with Spanish-American War hero, Richmond Hobson, Rudyard Kipling, or, of course, the Johnsons throughout this period. Katharine was caught up in her interest in spiritualism and tried an experiment in thought transference without the inventor knowing. In jest, Tesla wrote back:

> My dear Mrs. Johnson,
> There was no telepathic influence this time. I never thought of you even for a moment.
> Sincerely,
> Millionaire Kid![14]

Rooted in his materialistic philosophy, the superstitious "wizard who talks with other worlds"[15] continued to repudiate the notion that human minds could interact by extrasensory means, even though he had recently saved some friends from a train wreck because of a premonition.[16] Overtly, however, he would maintain that psychic phenomena was poppycock. Katharine would not only be teased for her mystical bent but also over possession of Hobson and over her looks.

> My dear Luka,
> When Mrs. Filipov is out of town I think of Mrs. de Kay as the most charming lady of my experience. It would be advisable, Luka, to keep both ladies in ignorance of this. A word to the wise is sufficient.[17]

Considered much like an uncle to the family, Tesla also expressed his continuing affection for the Johnson children. For Agnes, he would sign a New Year's card "Nikola Hobson," and for Owen, he took the time to read the youngster's first novel, *The Arrow of the Almighty*. He also congratulated Owen on his impending marriage. Agnes would also later marry, and she would inherit the vast Johnson correspondence, much of which was donated to Columbia University. According to the present Mrs. Robert Underwood Johnson, the wife of Owen's son (named after the grandfather), Agnes was "awful. I didn't like her at all. Her daughter, however, was very beautiful. Paderewski felt that Ann had great talent...Owen was very dashing and attractive, and had a lot of his mother's qualities. As a writer, he authored the Lawrenceville series and made a good living as a novelist."

The present Mrs. Johnson stated that Katharine had "an Irish personality." She could be "gay and lively and fun loving, but also depressive underneath." The present Robert Underwood Johnson lived with his grandparents. "They had two Irish servants, Josie and Norah. Katharine would go into one of her moods, and just stay in her room and wouldn't come down even for meals. Her depression became more severe...after World War I." She said that Robert was considered to be "boring, very formal with old world manners...a fine old gentleman. Katharine was attracted to Tesla because he was imaginative and exciting from a European point of view. He might have brought more gaiety into the house."[18]

Wardenclyffe

> "Inventor Nikola Tesla has purchased a 200-acre tract at Wardenclyffe on the Sound, nine miles east of Port Jefferson for the establishment of a wireless telegraphy plant. The land and improvements will cost $150,000."[19]

On March 1, 1901, Tesla officially signed his contract with Morgan. He was now able to begin construction of his laboratory and tower on Long Island, sixty-five miles from New York City. Two days later, Morgan officially announced the creation of U.S. Steel. No such announcement was made about the Tesla Company. The above article, which appeared in a local paper, the *Long Island Democrat,* was perhaps the only one to make reference to the correct figure of $150,000, which Morgan provided. When John O'Neill wrote his biography in 1944, he did not know the details of the Morgan-Tesla relationship, even though he had personally known Tesla for over thirty years. The inventor, whose papers were still under lock and key at the time of the completion of the biography, had told O'Neill that the financier provided the funds in his capacity as a philanthropist, although this was not the case at all. It was a simple business partnership.

Tesla celebrated the new connection by giving a large party at the Waldorf-Astoria. He had discussed with Oscar the details of the menu and participated in tasting the various sauces. Impeccably dressed, he reserved one of the smaller banquet halls, requesting his guests to arrive at seven-thirty *sharp.* White was probably there, along with the Johnsons, Hobson, and perhaps Miss Merrington or Vivekananda acolytes Miss Thursby or Miss Farmer or Anne Morgan. As legend tells it, when it came time for dinner to be served, the maître d' was forced to call Tesla aside to inform him that he owed the hotel back bills totaling more than $900. He was under orders. Dinner could not be served unless the matter was straightened out first. With an ace up his sleeve, Tesla nonchalantly welcomed his guests and then eased out to see the manager. Mr. Boldt was cordial but insistent, so Tesla made a call and put him on the phone with Morgan. Flustered, Boldt was nevertheless able to hold his ground. A check was sent immediately, and the inventor was saved from embarrassment.[20]

Soon after, Tesla met with real estate mogul Charles R. Flint, who arranged a meeting with James Warden, director of the North Shore Industrial Company. Warden, who was in control of an eighteen-hundred-acre potato farm on Long Island Sound, in Suffolk County, provided Tesla with two hundred acres adjacent to what is today called Route 25A. The inventor was also given the option to purchase the remaining parcel.

Perhaps to sweeten the deal or in lieu of certain other arrangements, the cite was named Wardenclyffe, after the owner, and a post office under that appellation was established on April 2. Five years later, in 1906, the name was officially changed to the Village of Shoreham.[21]

Electrical World & Engineer reported: "The company is offering its stock for sale at $100 per share, expecting to pay 15 per cent dividends.... The Wardenclyffe Building Company shall have the first right and privilege to do all building and make all constructional improvements...and shall have the first right of purchase of any additional land offered by it for sale." Warden, who was interviewed for the article, predicted that "large profits will be realized in the future." Describing Tesla as "the foremost electrician of the age, whose achievements in electrical science eclipse in practical importance all other discoveries of the century," Warden noted that the inventor "has just closed a contract to expend...a very large sum of money in constructing electrical laboratories and the main station for his wireless telegraphy system of communication with Europe and Australasia. This development will require a large number of houses for the accommodation of the several hundreds of people whom Mr. Tesla will employ."[22]

Tesla's ultimate plan was to construct a "World Telegraphy Center," with a laboratory, wireless transmitter, and production facilities for manufacturing his oscillators and vacuum tubes. He had negotiated with Morgan the first step, that is, to build the laboratory and a simple tower for reporting yacht races, signaling ocean steamers, and sending Morse-coded messages to England. Simultaneously, he discussed with McKim, Mead & White the construction of an entire metropolis, "a model city," using the eighteen-hundred acres available, with homes, stores, and buildings to house upwards of twenty-five hundred workers.[23] "Wardenclyffe will be the largest operation of its kind in the world," Tesla told the local newspapers. "The laboratory will draw men from the highest scientific circles and their presence will benefit all of Long Island....[24] With a staff of 75 draftsmen, the eminent architectural firm of McKim, Mead and White is well suited to the task," Tesla concluded. They billed him $1,168 for blueprints.[25]

White, it appears, was placed in a precarious position: Morgan still had reservations about his connection with the flamboyant engineer. In his capacity as interior decorator, White, in February, had located a statue in London which he knew the financier was interested in. "My dear Commodore," the engaging redhead wrote, "it is really like parting with a piece of my heart to give it up....[Thus] I would honestly rather [present it to you, than sell it]...as my only desire...is to please you." But Morgan insisted on compensating him for the statue, paying White "double the normal commission" in order to show that he, too, felt their affiliation transcended "business."[26]

April 26, 1901

Dear Nikola,

I send you with this a new revised plan for your power house, also a very close estimate which we have made ourselves in this office....The work could be done for about $14,000. I am very sure for a building of this size, that it would be impossible to get these figures down lower.

Affectionately,
Stanford White[27]

Bids were taken from Sturgis and Hill and also Mertz and Co., two contractors White often worked with. In May blueprints for the laboratory were sent over to Tesla for his approval, in June a contract with Sturgis and Hill was finalized,[28] and in July the land was cleared, and a road was put in.[29] White recommended one of his associates, W. D. Crow, as architect in charge. Crow would also do the actual construction of the tower.[30]

The sun was out bright one morning in the early spring as the inventor strutted down Peacock Alley and up Fifth Avenue toward Forty-second Street and Grand Central Station, to transfer to Pennsylvania Station and take an early train to Wardenclyffe. Golden shafts of light billowed through the upper windows of the cathedral-like edifice as Tesla crossed the mighty corridor and stepped aboard the luxury car. Ordering a pot of coffee, he began to peruse his mail. The train rattled out of the city, past Manhassett, Oyster Bay, where Vice President Roosevelt lived, St. James near Smithtown, where White had his family estate, Port Jefferson, and finally to Wardenclyffe. With stops at the various towns, the trip took about an hour and a half, the Connecticut shoreline sometimes visible across the Sound. It was when he reached page 280 of the *Electrical Review* that his jaw dropped and his cup of coffee spilled all over the prim white tablecloth.

Syntonic Wireless Telegraphy

Guglielmo Marconi

A large amount of inacurrate and misleading information is being published...even in the scientific press...[on] telegraphy through space....I shall endeavor to correct some of the[se] misstatements....

It is my intention to describe fully the efforts I have made in order to tune or syntonize the wireless system, efforts which, I am glad to say, have been crowned with complete success....

I first constructed an arrangement which consists of a Leyden jar or condenser circuit in which included the primary of what may be called a Tesla coil [and a] secondary [which] is connected to the earth or aerial conductor. The idea of using a

> Tesla coil to produce oscillations is not new. It was tried by the Post Office [i.e., with Preece] when experimenting with my system in 1898 and also suggested in a patent specification by Dr. Lodge dated May 10, 1897 (No. 11,) and by Professor Braun in... 1899.[31]

Tesla described to Morgan in a letter written three years later how this information affected him:

> When I discovered, rather accidentally, that others, who openly cast ridicule on what I had undertaken and discredited my apparatus were secretly employing it, evidently bent on the same task, I found myself confronted with wholly unforeseen conditions.... Your [Morgan's] participation called for a careful revision of my plans. I could not develop the business slowly in grocery-shop fashion. I could not report yacht races or signal incoming steamers. There was no money in this. This was no business for a man of your position and importance. Perhaps you have never fully appreciated the sense of this obligation.[32]

This passage displays misunderstanding by Tesla of Morgan's personality. Unlike the inventor, whose ideas existed in abstract and futuristic forms, the pragmatic financier's mind was on the present. Morgan loved sailing and yacht races and would resent another suggesting what a man in his position should or should not do.

Tesla reveals in this letter that he had to change his plans due to the "advantage shrewd competitors had" (e.g., due to Marconi's pirating and connections with Pupin, Edison, European investors, and sovereign rulers). He thereupon decided to abandon the agreed-upon scheme of building a modest-sized transmitter and replaced it with the idea of constructing a skyscraping tower six hundred feet high, designs for which he scratched out on his fancy Waldorf-Astoria stationery.[33] Ironically, for the ofttimes altruistic Tesla, it was his greed, vanity, and megalomania which drove him into the new venture. To have his ideas stolen from him was abhorrent. In his autobiography, Tesla would later refer to Marconi (although not by name) as a "parasite and microbe of a nasty disease." It was at this moment that the inventor decided to scrap the trivial idea of sending mere Morse codes across the Atlantic. He would inaugurate a world communications enterprise to pulverize the vermin as a pachyderm would crush a toad.

Having reached the pinnacle of the ruling class, Tesla's self-image swelled with the occasion, for he had conceived a telecommunications enterprise more efficient than the combined forces of today's radio, television, wire service, lighting, telephone, and power systems! His ultimate plan even included the production of rain in the deserts, the lighting of skies above shipping lanes, the wireless production of energy for automobiles and airships, a universal time-keeping apparatus, and a mechanism for achieving interplanetary communication. Having attained cosmic consciousness, he offered this creation to the king of the financial world, and the king had accepted. To the inventor, it was simply a detail that this vision was not in agreement with the specifics of the contract or that Tesla never really told Morgan his greater scheme; this work, like White's statue, transcended the traditional rules of business.

PANIC ON WALL STREET

It was just sixty days since Tesla had signed his contract with Morgan, thirty days after Morgan sailed for Europe. Yet already Tesla had irreversibly changed his plans. Since he had been a gambler and pool player in his salad days and was now living among the most audacious of high rollers at the Waldorf, these old tendencies resurfaced now that he had "hooked the biggest fish on Wall Street." He had calculated the odds based on certain assumptions about the stability of the economy and the quickness of his access to Morgan's $150,000, and he proceeded boldly with the completion of the masterwork.

How could the inventor know that on May 10 the stock market would crash and that the main culprit in the catastrophe would be his backer, J. Pierpont Morgan?

The collapse of the stock market occurred because of a bitter rivalry that existed between Morgan and Ned Harriman. Morgan, who was in charge of the Northern Pacific Railroad, having tossed Henry Villard out a decade earlier, had bought control of an enormous line called the Chicago Burlington. This concern stretched its tracks from Atlantic ports to Chicago and down the Mississippi all the way to New Orleans. Harriman, who controlled the Union Pacific, or southern route to the west, wanted access to the Burlington as well, and he tried to negotiate with Morgan for a seat on the board of directors. Unfortunately, due to bitter disagreements stemming from an old railroad deal in which the crafty Harriman had outwitted the commodore, Morgan came to detest the man. Therefore, he would not share the Burlington and became irrational whenever Harriman's name was mentioned.[34]

Thus, while Morgan enjoyed his art purchases in England and his mistress in France, Harriman, with the help of his broker, Jacob Schiff, clandestinely began purchasing Morgan's Northern Pacific. Instead of trying to outbid Morgan for the Burlington, Harriman boldly bought Morgan's own holding company out from underneath him! To accomplish this coup(!), it would take the adventuresome Harriman close to $100 million, which he raised by selling enormous blocks of Union Pacific bonds, and Harriman actually successfully carried out the operation. By the first week in May, Harriman owned more than 50 percent of Morgan's precious company, which had been affectionately called the Nipper. When Morgan received the fateful telegram from his underlings while in France, he tossed his paramour off his lap and wired back the order to buy Northern Pacific back *at any price,* for Harriman did not yet own a majority of the voting common shares.

On May 9, the stock rocketed from $150 to $1,000 per share! A panic ensued as people who bought in on the Nipper were unable to obtain possession of their shares, as neither Morgan nor Harriman would release any; most of the other stocks dropped when investors sold in order to cover their losses. The end product was the tumbling of the market and the creation of extreme economic and political times and also monetary chaos. Stanford White was one of many who lost heavily in the market. For Tesla, costs increased dramatically, and credit was nearly impossible to obtain. The front page of the *New York Times* reported the calamity: "The greatest general panic that Wall Street has ever known came upon the stock market yesterday, with the result that before it was checked many fortunes had been swept away...."[35]

Even Morgan's precious U.S. Steel dropped from its current price of forty-six to a low of eight dollars a share.[36] Numerous investors were financially ruined, and some purportedly committed suicide. (One rumor concerning this famous event was that Morgan won back the company because Harriman's broker, Jacob Schiff, was in the synagogue on the fateful Saturday when Morgan began buying more shares of Northern Pacific. Schiff, however, never intended to wrest the company from Morgan. His goal was simply to obtain a large enough percentage of the company to force Morgan to give Harriman a piece of the Chicago Burlington. Harriman, in his frenzy, wanted to change their plan so as to gain control over all three railroads, but Schiff overruled him. Thus, the skyrocketing of the stock and ensuing crash of the market were based solely on Morgan's buy orders.)

The economic upheaval created heavy financial burdens on Tesla. However, he would not completely realize the increasing financial difficulties immediately, for the affects on construction costs, wages, and incidentals would ripple throughout the summer and fall.

Before Pierpont's departure for England, in April, he had assured Tesla that now he "had no doubts" about the inventor's abilities,[37] and even if, by some fluke of misfortune, Morgan should not supply additional funds, Tesla still had his own money and the personality to attract additional investors. At forty-five years of age, wealthy, established as a leader of his field (however controversial), and rubbing elbows with the crème de la crème, the tall inventor was well equipped for the task he foresaw.

31

CLASH OF THE TITANS (1901)

Make a Tesla or buy a Tesla coil. I made one, find it.... Get books on wireless telegraphy.

FROM THE PRIVATE NOTEBOOK OF THOMAS EDISON[1]

Throughout the late spring and summer, Tesla commuted regularly to Wardenclyffe, often with a servant of Serbian extraction and a box lunch from the Waldorf-Astoria.[2] At night he returned to the city, where he could stop in at the Players' Club, attend a concert, or dine at Delmonico's or Sherry's. In June, having to forgo yet another "Johnson blowout," he apologized to Robert and Owen for being "unable to meet the lady who inspired the celebrated author of the *Arrows of the Almighty*."[3]

White was on his yearly fishing vacation in Canada,[4] so Tesla was on his own for the month. Some of the time was spent scouting out possible apartments to rent in Shoreham, and George Scherff also had to look for a place. In July the architect returned, and talks resumed concerning the construction of the tower. Having recently joined the Automobile Club of America, which had its headquarters in Locust Valley and counted Vice President Teddy Roosevelt among its members,[5] White would motor out from the city to the club or to St. James in his new "steam-powered Locomobile," which required a driver, or on his own in a fashionable "electric two-seater runabout."[6] With his estate just a few miles from Shoreham, the avid motorist could drive out past the endless flats of potato fields, along the same road that also led to the site of Wardenclyffe to oversee the work and perhaps take the inventor for a spin. The architect's son recalled: "I remember Tesla well, as he often came down to stay with us on Long Island. He used to wander around at night in the garden in the moonlight; and when my mother asked him why he wasn't asleep, he replied: 'I never sleep.' I also remember going to his laboratory [in the city] as a boy, and watching him put several million volts through his body lighting up two Crookes tubes which he held in his hand."[7]

White made Tesla realize that a six-hundred-foot-high transmitter (roughly two-thirds the height of the Eiffel Tower) was simply out of the question, so Tesla dabbled with harmonic ratios of one-half and one-quarter that size. With prices so unstable, it would be difficult to calculate the new costs.

The Johnsons were as excited as Tesla about the purchase of Wardenclyffe. By July, with the land cleared, it became an opportune moment to visit the site—just a few miles from a lovely beach on the Sound at Wading River and also not far from Southampton on the Atlantic side. A weekend was set aside to coax Tesla into spending some time swimming. It was a delightful moment, a time to enjoy the salt water and picnic grounds and pose for whimsical photographs of their heads placed in the notches of standard billboard pictures of bodies in striped bathing suits or, gaily adorned, sitting in the seats of an imitation automobile.[8]

In August, with the frame up on the lab and plans for the facing under way, Tesla once again refused a respite in Maine with the Johnsons, writing a teasing letter that as a member of the Four Hundred he could not meet with "people whose fathers have been fruit peddlers and grocers."[9] Perhaps Katharine's father had been such.[10] With the tower now planned as a much larger venture and Morgan delaying payment of promised funds, Tesla pondered a way not only to obtain owed funds but also to get Morgan to increase his investment.

The monarch with the bulbous snout returned from Europe on the Fourth of July. Leaving the ocean liner from the stern in order to avoid the throng of reporters, Morgan ignored his home and moved to the *Corsair*, his 300-foot-long yacht, living there for the month and staying in Bar Harbor, Maine, through part of August.[11] The art connoisseur was pleased with his recent acquisitions of paintings, gemstones, and rare manuscripts—he did not shorten his yearly European excursion on account of a Wall Street crisis—but he was irritable, irrationally perturbed with Harriman, and fearful that the press and public might threaten his empire, if not his life. Morgan was an opinionated man, fair-minded much of the time but dangerously stubborn at other times. Having detested Harriman for outwitting him not once but twice, he must have been furious when the world viewed *him* as the villain who had destabilized the economy because of a personal vendetta. A turbulent labor strike by the steel workers added to his unhappiness and the uncertainty of the times. Headlines such as the following caused him to seek armed protection:

The Rich Denounced by Socialist Labor

Thousands at Cooper Union
Cheer Wordy Assaults on Capital.
J. P. Morgan Accused of Trying to "Trustify the Earth."

> "This is the century," said chairman Lucien Sanial, "in which there is going to be social revolution."
>
> Whoop went the audience waving hats and yelling madly for a minute or so.... Charles Knoll [followed and] said he favored the adoption of such resolutions as would "chill and make to shiver the spinal column of the capitalists."[12]

To end the Northern Pacific fiasco, Morgan and Harriman allowed investors to settle their accounts at $150 a share. The public was not supposed to notice that this price would yield hefty profits to the giants who had purchased their stock for one-third less than that price only days earlier. Instead, they were perceived as noble in their attempts to restore order and sanity to the economy. At first, the government wanted Morgan to return stocks to the original investors at original prices. In response, Morgan was quoted as saying that it would be quite a feat to "unscramble the eggs and return them to their original hens!" When accused of avoiding his responsibility to society, Morgan responded in anger, "I owe the public nothing." For this remark, Morgan would be interrogated by the governmental investigative committees until his dying days, but he weathered the storm easily.

Before Morgan left for Maine, he met with the inventor. In a new satchel purchased for the occasion, Tesla brought his latest patent applications, drawings of the half-completed laboratory, and schematics for the tower. The secretary at 23 Wall Street ushered him in.

"Mr. Morgan, you have raised great waves in the industrial world and some have struck my little boat. Prices have gone up in consequence, twice, perhaps three times higher than they were and then there were expensive delays, mostly as a result of activities you excited."[13]

"We've all suffered, Mr. Tesla," Morgan said, already perturbed and testy because of the more important imbroglios he was enmeshed in.

Tesla pushed on, informing Morgan that he had decided to design a larger tower than agreed upon because of Marconi's piracy. Morgan looked on, at first in detached amazement, as Tesla continued.

"Suppose a plant is constructed capable of sending signals within a given radius, and consider an extension to twice the distance. The area being then four times as large, the returns will become more valuable. Approximately computed, the average price will be tripled. This means that a plant, with a radius of activity twice as large, will earn twelve times as much. But it will cost twice as much.... The greater the distance the greater the gain until, when the plant can transmit signals to the uttermost confines of the Earth, its earning power becomes, so to speak, unlimited.

"The way to go, [Mr. Morgan, is] to construct such a plant.... It [will] give the greatest force to my patents and ensure a monopoly.... [and] offer[s] possibilities for business on a large, dignified scale commensurate with your position in life and mine as a pioneer in this art, who has originated all essential principles."[14]

"Let me understand you, Mr. Tesla. You have not exploited the lighting enterprise?"

"Not yet, sir."

"You have not constructed a transmitting tower, but you have just about completed the construction of a laboratory?"

"Yes."

"You have purchased, is this 200 acres? with an option on 1,600 more, and you have run out of funds?"

"Only temporarily, sir. Once you supply the balance..."

"And if I relinquish these funds, will that be enough for the creation of your 'model city'?"

"No, as I explained..."

"If we double the size of the tower, I will earn twelve times as much. Is that it?"

"Precisely."

"Get out, Mr. Tesla."

"But, sir...,"

Raising his voice to a dull roar, Morgan reiterated his command. Quietly, Tesla packed his satchel and slipped away.

The inventor was in shock. One can only imagine which favored swear words Morgan, known "on Wall Street...for his gruffness and blunt expletives,"[15] unfurled upon the prima donna who had pranced into his office with his fantastic scheme and high and mighty demands. Tesla required a few days to regain his composure. Morgan still legitimately owed him a significant amount of funds. The banker was upset, with the problems, with the Northern Pacific, and with the ensuing denigration by the press. It will pass, Tesla reasoned. The best thing to do would be to reestablish credibility. He forwarded Morgan his most recent patent assignments and then backed off.[16]

The following week, White telephoned to suggest that they go with a rough stone facing on the laboratory instead of brick, and the inventor agreed. "Please make sure they also put in a fireproof roof," Tesla told his architect.[17]

"Let's not rush the tower," White cautioned. "I'm still calculating figures for you."

As White was courting during these very weeks on a daily basis

Evelyn Nesbit and also helping Tesla recalibrate the construction of the complex, it is possible that the tight-lipped inventor became privy to the architect's intimate liaison.[18]

In a financial predicament, Tesla reported to White that he had visited the "American Bridge people to ascertain whether they will be able to construct the cupola of my [tower] without much delay. As this item will consume the longest time, it is necessary to take all the preliminary steps, so that the work may be begun just as soon as you have passed upon the plans.

"I believe that the American Bridge Company is the best concern to deal with in this matter," Tesla continued, "but I beg you not to pay any attention to my suggestion, if you think otherwise. The Bethlehem Steel Company will furnish the sheets, but I cannot give the order until we have agreed upon all details."[19]

"You must enjoy parting with your money, if you are negotiating with American Bridge," White responded. "I implore you to let me handle the contracts. I should have the figures in a few weeks, but I can tell you right now, a 300 foot tower is out of the question so we do not know what size the cupola will be.... You must also consider the extra cost for designing the tower so that each individual strut can be replaced if need be, without toppling the entire edifice."

"[Please understand, Stanford,] I went to the American Bridge Company simply because of my anxiety to have the work pushed through as fast as practicable. I am only too glad to follow your advice, and beg you to consider yourself absolutely free in your choice and arrangements regarding this work."[20]

On Friday, September 6, 1901, President William McKinley journeyed to Buffalo to attend an exposition and see firsthand the remarkable project instituted at Niagara. Dwarfed by the colossal Tesla turbines, the president wove his way back to the train station shaking hands with many of the onlookers. While waiting on the platform, a crazed anarchist lurched forward and fired at point-blank range. As McKinley struggled throughout the week at the edge of death, Tesla decided to write to the president's old-time friend Morgan, sending the appeal on the inventor's favorite day, Friday the thirteenth, the day the president died. "McKinley's passing," Morgan cried, "is the saddest news I have ever heard."[21]

Beginning the letter with "I respectfully apologize for disturbing you at a time when your mind must be filled with thoughts of a more serious nature than usual," the inventor imprudently reiterated his recent proposal, suggesting that if Morgan doubled his investment, Tesla would be able to send messages across the Pacific as well as the Atlantic, or better yet, if Morgan tripled the amount, the wizard could send messages to any point on the globe, "no matter what the distance."[22] On the same day, he also wrote to White, who had finally given him precise figures on his proposed monster-sized tower.

> My dear Stanford,
>
> I have not been half as dumbfounded by the news of the shooting of the President as I have by the estimates submitted to you, which, together with your kind letter of yesterday, I received last night. One thing is clear we cannot build that tower as outlined.
>
> I cannot tell you how sorry I am, for my calculations show, that with such a structure, I could reach across the Pacific.

Tesla told White that because of the limitations on capital, he would have to "fall back on an older design...involving the use of two, and possibly three towers, but much smaller." The design would be the same, only the dimensions would be reduced. "I shall make some calculations," the inventor concluded, "to see how far I can reduce the height without impairing materially the efficiency of the apparatus, and will communicate with you as soon as practicable."[23]

The following day, Tesla wrote White again, agreeing to construct a tower with a height of approximately 150 feet.[24] Having worked with figures on a tower 600 feet high in May, Tesla probably reduced it by half by the time he saw Morgan in August, and then in half again after White told him that the costs of a 300-foot tower were prohibitive. Adding the figure of "1/6th larger," which he mentioned to White in his last letter (or 25 feet), we come to 150 + 25, or 175 feet, which was just about the height of the actual tower. (After construction, it was 187 feet.) However, Tesla also constructed a well beneath the tower, with an accompanying spiral staircase, which ran ten stories below the ground, to a depth of 120 feet.[25] Adding this figure to the total (i.e., 187 + 120), we come to a length of approximately 300 feet, or 1/2 the size of the original plan and therefore in a harmonic relationship to it. Even this tower, however, was too expensive to build, given the cost of the machinery, complex design of the housing, which required that it be fireproof, and the inflation produced by the crash.

In pondering these letters, it becomes apparent that Tesla was not too perturbed by the assassination of the president. Self-engulfed, he was utterly amblyopic when it came to negotiating with Morgan, a man enmeshed in two potentially epic crises and one history-altering tragedy. Theodore Roosevelt, who now became president, was not a man who was going to be particularly kind to big business.

To say that Tesla blundered here would be an understatement. His decision to alter his contract without telling Morgan and his resolution to proceed with the grand vision when he knew that his funds would be inadequate were addlebrained. One suspects that once Tesla had signed a contract with the greatest financial force on the planet, a deep-seated, subconscious complex was triggered involving an impatient egomaniacal streak that forced the inventor to place everything on the line when he should have proceeded in a more discerning way. Incapable of compromising and at the risk of self-obliteration, Tesla began construction of the tower *after* his falling out with Morgan. On the positive side, the inventor knew that he was in a race against pirates and for what he perceived as the "Holy Grail," his peerless notch in history. Undaunted, the courageous inventor moved ahead with the conviction that his path was right and that he could not fail.

Tesla's seemingly foolhardy decision must be understood in light of the fact that fortunes had already been reaped on his former inventions by Morgan and others. By 1901, for instance, Morgan's General Electric Company was actually producing *more* induction motors than the Westinghouse Corporation; Morgan was involved, along with Westinghouse, in instituting an electric subway system in the bowels of Manhattan based on Tesla's polyphase system; and then, of course, there was Niagara Falls. Every home in the world was going to be illuminated by Tesla's system. The revenues pouring into the electric power companies for this new technology were staggering, but Tesla received not a cent. One way or another, he felt that Morgan should give him carte blanche.

It was a bleak autumn for the inventor when the ground was broken for the eighteen-story edifice, Tesla now naming it his "magnifying transmitter." Although mostly constructed of wood, "50 tons of iron and steel" were also used, along with "50,000 bolts."[26] Taking into account the amount of lumber it took to line the well and build the staircase down, and the difficulty in digging it, one begins to visualize the enormous expenditures that were going to be involved. W. D. Crow stayed in charge of construction. Hoping for the best, Tesla wrote to Katharine on October 13:

> My dear Mrs. Johnson,
>
> 13 is my lucky number and so I know you will comply with my wish...[to] come to the Waldorf. And if you do—when I transmit my wireless messages across seas and continents you will get the finest bonnet ever made if it breaks me....
>
> I have already ordered a simple lunch and you must come en masse. We must exhibit Hobson....I know he likes me better than you.
>
> Nikola Tesla
> Electrical Engineer & Inventor[27]

In November, the inventor tried once again to approach Morgan, setting up a meeting at 23 Wall Street and bringing a succinct list of his latest patent assignments and his report on how the construction was going.

> Dear Mr. Morgan,
>
> Pardon me for trespassing on your valuable time.... The practical significance of my system resides in the fact that the effects transmitted diminishes only in a simple ratio with the distance whereas in all other systems it is reduced in preportion to the square. To illustrate, if the distance be increased 100-fold, I get 1/100th of the effect, while under the same conditions others can obtain at the very best, 1/10,000th of the effect. This feature alone bars all competition.
>
> In regard to [other advantages], there are only two ways possible of economically utilizing the energy transmitted...: either storing it in dynamic form as, for instance, the energy of well timed thrusts is stored in a pendulum, or by accumulating it in potential form, as for example compressed air is stored in a reservoir.... My rights [through patents] on both are fundamental.
>
> Referring particularly to telegraphy and telephone, I have still in the patent office two applications [pending].... In one I describe and claim discoveries relating to the transmission of signals through the earth to any distance no matter how great, and in the other a new principle which secures absolute privacy of messages and also enables the simultaneous transmission of any desired number of messages up to many thousands, through the same channel, be it the earth, or a wire or a cable. On this latter principle I have applied for patents in the chief foreign countries. I consider these inventions of extreme commercial importance.
>
> Hoping that I shall be able to satisfy you that your generosity and confidence in me have not been misplaced,
>
> I remain,
Yours very respectfully,
N. Tesla[28]

Morgan's continuing mistreatment of the inventor and lack of acknowledgment of the significance of the plans revealed were almost too much to bear. Tesla could not face the Johnsons or anybody for Thanksgiving and so declined their invitation. "Dear Luka," Tesla wrote, "Kindly excuse and remember me with kind regards," signing the letter, "Nikola Faraway."[29]

NEWFOUNDLAND

Signor Marconi had been commuting regularly between England and the States throughout this period, looking for sites to place his wireless stations. Prime locations included the eastern tip of Long Island, Martha's Vineyard, and Cape Cod. "In September 1901, the new equipment, including the immensely powerful transmitter, was installed at Poldhu [England] and a great 200-foot diameter ring of masts 200-feet high had risen like [a huge] skeleton...on the edge of the cliff. Test transmissions to other Marconi stations, in particular that at Crookhaven, Ireland, over two hundred miles away, had shown that the waves did—at least to this extent—follow the curvature of the Earth and not fly off into space. Across the Atlantic, at Cape Cod, the twin station was similarly nearing completion, and plans were made with quiet optimism for the experiment to take place in a few weeks' time."[30]

In September, gale-force winds leveled the aerial in Poldu, and in November the same thing occurred on the Cape. Tenaciously, Marconi pressed on, gambling with a less powerful but sturdier transmitter in England and abandoning the idea of constructing a twin in the States. Instead, the Italian would try simply to intercept the signals of the English transmitter by fashioning an aerial with kites, high-altitude weather balloons, and a sensitive coherer as a receiver.

On December 6, he landed, with a small crew, in Newfoundland, Canada, and floated up his receiving antenna on a spot designated appropriately as Signal Hill. December 12 was chosen as the day for the experiment, the beacon selected dot dot dot, the Morse code for the letter *S*.

On Friday the thirteenth, during a lull in a miserable storm of hail and rain, three faint taps were heard on his equipment. The world was rocked; Guglielmo Marconi's name was irrefutably carved into the history books; the age of mass communication had begun.

32

The Passing of the Torch (1902)

December 1901: Signor Marconi has scored a shrewd coup. Whether or not the 3 dots he heard came from England or, like those Tesla heard, from Mars, if I am aught a prophet, we will hear no more of trans-Atlantic messages for some time.

Lee De Forest[1]

Embittered, Tesla knew that Marconi's achievement was predicated on the use of his coil, oscillators, and general design which he had spelled out in lectures years earlier. Preece assumed partial culpability, as he had requested from Tesla the use of this equipment for the work, but Marconi had announced that the Tesla apparatus was unnecessary and ineffectual,[2] and this had caused a rift in the Italian's relationship with Preece. Fleming, on the other hand, having studied Tesla's work in earnest ever since he had received the inventor at his home in London in 1892, saw no such conflict; for it was he who "arranged for Marconi the transmitting plant at Poldhu."[3] Tesla revealed many years later: "[Marconi had] declared that wireless communication across the Atlantic was impossible because there was a wall of water several miles high between the two continents which the rays could not traverse. But subsequent developments showed that he had used my system in secret all the time, received the plaudits of the world and accepted stolidly even my own congratulations, and it was only a long time after that he admitted it."[4]

Thomas Commerford Martin arrived at his office on Monday, December 16 to review the astonishing report from Newfoundland. With only Marconi and one aide as a witness and the plans having been kept secret until the deed was accomplished, many doubted the Italian's proclamation. Prof. Silvanus Thomson, of Great Britain, suggested that Marconi had probably received static caused by severe weather conditions. One of Martin's colleagues concurred: "It's a fake. Such a thing cannot be done."

"I think I should seek another opinion," the editor said as he put in a call to Tom Edison.

"Very doubtful. How's it going to get around that blasted curve?" came Edison's hedging response. Martin dialed up Michael Pupin.

"Professor, do you believe Marconi's transmission is genuine?"

"I most certainly do."

"Then I think we ought to celebrate it."[5]

It was the dead of winter when Tesla ducked out of the Waldorf-Astoria as the new crowning electrical savant checked in. Tesla probably went to Wardenclyffe to stare at the first tier of the tower, which was finally under construction. With the temperature so cold, it was just one more annoyance to slow down progress.

With only a few days to prepare, Martin was able to book the Astor Gallery at the Waldorf for the banquet, on Monday, January 13, 1902. With three hundred guests arriving, the task of arranging all the particulars put him in a frenzied state. He brushed by fantastic pictures of a wizard's laboratory as he hurried out the door.

The hall was decorated with a large map of the Atlantic placed on the wall and a festoon of wires "with clusters of three lights" blinking *dot-dot-dot* strung between large tablets reading Signal Hill in Newfoundland and Poldhu in England. Around the room, each table had its own model transmission tower, nameplates, and "Italian olive green menus" on card stock with pen and ink drawings of the transatlantic accomplishment. At the upper dais "in the middle was a medallion with Mr. Marconi's portrait, draped with the Italian flag." American and British flags and pendants for the AIEE and the Italian coat of arms were also placed there.

"At fitting times, [the lights] were flashed" to the applause of the audience; and to cap off the dinner, for dessert, a "procession of waiters" came marching in with ice cream imbedded in ice carvings of incandescent lamps, ships at sea, electric vehicles, and wireless telegraph towers.[6]

The four-foot gnome trimmed his goatee and checked his gold pocket watch before smiling at himself once more in the mirror as he departed for the affair. Although he rocked from side to side when he walked, Charles Proteus Steinmetz developed a new swagger, for he had just been elected to the presidency of the AIEE. He was about to receive an honorary doctorate from Harvard and a professorship in engineering from Union College, located near GE headquarters in Schenectady, New York. The college appointment enabled Steinmetz to divide his time between academia and the corporate world.

During the six-hour train ride down to the city, the abstract mathematican carefully read through galleys of his opus on AC which was about to be republished in a larger format by McGraw-Hill. It was a small matter to the preeminent scholar that he had removed the name of his coauthor and had continued the practice of eliminating reference to the source of his work, *The Inventions, Researches and Writings of Nikola Tesla*. Electricians, he rationalized, would be more interested in his advanced concepts than "in knowing who first investigated the phenomena."[7] By 1907, Steinmetz would lead in establishing the AIEE Code of Ethics.[8] Who was Tesla, anyway? Marconi was the man of the hour.

At the upper dais sat a gaggle of Tesla adversaries. Aside from the new president, there was Professor Pupin, now financially tied to Marconi; Elihu Thomson, who claimed priority on the invention of the AC motor and Tesla coil; Carl Hering, who had backed Dobrowolski in priority discussions of the inventor of long-distance AC transmission; William Stanley, who had pirated the Tesla/Westinghouse induction motor and was now producing them legally for GE; Frank Sprague, who gained his reputation in part as the inventor of the electric railroad when, in fact, it was all part of Tesla's AC polyphase system; T. C. Martin, who was still angry about past moneys owed for sale of the inventor's collected works; and, of course, Guglielmo Marconi, the chap who had beaten him to the punch. Tesla's decision not to attend created an excellent atmosphere for gaiety—and for perpetuating Steinmetz's published position of relegating the pioneer to the category of nonperson.

Others present at the infamous upper dais included Alexander Graham Bell and general counsels from Great Britain and Italy, and around the hall were Josh Wetzler, D. McFarlan Moore, many of the men's wives, and Mrs. Thomas Alva Edison, who was representing her husband.

Martin presided over the occasion, opening up the period for lectures with readings of telegrams from those not in attendance. He began with a letter from the mayor and then read a communiqué from the Wizard of Menlo Park.

> To T. C. Martin:
>
> I am sorry that I am prevented from attending your annual dinner to-night, expecially as I would like to pay my respects to Marconi, the young man who had the monumental audacity to attempt, and succeed in, jumping an electric wave clear across the Atlantic. Thomas A. Edison[9]

Martin did not announce that at Christmas Marconi had sent Edison a "cheerful telegram" reiterating his success and offering to display personally to the master his transatlantic equipment or that Marconi was already tendering Edison offers for his early wireless patents.[10]

The *New York Times* reported "cheers when the toastmaster came to a letter from Nikola Tesla who said that he 'could not rise to the occasion.'"[11] No doubt, they masked the jeers. Grinning through his oversized mustache, Martin waited for the clamor to subside before continuing with the rest of the letter:

> I regret not being able to contribute to the pleasure of the evening, but I wish to join the members in heartily congratulating Mr. Marconi on his brilliant results. He is a splendid worker, full of rare and subtle energies. May he prove to be one of those whose powers increase and whose mind feelers reach out farther with advancing years for the good of the world and honor of his country.

Prof. Elihu Thomson followed. "I had received the news of Marconi's great feat from over the telephone from a reporter, who wished to know whether I believed that signals had actually been received across the Atlantic." Riding on the accolades of the audience, Thomson related his response: "As I told the reporter, if Marconi says that he received the signals, I believe they were received." And then the hero took the pulpit. Waiting gratiously for the applause to subside, he began.

Signor Marconi explained his syntonic wireless system and pointed out that "he had built very largely on the work of others and mentioned Clerk Maxwell, Lord Kelvin, Professor Henry and Professor Hertz." At this time, its most important use was for communication between ships. The Italian was pleased to announce that "over 70 ships now carried his wireless system, 37 for the British navy, 12 for the Italian navy, and the remainder on large liners, such as the Cunard Line, the North German Lloyd and the Beaver Line. There were also over 20 stations in operation...with more on construction." Marconi addressed the problem of selective tuning and suggested that he had such a system created so that "messages transmitted [from one ship] can in no way be received by any other, except that attuned to receive the message."[12] Stated as bravado, Marconi was bluffing, as he had no system for creating separate channels.

"It is my hope," Signor Marconi concluded, "that at no great distant date, I shall bring my system to the point of perfection of allowing friends and relatives to communicate with each other across the ocean at a small expense."

Professor Pupin concluded. "Referring to claims made that previous to Marconi wireless signals had been transmitted over short distances," Pupin said, glancing back to the upper dais, "any schoolboy by means of an Hertzian oscillator could transmit such signals over a short distance...but it required the engineer to make [such] work of avail to the world." In attempting to allay fears that Marconi's system would make obsolete

Atlantic cables, Dr. Pupin shrewdly "pointed out, as an illustration, how the completion of electric lighting has aided the gas industry and enhanced rather than decreased the value of [their] investments."[13]

On January 9, Tesla dispatched a letter to Morgan explaining that the patents of the "Marconi-Fleming syndicate" do not accurately reflect their actual apparatus but are covered "by my patents of 1896 [and] 1897." The balance of the letter describes the precursor to what became, a half century later, the major television networks:

> I need not tell you that I have worked as hard as I have dared without collapsing.... Hav[ing] examined and rejected hundreds of experiments...with the capital at command, I am glad to say that by slow and steady advances, I have managed to continue a machine...[which shall produce] an electrical disturbance of sufficient intensity to be perceptible over the whole of the earth.... [When] I throw the switch, I shall send a greeting to the whole world and for this great triumph I shall ever be grateful to you....
>
> [This system] will do away not only with the cables but with the newspapers also, for how can journals as the present [stay in business] when every [customer] can have a cheap machine printing its own world news?
>
> [The] beautiful invention which I am now developing, will enable me to spread our name into [every] home, and it will be [able] everywhere to [hear] the tune of my voice.[14]

This would be the inventor's last communiqué to the financier for fully nine months. He set himself the difficult task of completing construction of the eighteen-story transmission tower, knowing full well that he did not have sufficient funds. From bank records dating back to 1896, it is apparent that Tesla had nearly $50,000, some of which had been converted into land assets.[15] The last of Morgan's money had probably been received, so it was at this time, in the midst of 1902, that the trailblazer began to tap his personal reserves to keep the project going. Work would continue at a steady pace throughout the year.

OTHER COMPETITORS

Having received his doctorate in electrical engineering in 1899, Lee De Forest had tried once again to gain entrée into his idol's laboratory, but for a third time Tesla refused him. De Forest decided, therefore, to set out on his own. In 1901 he succeeded in sending wireless messages across the Hudson River over a distance of one or two miles and shortly thereafter sent impulses from State Street, in downtown New York, to Staten Island, seven miles away. By using "self-restoring detectors with telephone receivers instead of Morse inkers or sounders," De Forest succeeded in increasing substantially the speed of transmission. Now his apparatus threatened Western Union's local telegraph lines. Working with D. McFarlan Moore, who had "studied Tesla's monumental early volume," De Forest was able to decrease the problems of static interference. By 1903 he was reporting yacht races at a speed of twenty-five to thirty words per minute, or about as fast as a Morse-code operator could send them. By 1904 he could transmit messages "180 miles over land, between Buffalo and Cleveland," and by 1908 his signaling device was jumping continents.[16]

Perhaps it was because of Tesla's regard for the Yale graduate or because of the canniness of his patents that Tesla did not try to prevent De Forest from using his oscillators and general scheme. The same, however, could not be said for Reginald Fessenden, whom he sued for patent infringement in April 1902.

Fessenden, who had worked for both Edison and Westinghouse as far back as the early 1880s, is generally credited with having invented the means of sending voice over the airwaves. Although Marconi was using the electromagnetic frequencies to mimic the impulse patterns of the Morse code, "it occurred to Fessenden to send out a continuous signal with the amplitude of the waves varied (or 'modulated') to make the variation follow the irregularities of sound waves. At the receiving station, these variations could be sorted out and reconverted into sound. In 1906 the first such message was sent out from the Massachusetts coast, and wireless receivers could actually pick up the music. In this way radio, as we know it, was born."[17] A year later, using his patented audion, which was, in essence, a modification of Tesla's 'brush tube,' De Forest succeeded in transmitting the voice of Enrico Caruso, who was singing at the Metropolitan Opera House in New York.[18]

Interested for reasons of priority in obtaining Edison's grasshopper wireless patent (dating from the 1880s), Fessenden sought a job with GE in 1902 to inaugurate construction of a wireless transmitting station at Brant Rock, Massachusetts. Although he stayed friendly with Edison and looked after his wayward son, Tom Junior, who had been caught passing bad checks, Fessenden was unsuccessful in obtaining Edison's key wireless patent; the Menlo Park wizard had sold it to Marconi for $60,000.[19]

Legal entanglements were expensive, but Tesla felt that he had no choice but to protect as many fundamental aspects of his system as he was able to. How else could he prove to Morgan that his work in the field really was the basis of the systems that were succeeding?

In June 1900, Reginald Fessenden had applied for a patent on tuned circuits. The following month, Tesla filed for one as well. It was a matter of public record that Fessenden's application preceded Tesla's. What was at issue was whether or not Fessenden had compiled his invention from Tesla's earlier experiments. Although Fessenden claimed that he had conceived the idea in 1898, Tesla pointed out that Fessenden was (1) unable to provide documentation of this earlier date; (2) he did not create a working model of his apparatus; and (3) the machine had not been used commercially.

Whereas Fessenden's application was rudimentary, Tesla's delineated clearly a multiplicity of goals, for example, (1) operating distant apparatus; (2) controlling signals by using two or more idiosyncratic electrical frequencies; (3) producing a plurality of distinctive impulses onto a receiving apparatus comprising a manifold number of circuits; and (4) creating a combination transmitter-receiver arrangement set up to respond to a succession of impulses released in a given order. Whereas Fessenden could date his theoretical conceptualizations to, perhaps, 1898, Tesla dated the onset of his work to 1889 and provided his numerous publications as evidence. With specific reference to the operation of "tuned circuits," the inventor displayed his fully working telautomaton, which he introduced to the world in 1898. Without Tesla's AC oscillators, Fessenden's machinery could not operate. Unless he had lived in a hermitage, it would have been impossible for Fessenden to have conceived his invention without utilizing Tesla's attainments. Parker W. Page questioned his client for hours; Tesla's testimony would run seventy-two typewritten pages.

Throughout mid-April the testimony continued, and after Tesla was finished, his twenty-nine-year-old manager, George Scherff, took the stand. Scherff, who was living at Wardenclyffe by this time, was able to substantiate that Tesla's tuned circuits and long-distance wireless experiments were first conducted in his presence in 1895, when he began working for Tesla at his laboratory at 33-35 South Fifth Avenue (before it burned to the ground). Scherff remembered the inventor transmitting wireless impulses from the Houston Street lab to the roof of the Hotel Gerlach, which was one or two miles away.[20]

Fritz Lowenstein, just a year younger than Scherff, followed. Having returned from Europe in February and newly married, Lowenstein had gained reemployment with Tesla at Wardenclyffe. In a heavy German accent, Lowenstein thoroughly described the nature of the confidential experiments at Colorado Springs. "Mr. Tesla explained to me," Lowenstein stated, "that the chief feature of a practical wireless telegraph system was the secrecy, immunity and selectivity; at the same time he explained to me how two oscillations are secured from one oscillating apparatus... When I came to Mr. Tesla," the college-educated engineer revealed, "I didn't understand anything at all about it, but he soon showed me the great value [of] tuned circuits, and then I understood what tuning was."[21]

> Dear Mr.Scherff,
>
> Mr. Page has just told me that my opponent's attorney has admitted my priority.... [Mr. Fessenden] must be disappointed of course, and I am sorry for him although you know he has written some articles which are not very nice.... My honor as the originator of the principle is assured.[22]

Tesla may have won the case, but he was not about to celebrate. Foremost among his priorities was his wish to keep the details of the litigation secret. The last thing he wanted was publicity, for the transcripts of the trial revealed many technicalities that would aid his competition in numerous ways. In the short run, the inventor succeeded in protecting important aspects of his wireless scheme; but in the long run the testimony became an important source text for Fessenden, who now had a legal basis for fashioning a plethora of second-order patents. By the time of his death, Fessenden had compiled an astounding five hundred patents, which was nearly as many as Tom Edison. Obviously, this work also aided Lowenstein, who became a wireless expert valuable to different members of the new crop of engineers who were rapidly emerging.

33

WARDENCLYFFE (1902–1903)

> *While the tower itself is very picturesque, it is the wonders hidden underneath it that excite the curiosity of the little [hamlet]. In the centre of [the] base, there is a wooden affair very much like the companionway on an ocean steamer. Carefully guarded, no one except Mr. Tesla's own men have been allowed as much as the briefest peep....*
>
> *Mr. Scherff...told an inquirer that the [shaft entrance] led to a small drainage passage built for the purpose of keeping the ground about the tower dry; [but] the villagers tell a different story.*
>
> *They declare that it leads to a well-like excavation as deep as the tower is high, with walls of masonwork and a circular stairway leading to the bottom. From there, they say. the entire ground below has been honeycombed with subterranean [tunnels that extend in all directions].*
>
> *They tell with awe how Mr. Tesla, on his weekly visits...spends as much time in the underground passages as he does on the tower or in the handsome laboratory where the power plant for the world telegraph has been installed.*
>
> NEW YORK TIMES[1]

Just as Tesla had created a notebook for his experiments in Colorado, he also compiled a daily log of activities at Wardenclyffe. His records for 1902 reveal little activity for the first third of the year except for the month of March. Extended note taking really began again in May and continued uninterruptedly until July 1903. Each week, Tesla watched the tower attain a new height as he experimented by measuring the capacitance of his apparatus and constructing a prototype planet to calculate "his theory of current propagation through the Earth." On this sizable metal sphere, the inventor would transmit varying frequencies to measure the voltage, wavelength, and velocity of the transmitted energy and also assess various nodal points, such as along the equator and at the pole opposite the point of generation.[2]

In February 1902, along with Stanford White, Tesla entertained Prince Henry of Prussia who had come to New York to retrieve a royal yacht that had been built in America. The brother of Kaiser Wilhelm, it had been Prince Henry who had assisted in the performance of the inventor's famous experiments under Tesla's name in Berlin six years earlier. The yacht was being christened by Alice Roosevelt, daughter of the president.[3] In June out at Wardenclyffe, "two grey-haired East Hampton pilgrims drop[ped] in on their way to a Spring retreat."[4] Kate's eyes glowed as she broke off from Robert and Nikola to approach the tower and feel it with her own hands. A radiant warmth flowed through her being as she watched the lanky engineer converse with her husband.

By September, the transmitter had reached its full elevation of 180 feet. Exhausted of funds, with still the dome to place on top, the inventor had no choice but to wind down activities and lay off most of the workers. Tesla had sold his last major asset, a $35,000 land holding, but even that could not keep the operation fully afloat. Nevertheless, with this new source of revenue, he procured enough funds to keep a skeleton crew, cover Scherff's and his own lodging, and pay for a chef from the Waldorf to come out at regular intervals. Tesla also took this time to photograph the interior operations of his entire plant. These pictures not only included reproductions of all his machinery but also contained a representative sample of the myriad different types of radio tubes that the inventor had designed. They numbered nearly a thousand.[5]

The *Port Jefferson Echo* reported "War between Marconi and Tesla" in their headlines in 1902. According to the paper, the United States Marconi Company had purchased land west of Bridgehampton and was planning on constructing its own competing 185-foot tower, with New York City connections to Western Union. "It should become the most important wireless center in the country." Whereas Marconi was going to send his signals through the air, the paper said, Tesla plans, through his "500-foot deep" shaft, to send messages through the earth as well. Though the hollow was 120 feet, the gist of the report was accurate.

Having first sent an emissary, Marconi himself hired a horse and buggy one rainy morning to go out to Wardenclyffe to meet with Tesla and see the operation with his own eyes.[6] Marconi's limitations became apparent in conversation, and this gave Tesla the courage to return to see Morgan. He had stayed out of Morgan's way for nearly a year, but now it was clear to the inventor that once his partner realized the work accomplished with so little funds at hand, he would reconsider his position and agree to resume investment. All the inventor had to do was change the man's mind.

On September 5, Tesla wrote the financier to inform him that the foreign patents had been assigned to the company. The letter explained clearly that his wireless system ensured privacy and had the ability to create a virtually infinite number of separate channels that were dependent not only on particular combinations of differing frequencies but also on "their order of succession." In essence, the inventor had explained to Morgan the concepts inherent in such devices as the protected channels on cable TV, digital recording, and the wireless telephone scrambler.

In the letter, Tesla explained that he was forced to increase the power of his apparatus because of "bold appropriations" of his equipment (Marconi's piracy), but his *mea culpa* also betrayed paranoid tendencies, even though he was attempting reconciliation: "The only way to fully protect myself was to develop apparatus of such power as to enable me to control effectively the vibrations throughout the globe. Now, if I had received this necessity earlier, I would have gone to Niagara, and with the capital you have so generously advanced, I could have accomplished this easily. But unfortunately my plans were already made and I could not change. I endeavored once to explain this to you, much to my sorrow, as I impressed you wrongly. Nothing remained then but to do the best I could under the circumstances."

Morgan's response was utter astonishment. Tesla had not only reiterated the breach of contract but had also revealed a striking flaw in his plan. Whereas he would have to haul in continuing truckloads of coal to provide the energy required to fire up his transmitter on Long Island, if he had set up operations at Niagara, there would have been easy access to unlimited amounts of power, and thus the operation would have been considerably less expensive to initiate. Furthermore, there was a standing offer by Rankine and his associates to provide power at the Falls for little or no cost, thereby reducing potential expenses even more. Nevertheless, Tesla incredibly informs Morgan that even on Long Island he could outdo the power of the giant cataract.

"By straining every part of my machinery to the utmost," Tesla wrote, "I shall be able to reach...a rate of energy delivery of 10 millions of horse power." This production, he proclaimed, would be equal to "more than twice that of the entire Falls of Niagara. Thus the waves generated by my transmitter will be the greatest spontaneous manifestation of energy on Earth...the strongest effect produced at a point diametrically opposite the transmitter which in this instance is situated a few hundred miles off the western coast of Australia."

Whether or not this statement was true, it was counterproductive to divulge this. (Most likely Tesla is discussing the production of a single massive burst of electrical energy rather than a continuous, never-ending flow. In either case, having spent a year in the outbacks of Colorado, the bon vivant was not about to give up his lifestyle at the elegant Waldorf for another lonely excursion to the dreary location near Buffalo.) Neurotically impelled , the inventor was stirred to impress the financial magnate when all Morgan wanted was to signal ocean liners and send Morse code to Europe.

Later in the letter the inventor did address this more modest task of broadcasting simple messages. Over transmission lines, dispatches from New York Telegraph & Cable could be delivered to Wardenclyffe, where they could then be distributed to Europe by means of wireless to a central receiving station.

> Since your departure, Mr. Morgan, I have had time to reflect...[on] the importance and scope of your work, and I now see that you are no longer a man, but as a principle and that every spark of your vitality must be preserved for the good of your fellowmen. I have therefore given up the hope that you might aid me in establishing a manufacturing plant which would enable me to reap the fruit of my labors of many years. But some ideas which I have not simply conceived—but worked out—are of such great consequence that I honestly believe them to deserve your attention...
>
> I have no greater desire than to prove myself worthy of your confidence, and that to have had relations, however distant, with so great and noble man as you will ever be for me one of the most gratifying experiences and most highly prized recollections of my life.
>
> Yours most devotedly,
> N. Tesla[7]

Perhaps it was the finale or the inventor's rank or maybe the realization of the value of the fundamental patents that continued to flow into his office, but for whatever reason, Morgan was moved to the point of allowing another meeting—as long as it was expressly understood that the liaison would still be kept unpublicized.[8]

Tesla's plan was to seek new investors by selling bonds and capitalizing the company at $10 million. What he did not realize was that his silent partner wanted to maintain his 51 percent control over the patents, take in "about 1/3 of the securities," and also be reimbursed for his initial investment.[9] In other words, whereas Tesla required about an additional $150,000 to complete the tower, pay off his debts, construct the receiving equipment, and so on, the Wall Street hydra wanted his money back in full, wanted to maintain his large share of the concern, wanted the inventor to raise the funds on his own, and wanted no one to know of his connection to the project! If Tesla were to agree to all that, they would have a deal.

Kate gazed out her window at the changing colors of the leaves as she went over the Thanksgiving menu with Agnes.

"Do you think he'll come, Momma?"

"Of course he will."

"He didn't last year."

"Mr. Tesla was not himself," Katharine stated matter-of-factly. Sitting by a fizzling Edison lightbulb, she penned an urgent notice and hired a special messenger to deliver it.

Tesla was working on his list of prospective investors when a loud banging at his door interrupted him. "Sorry, sir, it is important."

The inventor grabbed the letter and ripped it open. Thoughts of accident and death flashed through his brain as he reflected upon his relationship with the Johnsons and his inability to deal with them now. He wrote back:

> Some day I will tell you just what I think of people who mark their letters "important" or send dispatches at night.
>
> You know I would travel 1000 miles to have such a great treat as one of Mrs. Filipov's dinners, but this Thanksgiving day I have a great many hard nuts to crack and I will pass it in quiet meditation. The rest of the holidays I propose to pass in the same good company.
>
> Never mind my absence in body. It is no consequence. I am with you in spirit.
>
> With love to all and Agnes in particular,
>
> Nikola[10]

With this painful letter, Tesla succeeded in transferring some of his anguish to his friends. Not only would he not attend the November holiday feast; he would also not partake of the Christmas festivities. Unable to escape himself or dig his way out of the deep hole he was in, he passed the time cranking out ways to enlist new investors.

When Katharine realized the depth of the inventor's despair, she became alarmed. The emptiness at such an important occasion was almost too much to bear. "Agnes, please write Mr. Tesla once again and tell him that he may stop by at any time that he likes, and forever long he wants."

"Don't you think he knows what he is doing," Robert interrupted. "He needs to be by himself right now."

"Don't tell me what he needs!" Kate exploded, her Irish temper flaring. "Do it," she commanded her daughter. Robert moved quietly into the living room to read a book of poems.

As Agnes sat down to write, a peculiar dark and twisted expression passed over her mother's face. Kate retired to her boudoir to sit there in her gloom. Tesla wrote, "My dear Agnes," back two days after the new year.

"I have no time but plenty of love and friendship for all of you. I would like very much to see you, but it is impossible. Even kings are beginning to infringe my patent rights, and I must restrain myself."[11]

Katharine could take solace in the fact that her platonic aficionado had used the word *love* twice in two letters. In his own peculiar way, he had transformed the pain of his career to their linkage, and in that sense, he had reached out, as never before, but it was by withdrawing, and this made her love him more.

Thomas Commerford Martin may have had mixed feelings about having supported so vigorously a man he knew had infringed on Tesla's patents, but Martin had warned Tesla nearly five years earlier that whereas Marconi was succeeding on the physical plane, Tesla appeared to advance mostly in theory. A social gadfly, Martin had also upped his prestige considerably by hosting the gala Marconi affair, and he continued to elevate his standing among the upper echelon of the electrical community. *Harper's* decided to do a feature on him, and Edison was becoming amenable to the idea of allowing him to write his biography.[12] As a precursor to the new venture, the editor prepared for the weekly a short piece on the "volcanic lifetime of a master who has produced a patent every fortnight for over thirty years."

> An "Edison Man" remains an Edison man to the end of the chapter, and is proud of the stamp left upon his career by the great spirit with whom trials and triumphs have been shared....[Although] Edison has always been surrounded by a willing host of workers, [he] has always held easily his leadership among them. This is by no means true of other[s]....Some powerful thinkers, whether from instinctive distrust or unavowed jealousy, endeavor to hammer out their conceptions in lonely struggle, and names could be mentioned here of electrical inventors whose curse seems to be this sterile seclusion. In Edison's case, the sunny, kindly temperament of the man makes for friendship.[13]

Martin was almost certainly talking about Tesla here, and in so doing, he highlighted ponderous debilities in the inventor's personality. Tesla could be inordinately withdrawn, private, "distrustful," an elitest, and yet envious of others, unable to share in the development of his ideas lest he would have to share credit. Later in this article, Martin wrote that "a great many first-class inventors are sharply concentrated along one line," while Edison had "many irons in the fire." As versatile and incredibly prolific as Tesla was, he stuck, until the end of his days, to the single monumental Wardenclyffe idea when any aspect of the grand plan, in and of itself, would have been a revolutionary accomplishment.

Underneath it all, however, Martin was, without doubt, the one individual who had accomplished the most toward explicating the wondrous achievements of the sequestered genius, and his connection to Tesla would remain sacred to him for the duration of his life. Martin took the opportunity to let Tesla know this when he gave the inventor an updated version of the collected works.

"Many thanks for the book," Tesla wrote his old-time friend. "It was a pleasure to read the dedication which tells me that your heart is true to Nikola."[14] This occasion served to break the ice again, and the two resumed their friendship, albeit not with the same intensity as before.

In a dreadful predicament, the inventor was beginning to stare down a canyon of doom, writing his letters now in pencil, abandoning the certainty of the pen.

> Ever since I was a boy I was desirous of drawing on the Bank of England. Can you blame me? I confess my low commercial interests dominate me.... Will you please give me a list of people almost as prominent and influential as the Johnsons who desire to get into high society. I will send them my letter.
>
> Nikola Busted[15]

Forced to approach both the Johnsons and his manager, George Scherff, for funds, both lent him thousands over the next few years.[16] Simultaneously, he returned to former enthusiasts such as Mrs. Dodge and Mrs. Winslow, and new investors, such as Mrs. Schwarz, wife of toy store owner F. A. O. Schwarz.[17] The price for investment was $175 for each share.

With his back to the wall, he approached again his Wall Street benefactor. "Will you now let me go from door to door to humiliate myself to solicit funds from some jew or promoter and have him participate in that gratitude which I have for you?," he wrote Morgan. "I am tired of speaking to pusillanimous people who become scared when I ask them to invest $5000 and get the diarrhoea when I call for ten."[18]

MAD SCIENTIST

> "It is a mighty fine tower," said one good farmer...last week. "The breeze up there is something grand of a Summer evening, and you can see the Sound and all the steamers that go by. We are tired, though, trying to figure out why he put it here instead of Coney Island."[19]

Although operations were suspended, there were many avenues the inventor would explore in his attempts to complete the vision. One of his first decisions was to comply with George Scherff's plan to become more pragmatic. Throughout the balance of the year and much of 1903, Tesla

HIS HANDS FULL.

THE OCTOPUS: "Guess I'll have to grow some more arms." —*Minneapolis Journal.*

J. Pierpont Morgan: one of numerous political cartoons poking fun at the most powerful financier in the world, circa 1901

began to manufacture oscillators and continue development of his fluorescent lamps. Revenues began to trickle in, and by midyear he had compiled enough savings to hire back a half-dozen workers and pay for the cupola, which was placed at the apex of the spire.[20] Over fifty feet in diameter, ten feet tall, and weighing fifty-five tons, this iron-and-steel crown, with its specially designed multitude of nodal points, would serve the purpose of storing electrical charges and distributing them either through the air or down the metal column and into the hollow. The cupola was linked to four large condensers behind the laboratory, which also served the purpose of storing electrical energy, and these in turn were coupled with "an elaborate apparatus" which had the ability to provide "every imaginable regulation...in the control of energy."

At the base of the edifice, deep below the earth, along the descending spiral staircase, was a network of catacombs that extended out like spokes of a wheel. Sixteen of them contained iron pipes which protruded from the central shaft to a distance of three hundred feet. The expense for these "terrestrial grippers" was notable, as Tesla had to design "special machines to push the pipes, one after the other" deep into the earth's interior.[21] Also in the well were four stone-lined tunnels, each of which gradually rose back to the surface. Large enough for a man to crawl through, they emerged like isolated, igloo-shaped brick ovens three hundred feet from the base of the tower.

Although the exact reason for the burrows has not been determined, their necessity was probably multifaceted. Tesla had increased the length of the aerial by over a hundred feet by extending the shaft into the earth. Simultaneously, he was able to more easily transmit energy through the ground with this arrangement. It is possible that he also planned to resonate the aquifer which was situated slightly below the bottom of the well. The insulated passageways which climbed back to the surface may have been safety valves, which would have allowed excess pressure to escape. They also provided an alternative way to access the base. Tesla may have planned to fill other shafts with salt water or liquid nitrogen to augment transmission. There may have also been other reasons for their construction.[22]

Just as the inventor prepared to test his new equipment, creditors began to encroach more vigorously and Tesla was never able to put the final fireproof protective outer facing on the cupola and tower. To Westinghouse he owed nearly $30,000,[23] the phone company was billing him for the telephone poles and lines which they erected to connect him with civilization,[24] and James Warden was suing for taxes owed on the land.[25] With the exigency of time inexorably pressing down, the inventor worked furiously to link his transmitter to the power source and test its potentialities.

Throughout the early part of 1903 the engineer "performed many measurements of ground resistance and insulation resistance of the tower. He even considered temperature [increases] caused by ground losses, differences when salty water was spread around the base, weather conditions and time of day."[26]

In the last week of July, just days before men came to cart away part of his equipment, the inventor fashioned a way to couple his behemoth and fire it up. As pressures reached their maximum with the cupola fully charged, a dull thunder rumbled from the site, alerting the hamlet that something was about to happen.

Strange Light at Tesla's Tower

> From the top of Mr. Tesla's lattice work tower on the north shore of Long Island, there was a vivid display of light several nights last week. This phenomena [*sic*] provoked the curiosity of the few people who live near by, but the proprietor of the Wardencliffe [*sic*] plant declined to explain the spectacle when inquiries were addressed to him.[27]

Tesla's mushroom-shaped citadel spewed forth a pyrotechnic eruption that could be gleaned not only by those who lived nearby but also by the populace inhabitating the shores of Connecticut, across Long Island Sound. But by the end of July the tower fell silent, never to raise its radio cry again.

It was a foggy morning when the Westinghouse crew appeared with their horse-drawn wagon and court order granting them permission to cart away the heavier equipment. The gargantuan edifice loomed as a specter of what could be, its head still obscured by the low-hanging clouds. Except for a guard, George Scherff, and a handyman, the entire crew was let go. His dream now hobbled, the sullen wizard crept back to the city to weep alone in his Waldorf suite.

34

THE WEB (1903–1904)

July 14, 1903

Dear Sir:

I have received your letter... and in reply would say that I should not feel disposed at present to make any further advances.

Yours truly,
J. Pierpont Morgan[1]

Mr. Boldt removed his monocle from his top vest pocket and eyed the man suspiciously. The foreigner was sweating profusely from the August heat. "Niko Tezlĉ," he said in a heavy accent as he handed the hotel manager a crinkled envelope.

Momentarily sidetracked by a spot of dust, Boldt retrieved the document to inspect Tesla's letterhead. "You may go up," he said disdainfully as he slammed his hand down on the clangor to call a bellhop.

The man entered what appeared to him to be a palatial suite. "Jovan, so good of you to come," the venerated engineer said in his native tongue. "You must find Uncle Petar. It is a matter of utmost urgency."

"He may be in Bosnia."

"Then go there." Tesla handed the man a round trip boat ticket, a communique sealed with wax, and a billfold of spending money. "I am depending on you."[2]

Tesla's three sisters, Angelina Trbojevich, Milka Glumicic and Marica Kosanovic, their husbands, all Serbian priests, and all of their children sat in the rectory to listen to the spell-binding account the courier brought of Niko's laboratory and world telegraphy tower. "Its head reaches up to the clouds," Jovan said, spreading his arms to their full extent, "and some day it will send messages to the whole world—right here to this town." Jovan's cousins gasped in amazement as he passed around a photograph taken just three months earlier by Dickenson D. Alley, the photographer also responsible for the Colorado pictures.[3]

"How will we know when a message is sent?"

"Everyone will have a little device, the size of"—Jovan looked around the room and spotted a prayer book—"the size of this book, with a wire attached which you will stick into the ground to receive the message!"

Two first cousins, Nicholas Trbojevich and Sava Kosanovic, both just emerging from boyhood, listened intently. Fully captivated by such news from America, Nicholas announced proudly, "Some day I will be an inventor too." Sava, destined forty years later to be Yugoslavia's first ambassador to America, smiled back and nodded in agreement.

Uncle Petar stepped from the room to open the dispatch in private. Niko had told his uncle that the Panic had hit him hard, that he was out of funds. He would have to close down the World Telegraphy Center if assistance was not received immediately. He asked Petar to go to a local bank and borrow funds using shares in Wardenclyffe as collateral. The bank, of course, would not comply,[4] so Petar called a meeting with family elders for a pooling of resources. They then returned to Belgrade, where the transfer of funds could be accomplished. "You wish Niko well for us," Petar said, grasping Jovan with both arms. Tesla received the money by the end of the month, but it was really only enough to keep a door open for the balance of the year.

September 13, 1903

Dear Mr. Morgan,

Many years I was at your door with this invention, but I did not go in thinking it would be useless.... My last undertaking has returned more that two dozen times the original investment and this in strong hands should do better still.

Help me to complete this work and you will see.

Yours most faithfully,
N. Tesla[5]

Altering his schedule, the inventor began visiting the lab only on weekends. Having thought the matter through, he abandoned the idea of meeting with small-time investors and placed his efforts in two separate directions. He would step up the manufacture of the oscillators and enlist other tycoons.[6] With the money from his relatives, Tesla was able to hire back a few workers. Operations, however, were just about nil, and the rest of the crew were angry about not being paid.

"You know, of course," he told Scherff, "that when this panic came, a great many manufacturers simply dismissed their men. Our employees should understand that I have tried to treat them generously in the hardest times this country has known, and they should be grateful instead of impatient. Although the panic is practically over, there is still a general feeling of apprehension on the street. However, I have a few irons in the fire, and at any moment, I may come up with the solution which confronts me. I am more than ever assured that nothing can prevent my ultimate success."[7]

Tesla's first stop was the home of John Jacob Astor. Taking with him Alley's new dramatic glossies, the inventor tried his best to reignite the flame. Astor, who was earning approximately $3 million per annum, was nevertheless not a frivolous man. Most of his money these days was being spent on his yacht, the *Nourmahal.* Whereas Tesla had had an ally in Astor's wife, the marriage was now in decline, with Ava spending most of her time in Europe with her two small children and her husband, Jack, continuing his practice of straying at home.[8] "While wishing you all possible luck," Astor wrote the inventor, "do not care to go into the company myself."[9]

Deciding that there were other fish in the sea worth hooking, Tesla compiled a list of the biggest ones and worked with a graphic designer to produce a flashy mailer. Known as the "Tesla Manifesto," the pamphlet boldly announced the expectations of his world telegraphy enterprise.[10]

Folded into a maroon vellum binder, the brochure contained lavish prints of the Colorado Station and also the imposing laboratory and tower at Wardenclyffe, a list of relevant patents, past and future accomplishments, a statement of his availability to be hired as a consultant, and a declaration of the breadth and scope of his plan, all of this surrounded by a scalloped design of pen-and-ink drawings of his many other inventions. Atop the new magnifying transmitter, drawn as part of the frame in fancy cursive, were the following words: "Electrical Oscillator Activity Ten Million Horsepower." People on his list included numerous moguls, each worth anywhere between $20 and $200 million, just about all personally known by the inventor.

"Luka," Tesla wrote to Robert Johnson, "Rockefeller and Harriman are now taking up every moment of my time, but I think I shall get through with them very shortly."[11] Written partly in jest, this letter was not far from the truth, for Tesla was becoming more successful in penetrating a number of other wealthy enclaves.

An Interview with Wizard Edison

> "Do you believe, [Mr. Edison] with Tesla that we should be able to talk around the world one of these days?"
>
> "No, I do not look for developments in that line. The wonderful thing that will be more and more developed is wireless telegraphy. Marconi is all right and is bound sooner or later to perfect his system.[12]

With Edison leading the charges against his credibility, Tesla was placed in a delicate situation. The public at large had yet to know of the Morgan connection, and yet Tesla had to inform potential investors of his interest. Due to animosities that Morgan still held for Harriman, a contract with him was out of the question, but there were still many other financiers to consider.

On October 12, 1903, Tesla met with Thomas Fortune Ryan. A stalwart, corpulent man five years Tesla's senior, Ryan, whose real middle name was Falkner, had gotten his start as a dry-goods clerk in Baltimore. His break came after he moved to Wall Street, where he became a stockbroker and investor in large financial institutions. By 1905 his position had grown so extensively that he gained control of nearly $1.4 billion, which was equivalent to almost half of the entire public debt of the United States. Almost one-third of this figure stemmed from his acquisition of the controlling shares of Equitable Life Assurance Society,[13] and this came about a little over a year after his initial meeting with Tesla. In and of itself, the connection would appear to be incidental; however, it was Tesla who arranged for a meeting between Ryan and Morgan in attempts to iron out an amenable agreement, and it is well known that Morgan was the secret power behind the famous Equitable Life insurance scandal which erupted in 1905.

Aside from the handful of diamonds he perpetually fondled in his palm, Ryan's other assets included control over Mutual Life and Washington Life Insurance, New York City Railway, American Tobacco, Morton Trust, Metropolitan Securities, and Mercantile Trust. Ryan also sat on the board of a dozen other insurance, banking, rail, and utility concerns.[14] He also owned an immense estate covering hundreds of acres north of the city, near Monticello, where he had erected a $500,000 mansion. From this retreat, his wife, Mrs. T. F. Ryan made a name for herself as a philanthropist, founding a theological institution, erecting "a magnificent Catholic Church [and] a public hospital, [purchasing] fire company equipment, and [making] scores of minor contributions for the good of the village."[15]

Known for his "striking [proficiency] of systematic organization...decisive action, secrecy and the art of using great power behind the throne," Ryan's most effective gift was his persuasive ability. "Mr. Ryan makes his headquarters in the Morton Trust Company's office, where he is Vice-President. He does not bar the door like John D. Rockefeller....Any caller gets to his secretary and very often to the inner room. There is none of the gruffness that so characterizes Mr. Morgan...Mr. Ryan is suavely silent....He never does and never says an important thing without consulting lawyers....Yet he always is polite, and never shows anger."[16]

Tesla had calculated that he required approximately $100,000 to complete his project, so his plan was to enlist "ten subscribers at $10,000 each."

"What is the use of going to so many people," Ryan suggested, staring at the contents of the vellum binder. "I shall take one-fourth. Where do I sign?"[17]

Obviously, this was a great opportunity, but it was not a simple matter of just signing a paper. Tesla had to go back to Morgan to confer, and $25,000 would still leave him way short of his goal. "Would you consider underwriting the entire $100,000?"

"That's a possibility."

"Let me speak to my partner about your generous offer and get back to you."

"Is it anyone that I know?" Ryan inquired.

"I am under orders to keep his name confidential."

"We are talking about a significant amount of money, Mr. Tesla. I want to know who I am going into business with."

"It is Pierpont Morgan," the inventor said cryptically.

"Morgan! Well, then, you couldn't be in better hands."

Tesla wrote Morgan on the following day to arrange a meeting. "Mr. Ryan is a great admirer and a loyal friend of yours and for this reason as well as on account of his ability, I am very anxious to enlist his cooperation. I have told him that $100,000 would be sufficient to reach the first commercial results, which will pave the way to other greater successes. Knowing your generous spirit, I have told Mr. Ryan that any terms you may decide upon will be satisfactory to me."[18]

For Thanksgiving, Tesla joined the Johnsons. He had good news to report. Just back from a two-month stint in Europe, Robert and Kate were eager to describe their meeting with Queen Elena of Italy. Having been decorated by King Humbert in 1895 for his work on the international copyright, Robert was received by the widowed queen as a distinguished personage. He read to her and the queen mother a selection from his most recent book of poems.[19] They also stayed a few extra weeks to witness the jubilee at the Vatican, celebrating twenty-five years of the reign of Pope Leo XIII.

"Mr. Tesla, do I detect a gleam in your eye?" Kate inquired.

"You're not about to become one of the detested wealthy elite, are you?" Robert teased.

"My dear Luka, I do not want you to despise millionaires, as I am hard at work to become one. My stocks have gone up considerably this week."

"Morgan?" Katharine whispered hopefully.

"Fortune Ryan," Tesla said, taking out a bank note to display proudly. He would receive a total of $10,000 from the financier. "If it continues for a few weeks like this, the globe will be girdled soon. Now Kate, where's that turkey?"[20]

At this time, Morgan met with Ryan. Obviously, the Tesla deal did not go through. The question is, why?

Every indication suggests that the meeting went well. Ryan had been described by Wheeler (one of Morgan's harshest critics) as "a most adroit, suave and noiseless man,"[21] and this makes some sense, for a few years later it would become evident that Ryan was, in essence, a Morgan puppet.

In 1899, Henry Hyde passed away. He had been controller of the half-billion-dollar Equitable Life Assurance Society, "comprising the scrapings of the poor."[22] Fifty-one percent of the company was left to his son James, who was twenty-three at the time. A rather eccentric and naive millionaire, James charged the company for petty extravagances, such as importing barbers from France and placing his personal chefs at all of his favorite restaurants. The muckrakers were unable to tolerate the extravagent lifestyle of James, especially after it was learned that on January 31, 1905, he had spent $200,000 on a Louis XV costume ball at Sherry's. They demanded his resignation.

No doubt, ever since Hyde's father had passed on, Morgan had had his eye on the company, but because of the Northern Pacific debacle, he had to proceed cautiously. It is plausible to consider that at the meeting concerning the Tesla project Morgan might have suggested that Ryan's money could be better spent in a different area. Whatever he said, it appears to have been stated in such a way as to not completely turn off Ryan, as he did invest and maintain interest in the enterprise from the years 1903 to 1906, although he never paid Tesla more than the initial modest subscription.

In a situation that paralleled the Northern Pacific fiasco, Morgan fought for the Equitable on one side against Harriman and his broker, Jacob Schiff, another financier who almost went into the Tesla deal. The end result was an obscure maneuver whereby Thomas Fortune Ryan was able to purchase the controlling shares of Equitable for a paltry $2.5 million. Young Hyde moved to France, where he remained for a quarter century.

Rumors of mismanagement of the funds of a few million investors continued, however, and Ryan's picture, throughout the summer of 1905, was slapped on the front pages of every newspaper in town. Drawn as a rapacious spider in his web, clutching his prey of the Equitable in one political cartoon, Ryan was described by southern banker and critic John Skelton Williams in a scathing article that appeared in the *World:* "Mr. Ryan has the tendencies which, if his lines had been cast in a humble and contracted sphere probably would have made him a kleptomaniac. His strongest impulse is to acquire money, and his one robust passion is to keep it.... Ryan is simply an acquiring machine and operates himself for the purpose of getting what others have."[23]

Although Morgan, as head of a rival insurance company, supposedly had no interest in Equitable, Herbert Satterlee, his son-in-law, uncharacteristically revealed that Ryan was actually controlled by Morgan, stating that Morgan "was entirely conversant with Mr. Ryan's plans.... They had his thorough approval and possibly his financial backing."[24]

During the government investigations that continued relentlessly throughout the first decade of this century, it was discovered that dummy loans totaling $1.8 million were made to a fifteen-year-old Negro messenger boy in Morgan's New York Life Insurance Company.[25] Ryan also sold his 51 percent interest in Equitable to Morgan for $3 million in 1910. Testimony from the investigation explains quite clearly Ryan's role as a Morgan puppet.

Q: Untermeyer: Did Mr. Ryan offer this stock to you?
A: Morgan: I asked him to sell it to me.
Q: Untermeyer: Did you tell him why you wanted it?
A: Morgan: No; I told him it was a good thing for me to have.... He hesitated about it, and finally sold it.

Taken from Wheeler's *Pierpont Morgan: Anatomy of a Myth,* the author ended the passage as follows: "During the Panic of 1907, Morgan, in his 71st year, truly manipulated vast sums and groups of men [including President Theodore Roosevelt] in commanding fashion."[26]

As "the power behind the throne," clearly Morgan dominated the conversation when Ryan came to invest in Wardenclyffe. Morgan redirected his new chess piece into the insurance arena, but this still does not answer the question as to why. If Tesla's enterprise were perceived to have become potentially profitable, having Ryan put in another $100,000 to complete it would have been an easy thing to do. We can therefore conclude that Morgan deliberately scuttled the Tesla venture.

THE GUGGENHEIM CONNECTION

Aside from being annoyed with the difficult inventor, Morgan was concerned about Tesla's suggestion that he could transmit "unlimited power" by means of wireless. Tesla was boasting here, since he assured Morgan in future letters that Wardenclyffe would only be able to transmit "feeble amounts" of energy.

We come to a juncture here where the decisions of individuals redirected the course of history. As John Stewart Mill has noted, individuals make the history, and here is a clear instance of this proposition. For all intents and purposes, Morgan made the decision in October 1903 to do his best to ensure the inventor's defeat.

Morgan, however, was opposed philosophically to Tesla's overall scheme. It has been suggested that Morgan felt pressure from other Wall Street moguls, such as Bernard Baruch, Thomas Fortune Ryan's young and highly successful stockbroker. One day, Baruch had mistakenly suggested that Morgan, like him, was a gambler. Morgan was reported to reply, "I never gamble," and, indeed, may have backed down from the Tesla deal for fear of taking a chance that the controversial inventor might fail or that he might succeed in a way detrimental to existing corporate structures.

The following quotation is from physician and inventor Andrija (Henry) Puharich, a man of Yugoslavian heritage who was tangentially involved in helping ship Tesla's papers to the Tesla Museum in Belgrade in the early 1950s[27] and who personally knew John O'Neill, Tesla's first major biographer: "Now, I [Puharich] always got this second hand; you won't find it anywhere in print, but Jack O'Neill gave me this information as the official biographer of Nikola Tesla. He said that Bernard Baruch told J. P. Morgan, 'Look, this guy is going crazy. What he is doing is he wants to give free electrical power to everybody and we can't put meters on that. We are just going to go broke supporting this guy.' And suddenly, overnight, Tesla's support was cut off, the work was never finished."[28]

From a technical and economic point of view, Morgan could not understand how free information and/or power could yield returns. And whether it was Baruch or not who warned Morgan, Tesla himself had voiced the opinion boldly, a decade earlier in the *Sunday World,* that by providing a reservoir of electrical energy throughout the earth through his apparatus, "all monopolies" that depend on conventional means of energy distribution—that is, through wires, "will come to a sudden end."[29]

As the ultimate capitalist, Morgan's existence was greatly defined by controlling the price and distribution of energy and maintaining a working class to support the giant corporate monopolies (called "public" utilities). Thus, he simply could not support a system where wireless information and power could be tapped by anyone with a receiving instrument and machines would replace the work force. Reorganizing existing power, lighting, and telephone industries to please the vision of a somewhat eccentric inventor was certainly an unlikely undertaking for the cautious Wall Street financier. "All those businesses who would [no longer] need loans [would] then [no longer] deposit the[ir] profits in his bank."[30] Tesla, as the quintessential iconoclast, had struck a bargain with the wrong king.[31]

In 1903, Bernard Baruch, just thirty-three years old and one of the wealthiest men on Wall Street, retired from the firm he had worked for, for over a decade, to set up his own office. One of his first major clients were the Guggenheim brothers. Their interest lay in metals, so Baruch met with the miners Darius Ogden Mills and John Hays Hammond Sr. to seek advice and gain investments. Mills suggested that Baruch go west and purchase mines himself, and Hammond was hired by the Guggenheims to act as an adviser and acquire silver mines in Mexico. One of Baruch's first acquisitions was the Utah Copper Company, for "he knew that there would always be a need for copper in the world."[32] Tesla's wireless enterprise clearly threatened the Baruch-Guggenheim investment that Morgan was trying to involve himself in.

By 1905 the Utah Copper Company was producing copper and other metals at a rate in excess of $100 million per year, and this was a pace which maintained itself for another twenty-five years![33] In later years, John Hays Hammond would claim that "the development of the electrical and automotive industries were possible... 'only with assured large supply of copper.'"[34] Unfortunately for Tesla, his ultimate world plan was perceived as a threat on quite a number of key fronts.

Tesla waited through the first weeks of December for a positive sign from the meeting between Ryan and Morgan, but none came. With no choice left, he was forced to confront the financial potentate eyeball to eyeball. He decided to take a pragmatic approach. "Will you permit me to call this or any other evening," Tesla wrote, "[I wish] to bring a small instrument along to show you one or two experiments with my 'daylight'?...The shark who will come after me will get the contract for lighting your home."[35]

Timing his visit to coincide with the holiday, Tesla purposefully chose this opportune occasion to try and pierce the financier's stalwart exterior. But at the same time, he used the symbol of the shark in his letter, as this was an animal which ate big fish. Anne Morgan was one of his allies. She met him at the door. Having recently become "a founder of the Colony Club, the first American ladies' [society], patterned after a British gentlemen's club" and designed by Stanford White, Anne had entered into the midst of an androgenous theater crowd with ties to the infamous homosexual author Oscar Wilde. With Tesla's sexual orientation perpetually an enigma and Anne about to dabble in a lesbian fling,[36] their bond transcended the surface amenities. Before meeting with her father, Anne was able to corral the avowed celibate and discuss with him the emerging view of the new women.

"Mr. Tesla, I believe that it is absurd that in this day of enlightenment, women do not yet have the right to vote."

"I concur wholeheartedly, Anne. I also believe that this struggle of the human female toward sex equality will end up in a new sex order, with the females superior."

"Indeed?" Anne responded, the pupils of her eyes widening into large black pools. "I would have considered the sexes equal," she said as her hand reached out to touch the inventor.

"The modern woman, who anticipates in merely superficial phenomenon the advancement of her sex, is but a surface symptom of something deeper and more potent fomenting in the bosom of the race. It is not in the shallow physical imitation of men that women will assert first their equality and later their superiority, but in the awakening of the intellect of women. As generations ensue, the average woman will be as well educated as the average man, and then better educated, for the dormant faculties of her brain will be stimulated into an activity that will be all the more intense because of centuries of repose."[37]

"Mr. Morgan will see you now," the butler interrupted.

"Mr. Tesla."

"Mr. Morgan, thank you for seeing me. My enemies have been so successful in representing me as a poet and visionary that it is absolutely imperative for me to put out something commercial without delay. If you will only help me to do this you will preserve a property of immense value."

"I am sorry, Mr. Tesla, as I said before..."

"Won't you enable me to complete the work and show you that you have not made a mistake in giving me a checkbook to draw on your honored house. If you will imagine that I have found the stone of the philosophers you will be not far from the truth. My invention will cause a revolution so great that almost all values and all human relations will be profoundly modified."

"Had you simply achieved what I had asked, you would not be in this predicament."

"Mr. Morgan, I beg to call again to your attention that my patents control absolutely all essential features and that my work is in such a shape that whenever you tell me to go ahead, I shall girdle the globe in three months as surely as my name is Tesla. I have promised to the St. Louis people to open the door of the Exposition with power transmitted from here. It is a great opportunity, Mr. Morgan. I can easily do it, but if you do not aid me soon, it will be too late. Please think for a moment what this means for me. What I have told you long ago has happened. My competitors have collapsed, since their wholesale attempts [at practicable wireless transmission] has not succeeded. Now is the time to aid me. You know this better than anyone else."

"I have done my part. There are many other businessmen out there with the capital to complete your project."

"But you, sir, are the controlling partner. Should I obtain a commitment from another, will you consider a renegotiation."

"I will think it over."[38]

Having met with Morgan in December, Tesla not only came away empty-handed; he also came away with the distinct feeling that his partner was not going to facilitate finding new investors. Avoiding Katharine at

Christmastime was one way to telegraph the bad news. Using a full range of her feminine charms, the coquettish Mrs. Filipov issued one more of her provocative missives.

December 20, 1903

Dear Mr. Tesla,

You are most unkind...dear friend! Why do you not come to see ME instead of always dropping into The Century to see Robert.[39] I must have done something to offend you, but what?

How can you be indifferent to such devotion?...If you are unhappy and disappointed and down on your luck, then all the more reason why you should seek the companionship and support of your loyal friends.

Indeed, if the whole world were against you, the more firmly would they cling to you.

Faithfully yours,
Katharine Johnson[40]

Aware of Morgan's trepidation concerning the potential for his equipment to usurp the need for existing electrical power companies, the inventor wrote explicitly in attempts to allay these fears:

January 13, 1904

Dear Mr. Morgan,

The Canadian Niagara Company will agree in writing to furnish me 10,000 H.P. for 20 years without charge, if I put a plant there to transmit this power without wires to other parts of the world....As I outlined [earlier] I would use the energy not for industrial purposes, but for operating clocks, stock tickers and other apparatus which there are millions now in use, requir[ing on average] not more than 1/10 of the H.P. for each instrument....

Will you help me on any terms you choose and enable me to insure and develop a great property which will ultimately yield hundredfold returns. Please do not do me an injustice in believing me incapable simply because a certain sum of money was not sufficient to carry out my undetaking....You may see that my work remains uncompleted because of a lack of funds, but you will *never see* that machinery which I construct does not fulfill the purposes for which it was designed.

Tesla ended the letter with "hearty wishes for the new year."[41] How could he know that Morgan did not doubt that Tesla could succeed. He feared it.

January 13, 1904

My dear Sir,

In reply to your note I regret to say that I should not be willing to advance any further amounts of money as I have already told you. Of course I wish you every success in your undertaking.

Yours truly,
J. Pierpont Morgan

Close to making deals with other investors but hampered by his relationship to Morgan, Tesla was particularly upuset that his partner replied on the same day his own letter was sent. Surmising that Morgan was not even considering the situation, Tesla became enraged. For the first time he dropped all pretense and told Morgan what he really thought of him.

January 14, 1904

Dear Mr. Morgan,

You wish me success! It is in your hands, how can you wish it?

We start on a proposition, everything duly calculated; it is financially frail. You engage in impossible operations, you make me pay double, yes, make me wait 10 months for machinery. On top of that you produce a panic. When, after putting in all I could scrape together, I come to show you that I have done the best that could be done, you fire me out like an office boy and roar so that you are heard six blocks away: Not a cent; it is spread all over town. I am discredited, the laughing stock of my enemies.

It is just 14 months that the construction work on my plant was stopped.... Three months more with a good force of men would have completed it and now it would be paying $10,000 a day. More than this, I would have secured contracts from governments for a number of similar plants....

Now, when I have practically removed all obstacles skillfully put in my way and need only a little more to save a great property, which would pay you 10 million dollars as surely as one cent, you refuse to help in a trouble brought on by your own doings.

Tesla suggests in the balance of the letter that a subscription of $25,000 would enable him to start up operations of the production of oscillators and the fluorescent light and that eventually, "in a slow and painful way," he would be able to obtain the necessary funds to complete the tower.

I am anxious to succeed on your account as mine. What a dreadful thing it would be to have the papers come with your name in red letters [A MORGAN DEAL DEFAULTS]. It would be telegraphed all over the globe. You may not care for it Mr. Morgan. Men are like flies to you. But I would have to work 5 years to repair the damage, if repairable at all. I have told you all. Please do not write to refuse. I am pained enough as it is.

Yours sorrowfully,
N. Tesla[42]

Having received no response, Tesla shot off another letter the following week:

January 22, 1904

....Are you going to leave me in a hole?!!

I have made a thousand powerful enemies on your account, because I have told them that I value one of your shoestrings more than all of them....

In a hundred years from now, this country would give much for the first honors of transmitting power without wires. It must be done by my methods and apparatus and I should be aided to do it first myself.

April 1, 1904

...Will you aid me to complete this great work?

April 2, 1904

Have you ever read the book of Job? If you will put my mind in place of his body you will find my suffering accurately described. I have put all the money I could scrape together in this plant. With $50,000 more it is completed, and I have an immortal crown and an immense fortune.

Unable to understand why the Ryan deal went sour, Tesla nevertheless deduced that it had been Morgan's doing. Retaliating in any overt way would have been suicide. And although Tesla had self-destructive tendencies, breaching a contract with J. Pierpont Morgan being one example, the inventor wanted desperately to succeed. His goal was not so much to line his own pockets, although surely he sought to get rich from the invention, but rather to help society. Tesla was well aware of his potential role in reshaping the course of human events.

Seeing no other choice, he brashly decided, in early 1904, not to hide the Morgan connection anymore but rather to publicize it and to maintain the front that everything was okay. To one of his worried investors, William Rankine of the Niagara Falls project, he wrote on April 10, "Doubt the light of the sun, doubt the brightness of the stars, but do not doubt the existence of the Nikola Tesla Company."[43] Landing Morgan in the first instance had been a feather in his cap. Now the bon vivant decided to exploit the connection and also brazenly rebel against a man who seemed bent on sinking his ship. In quintessential Teslaic fashion, he published a spectacular article simultaneously in *Scientific American* and *Electrical World & Engineer.* In it he outlined his work to date and plans for the future, adorning the piece with breathtaking photographs of his transmission stations in Colorado Springs and Wardenclyffe.

> The results attained by me have made my scheme of "World Telegraphy" easily realizable. It constitutes a radical and fruitful departure from what has been done heretofore. It involve[s] the employment of a number of plants each of [which] will be preferably located near some important center of civilization and the news it receives through any channel will be flashed to all points of the globe. A cheap and simple [pocket-sized] device, may then be set up somewhere on sea or land, and it will record the world's news or such special messages as may be intended for it. Thus the entire earth will be converted into a huge brain, as it were, capable of response in every one of its parts. Since a single plant of but one hundred horse-power can operate hundreds of millions of instruments, the system will have a virtually infinite working capacity....
>
> The first of these central plants would have been already completed had it not been for unforseen delays which, fortunately, have nothing to do with its purely technical features. [This delay may prove] after all to be blessing in disguise....
>
> For the work done so far I am indebted to the noble generosity of Mr. J. Pierpont Morgan, which was all the more welcome as it was extended at a time when those, who have since promised most, were the greatest of doubters. I have also to thank my friend, Stanford White, for much unselfish assistance. This work is now far advanced, and though the results may be tardy, they are sure to come.[44]

35

DISSOLUTION (1904–1906)

I have observed in the House of Morgan a largeness, nobility and firmness of character the like of which is very scarce indeed. I can only smile when I read of the attempts to find something discreditable in the transactions of J. P. Morgan & Co. Not a hundred of such investigations will ever uncover anything which an unprejudiced judge would not consider honorable, fair, decent and in every way conforming to the high ideals and ethical standards of business. I would be willing to stake my life on it.

NIKOLA TESLA[1]

Sociologist Karl Mannheim suggests that the psychohistorian should attempt to reconstruct both the subject's *Weltanschauung* and the spirit of the age in question. Irrational components should be recognized. Thus, and in this sense, history is paradoxical; it is contradictory, dynamic, multileveled, and dialectic.[2] Tesla's worldview involved a philosophy based on the work of Wolfgang von Goethe. His inventions were for him not true creations in the sense that they stemmed from nothing. They evolved from the work of others and from uncovering secret mechanisms lying within hidden laws of nature.

Was it God who wrote each sign?
Which, all my inner tumult stilling,
And this poor heart with rapture filling,
Reveals to me, by a force divine,
Great Nature's energies around and through me thrilling?
Am I a God? It grows so bright to me!
Each character on which my eye reposes
Nature in act before my soul discloses.

This idea can clearly be found in Goethe's *Faust*, Tesla's favorite poem, which he memorized in its entirety and which he referred to throughout his life.[3] It was *Faust* which he recited in Budapest during his salad days, when he uncovered the secret to the rotating magnetic field, and it was a Faustian paradigm to which he adhered when he linked the invention of the world telegraphy system to the discovery of the Holy Grail.

> There manifests itself in the fully developed being—Man—a desire mysterious, inscrutable and irresistible: to imitate nature, to create, to work himself the wonders he perceives.... He subdues and puts to his service the fierce, devastating spark of Prometheus, the titanic forces of the waterfall, the wind and the tide. He tames the thundering bolt of Jove and annihilates time and space. He makes the great Sun itself his obedient toiling slave....
>
> Can man control this grandest, most awe-inspiring of all processes in nature? Can he harness her inexhaustible energies to perform all their functions at his bidding?... If he could do this, he would have powers almost unlimited and supernatural....
>
> [This] would be the supreme manifestation of the power of Man's mind, his most complete triumph over the physical world, his crowning achievement, which would place him beside his Creator, make him fulfill his ultimate destiny. Nikola Tesla[1]

Two major themes which run through Goethe's poem are that (1) secrets of nature can be revealed and harnessed to human needs and that (2) humans are enticed by satanic forces. Clearly, Tesla was driven by both tenets. In the second case, consciously or unconsciously, Morgan was sought out for the very reason that he was a demigod, a superhuman, whose life transcended that of mere mortals. Just as Faust was tempted by Mephistopheles, Tesla was lured by the House of Morgan. In the financier's "strong hands" Tesla willingly, and alas irrationally, handed over the 51 percent control, insisting on it. Knowing that the contract involved relinquishing his cornucopia of past and future patent applications, the inventor still sealed the Faustian pact, as "the terms were immaterial" to him.

> MEPHISTOPHELES: *Whatever promise on your Books find entry, we strictly carry into act.... for the present I entreat Most urgently your kind dismission.*
>
> FAUST: *Do stay but just one moment longer then, tell me good news, and I'll release thee.*

Dear Mr. Morgan,

Since many years I have known one side of your character intimately. I believe that in my first approach to you I have given you evidence of this knowledge.... You have already put aside the money necessary to complete the work begun—in your thoughts—and that is as good as done. But I did not understand you as a businessman until lately.

I have worked for results carrying with them a dignity and force such as to deserve your attention. What you wanted was a simple result. Will you let me profit by this later knowledge and give me an opportunity to rehabilitate myself in your opinion as a businessman?[5]

In October 1903, fully two months before Orville and Wilbur Wright made aviation history, Prof. Sameul P. Langley launched a heavier-than-air ship from the roof of a houseboat situated on the Potomac. With photographers from the Smithsonian Institution present, the craft was slingshot "over the 70 foot rails and in a moment was free upon air. Then it wavered. Down the aerodrome sank...with its daring navigator [emerging from]...the disaster...suffering only a ducking." The press called Langley's airplane "a failure,"[6] but Tesla rallied to Langley's defense. "Langley has perceived a great truth," Tesla wrote in the *New York Herald*, namely, that "a machine heavier than air could be made to fly....Such a man should be provided with the necessary means to complete his work, great honor attaches to this achievement, [and] also great practical utility which this country can ill afford to lose."[7]

Tesla started off 1904 on the offensive with his manifesto and striking publications. Work on the wireless operation was suspended, and essential components were stripped from the tower and returned to digruntled creditors. A skeleton crew kept up appearances and continued the development of his lamps and oscillators. But as his Wardenclyffe notes reveal, there were no theoretical writings during this period;[8] all of Tesla's energies were concentrated in one avenue: raising the funds to resurrect the project. Tesla's attorneys had located a manufacturer in Connecticut to "make all the metal parts" of his oscillators, but there was still the problem of distribution, and the revenues would not enable him to reopen the Wardenclyffe plant. Another predicament had to do with Mr. Warden himself. Apparently he had not conducted a title search when he purchased the land, and a legal shadow had fallen on the property. Tesla used the entanglement to further delay payment of the mortgage.[9]

"One consideration," he wrote Scherff, "is that the Edison-Pupin-Marconi combination, who have given me so much trouble, are in a worse fix."[10]

In February, Tesla attended a musical recital and party in Gramercy Park hosted by Stanford White and his wife, Bessie, for 350 of his friends, with dinner afterward at Sherry's.[11] Most likely the inventor crossed paths with Morgan as well as other potential investors. The following month, the inventor conferred with the CEO (chief executive officer) of GE, Charles Coffin. "If [the GE people] refuse they are simply snoosers," he wrote Scherff.[12] Nothing came of the meeting, but in April a solid lead came via John S. Barnes, a well-connected financier who had read Tesla's article in *Electrical World & Engineer*. An associate of Col. Oliver Payne's from the Rockefeller clique, Barnes had invited Tesla to his home for dinner and to discuss the inventor's plans.

"I have always had the highest regard for Col. Payne, and would be happy indeed should he ever deem me worthy of his association."

"We are curious as to the details of the Commodore's bestowal," Barnes interjected.

"Mr. Morgan has not made a generous donation as you might have inferred from my article," Tesla craftily retorted. "He is a man with a great brain and has seen that [by forming a business partnership instead] he can make an extremely profitable investment."[13]

Although hesitant to make a commitment, Barnes nevertheless suggested that Tesla have his lawyers write up their evaluation of his patent applications.

Because of his link to Colonel Payne, Tesla took the suggestion very seriously. A multimillionaire from the city of Cleveland, Payne had made his fortune as a partner of John D. Rockefeller; the duo had earned fifty cents per barrel for every barrel of oil that was shipped by rail. This enormous kickback had been set up as a rebate for their own crude and as a tariff for every competitor. With their vast holdings and Rockefeller's ferretlike spirit, they had simply bullied the railroads into this contractual arrangement.[14]

Known as a haughty fellow and "kin of God," Payne never did take to John D., but he did maintain the partnership. He had a residence in New York and was a friend and financial benefactor of Stanford White, whom he commissioned to purchase art for him while in Europe and design his nephew's mansion, Payne Whitney, in the city.

White, who was in an awful fix, nearly three-quarters of a million dollars in debt, mainly from the Northern Pacific fiasco, informed Tesla that Payne had provided him with notable assistance to help ease his burden. White was also disheartened because his girlfriend, young Evelyn Nesbit, had now begun dating a deranged multimillionaire from Pittsburgh, Harry Thaw. "I have heard stories from the Floradora girls that he whipped one of them in bed with a cat-of-nine tails."

Ironically, White held no ill feelings for Morgan, even though the financier was directly responsible for the market crash. At the close of 1903, White and his wife, Bessie, joined the commodore on the *Corsair* to watch the yacht races, and White's partner, Charles McKim, was still busily involved in constructing the Morgan Library. One wonders how Morgan might have felt when he boated up to Newport or Bar Harbor and looked to the eastern horizon and saw Tesla's mushroom-shaped behemoth. "Do you think he will ever reconsider?" the inventor inquired.

"With Morgan, anything is possible" came the architect's reply. "However, I think at this stage, Colonel Payne is a safer bet."

A Yankee cornerstone, the good colonel was connected to the highest echelons in government. Through the marriage of his sister, he was linked to William C. Whitney, secretary of the navy, and also to John Hay, secretary of state; Payne's father, Henry Payne, was a well-known senator, often discussed as a potential candidate for president of the United States.[15] This was not a lead to take lightly.

As a favor to this Ohio nobleman, Tesla opened up his storehouse to the *Cleveland Leader* and conferred with Kerr, Page & Cooper for a way to create a legal document which would delineate the scope and fundamental might of his arsenal of patents. In a comprehensive article entitled "Harnessing the Lightning," journalist Alfred Cowles noted that the inventor's prognostications "were so startling that had they come from another source, one would naturally consider them the vagaries of a wandering mind. If he can accomplish what he is undertaking, his fame will, in future centuries, overshadow the greatest names of the past."

Echoing sentiments Tesla had expressed during their interview, Cowles concluded, "Real inventions are only possible when the mental creation of the inventor proves to be in harmony with natural law; and such inventions, when they are necessities, are in themselves a part of the evolutionary process, where development is an adjustment to environment."[16]

How all one whole harmonious weaves,
Each in the other works and lives!
See heavenly powers ascending and descending
The golden buckets, one long line extending.

Tesla presented to Barnes and Payne a comprehensive lawyers' brief delineating essentially every feature of his master plan. Included were patent specifications and plans for "distributing electric energy without wires for telegraphic, telephonic and industrial purposes," for storing the energy, localizing transmissions, insuring noninterferability, and for creating separate channels. Also included was Tesla's work in telautomatics, means for creating high frequencies, his oscillators, and "a method for insulating electric mains by refrigeration to very low temperature.... By

[this] means, power can be conveyed to great distances cheaply, and literally, without any loss." The plan also suggested the "perfect solution of the problem of underground distribution in cities and populated districts." Thus, the ultimate scheme would involve both wireless and conventional means of distributing the electrical energy. Analyzing the viability of each of the twenty-three patents, lawyers Kerr, Page & Cooper concluded, "We know of nothing to anticipate the claims and are of the opinion that they are valid."[17]

The report was also circulated to other major players, including Fortune Ryan and Pierpont Morgan. "I SWEAR," Tesla wrote to Scherff, "if I ever get out of this hole, nobody will catch me without cash!" Simultaneously, he haggled with the coal company to maintain fuel deliveries and with the telephone company to keep his line at Wardenclyffe open. "I am now sure that the two lamps as proposed will be a *perfect success* and you know that after that I can draw on the U.S. Treasury."[18] Problems with the lamp persisted, however, and it would never be marketed under Tesla's name.

October 28, 1958
Westinghouse Announcement
Lamp Division

> The Westinghouse Corporation [is pleased to] announce [that] a "flat light bulb" which has no filaments, which produces no heat, which is glare free, and which will burn night and day for a year for less than a penny...has been introduced....This marks the first time the public has been able to purchase an electroluminescent lamp as a light source for the home.[19]

As the Colonel Payne deal fell through, Tesla wrote to Morgan from Wardenclyffe, "I hope the unfortunate misunderstanding, the cause of which I have been vainly trying to discover will be removed...and that you will recognize that my work is the kind that passes into history and worthy of your support."[20]

Throughout the spring and summer, Tesla would visit the plant again and again for strength and confirmation. In June he instructed Scherff to make sure that the lawn was manicured at Wardenclyffe, as he was arriving with another potential investor.[21] Yet his resolve was beginning to waver; everywhere he turned, he met rejection. He became convinced that his success rested on changing the mind of a single man. In September a dispatch was hand-delivered to Morgan "assur[ing] contracts for several such plants in England and Russia,"[22] but no response came from the financier.

During the height of autumn, Morgan conferred with the archbishop of Canterbury.[23] Taking this as a mystical sign in the aid of his quest, the engineering cognoscente authored, on October 13, his favorite day of the month, a thirteen-point letter spelling out the entire chain of events to the omnipotent capitalist. The letter began with a discussion of his patent applications, the development of their liaison, and Tesla's decision to change the nature of the agreement because of Marconi's piracy. Simply transmitting mere Morse-coded messages was beneath consideration for the pompous conceptualist. As he had aligned himself with the greatest economic force on the planet, this confirmed the necessity of engaging in the larger endeavor.

> Your participation called for a careful revision of my plans.... Perhaps you have never fully appreciated the sense of this obligation....
>
> Once I lost your support I could not because of your personality and character of our agreement interest anybody else, at least not for several years, until the... commercial value of my patents [was] recognized.
>
> [By increasing the size of the transmitter] until... the plant can transmit signals to the uttermost confines of the Earth, its earning power becomes, so to speak, unlimited... but it will cost scarcely twice as much [e.g., $300,000]. ...[This] offered possibilities for a business on a large, dignified scale, commensurate with your position in life and mine as pioneer in this art, who has originated all its essential principles....
>
> You had told me from the outset that I should not ask for more, but the work was of such transcending importance... that I undertook to explain to you the state of things on your first return from abroad. You seemed to misunderstand me. This was most unfortunate....
>
> The audacious schemers who have dared to fool the crowned heads of Europe, the President of the United States, and even His Holiness the Pope have discredited the art by incompetent attempts [far more than they ever could by success] and spoiled the public by false promises which it cannot distinguish from legitimate right....and skill....
>
> I know you must be sceptical [*sic*] about getting hundredfold returns, but if you will help me to the end, you will soon see that my judgement is true.... I have expended about $250,000 in all and a much smaller sum separates me from a giant triumph....$75,000 would certainly complete the plant....

This letter (greatly condensed here) was a fair and accurate assessment of what had occurred and why. Clearly, it was written by a lucid savant, one who had proved himself in industry on many other occasions and one who was on the verge of altering the course of civilization in a dramatic and revolutionary way. Tesla was operating at the level of *soul consciousness,* and so he removed all defenses and revealed the very depths of his being with the following salutation and sacred vow:

> Since a year, Mr. Morgan, there has been hardly a night when my pillow is not bathed in tears, but you must not think me a weak man for that. I am perfectly sure to finish my task, come what may. I am only sorry that after mastering all the difficulties which seemed insuperable, and acquiring a special knowledge and ability which I now alone possess, and which, if applied effectively, would advance the world a century, I must see my work delayed.
>
> In the hope of hearing from you favorably, I remain,
>
> N. Tesla[24]

> October 15, 1904
>
> Dear Sir,
>
> Referring to your letter of the 13th October, Mr. J. P. Morgan wishes me to inform you that it will be impossible for him to do anything more in the matter.[25]

This cavalier dismissal tore the inventor apart and opened up a tendril that bared not only his wrath against a force that had blocked his crusade but also a poetic eloquence.

> October 17, 1904
>
> Dear Mr. Morgan,
>
> You are a man like Bismarck. Great but uncontrollable. I wrote purposefully last week hoping that your recent association [with the Archbishop] might have rendered you more susceptible to a softer influence. But you are no Christian at all, you are a fanatic musoulman. Once you say no, come what may, it is no.
>
> May the gravitation repel instead of attract, may right become wrong, every consideration no matter what it may be, must founder on the rock of your brutal resolve.
>
> It is incredible, a year and a half ago, I could have delivered a lecture here which would have been listened to by all of the academicians of the world.... That would have been the time to thank you. [But] you let me struggle on, weakened by shrewd enemies, disheartened by doubting friends, financially exhausted, trying to overcome obstacles which you yourself have piled up before me....

> "If this is a good thing, why does not Morgan see you through?" "Morgan is the very last man to let a good thing go." So it has been going on for two years. I advance, but how? Like a man swimming against a stream that carries him down.
>
> Will you not listen to anything at all? Are you to let me perhaps succumb, lose an immortal crown. Will you let a property of immense value be depreciated, let it be said that your own judgement was defective, simply because you once said no. Can now I make you a new proposition to overcome the difficulty? I tell you I shall return your money a hundredfold.[26]

The letter was followed up with testimonials to his abilities as espoused by various leaders in his field. It also explained in detail how this operation advanced the work performed at Colorado Springs. On December 16, Tesla sent an ultimatum. He requested either $100,000 to complete the plant or $50,000 "to finish the indispensable parts, make everything perfectly fireproof...and take out insurance," or "if you do not want to do this, only one thing remains. You release me of all obligations, give me back my assignments and consider the sum you have invested as a generous contribution leaving it all to my integrity and ability to work out the best results for you and for myself." Tesla suggested that he could go on a lecture tour to raise the funds; then it would take him "not more...than a week to get a few million in Wall Street."[27]

On the seventeenth, Morgan wrote back:

> I am *not* willing to advance you any more money as I have frequently told you. As to your third proposition, I am not prepared to accept this either. I have made and carried out with you in good faith a contract and having performed my part, it is not unreasonable that I expect you to carry out yours.[28]

> December 19, 1904
>
> Dear Mr. Morgan,
>
> Owing to a habit contracted long ago in defiance of superstition, I prefer to make important communications on Fridays and the 13th of each month, but my house is afire and I have not an hour to waste.
>
> I knew that you would refuse....What chance have I to land the biggest Wall Street monster with the soul's spider thread.
>
> ....You say that you have fulfilled your contract with me. *You have not.*
>
> I came to enlist your genius and power, not because of money. You should know that I have honored you in so doing as

> much as I have honored myself. You are a big man, but your work is wrought in passing form, mine is immortal. I came to you with the greatest invention of all times. I have more creations named after me than any man that has gone before not excepting Archimedes and Galileo—the giants of invention. Six thousand million dollars are invested in enterprises based on my discoveries in the United States today. [This is not a boast, Mr. Morgan; only my credentials.] I could draw on you at sight for a million dollars if you were the Pierpont Morgan of old.

At this point, Tesla refers to what he perceives as Morgan's breach of contract:

> When we entered our contract I furnished: (1) patent rights; (2) my ability as an engineer and electrician; (3) my good will. You were to furnish (1) money; (2) your business ability; (3) your good will. I assigned patent rights which in the worst case are worth ten times your cash investment. You advanced the money, true, but even this first clause of our contract was violated. There was a delay of two months in furnishing the last $50,000—a delay which was fatal.
>
> I complied conscientiously with the second and third obligations. You ignored yours deliberately. Not only this, but you discredited me.
>
> There is only one way to [go], Mr. Morgan. Give me the money to finish a great work...Or else, make me a present and let me work out my salvation. Your interest is sacred to me and my hearty wishes for your happiness and welfare will always be with you.
>
> Faithfully yours,
> N. Tesla[29]

To demonstrate goodwill, Tesla enclosed a royalty check on one of his patents and an advance copy of his theoretical masterpiece "The Transmission of Electrical Energy Without Wires as a Means of Furthering Peace." A fortnight later, on January 6, 1905, Morgan sent the 49 percent balance legally due the inventor.

THE WARDENCLYFFE PEACE PLAN

Published in *Electrical World & Engineer,* Tesla's treatise ran nearly six thousand words. It began with a discussion of how "philanthropy" and "the practical utilization of electrical vibrations," that is, a mass communication system, might bring about "universal peace." In defining and analyzing this theme, Tesla noted that it may come about suddenly, as the result of a slow accumulation of past efforts through history. "We must think cosmically.... The race enmities and prejudices are decidedly waning.... So far, however, universal harmony has been attained only in a single sphere of international relationship: that is the postal service....

"A few strong countries [might] scare all the weaker ones into peace," Tesla suggests, "[but] to conquer by sheer force is becoming harder and harder every day." Just as cruise missiles, CNN, and the world news organizations have altered markedly the way war is conducted today, by having machines fight instead of human (Nintendo war), and by usurping the old conventional chains of command, Tesla prophesied similarly in 1905: "Had only a few [of my] 'telautomatic' torpedoes been constructed and adopted by our navy [instead of rejected], the mere moral influence of this would have been powerfully and most beneficially felt in the present Eastern [Russian-Japanese] complication. Not to speak of the advantages which might have been secured through the direct and instantaneous transmission of messages to distant colonies and scenes of the present barbarous conflicts."

His treatise went on to describe a new "quasi-intelligent" missile guidance system that he was developing which would have a "greater range and unerring precision," but also that "misunderstanding" is the basis of wars. Speaking to Morgan in veiled terms that help to explain why so many of Tesla's most important theoretical writings can be found in the pages of the *New York Times*, the *Herald Tribune*, the *Sun* and the *World*, the inventor wrote: "Mutual understanding would be immensely facilitated by use of one universal tongue.... Next to speech we must consider permanent records of all kinds.... Here the newspapers play by far the most important part.... Disregarding the force of electrical invention, that of journalism is the greatest in urging us to peace.... That which is most desirable... in the establishment of universal peaceful relations is—the complete ANNIHILATION OF DISTANCE. To achieve this wonder, electricity is the one and only means."

If Morgan were to fund him, universal peace would ensue. This was quite a responsibility he placed on his benefactor's shoulders. Continuing for five more fact-filled pages, Tesla described in vivid detail his entire world telegraphy operation, precisely how it worked, and what it attempted to accomplish. Referring to the thunderstorm he had witnessed that summer night in the Colorado Rockies and his discovery of stationary waves, Tesla concludes: "On that unforgettable day, the dark God of Thunder mercifully showed me in his vast, awe-sounding laboratory [the geomagnetic pulse]. I thought then that it would take a year to establish commercially my wireless girdle around the world. Alas! my first "world telegraphy" plant is not yet completed, its construction has progressed but slowly during the past two years. And this machine I am *building* is but a plaything, an oscillator of a maximum activity of only ten million horsepower, just enough to throw this planet into feeble tremors, by sign and word—to telegraph and to telephone."

Although trying to assure Morgan that Wardenclyffe would *not* usurp the power companies, Tesla had to go on: "*When* shall I see completed that first power plant, that big oscillator which I am *designing?*...Which will deliver energy at the rate of one thousand million horse-power—one hundred Falls of Niagara combined in one, striking the universe with blows—blows that will wake from their slumber the sleeping electricians, if there be any, on Venus or Mars!...It is not a dream, it is a *simple feat of scientific electrical engineering,* only expensive—blind, faint-hearted world!"

Had he shown restraint, would this treatise have served its purpose, transforming the capitalist into the philanthropist? Probably not. In any event, Tesla further injured his position by ending the essay with a final stab: "Perhaps it is better in this present world of ours that a revolutionary idea or invention instead of being helped and patted, be hampered and ill-treated in its adolescence—...by selfish interest, pedantry, stupidity and ignorance; that it be attacked and stifled; that it pass through bitter trials and tribulations, through the heartless strife of commercial exis[t]ence. *So* do we get our light. So all that was great in the past was ridiculed, condemned, combatted, suppressed—only to emerge all the more powerfully, all the more triumphantly from the struggle."[30]

February 17, 1905

Dear Mr. Morgan,

Let me tell you once more. I have perfected the greatest invention of all time—the transmission of electrical energy without wires to any distance, a work which has consumed 10 years of my life. It is the long sought stone of the philosophers. I need but to complete the plant I have constructed and in one bound, humanity will advance centuries.

I am the *only man* on this earth *to-day* who has the peculiar knowledge and ability to achieve this wonder and another one may not come in a hundred years....Help me to complete this work or else remove the obstacles in my path.

I was heartily glad to see you in such splendid health yesterday. You are good for another 20 years of active life.

Faithfully yours,
N. Tesla[31]

Pressures continued to mount. Warden's lawyer pressed for a mortgage payment, and one of Tesla's previous workers, a Mr. Clark, was suing for past wages. When some of the bad news arrived at his Waldorf suite, the inventor "tore the letter sufficiently to prevent any undesirable person

[from] read[ing] the terrible open secret of Wardenclyffe." Tesla tried to maintain a balance by persevering with the development of his oscillators and other inventions, such as a transformer, a condenser, and a steam turbine, but these operations were long-range projects that would not provide immediate payoffs on the order needed to reopen the wireless operation. "The obstacles in my way," Tesla wrote Scherff, "are a regular hydra. Just as soon as I chop off a head, two new ones grow."[32]

Tesla began to write exclusively in pencil during this period, his writing becoming less distinct, lacking the boldness and clarity which had characterized it at earlier periods. Beginning to feel run-down on a more regular basis, Tesla likened his task to that of a weight lifter. "Every ounce counts now," he told his manager.

> March 10, 1905
>
> My dear Luka,
>
> Won't go to dinner [with you]. I am hard at work to get [Mrs. Filipov] that fine automobile.[33]

At the end of March, Tesla was "thunder struck at the Waldorf" when Warden's lawyer stormed the premises demanding immediate payment. The raising of even one of Mr. Boldt's eyebrows was enough to unsettle the struggling entrepreneur. "Hope to get out [to Wardenclyffe] Sunday." Tesla wrote Scherff. "Need it badly."[34]

Tesla set to work with his lawyers in April, finalizing patent applications for England, France, and Italy. But his inability to adequately compensate his workers on Long Island was creating a "[de]moral[izing] effect at Wardenclyffe. Perhaps we are nearing a revolution down there?" he inquired of Scherff. "Disappointments...and dangers...troubles and troubles again" continued to plague him.[35] He was beginning to crumble.

> May 1, 1905
>
> My dear Tesla:
>
> I know it will please you to hear of the great happiness that has come to me. Miss Grizelda Houston Hull...has consented to become my wife and the wedding has been [planned] for May 25th....Do you know, my dear Tesla, you are the very first person, outside of my family that I thought of and which the ceremonies will be very simple, I wish to feel you present in standing close to me on this occasion so full of incoming in my life.
>
> Indeed, I could not feel the occasion complete without you.
>
> Sincerely yours,
> Richmond Pearson Hobson[36]

The joyful occasion was a needed respite, Tesla "chatting" with Hobson's mother-in-law, hiding his worries, and characteristically teasing his friend. "Hobson," he declared, "now that you are married, your career is over."[37] This, of course, was far from the truth, as rumors had long been circulating about the possibility of grooming the charismatic lieutenant for the presidency.

"I must do something for our dear Mr. Tesla," Katharine told Robert on the ride home.

"What could you do that he hasn't?"

"Appeal to the king," she answered wistfully. Upon their arrival home, Katharine ran to the house to hide her tears.

While Morgan was away in Europe for his summer sojourn, Tesla met with Jacob Schiff, a man in the midst of lending the Japanese large sums of money in their war against the Russians. "S. said that perhaps he may take it himself," Tesla wrote hopefully to Scherff. "I believe that he will be a valuable man to me."[38]

In August, Morgan returned. Tesla sent Scherff to hand-deliver his newest list of patents granted, which the octopus grabbed in his tentacles and tossed into the wastebasket.

November 11, 1905

Dear Mr. Scherff,

Thirteen seems to be my lucky number. First of all, I met with Mr. F. just for a moment as he was going out of his office. He was most friendly and said that he was sorry he had to go out, but he will talk with me some other day. I HAVE MY MAN as sure as the law of gravitation. I know it.[39]

December 14, 1905

My dear Mr. Tesla,

I have received your letter of the 13th and in reply would state that I am *not* willing to invest any more money in the enterprise. I should be very glad if Mr. Frick would join you. You could have no better associate and I should be very glad to work with Mr. Frick in the matter, putting in what I have against his $100,000 to which you allude.

Yours very truly,
J. Pierpont Morgan[40]

Christmas was approaching, and Tesla had apparently struck a deal with another of the superrich. Frick, we remember, had earned upward of $60 million in 1901, when U.S. Steel was created. Ryan and Schiff were also involved in this new potential syndicate. Tesla wrote Morgan to thank him for allowing the liaison to come about. "You and Mr. Frick can take whatever you like. [I shall be satisfied] with a very small interest," Tesla wrote in a first draft. "I understand your attitude perfectly. You adhere strictly to principles. Never in my life have I [met] a man who even in a small [way can approach] the state [as described] by Goethe."[11]

Yet again Tesla reviewed their relationship, his decision to change the contract, and his ultimate wish to transmit power for industrial purposes. Why? Morgan sidestepped a meeting with Frick in December. An ill wind was blowing. Tesla began to crumble, his handwriting now barely legible, written as a meager, wispy stroke in pencil.

> December 24, 1905
>
> Dear Tesla,
>
> I have been sorry to hear of your recent illness—well concealed from your friends and the public—and I am also now very glad to hear of your recovery. Please stay well and strong.
>
> Yours sincerely,
> T. C. Martin[12]

> January 24, 1906
>
> Dear Mr. Morgan,
>
> I have just learned that the Germans have commenced the construction of a plant in all respects similar to mine which they expect to complete in a year.... Now, Mr. Morgan, you do not wish such a horrible thing to happen. If Frick will aid me, I can without delay, put my plant in operation by July next. Please see him at your earliest opportunity. I have not much time to lose.[13]

With the Johnsons just back from Europe and a meeting with "His Majesty in Rome," Kate decided to seize the moment. Completely on her own, Mrs. Robert Underwood Johnson called a hansom. "Twenty-three Wall Street," she said boldly. Morgan would not see her.[14]

> February 2, 1906
>
> Dear Mr. Morgan,
>
> Please see Mr. Frick.... He is going to call on you. Time is flying.[15]

A fortnight later, Tesla wrote again, pleading with Morgan to allow the formation of "a reasonable foundation upon which I am justly entitled." Tesla requested that Morgan agree to take one-third of the enterprise, thereby reducing his share by approximately 20 percent. "Please do not spoil the letter by unnecessary reference to your unwillingness of furnishing more money. The whole town knows it."[16]

Morgan did his best to scuttle the deal, and with it Tesla fell. His body shook violently, and his eyes began to bulge out of his head. Forgetting to shave or even shower, the ailing engineer grabbed the first train to

Shoreham and ran from the station to his precious tower, just a few hundred yards away. Clutching the girders for support as he climbed, the pulverized sorcerer ascended fifteen stories to the apex and looked out at the flat land which lay undisturbed for miles in every direction.

April 10, 1906

Dear Mr. Tesla,

I have received your letter and am very glad to know that you are vanquishing your illness. I have scarcely ever seen you so out of sorts as last Sunday; and I was frightened.

Yours sincerely,
George Scherff[17]

In May there was a portentous explosion in Bridgeport which caused a shock wave that was felt in Shoreham. "I hope my tower will not be subjected often to such tests," Tesla wrote Scherff. It was less than a month after the great San Francisco earthquake, but the jolt served to reawaken a new reserve of energy. Working with Scherff and a few men at the plant, Tesla continued development of his condenser and steam turbine which he was planning on placing inside a torpedo.

Concerned about the stability of the buildings America was constructing, Stanford White wrote the city of San Francisco, telling them to pass "stringent laws.... The hot riveted steel building stood the shock wonderfully," he concluded.[18]

White, like Tesla, was a victim of financial misfortune and he began drinking heavily. Only fifty-four, his health was in decline, and he was suffering from tuberculosis. In February the architect had planned to auction off $300,000 worth of his tapestries, sculptures, and paintings in order to reduce his debt by half, but a fire wiped out the uninsured holdings just two weeks prior to the sale.

Harry Thaw, having married Evelyn Nesbit, was having White followed day and night. On June 25, 1906, with Evelyn present, Thaw snaked his way down the aisle of the Roof Garden restaurant at Madison Square Garden, pearl-handled pistol in hand. While entertainer Harry Short sang "I Could Love a Thousand Girls," Thaw spotted and shot the "Beast" between the eyes. Stanford White, designer of Madison Square Garden, the agriculture building at the Chicago World's Fair, the Niagara Falls power plant, the Capitol in Providence, the Washington Arch in New York, Rosecliffe and the Tennis Hall of Fame in Newport, the Casino in Narragansett, the Boston train station, the Players' Club, numerous churches and other mansions, the new extension on the White House, and Wardenclyffe was dead.

Few attended White's funeral for he had been accused of raping a sixteen-year-old girl. But Tesla came.[49] The dream vanquished; the Gilded Age ended.

Throughout the year, Tesla's handwriting began to unglue, and by August it completely disintegrated, supporting the hypothesis that he suffered a nervous collapse at that time.[50] Entering his own private hell, the inventor was forced to endure an emotional enervation which caused a corresponding dysfunctional shift in his personality. Self-alienation took hold, bitterness and displaced anger became manifest as the quirks in his nature became more pronounced. Letters even to his closest friends would be signed N. Tesla, not Nikola. The untimely death of William Rankine at the age of forty-seven in September was another nail in the coffin of his dreams.

In a barely legible letter dated October 15, 1906, the last in this incredible series to Morgan, Tesla informed the Wall Street monarch that Messieurs Ryan, Schiff, and Frick were all willing to enter the partnership.

> Every opportunity is there.... I have high regards for you as a big and honorable man.... There is greater power in the leaf of a flower then in the paw of a bear. That is as much as I'll ever say.... You are reputed as a builder of properties, but if you prefer in this instance to chop down poles... go ahead.[51]

FAUST: *Gnash not so thy greedy teeth against me!—*
Great and glorious spirit, thou that deniest to appear to me,
who knowest my heart and soul, why yoke me to this shame-fellow,
who feeds on mischief and feasts on ruin?
MEPHISTOPHELES: *Who was it that plunged her into ruin? I or thou?*

36

THE CHILD OF HIS DREAMS (1907–1908)

I do not hesitate to state here for future reference and as a test of the accuracy of my scientific forecast that flying machines and ships propelled by electricity transmitted without wire will have ceased to be a wonder in ten years from now. I would say five were it not that there is such a thing as "inertia of human opinion" resisting revolutionary ideas.

NIKOLA TESLA, MAY 16, 1907[1]

"It's three o'clock in the morning, Mr. Tesla," George Scherff rasped into the phone as his wife grumbled in her sleep.

"The sheriff's seized the land."

"You owed Warden a hundred ninety-nine dollars!" Scherff said in amazement.

Fighting back a flood of tears, the inventor rasped, "I don't have it."

"I'll take care of it, Mr. Tesla."[2]

"Thank you," Tesla said as his hand limply hung up the line. His hair disheveled, his clothes scattered about, the former man of the hour was going to have to let the maid in soon. What would she say about the drapes he had placed over the mirrors? And then there was the tower. He had to go back *there* to seal up the property. Would he have the strength to make the journey?

His appetite all but gone, Tesla hadn't seen his friends for months. He managed a letter to Katharine as he rang room service to send up breakfast. "I'm ever in so much greater trouble," he scratched out on his letterhead.[3] But he would allow no one to truly know the hell he had entered. No sunlight must enter his room. He sat in the shadows and petted a wounded pigeon he had found floundering by the New York Public Library. If Boldt ever knew, the bird would have to be smuggled back out.

The withered man reached over to the envelope addressed to him in a feminine hand. Carefully he removed the letter and theater ticket. Marguerite Merrington had invited him to her new play *Love Finds a Way*. He stared at the title and broke down once again into uncontrollable sobs.[4]

Reconciling the torment, Tesla eased himself back into the social net as 1907 commenced. As part of his therapy, the recluse would surreptitiously board a moonlight train to Wardenclyffe. There, in the wizard's chambers, the Balkan genie would hook up high-frequency apparatus to his skull and thereby impress macabre waves of soothing electrical energy through his brain. "I have passed [150,000 volts]...through my head," Tesla told the *New York Times*, "and did not lose consciousness, but I invariably fell into a lethargic sleep some time after."[5]

In May, Tesla was inducted as a member of the New York Academy of Sciences.[6] Slowly, he began to see once again that perhaps his grand plan could be resurrected. To raise the capital to keep his ship afloat, the inventor took out a series of mortgages, subdividing the enterprise into a string of hypothetical parcels. In the spring of 1904 he had borrowed $5,000 from Thomas G. Sherman, a law partner of Stanford White's brother-in-law, and in the winter of 1906 he obtained $3,500 from Edmund Stallo, a son-in-law of one of Rockefeller's Standard Oil partners; but those funds had long disappeared. Having dodged the Waldorf management for nearly three years, he took out another mortgage for an additional $5,000 against the rent he owed with the proprietor, George Boldt.[7] And thus began a fresh plan for continuing to live in the lap of luxury without laying out another dime.

Boldt had done exceptionally well for himself. Having hobnobbed with the megarich for many years, the Waldorf manager had been able to take advantage of a number of inside opportunities. By 1907, a millionaire in his own right, he had expanded his base to become a banker, orchestrating the creation of the Lincoln Trust Company, which was located across the street from Madison Square Garden.[8]

Everyone except Tesla seemed to be flourishing. Morgan, through Jacob Schiff, had finally iced his deal with the Guggenheims to form the "Alaska Syndicate," an enormous corporation which had been set up to exploit a copper find in the inviolable northern wilderness. Whereas the Guggenheim mountain in Utah contained only 2–3 percent ore, this lode, according to John Hays Hammond's report, was 75 percent pure copper! A site of incalculable wealth, it would take a fleet of steamships, a thousand-man crew, and an up-front capital investment of $25 million to construct a railroad just to reach the find.

But copper was not all the "Morganheims" had their eye on. They also began purchasing coal and iron reserves and hundreds of thousands of acres of forestland. "Thus the press, the few environmentalists active at the time, and a significant portion of the American people began to vigorously oppose the Guggenmorganization of Alaska."[9]

With the growing need for copper wire came also a demand for insulation. Seizing the opportunity, Thomas Fortune Ryan and Bernard Baruch went to Europe to sign a contract with the king of Belgium (the former Prince Albert, an acquaintance of Tesla's). Their plan was to take over the rubber industry in the African Congo. The financiers negotiated an even split, with the king allocating 25 percent for his country and retaining 25 percent for himself. As Baruch returned to Wall Street to handle marketing, Ryan traveled to Africa to oversee the product's manufacture. Naturally, the tire companies were just as interested as the electrical concerns.

Once it was realized that Tesla's plans to do away with transmission lines had been abolished, it seemed as if there began a feeding frenzy on copper stocks, as this market was now assured a continually increasing demand.

PANIC OF 1907

The first signs of economic distress was heralded in August, when John D. Rockefeller of Standard Oil was fined the staggering sum of $29 million for price gouging and illegal tariffs. Suddenly, Wall Street became edgy. In October, F. Augustus Heinze, a well-known speculator and enemy of the Guggenheim syndicate, began dumping large blocks of United Copper onto the market. Heinze miscalculated in his attempts to buy back the stock at a much lower price, and his shifty scheming resulted in a drop in the market and a run on his bank, the Mercantile Trust Company. Due to Heinze's links to other financial institutions, the hysteria spread, and the Panic of 1907 began. Depositors emptied out every bank they could get into.

J. Pierpont Morgan called an emergency conference of all the bank and trust presidents, gathering them together in his newly constructed library in an all-night vigil. Sitting among his tapestries, original manuscripts, paintings, and jewels, the Wall Street monarch did his best to orchestrate a bailout of those institutions that were salvageable. Some, however, were beyond repair, and the stronger banks would go only so far in dipping into their reserves. Charles Barney, director of the Knickerbocker Trust Company, and father of "two awfully pretty sisters," pleaded for assistance, but he was rebuked. Barney went home and put a pistol to his head. This act produced a wave of suicides, particularly among the

Knickerbocker's eighteen thousand depositors. With Henry Clay Frick acting as liaison, President Theodore Roosevelt would transfer $25 million into Morgan's control. Although this figure matched the pledge of the stronger institutions, the new influx could only stretch so far. Boldt's bank, the Lincoln Trust, along with the Knickerbocker, the Mercantile, and half a dozen others, had gone under by the end of the week.[10] Now Tesla's chances of resurrecting his own enterprise became even more remote.

"These are simply awful times," Tesla told Scherff. "I cannot understand at all how Americans who are so daring and reckless in other respects can get scared to such a degree. My ship propulsion scheme is really great, and I feel sure that it will pull me out of the hole. Just how, I do not see as yet because it seems almost impossible to [amass] any money at all."

"We're still waiting to hear from the International Mercantile Marine Company," Scherff said.

"Be patient, my man. They are certainly interested, but make conditions which I am unable to accept for the present. If I had just a little capital I would not worry about finishing my place."

"What about Astor?"

"He told me over the phone, that he would see me as soon as possible, but up to present, nothing has materialized. I know now, that if I am to get capital, I can only get it from some fellow who has not less than a hundred million."

"Then, Mr. Tesla, let's hope for the best."[11]

"Dulled by [his] own suffering,"[12] Tesla began edging himself out of his depression by producing a number of acerbic essays for the electrical journals and local newspapers. Covering a wide range of topics, the inventor sought to vindicate himself and thereby try and make sense out of an absurdly ironic situation. Simultaneously, he sought to explain the Wardenclyffe vision yet again in the vain hope that some financier with a transcendent vision would come to his rescue. He was searching for a hero, not only for selfish desires but, in his eyes, for the future of the planet.

Under the guise of commenting on Commodore Perry's exploration of the North Pole, Tesla explained in detail the modus operandi of his world wireless scheme.[13] For *Harvard Illustrated*, he discussed Lowell's Martian discoveries and the way to signal the nearby planet;[14] for the *World* and *English Mechanic & World of Science*, he described how a tidal wave could be created by using high explosives to set the entire earth in oscillation and discussed how this wall of water could be harnessed to "engulf" an advancing enemy;[15] and for the *New York Sun* and *New York Times* he drafted a flurry of letters to the editor on such topics as his dirigible wireless torpedo,[16] the transmission of voice by means of wireless, the "narcotic influence of certain periodic currents" when transmitted through the body for therapeutic reasons, the inefficiency of Marconi's system, and the piracy of his oscillators by Marconi and another wireless inventor Valdemar Poulsen. Tesla also declared that the telephone was invented by Philip Reis before Bell and the incandescent lamp by King and J. W. Starr before Edison.[17]

Unlike Bell and Edison, Tesla wrote, "I had to cut the path myself, and my hands are still sore." After reviewing his bitter battle for vindication as the true author of the AC polyphase system against such "feeble men" as Professor Ferraris, the wizard went on to discuss his seminal work in wireless telegraphy. "It will never be possible to transmit electrical energy economically through this [planet] and its environment except by essentially the same means and methods which I have discovered," he declared, "and the system is so perfect now that it admits of but little improvement.... Would you mind telling a reason why this advance should not stand worthily beside the discoveries of Copernicus?"[18]

This was a new Tesla—resentful, indignant, defiant, petulant. He was the discoverer of the AC polyphase system, the induction motor, fluorescent lights, mechanical and electrical oscillators, a novel steam propulsion system, wireless transmission of intelligence, light, and power, remote control, and interplanetary communication. He was an original discoverer, whereas Bell and Edison had merely modified the works of others. How dare the world deny him his due?

Tesla's inventions were even at the heart of the new electric subway system which had just opened its doors beneath the thriving metropolis. Flooding, however, was a continual problem which marred this newest Tesla spin-off. The public had to be warned lest water cause corrosion of vital components, thereby increasing the risk of causing an explosion, and so another article advised the authorities on ways to cure the problem.

After one of his biweekly trips to his esteemed tonsorial artist for the warm compresses on his face and vigorous head massage to stimulate brain cells,[19] Tesla picked up his walking cane and strolled out in his green suede high-tops to Forty-second Street, to the entranceway of the freshly tiled Interborough catacombs. He was looking for new office space. Descending the staircase, the creator was overtaken with a pompous sense of pride as he stood by the tracks to await the next train. It was an almost magical experience for him to drop down in one part of the city, only to pop up majestically in another spot a few minutes later.

While waiting at a stop one ordinary day in 1907, he was approached by a lad and asked if he were the great Nikola Tesla. Catching a gleam in the inquirer's eye, the inventor answered in the affirmative.

"I have many questions to ask you," the youngster said as Tesla moved forward to step aboard the train.

"Well, then, come on," Tesla responded, unable to understand why the boy hesitated.

"I do not have enough money for the fare" was the embarrassed reply.

"Oh, is that all," the electronic savant chuckled as he tossed the youngster the required sum. "What's your name?"

"O'Neill, sir, Jack O'Neill. I'm applying for a job as a page for the New York Public Library."

"Good. We can meet there and you can help me research the history of some patents I am investigating."

O'Neill, who also had a keen interest in psychic phenomena, would go on a decade later to become a science reporter for the Long Island paper, the *Nassau Daily Review Star*. Eventually he took a position at the *Herald Tribune*, where he won the Pulitzer Prize before penning *Prodigal Genius*.[20]

In June came yet another legal suit, again from Warden, only this time from his heirs, as he had passed away. The amount was for $1,080, for money owed on an option Tesla had on four hundred acres adjacent to the two hundred he controlled.

"This is an old case which has been dragging in the courts for years," Tesla told the *Sun* reporter. "I [had] intended to use this land for an agricultural experiment in fertilizing soil by means of electricity. I thought that by the use of certain electrical principles [in producing nitrogen], the soil could be increased very much, [and thus I had] agreed to take a certain option. But subsequently [I] discovered that the person who entered into the agreement had no right to make such a disposition.... I told him the option was off... [but] the heirs of the owner had simply pressed the claim, and it is very likely that it will have to be paid."[21]

VTOL'S: A HISTORY OF VERTICAL & TAKE-OFF LANDING AIRCRAFT

June 8, 1908

My dear Colonel,

I am now ready to take an order from you for a self-propelled flying machine, either of the lighter or heavier-than-air type.

Yours sincerely,
Nikola Tesla[22]

Astor was particularly interested in flying machines, but as would become his habit, Tesla would be working at cross-purposes. He wanted the good colonel to fund this work in aeronautics, but, in actuality, his ultimate goal was to earn enough money so that he could return to Long Island and reopen his world telegraphy plant. Thus, any potential profits were always threatened by the greater plan. This problem would continue to encumber any possible deal, especially with someone like Astor, who knew full well the inventor's primary intentions.

One of Tesla's most confounding prognostications came at the onset of 1908. Having finally located a new work space at 165 Broadway, Tesla felt that he was getting back on track. Shortly after he moved in, he received an invitation to speak at a Waldorf-Astoria dinner in honor of himself and Rear Adm. Charles Sigsbee. "Th[is] coming year will dispel [one]...error which has greatly retarded aerial navigation," Tesla prophesied. "The aeronaut will soon satisfy himself that an aeroplane...is altogether too heavy to soar, and that such a machine, while it will have some use, can never fly as fast as a dirigible balloon....In strong contrast with these unnecessarily hazardous trials are the serious and dignified efforts of Count Zeppelin, who is building a real flying machine, safe and reliable, to carry a dozen men and provisions, and with a speed far in excess of those obtained with aeroplanes."[23]

Assuming that the viscosity of the atmosphere exceeded that of water, Tesla had calculated that an airplane could never fly much faster than "an aqueous craft." The inventor further reasoned that for highest velocities "the propeller is doomed." Not only was its rotational speed restrictive, it was also subject to easy breakage. The prop plane, according to calculations, would have to be replaced by "a reactive jet."[24]

In the short run, that is, for the next thirty years, the airship was the preferred method of passenger travel.

> BERLIN, May 30 [1908]. Count Zeppelin, whose remarkable performances in his first airship brought such signal honors, today accomplished the most striking feat in his career so far. He guided his Zeppelin II, with two engineers and a crew of seven aboard, a distance of more than 400 miles, without landing....
>
> All through the night the vessel...sped over Wertenberg and Bavaria, passing over sleeping countryside and villages and cities hardly less asleep....
>
> It was announced and widely published...that the Count would come to Berlin and land at the...parade ground. In expectation of the event...the Emperor and Empress...and hundreds of thousands gathered there.[25]

It would be two decades before Lindbergh would capture the imagination of the public by flying solo in a propeller-driven airplane across the high seas, but airships were already close to accomplishing that feat. In 1911, Joseph Brucker formed the Transatlantic Airship Expedition, but he was beaten in the quest by the British air force, which succeeded in crossing the Atlantic eight years later.[26] During World War I, the zeppelin ran frequent bombing missions from Berlin to London; Robert Underwood

Johnson flew with fifty other passengers in a similar "leviathan" over Rome just two years later, in 1919.[27] However, by the late 1920s this infamous legacy was all but forgotten, as these great airships were flying regularly across the Atlantic from Europe to both North and South America, and Germany was enjoying a reputation as the new leader in futuristic technology.

One curious and unfortunate footnote to history was the senseless choice to fill those blimps with hydrogen, a highly explosive gas, instead of nonflammable helium. Had engineers insisted on the much safer medium by heeding Tesla's 1915 warning, the great *Hindenburg* disaster of 1937 would never have taken place, and the use of zeppelins would probably have continued for many more years to come. The problem stemmed all the way back to the late 1700s, when Jacques Charles, a French scientist, discovered that hydrogen was fourteen times lighter than air and filled a balloon with it. Monsieur Charles, like Count Zeppelin of Tesla's time, gained great notoriety by traveling in his balloon fifteen or twenty miles at a stretch.

Today airships create stable platforms for TV sports cameras, advertisers use them because of their unique ability to generate "brand-name recognition," and the military likes them because they offer singular advantages over the helicopter. They can be used for low-flying rescue missions without creating hazardous turbulence; they can be used to detect the submarine launching of cruise missiles by positioning themselves in a single area for hours or days on end; and they are extremely difficult to locate by ground surveillance. "Why don't they show on radar?" a recent *Popular Mechanics* article asks. "Because the Skyship's gondola is made of Kevlar, the envelope is of polyurethane fiber and it's filled with helium. They all have little or no radar register.... The next-generation [air] ships would sip fuel. And they would stay operational for months at a time. As these are developed for military use, it is not too farfetched to predict that airships of the [21st century] may even be used for trans-Atlantic passenger service."[28]

Tesla reveals in his Waldorf-Astoria speech his prophecy of the inevitable development of the jet plane, which would be about as close as he would come to explaining his highly novel and still obscure invention of an airplane that operated much like today's VTOL "vectored thrust" aircraft. Tesla had played with a number of airship designs since his college days. One of his models, drawn up in 1894, was a traditionally shaped hot air balloon. Inspired by those he had seen at the World Fairs in Paris and Chicago, this dirigible received its continuing supply of heat from a gigantic induction coil that was placed high above the gondola, in the center of the hot-air container.[29]

The more recent model, which resembled a gigantic teardrop, took into consideration aerodynamic principles uncovered by such researchers as Leonardo da Vinci, Count von Zeppelin and Lawrence Hargrave, an Australian who, in 1890, fashioned rubber-band powered prop planes which traveled through the air over distances exceeding a hundred yards. This design was schematically prepared in the shape of a conventional airfoil by one of Tesla's draftsmen in 1908.[30]

> My airship will have neither gas bag, wings nor propellers.... You might see it on the ground and you would never guess that it was a flying machine. Yet it will be able to move at will through the air in any direction with perfect safety, higher speeds than have yet been reached, regardless of weather and obvious "holes in the air" or downward currents. It will ascend in such currents if desired. It can remain absolutely stationary in the air even in a wind for a great length of time. Its lifting power will not depend on any such delicate devices as the bird has to employ, but upon positive mechanical action.... [Stability will be achieved] through gyroscopic action of my engine.... It is the child of my dreams, the product of years of intense and painful toil and research.[31]

Tesla's vehicle had the "reactive jet" placed at its "leading edge," or bulky end, and the fifty steering escape valves placed at the opposite, "trailing edge," or tapered end. If fashioned as a lighter-than-air dirigible, the ship would have been modeled, in part, after the work of Henri Giffard, a Frenchman who invented the first dirigible in 1852, as well as Count von Zeppelin, the inventor who had been the first to construct a successful prototype with a rigid metal framework "within the bag."[32] Zeppelin was also one of the first to take into consideration wind resistance; his ships could travel at speeds of more than forty miles per hour.

A well-designed airfoil can develop "a lift force many times its drag. This allows the wing of an airplane to serve as a thrust amplifier.... [If] the thrust is directed horizontally, a vertical lift force large enough to overcome the vehicle's weight can be developed."[33]

Thus, it appears that Tesla's reactive-jet prototype could have also been fabricated in a heavier-than-air design. Oliver Chanute, M. Goupil, and O. Lilienthal were other Gay Nineties aeronauts whose patented works Tesla had studied. Naturally, he was also influenced by Samuel Langley and the Wright brothers, both of whom had produced heavier-than-air models that had actually flown.[34]

THE HOVERCRAFT

Another horseshoe crab–shaped VTOL designed by Tesla was called a hovercraft. This vehicle, which resembled a Corvette, placed the powerful turbine horizontally within its center. Operating much like a great fan, the engine created a heavy downdraft which caused the vehicle to rise up and ride along the ground on a layer of air.[35] This invention, which apparently worked much like the hovercraft depicted in the original *Star Wars* film, was the early precursor of the army's car-sized "aerial jeep," which "derive[d] its thrust from ducted fans mounted rigidly in the airframe. To fly horizontally, the entire craft [was] tilted slightly [by the leaning motion of the driver]." In 1960, *Scientific American* could write that "this design is being explored because of its simplicity and...adaptability for flying at very low altitudes."[36]

It is doubtful that Tesla ever constructed any of the heavier-than-air hovercrafts, although he may have built a hydrofoil model to skim over the Hudson. There is no doubt that he also constructed "lighter-than-air" vessels which could be operated by means of remote control.

Ideas inherent in Tesla's hovercraft and paramecium-shaped reactive-jet dirigibles evolved into today's Harrier fighter plane, a supersonic aircraft considered one of the military's "most potent fighting machines," and the new, yet-to-be-built Lockheed Martin X-33, which is a lightweight VTOL replacement of the space shuttle having a new experimental engine, the plane itself being shaped like a "flat-flying wing."[37]

The seeds of this technology can also be traced to the work of "A. F. Zahm, a prominent aeronautical engineer who patented [in 1921] an airplane with a wing that would deflect the propeller slipstream to provide lift for hovering." Although Zahm did not actually construct his plane, his concept, which may have been influenced by Tesla's work, evolved into the English Hawker, a British fighter developed in the 1960s. This airplane utilized nozzles to deflect a slipstream downward for vertical takeoff or for hovering and horizontally for normal flight. Utilizing "thrust vectoring," this apparatus became more workable with the development of the Pegasus engine, an extremely powerful turbojet found in the Harrier, which was unveiled in 1969.[38] "From the pilot's point of view, there is only one extra control in the cockpit: a single [lever] to select the nozzle angle."[39] "AV-8B Harrier: The U.S. Marines' ground-support jet can take off vertically, hover close to a battlefield and let loose missiles, cluster bombs or smart bombs."[40]

FLYING ON A BEAM OF ENERGY

Whether or not Tesla was able to perfect his design for aircraft that operated without any fuel—by deriving energy from wireless transmitters—is unknown. This concept, however, has been adopted by the military. In 1987, the *New York Times* and also *Newsweek* reported large glider planes "powered without fuel." Their energy is derived from microwaves beamed up from ground transmitters to large, flat panels of "rectennae" on each wing's underbelly. These "special antennas, laced with tiny rectifiers that turn alternating current into direct current, power an electric motor to run the craft's propeller."[11] This concept is also utilized as solar panels onboard spacecraft as well as on solar-powered automobiles.

The Flivver Plane

Tesla Designs Weird Craft to Fly Up, Down, Sideways

Craft Combines Qualities of Helicopter & Plane

> Detailed descriptions were available yesterday of the helicopter airplane, the latest creation of Nikola Tesla, inventor, electrical wizard, experimenter and dreamer.
>
> It is a tiny combination plane, which, its inventor asserts, will rise and descend vertically and fly horizontally at great speed, much faster than the speed of the planes of today. But despite the feats which he credits to his invention, Tesla says it will sell for something less than $1000.[12]

Although this article was written in 1928, Tesla first applied for patents on his new "method of aerial transportation" in 1921.[13] Nevertheless, designs for propeller-driven VTOL aircraft dated back even before the turn of the century. One of Tesla's earliest and most primitive helicopters looked much like a washbasin, with vertical shaft rising from its center. Flailing out, like the skeletons of two umbrellas stacked above one another, were its dual horizontal propellers. This vehicle evolved into the flivver plane, which took off vertically like a helicopter and then flew like a normal airplane, when the propeller and craft were rotated 90 degrees into the horizontal position. The concepts found in Tesla's flivver plane can be found in another advanced military VTOL aircraft called the V-22 Osprey. In this design, the body of the vehicle resembles a normal military transport plane. It is the propellers, at the ends of each wing, which rotate ninety degrees from the helicopter position, for vertical takeoff, into the normal airplane position for forward flight. Used in the recent war with Iraq (February 1990), this vehicle, like the aerial jeep and VTOL Harrier fighter jet, evolved directly out of Tesla's designs. As Tesla's work in aeronautics has never received much publicity, it is quite possible that the military adopted it in secret.

VTOLs can be grouped into four general categories. The aircraft could be tilted, the thrust could be deflected, the propeller or turbojet engine could be tilted, or a dual propulsion system could be utilized. Bell Labs began constructing propeller-driven VTOLs in the 1940s. Early models included the wing-tilted XC-142A, developed by Vought, Hiller & Ryan, and the X-19 propeller tilted craft, developed by Curtiss & Wright.

The New Weapons

Every service has its favorite new weapon, and the Marine Corp's favorite is the V-22 Osprey, an aircraft that can take off like a helicopter and fly like a plane. Just the craft to ferry Marines quickly and far into the desert, argue its manufacturers, Bell Helicopter Textron Inc. and Boeing Vertol Co.... [The vehicle can carry] 24 men and costs $40 million.[44]

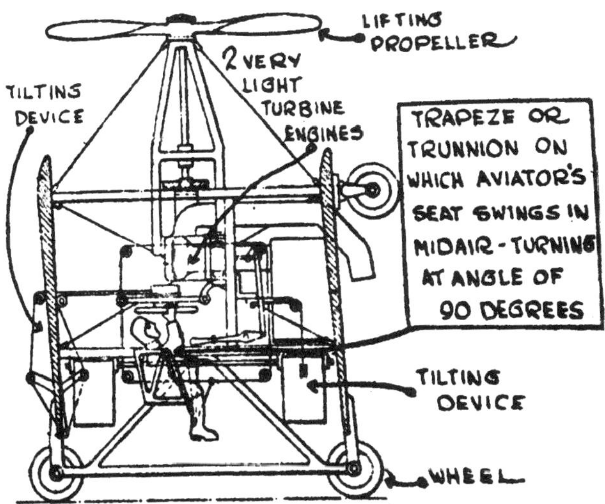

*Tesla's invention of the helicopter-airplane, which he called the flivver plane. (*New York American, *February 23, 1928)*

37

BLADELESS TURBINES (1909–10)

March 22, 1909

My dear Col. Astor:

I was very glad to know from the papers that you have returnedto the city and hasten to tell you that my steam and gas turbine, pump, water turbine, air compressor and propeller have all proved a great success. In the opinion of very competent men these inventions will create an enormous revolution. My gas turbine will be the finest thing in the world for a flying machine because it makes it possible to attain as much as 4 or 5 HP for each pound of weight. I have been hard at work on a design of the flying machine and it is going to be something very fine. It will have no screw propeller or inclined plane, rudder or wane—in fact nothing of the old, and it will enable us to lift much greater weights and propel them in the air with ever so much greater speed than has been possible so far. We are making up an automobile in which these new principles are embodied and I am also designing a locomotive for a railroad and am adapting my new propulsion scheme to one of the biggest Atlantic liners. All this information is confidential. I am merely writing knowing that you will be pleased with my success.

With kind regards, I am,

Sincerely yours,
Nikola Tesla[1]

With the death of the fragile poet Robert Watson Gilder in November 1909 came the advance in position from associate editor to editor in chief by Robert Underwood Johnson. Numerous dignitaries attended the somber affair, including Mark Twain and the latest rising star in the world of poetry, the twenty-four-year-old "wonder child" George Sylvester Viereck. It was an undesirable way for Luka to gain the promotion, but clearly the trustees had never considered anyone else for the post. Gilder's passing was yet another tangible sign of the end of an important era for the *Century*.

Tesla came over for Christmas dinner, and the discussion drifted to the problems Robert would now have in boosting a steadily declining circulation. In competition with a new crop of plebeian journals, Luka was forced to lower his standards by allowing the introduction of such four letter words as "hell" to grace the *Century*'s pages.

Katharine was interested in discussing Sir Oliver Lodge's recent contention that he had located a medium that had spoken to "dead members of the Society of Psychical Research," but Tesla thought such form of "wireless communication" poppycock. He was more interested in tearing apart Professor Pickering's supposition that he could erect a set of mirrors in Texas with $10,000 to signal the Martians.

"The idea that mirrors might be manufactured which will reflect sunlight in parallel beams, for the time being is beyond our range of ability. But there is one method of putting ourselves in touch with other planets," Tesla said as the eyes of his hosts lit up once again with the idea of Wardenclyffe. Capital, of course, was the problem, so Tesla began to describe his newest moneymaking scheme; it was his latest invention.[2]

Accused of being a visionary and dreamer, the consummate inventor "taxed his powers of concentration in the calm retirement of the night" to cultivate a way to bail himself out.[3] Often he would leave his Waldorf suite to walk the streets after hours and cogitate. His favorite sanctum was the colossal hall at Grand Central Station.[4] There, in the slumbering chapel, at four in the morning, he could follow the echo of his solitary thoughts down and into the tunnels to where the trains were berthed or up and around the majestic marble staircases which overlooked the vast commuters' arena and skyward to the starry dome, where the constellations and corresponding mythical gods were painted on the ceiling. This was *his* grand station for bouncing ideas off Pegasus or Hercules, Virgo, Centaurus, Gemini, Hydra, or Orion. Perhaps *Argus* (the ship) could provide a clue.

Wardenclyffe had become his obsession, and unless he was able to resurrect it in toto, he would never feel fulfilled. In-between measures were out of the question. Either he launched the entire edifice, or he would launch none of it. Scherff would visit the plant periodically with his wife, his father, and newborn baby and handle the money on the taxes and salary for Mr. Hawkins, who was retained as a guard.

But Tesla's competition had now caught up and in some ways was surpassing, if not replacing his vision. Airplanes and zeppelins were dotting the skies, the powers opposing illuminants without filaments were becoming more entrenched, and wireless transmitters were springing up like mushrooms at the banks of a woodland stream. In January 1908 the French placed a broadcast station atop the Eiffel Tower for the purpose of transmitting messages to Morocco. The director of operations predicted that such impulses "should theoretically go around the world [and] return to the tower."[5] Lee De Forest began to gain momentum in the States and soon began contracting with the government and the millionaires for the erection of "radio-telephones," which he placed on the roofs of the tallest structures in Manhattan. In 1907 he had aired the voice of Enrico Caruso, who was singing at the Metropolitan Opera House. Most of the listeners were in nearby boats. Concurrently, De Forest had refined a way to boost the speed of Morse-code transmissions. He could now direct telegrams at the astonishing rate of six hundred words per minute.

"I can confidently predict," De Forest proclaimed, "that within the next five years, every ship...will be equipped with the wireless telephone....I look forward to the day when by this means, the opera may be brought into every home. Some day, the news and even advertising will be sent out to the public over the wireless telephone." De Forest went on to criticize Marconi's devices, which still had not solved the problem of static interference, and predicted that his new system for tuning would eventually become standard.[6] The following year, he signed a contract for the "radio wireless" with Bell Telephone and installed operations between Philadelphia and New York.[7] Tesla was becoming a footnote to the field, Mr. Boldt hurling his own insult by hiring United Wireless to place two forty-foot wireless transmitters on the roof of the Waldorf and paying them $3,000 for the work.[8]

Marconi, however, was still the man of the hour, his name a household word, as the *New York Times* boasted in every Sunday supplement, a crest banner above their masthead boldly depicting Marconi wireless transmitters traversing continents and seas.

The Prime Mover

> The Tesla turbine is the apotheosis of simplicity. It is so violently opposed to all precedent that it seems unbelievable.[9]

With his wireless "project...evidently far in advance of the times," Tesla devoted himself "to other inventions which appealed more to practical men. After years of careful thinking, I found that what the world needed most...was an efficient prime mover." Tesla was referring to his new invention of a powerful and lightweight turbine, one that could be used to replace the gasoline engine in the car, fitted on airplanes, torpedoes, or ocean liners, or converted into a pump for transporting air, solids, or fluids. This remarkable machine could be used to create liquid oxygen or even be placed above incinerators to convert wasted heat into electricity. Born of Dane and Niko's childhood play with waterwheels in Smiljan, the multifaceted and revolutionary device first became manifest in 1906–07. It was called the bladeless turbine.[10]

Latest Marvel From the Monarch of Mechanics

Frank Parker Stockbridge

> "You have got what Professor Langley was trying to evolve for his flying machine, an engine that will give a horse power for a pound of weight," I suggested.
>
> "I have got more than that," replied Dr. Tesla. "I have an engine that will give ten horse power to a pound of weight. This is twenty-five times as powerful as the lightest weight engines in use today. The lightest gas engine used on aeroplanes weighs two and one half pounds [and produces one] horse power. With [that weight] I can produce twenty-five horse power.
>
> "This means the solution of the problem of flying," I suggested.
>
> "Yes, and many more," was the reply. "It is the perfect rotary engine. It is an accomplishment that mechanical engineers have been dreaming about ever since the invention of steam power."[11]

The inventor thereupon proceeded to explain its principles. Having studied the properties of water and steam as they passed through a propeller, Tesla explored the relationship of *viscosity* and *adhesion* to the blade's corresponding spin.

"The metal does not absorb any of the water, but [some of] the water adheres to it. The drop of water may change its shape, [yet its] particles remain intact. This tendency of all fluids to resist separation is viscosity," the inventor explained. By exploiting these principles, Tesla had patented an entirely new kind of turbine which did away with the blades of an everyday propeller and replaced them with a series of disks thinly spaced apart like a stack of pennies on their side. Each disk had a hole in its center for removing the incoming fluid and for turning the central shaft. Whereas "skin friction impedes a ship in its progress through the sea or an aeroplane through the air," Tesla exploited this seeming obstacle so that the spin of the turbine would be enhanced rather than retarded by the adhesion and viscosity of the medium. It was another stroke of genius from the master.

Spiral action was initiated at the periphery of each disk as the water formed a tighter and tighter corkscrew pattern as the center hole was approached thus augmenting spiral action. In this way, a fluid under pressure, such as steam, could enter the sealed chamber which housed the horizontal stack of disks and cause them to rotate. Following the natural tendency to create a whirlpool (like water exiting a drain), the fluid would naturally spin faster and faster as it moved toward the center. Simultaneously, its property of adhesion would carry, or drag, the corresponding disk around and around at a faster rate, and this spin could be used, for instance, as a turbine to generate electricity; reversing the entire process would turn the instrument into a pump; and hooking it up to an induction motor could transform the instrument into a jet engine.

> "One such pump now in operation, with eight disks, eighteen inches in diameter, pumps four thousand gallons a minute to a height of 360 feet....
>
> "Suppose now we reversed the operation," continued the inventor....Suppose we had water, or air under pressure, or steam under pressure...and let it run into the case in which the disks are contained—what would happen?"
>
> "The disks would revolve and any machinery attached to the shaft would be operated—you would convert the pump into an engine," I suggested.
>
> "That is exactly what would happen—what does happen," replied Dr. Tesla....
>
> "Then too," Dr. Tesla went on, "there are no delicate adjustments to be made. The distance between the disks is not a matter of microscopic accuracy....Coupling these engines in series, one can do away with gearing in machinery....The motor is especially adapted to automobiles, for it will run on gas explosions as well as on steam....
>
> "With a thousand horse power engine, weighing only one hundred pounds, imagine the possibilities. In the space now occupied by the engines of the *Lusitania* twenty-five times her 80,000 horse power could be developed, were it possible to provide boiler capacity sufficient to furnish the necessary steam....Here is...an engine that will do things no other engine ever has done."[12]

In January 1909, George Scherff, who was now working for a sulphur company, sent off a pleading letter for financial aid to Tesla. "My creditors are hounding me hard. Anything you can do for me will be much appreciated," Scherff wrote.[13]

Instead of sending him money, Tesla sent a check for Mrs. Schwarz, yet another disgruntled investor. In need himself, Scherff tried to divert the funds, but Tesla, having been in Scherff's position numerous times, wrote back lightheartedly, "I am sorry to note that you are losing your equanimity and poise. Mrs. Schwarz is weak and you are fully able to fight your own battles. You must pull yourself together and banish the evil spirits."[14] Shortly thereafter, Scherff sent off another note informing Tesla that he had prepared the taxes for Wardenclyffe. "A few nights ago," Scherff added, "a burglar entered my house and cleaned all the cash out of my pockets." Tesla took the hint and began recompensing his former secretary, sending a check in November.

November 11, 1909

Dear Mr. Tesla,

Thank you for the $200.... What gives me more pleasure than the money is the concrete evidence it furnishes of your progress towards the success for which you have battled so long and hard.

Sincerely,
George Scherff[15]

In March 1909, Tesla had formed the Tesla Propulsion Company with Joseph Hoadley and Walter H. Knight. With stock capitalized at $1 million, it was announced in *Electrical World* that turbines were being sold to the Alabama Consolidated Coal & Iron Company.[16] Tesla also set up other firms: the Tesla Ozone Company, capitalized at $400,000, which produced ozone, and the Tesla Electrotherapeutic Company, which marketed electrotherapeutic machines with Colonel Ray.

PRESENT-DAY OZONE THERAPY

During a recent symposium marking the hundred years since Tesla's arrival in America, G. Freibott, a medical doctor who used Tesla's ozone-producing equipment, stated that by injecting pure ozone directly into the bloodstream of a man afflicted with colon cancer, "thirty tumors were released." According to Freibott, this form of oxygen, which is naturally produced by the sun's action on the upper reaches of the atmosphere, contains "oxidizing, antiseptic and germicidal power...bringing palliative and curative results to many individuals." When questioned about the dangers of creating embolisms, Freibott noted that "air embolisms" are not caused by bubbles of oxygen in the bloodstream, as is commonly believed, but rather by impurities carried by the oxygen. This work is new and controversial, although verification of these findings has been established by physicians.[17]

Tesla, of course, did not inject ozone into people; however, he did construct an electrotherapeutic device for Scherff's wife, who was suffering from an ailment at that time. "I believe that it will do you and Mrs. Scherff a lot of good," Tesla wrote, adding for the sally, "unless you have no electric supply circuit in your home, in which case, it will be necessary to move into other quarters."[18]

Throughout 1909 and 1910 the inventor shuffled back and forth between Providence, Bridgeport, and New York City, where he had installed various renditions of his turbines. Most of the development work was at Bridgeport.

"I am now at work on new ideas of an automobile, locomotive and lathe in which these inventions of mine are embodied and which cannot help [but] prove a colossal success," he wrote Scherff. "The only trouble is to get the cash, but it cannot last very long before my money will come in a torrent and then you can call on me for anything you like." He added optimistically in another letter, "Things are developing very favorably, and it seems that my wireless dream will be realized before next summer."[19]

In March 1910, Owen's wife gave birth to their first son, Robert Underwood Johnson Jr., but in the spring a potential disaster of foreboding invaded the Tesla circle when it was announced that John Jacob Astor and his son Vincent were lost at sea. The inventor was one of the many who rejoiced when the news arrived that one of the ten richest men in the world (and his son) emerged unscathed. It is uncertain to what extent Astor contributed to Tesla's work with the turbine, however, there is some evidence that the inventor installed a hydrofoil jet engine in a "mysterious craft" Astor had docked in the Harlem River. The *New York Times* reported that the vehicle "seemed to embody an airship with a practical water craft."[20] If this was a radical flying machine which Tesla was working on, both he and Astor made sure that the reporters were kept away. One of the advantages of such a prototype was that the danger of death resulting from experimental flights could be minimized, since the craft was theoretically set up to hover only above water.

Feeling well on the way to success, Tesla wrote his friend Charles Scott at the Westinghouse Corporation for an order of a million induction motors to drive his turbines. "But as I have learned to go slow," he added to the letter, "I shall take only one at first."[21]

In November 1910, with his newfound tidal wave of momentum, he moved his headquarters to the prestigious forty-eight-story Metropolitan Towers, located at 1 Madison Avenue, just across the way from the Garden. With an office suite on the twentieth floor, right beneath the skyscraper's famous tower clock, the inventor could look out on the burgeoning metropolis from the tallest building in the world to plan his next move for the recapture of his holy grail, his world-telegraphy scheme.

38

THE HAMMOND CONNECTION (1909–13)

November 8, 1910

My dear Mr. Hammond [Jr.],

I was glad to read the enclosed newspaper reports. This is water on my mill. Just go ahead and make a lot of money and I will sue for infringement and we will divide.

Yours sincerely,
N. Tesla[1]

It is unclear as to exactly when, and in what capacity, John Hays Hammond Sr. became involved financially with Nikola Tesla. John O'Neill, who knew the inventor for nearly forty years, wrote in his biography that Hammond Senior gave Tesla a gift of $10,000 for the development of the telautomaton, which was unveiled in 1898.[2] John Hays Hammond Jr., or "Jack" Hammond, contradicted this assertion, writing twelve years after the book's publication, "My father was financing one of his later inventions and in this way, I had the opportunity of meeting him even while I was at Yale (1907–1910)."[3] Thus, based on Jack's letter, Hammond Senior most likely helped finance Tesla's bladeless turbine, although he may have invested in Wardenclyffe or some other enterprise.

In either case, it is unlikely that Hammond presented Tesla with an outright "gift," so it is clear that at least part of O'Neill's statement is incorrect. One of Hammond Senior's oldest friends stemming from childhood was Darius Ogden Mills. Both men grew up as California gold miners.[4] Mills, a long-standing friend of Stanford White's, became a principal in the Edison Illuminating Company back in 1883, along with J. Pierpont Morgan.[5] As a business associate of John Jacob Astor in the late 1890s, Mills was involved in the financing of the Niagara Falls enterprise and probably invested in Tesla's company as well. Tesla also knew Hammond's brother Richard, who had been to Niagara Falls to hear the inventor's invocation.

Having correctly anticipated a "depression" resulting from Grover Cleveland's 1892 election, Hammond had traveled with his wife and family to South Africa to run the Bernarto Brothers' gold and diamond mines. Thus, he was on the other side of the globe at the time of Tesla's work in telautomatics. Nevertheless, it is entirely possible that through Mills, Hammond participated in the venture. Jack Hammond, who would have been ten years old in 1898, would have therefore learned about this technology at an impressionable age. As the focus of Jack's extraordinary career revolved around his work in radio-guided weaponry systems, this early Tesla connection would help explain his ardent interest. Although Jack made no secret of crediting Tesla as being the primary inventor of telautomatics, he may still have wished to suppress Tesla's ultimate role in influencing so greatly the direction his life would take.

According to Jack Hammond's research, "Prof. Ernest Wilson in 1897 controlled a torpedo on the Thames by Hertzian waves. He is the pioneer inventor in this art."[7]

JOHN HAYS HAMMOND SR.

John Hays Hammond Sr., whose life became fictionalized as the "heroic Clay in *Soldiers of Fortune*,"[8] was the ultimate daredevil. Born in 1855, Hammond's maternal grandfather, Col. John Coffee Hays, was a Texas Ranger and the first sheriff of the "wickedest city in the world," the seaport and bonanza town of San Francisco. Raised in California during the gold rush, Hammond's father, Richard Pindell Hammond, was a West Point graduate and a friend of Robert E. Lee's and also Franklin Pierce. Hammond was also a gold miner and federal tax collector for the port of San Francisco.

Schooled at Yale University with a major in mining, Hammond continued his studies in the mid-1870s in Europe. After his return, the energetic adventurer set out for the Sierra Madre in his search for silver and gold. Traveling with his family and brother Richard, Hammond encountered Apache Indians on the warpath and Mexican desperadoes in his quest for buried treasure. "By way of encouragement," Hammond stated, "my wife frequently declared that in case Dick and I should be killed, she would faithfully promise to shoot: first the women,...then her child and then herself, rather than have them fall into the hands of the Indians."[9]

Other excursions included travel through alligator-infested swamps in Central America and "the cannibal country of Columbia."[10] Successful in finding gold in Guatemala, Hammond also opened up lead and silver mines throughout Mexico and the Midwest. In 1891, with a six-gun strapped to each hip, he helped quell a violent mining strike in Montana; but in 1893, unhappy with the new Democratic administration, he decided to leave America, taking his family with him, to fulfill his childhood dream of searching for diamonds in the depths of the Dark Continent.

Placed in charge of the British Consolidated Gold Fields, Hammond made his fortune when he realized that searching for diamonds twenty-five hundred feet under the ground would be much more lucrative when this type of land was selling for $10 per acre, whereas shallow mining stakes were going for $40,000 per acre.[11]

Among his children, the most precocious was a five-year-old named John Hays Hammond Jr., or Jack. There was also Harris, six years Jack's elder, Richard, a younger brother, and Nathalie, a little sister.

Swept into the Boer War in 1896, Hammond was arrested by the Transvaal government. Captured with Cecil Rhodes and the infamous Dr. Jameson, who had led a revolt against the Dutch, the elite members of the mining syndicate were sentenced to death by firing squad. With a plea from the U.S. secretary of state and perhaps a nudge from Mark Twain, who was in South Africa at the time, they were finally able to buy their way out. According to Hammond, Twain had informed the Dutch that they had captured "some of the wealthiest bugs in the world." President Krueger placed the ransom at $600,000, or $125,000 a piece. With Rhodes fronting the booty, the deal was struck, and they were released. Hammond, with his wife and family, were free to return to the States. He would pay back his share with future profits from new mining ventures.

Considered one of the wealthiest industrialists in the world, with a list of friends that included three presidents and former Yale classmate William Howard Taft, John Hays Hammond Sr. became a natural choice for vice president. Resigning from the Guggenheim copper coalition, Hammond sought the position as Taft's running mate with full vigor in 1908,[12] during the initial years of Tesla's partnership with his son Jack.

JOHN HAYS HAMMOND JR.

After a short stay in England in 1900, the Hammond family returned to the States and took up residence in Washington, D.C. Hammond Senior also had an office on Wall Street and a summer home in New Jersey. Having a keen interest in inventors, the mining engineer invited many of them to his home. Included on the list were Alexander Graham Bell, Guglielmo Marconi, Tom Edison, Nikola Tesla, and the Wright brothers.[13] In 1901, when Jack was just twelve years old, he was invited with his father to Menlo Park. There Edison, who was working on "a new process to extract gold from South African ore, showed Jack models of his first phonograph, and gave the youth some original sketches. It may have been this contact," Hammond speculated, "that stimulated my son's interest in the study of electricity."[11]

Shortly after Jack entered Yale in 1906, he began to study Tesla's inventions. He also worked for Alexander Graham Bell. Thus, it was during his college years that his interest in remote control became (re)awakened. "Tesla and Bell were, so to speak, my scientific god-fathers," Jack wrote in his diary. "I found them deeply inspiring."[15] Jack's "experiments started in early 1908, when he developed an electric steering and [also an] engine control for a boat...[finding] that he could control this mechanism over short distances with a radio impulse."[16]

It was at this time that the Hammonds set up permanent residence on the harbor at the fishing village of Gloucester, Massachusetts, and it was there that the enthusiastic engineering student performed most of his investigations. Destined to have more patents than any American inventor except Tom Edison, Jack began his interest in inventing during his New Jersey prep-school years. His first significant creation, at age sixteen, was a reverse switch which automatically turned off his night light when the headmaster opened his dorm door to check to see if he was reading after curfew.[17] After this, the floodgates were opened, and by the end of his career, John Hays Hammond Jr. had amassed an astounding array of over eight hundred patents, including inventions in the fields of military warfare, music and sound (no relation to W. H. Hammond of electric-organ fame), and home appliances. Some of Jack's most unique contributions include a cigarette case which "popped out a lit cigarette when opened," a microwave oven, a push-button radio, a superheterodyne (which greatly amplified radio waves and was coincident with Edwin Armstrong), aircraft guidance systems, a time-controlled gas bomb, a magnetic bottle cap, a combination piano-radio-phonograph, a windshield washer, a mobile housing unit, and a "telestereographer," or "mechanism for projecting three-dimensional images via wireless."[18]

In September 1909, during his senior year, the budding wunderkind wrote to his father to arrange for a meeting with the "Serbian High Priest of Telautomatics."[19] "Father, I have some important information that I desire to get from Mr. Tesla."[20]

Hammond Senior, who had just lost his bid for the vice-presidential slot, made the arrangements. Jack met the fifty-three-year-old inventor at his Metropolitan Towers office in New York the week of September 26, and it is most likely that Tesla reciprocated by visiting Gloucester shortly thereafter. Hammond requested that Tesla send his patent information on wireless control of machinery, and Tesla did so before the end of the month.[21]

Moonlighting at the patent office in Washington D.C., Jack had already rigged, by this time, a forty-foot vessel to be maneuvered by means of wireless. His broadcasting system, in part based on a Marconi design, also utilized Tesla oscillators and contained "two 360-foot radio control towers near the laboratory overlooking Freshwater Cove.... With these devices, a man standing at a shore lookout station could steer an empty boat in the water."[22] Jack also asked Tesla to speak at his Yale graduation.[23]

This period in Tesla's life was marked by extreme bitterness because Marconi was rewarded for his piracy with the Nobel Prize in December. Tesla informed Jack that the Italian tinkerer had "abandoned the old devices of Hertz and Lodge and substituted mine instead. In this manner the transmission across the Atlantic was effected."[24] Jack, however, held no ill will toward Marconi and included him prominently in the four-volume compendium he was writing on the history of wireless communication. He also invited Marconi to the Gloucester compound and formed a friendship that would last well into the 1930s.[25]

Having returned from a European excursion, where he had visited electrical engineers (and psychic researchers) in London, Paris, and St. Petersburg, Jack was able to complete his master's dissertation:

> Mr. Tesla in 1892 showed that the true Hertzian effect was not a means by which it was possible for a sending station to communicate with a receiving station at any great distance. He demonstrated furthermore, that waves propagated at a transmitting station travelled along the ground as a conductor. Today [1912] it is acknowledged that these views are correct. It was, however, left to the splendid enterprise of Marconi to crystallize the results of previous investigators into a complete and practical system of space telegraphy.... In 1897 Mr. Marconi transmitted messages to a distance of 8.7 miles. Today Mr. Marconi says that the maximum effective distance of transmission is 6,000 miles.[26]

Perturbed with Hammond's decision to highlight Marconi's dubious achievements and proceed in his field of remote control, Tesla sought compensation. Working with Fritz Lowenstein and Alexander Graham Bell, Jack invented a "mechanical dog" which followed its "master" when a lantern was beamed at it. Created in the shape of a milk carton on wheels, the "critter" made use of selenium cells for "eyes" to receive the glimmering command signal. Hammond assured Tesla that he was not infringing upon his work in telautomatics, but Tesla remained unconvinced, especially after an article in the newspapers reported that Hammond was in the midst of displaying remote controlled torpedoes to the military.

> My dear Mr. Hammond,
>
> Judging from the enclosed, I think that you are playing a

> wireless possum. Notwithstanding your assurances, I will watch your progress and bring a friendly suit for infringement as soon as I ascertain that you are in funds.[27]

Jack wrote back to reconfirm that he would give proper credit, but Tesla wanted a contract and a percentage of any profits.

"My dear Mr. Tesla," the twenty-two-year-old wrote back, "I am very agreeable to share the profits with you, but I shall only on the condition that you share our liabilities also."[28]

"As I naturally surmise that your Papa would pay all our liabilities," Tesla replied, "I am willing to share in these."[29]

Aside from the banter, Tesla was hoping that Hammond would succeed in his interface with the military, for now he would have a market for selling his new bladeless engines. Soon a partnership was formed, Hammond Senior footing the bill.

"Go in on this with your brother Harris," Jack's father cautioned. "He is older than you and more experienced. And be careful with Mr. Tesla. He tends to spend gold as if it were copper."

Having studied Tesla's method of selective tuning, Jack came to call it the "1903 prophetic genius patent."[30] Tesla had created this invention due to a recurrent problem which he had noticed in 1894–95, namely, that he had been having difficulty illuminating particular bulbs in his laboratory without illuminating others. After studying the work of Herbert Spencer on the combined action of two or more nerves in the human body, the inventor came upon a plan whereby bulbs would illuminate only when a combination of more than one frequency was transmitted. Jack noted that "Mr. Tesla resembles his system to a combination lock."[31] Explaining the particulars to the initiate, Tesla showed that devices could be made to respond not only to one frequency but to two, three, or even more. This combined arrangement, analogous to today's TV and telephone scramblers, would not only ensure privacy, it would also allow for a system with a virtually unlimited number of separate channels.[32]

The corresponding patent, along with Tesla's method of utilizing resonant earth frequencies for transmission, that is, Tesla currents, became the backbone for a plethora of inventions ranging from military guidance systems to radio and telecommunications. The further development and refinement of this foundation would also eventually make Jack and a few other inventors millionaires, for example, Edwin H. Armstrong.

In 1911, writing from the patent office, where Jack was still employed, the willing prodigy informed Tesla that he had contacted the War Department with the hope of selling ship-to-shore communication systems capable of transmitting twenty words per minute. Jack had also begun the construction of a military-linked think tank at Gloucester, where he hired such competent engineers as Fritz Lowenstein and Benjamin Franklin Meissner. Born in 1890, Meissner, who would come to author a textbook on radiodynamics with some help from Tesla,[33] became chief assistant at the lab. Having worked for the U.S. Navy in 1908, Meissner had aided in the development of the electric dog and superheterodyne. He is also credited with inventing the "cat whisker" which was a detector on the crystal radio set.[34] Jack was also conferring with Reginald Fessenden, Lee de Forest, John S. Stone, and Guglielmo Marconi.

> Washington, DC February 16, 1911
> Dear Mr. Tesla,
>
> Let us create an unpretentious company and call it the Tesla-Hammond Wireless Development Company. In thinking of this name, I have followed Emersonian advice, and as you see, have attached my chariot to a star....
>
> The purpose of this company would be to perfect an automatic selective system, to perfect the [submersible] torpedo, and eventually to carry out your magnificent projects that will wirelessly electrify the world.
>
> I am most sincerely yours,
> John Hays Hammond Jr.[35]

> 202 Metropolitan Towers February 18, 1911
> New York City
> Dear Mr. Hammond:
>
> The Tesla-Hammond combination looks good to me, but we should have to go at it with some circumspection. I have already interested a gentleman who signs himself J.P.M. in a part of my wireless inventions and my friend Astor is now waiting for the completion of my plant to go into the wireless power transmission business which should be a colossal success.
>
> In the art of Telautomatics, however, I am perfectly free and would be glad to go into any fair proposition to exploit the field. I think that in a few years this departure will command the attention of the world.
>
> I have just completed my turbines and am starting Monday to install them at the Edison plant where I expect to show them to you in operation on your next visit to the city.
>
> With kind regards,
> N. Tesla[36]

Writing on his classy Wardenclyffe letterhead, with the magnifying transmitter posing at the top of the page, the inventor pens, in the estimation of this author, a most exasperating dispatch. The tiff that Tesla had with Morgan had been held in secret. Only a handful of people even knew the details of their contract. Even Tesla's closest friends and latter-day biographers were kept in the dark. However, on another level, it had become obvious by 1911 to all but Tesla that Wardenclyffe was a ship with a lead hull.

Still intoxicated with the world-telegraphy idea, filled with hope that his new bladeless turbines would cause a revolution, the perennial iconoclast embarked foolhardly, albeit courageously, on the best-case scenerio: that he would raise enough capital with new inventions to finally return to Long Island to complete the tower.

Perhaps it was still possible at this juncture of his life if his motor, for instance, replaced the gasoline engine in the automobile or the prop engine in the airplane. However, what was not possible was the intimation in the letter that Morgan was still "interested." This was a blatant display of disinformation which the pompous Waldorf dandy conceitedly proclaimed in order to hide the very fact, maybe even to himself, that his optimism was possibly delusion.

Here was a chance to develop a concrete wireless system with the backing of the wealthy and powerful Hammond lineage, but Tesla turned the opportunity away because of arrogant, tunnel-minded, and narcissistic proclivities—and possibly because of contractual limitations imposed by the Morgan contract. Had he developed the wireless scheme with Jack, he may have had to legally compensate Morgan with 51 percent of any developments he achieved. Hammond hadn't figured that the star he had hitched himself to was a half-baked comet.

THE APOSTLE OF FIGMENTS

In May 1911, T. C. Martin invited Tesla to address thirteen hundred members of the National Electric Light Association, which was holding its annual symposium at the Engineering Society Building on Thirty-ninth Street.

"There is no enjoyment that I could picture in my mind so exquisite as the triumph which follows an original invention or discovery," the inculcation began. "But the world is not always ready to accept the dictum of the inventor, and doubters are plentiful, so that discoverers have often to swallow bitter pills, along with their pleasure."

But what magniloquent pills would this mad scientist force his congregation to ingest! Tesla proceeded to dazzle the audience with slides of his AC polyphase system, telautomaton, and world wireless experiments, flashing pictures from Colorado with streamers extending sixty-five feet.

In discussing his method of individuation, he stated that the broadcasting of combination and multiple frequencies was benefited in a system that did *not* use wires. "All the statements that you read in the newspapers that wireless messages are interfered with," the inventor explained, "are because the workers in that field are laboring under delusions—they are transmitting messages by Hertz waves, and in this way no secrecy is possible."

Having skyrocketed his vision to a world which was not of this earth, the wizard stepped into the shoes of Prometheus. "Now, the discovery [of standing waves that] I have made upset all that has gone before, for there was a means of projecting energy into space, absolutely without loss from any point of the globe to another, to the antipodes, if desired. In fact, a force impressed at one point could be made to increase with distance.... You can imagine how profoundly I was affected by this revelation. Technically, it meant that the earth, as a whole, had a certain period of vibration."[37]

Set against a sky of thunderclouds, the Wardenclyffe citadel was flashed on the screen, its mushroom-shaped vertex looming.

"I have annihilated distance in my scheme," deus roared, "and when perfected, it will not be one mite different than my present plans call for. The air will be my medium, and I will be able to transmit energy of any amount to any place. I will be able to issue messages to all parts of the world and send [forth] words which will come out of the ground in the Sahara Desert with such force that they can be heard for fifteen miles around."[38]

"It would be possible by my powerful wireless transmitter, to light the entire United States. The current would pass into the air and, spreading in all directions, produce the effect of a strong aurora borealis. It would be a soft light, but sufficient to distinguish objects."[39] Naturally, the tower would also be powerful enough to send signals to nearby planets, especially if there were any Martians out there to receive them.

And this was just the introduction to the topic he came to divulge that night: the Tesla bladeless turbine.

Tesla began to work on his new engines in earnest, shuffling between Providence and Bridgeport, with most of his operations now shifted to the New York Edison Waterside Station. He also looked for prospective clients. One of his plans was to sell, probably through Jacob Schiff, five hundred engines to the Japanese. "By applying my turbine to their torpedo," Tesla wrote Jack's brother Harris, "I can double the power. We should negotiate royalties on the basis of horsepower."[40] Tesla also conferred with GE and the exuberant Seiberling Company, leaders in the development of high-speed power boats.[41]

Promising "great success," the vulcan worked overtime, forging his revolutionary equipment, as Jack continued perfecting a prototype remote-operated boat and a wireless broadcasting station. With a range of two thousand miles, the Hammond transmitter became "the most important private sending station in the world." Jack also studied telephotography, and he worked on perfecting his electric dog. "And, if you reverse the motor by pushing over the tail switch," Jack announced to the press, "you can make the dog back away most surprisingly in either direction when you advance upon him with the light."[12]

Expanding his market, Tesla designed prototypes that could transmutate the gasoline engine in the automobile; he began to make overtures to Ford Motor Company and also Kaiser Wilhelm of Germany, who was planning on putting them in tanks. As with any new creation, there were problems. For instance, because the ball bearings were wearing down too quickly, the disks were not maintaining their spins at optimum accelerations. As a "sun dodger," Tesla also had a penchant for working throughout the night, and therefore his labor costs were often double. Naturally, there were other expenditures.

Jack suggested more publicity and sent over the well-known journalist Waldemar Kaempffert to interview, in Kaempffert's words, the "temperamental genius" for *Scientific American.*[13] But in Tesla's estimation he had enough publicity. He needed more capital.

Throughout the latter part of 1912 and through the first months of the new year, Tesla sent urgent pleas to his partner. He had expended $18,000, had worked without salary for all this time, and required $10,000 back immediately.

> Dear Mr. Hammond,
>
> ...in desperate need of money. I am unable to hold out any longer.[14]

But Hammond, who was helping Lowenstein install his wireless equipment aboard navy ships and competing against De Forest for a $50,000 amplifier deal with AT&T, ignored the request, his brother Harris taking a full quarter of a year to respond:

> June 10, 1913
>
> Dear Mr. Tesla,
>
> As you know, we have advanced a great many thousands of dollars in the development of this turbine and have expected each week the past year to be in a position to have tested it...[Now] we find that the turbine is only partially set up at the Edison plant....[We are missing] a splendid opportunity of having it thoroughly and honestly tested by people who would be the greatest benefit to us should these tests be successful.
>
> Sincerely yours,
> Harris Hammond

The High Priest of Telautomatics was incapable of believing that the son of one of the richest men in the world was scorning his entreaty. "Since my notice, I have done the best I could to save what was possible, the sacrifices which I have been compelled to make and the losses which I have suffered are such that if I were dealing with a man less attractive to me than yourself, I would disdain to answer." Tesla also enclosed glowing testaments to the turbine from professors and chief engineers, but the partnership was over.[15] Hammond would not come through.

THE CASTLE THAT JACK BUILT

Jack Hammond traveled to Europe only months before World War I to confer with various scientists in order to perfect a better receiving instrument than the Marconi coherer. It appears that Tesla and Hammond were working at cross purposes. As would become obvious, the $10,000 Tesla requested would not have been sufficient to complete his work on the turbines. He probably needed forty or fifty times that figure, and Jack's main interest lay in the perfection of wireless transmitting and receiving apparatus. The torpedo propellants were really secondary.

In retrospect, it seems that Tesla might have been better off abandoning the turbine for the time being and working with Jack to perfect the guidance system; but he was too close to a potential major success to sink more time into an invention he had already perfected fifteen years earlier. Jack would go on in 1913 and 1914 to demonstrate his remote-controlled boat before the U.S. military elite. General Weaver, chief of the U.S. Coast Artillery, and a small entourage traveled up to Gloucester to witness the *Natalia*, the prodigal son's newest success, the general even taking the controls himself. "Again and again the flashing craft shot forth and manoeuvered [*sic*] about the harbor under invisible control, while natives of Gloucester gasped in amazement.... They saw her headed for a definite mark a mile away, two miles, three miles away, and strike it with precision every time."[16]

A few weeks later, Hammond Junior demonstrated the long-range capabilities of the vessel. It could operate while twenty miles away from the Gloucester radio transmitter and in one way or another was directed the full distance of sixty miles to the naval base at Newport by means of wireless. Jack had also perfected the problem of static interference and selective tuning. In December he wrote:

> My dear father,
>
> We are now drawing up as systematically as possible the whole proposition to the present to the Board of Ordinance. This work means a good deal in the future financial success of the proposition.

I am your affectionate son,
John Hays Hammond Jr.[17]

It would be many years, however, before the U.S. government recompensed Jack for his remote-controlled guidance system. He would expend over the next decade in the neighborhood of three-quarters of a million dollars on the operation, expanding the system of radiodynamic control to include aircraft as well submersibles.[18] Problems in the creation of secret channels were made apparent in 1915 and 1916 when the USS *Dolphin* successfully interfered with a torpedo launched by Hammond over distances of two hundred to three hundred feet, but the Hammond system was successful when the torpedo was launched farther away.[19] The War Department also wanted to sustain visual contact of the weapon, so Jack began working on a device to be directed from flying machines. Every problem he encountered he was able to overcome.

The technology was still too new to be used during World War I, and the military kept avoiding expenditures of any funds. Ostensibly waiting for Tesla's fundamental telautomatic patents to expire, Jack finally presented his case before members of Congress. He stated that he had overtures from foreign governments, but he would refuse to negotiate with them because of the importance of the work and loyalty to his country. And so, in 1919, while his father, John Hays Hammond Sr., continued to gain publicity for his idea of a World Court to prevent war, the U.S. Congress and President Wilson approved an appropriation of $417,000 for the war patents for Hammond's son; however, still no monies changed hands.[50]

In the 1920s, Jack began working with David Sarnoff, who, along with Guglielmo Marconi and Edwin Armstrong, was forming the seeds of the Radio Corporation of America (RCA). And, in 1923, the fruits of his labor paid off when Jack sold a series of wireless patents to RCA for $500,000,[51] but he was still to be compensated by the U.S. government. In 1924, Hammond sent yet another dispatch to the War Department to gain the release of the appropriated funds, which were now up to $750,000. "I have brought the development to a state where we have demonstrated the feasibility of control of standard naval torpedoes while running at depths of 6 feet or over, submerged at speeds of 27-30 knots/hr," Hammond wrote.[52] Finally, in December 1924, with the help of Curtis Wilbur, secretary of the navy and admirer of Hammond's father, the government made good, recompensing Hammond and assigning his work to a secret file in the patent office. Assurances were also given that their exclusive patents did not compete with those sold to RCA. All this took place a full decade after his breakup with his Serbian mentor, who "now [had] the pleasure of simply looking on when others are using my inventions."

"I wish him luck," Tesla said. "But still, I ought to have had something for it." Tesla also pointed out that Hammond sought his patents just a few months after Tesla's had run out.[53]

Now a millionaire on his own merits, Jack set out to fulfill a dream that began in his youth, when the family moved to England, namely, to live in a castle. He also fell in love with an artist, Irene Fenton, a lovely daughter of a shipbuilder who unfortunately was married to a shoe merchant. Irene, forty-five, got divorced and married Jack, thirty-seven, clandestinely in 1925, as he began construction of the medieval dwelling which was situated on the cloistered and treacherous coastal cliffs where Longfellow wrecked the fictitious schooner *Hesperus* in Gloucester. The site was less than a mile from his parent's home.[54]

Jack's passion was music. Although not a musician, he had many acoustical patents and also an organ that was so enormous, comprising eight thousand pipes, that a palace would be the only edifice capable of housing it. Hammond designed the all-stone building around the instrument, complete with parapets, a moat, and a chain-link drawbridge. Inside could be found dark, winding corridors, hidden doorways, and breakaway walls at the entrance of the great room for moving organ pipes into it and out. In the center of the castle, which today is a museum, Hammond placed an indoor swimming pool and an atrium filled with plants and tropical birds. Ancient artifacts from Europe were purchased, and a nude statue of the celebrated innovator was sculpted, Irene designing a metal fig leaf to subdue the piece.

Jack continued to work on a long series of top-secret inventions for the War Department and for himself as he lived the life of the bon vivant. Visitors to his estate and to his pipe-organ concerts in the 1930s included the Hearsts, George Gershwin, Helen Hayes, David Sarnoff, Ann and Theodore Edison, the Marconis, J. Pierpont Morgan's daughter Louisa and her husband, Herbert Satterlee, Helen Astor, Marie Carnegie, David Rockefeller, the Barrymores, Noël Coward, and Leopold Stokowski.

It is doubtful that Tesla ever visited the castle, although he might have, but on March 30, 1951, nearly a decade after the Serb's death, another Slav and Teslarian, Andrija Puharich, stopped by.[55] Still interested in extrasensory perception, Jack had invited Puharich, a medical doctor and inventor of hearing aids, along with psychic Eileen Garrett, to his citadel for the purpose of testing her telepathic abilities. Placed in a Faraday cage so that electromagnetic waves could be screened out, Garrett performed at a level that astounded the experimenters.[56]

A world traveler and prodigious innovator throughout his life, Jack spent much of his time in the latter years traveling, with his wife, across the country in a mobile home he had designed. One day on a trip to see his friend Igor Sikorsky, in Bridgeport, the inventor of the helicopter asked if

Jack wanted one as a present. "Only if it can take my mobile home," Jack responded. After a long pause, with time out to stare at the mobile monstrosity, Igor responded, "It can be done."

John Hays Hammond Jr. died in 1965 at the age of seventy-seven. A genius in his own right, it is unfortunate that his relationship with Tesla was thwarted. Together they invented a rather sizable chunk of the appliances of the modern era.

39

J. P. MORGAN JR. (1912–14)

April 18, 1912

My dear Mr. Tesla:

I attended the Marconi meeting last night, in company with illustrious society. The everlasting toast-master, [T.C.] Martin, read, in most theatrical fashion a telegram; and after a pause of fully three minutes, announced its author as: "THOMAS A. EDISON!"

Mr. Marconi gave the history, as he sees it, of wireless up to this date. [He] does not speak any more of Hertzian Wave Telegraphy, but accentuates that messages he sends out are conducted along the earth. Pupin had the floor next, showing that wireless was due entirely to one single person....

The only speaker of the evening who understood Mr. Marconi's merit, who did not hesitate to vent his opinion was Steinmetz. In a brief historical sketch, he maintained that while all elements necessary for the transmission of wireless energy were available, it was due to Marconi that intelligence was actually transmitted....

That evening was, without any question, the highest tribute that I ever have heard paid to you in the language of absolute silence as to your name.

Sincerely yours,
Fritz Lowenstein[1]

Tesla was in the midst of working for W. M. Maxwell, superintendent of New York's public schools, on a controversial project to electrify classrooms with high-frequency currents. Following in the footsteps of a highly publicized experiment in Stockholm whereby a group of children in such an environment apparently showed increased scores on aptitude tests and accelerated growth, Maxwell was hoping to boost the health and intelligence of students in America. With the prestigious inventor wiring the walls with his "Tesla coil," guaranteeing complete safety, and concurring with the general conclusions of the hypothesis, Tesla helped Maxwell set up a pilot study with "fifty mentally defective school children." If the study proved successful, the superintendent boldly stated, "the new system would overturn all methods hitherto applied in its schools and introduce a new era in education."[2]

Marconi's talk, before eleven hundred members of the New York Electrical Society, was held on the very day the *Titanic* sank. Frank Sprague, who gave the eulogy, "visibly affected Mr. Marconi when he credited him with saving the lives of from 700 to 800 persons."[3] Unfortunately, Marconi was unable to save fifteen hundred other individuals, including Colonel Astor, who went down with the ship after helping his new bride board one of the remaining lifeboats.

If ever an event epitomized the loss of innocence, the myopic condition of humanity, it was the sinking of the *Titanic*. Reminiscent of Tesla's own voyage, this watershed recapitulated the story of Icarus, the prideful aeronaut who crumbled because of lack of respect for his limitations. With Tesla's outlandish wish to transmit unlimited energy to the far reaches of the world and bring rain to the deserts, to become a master of the universe, it was inevitable that he, too, would succumb. The tragedy prompted Congress to pass into law an act requiring the use of wireless equipment on all ships carrying fifty or more passengers. It also focused national attention on twenty-one-year old David Sarnoff, latter-day head of RCA, who was credited as the first wireless operator to pick up the *Titanic* distress signal.[4]

Tesla was not the only casualty in the wireless game. Reginald Fessenden's concern "all but ceased functioning in 1912" due to his erratic nature, infighting, and "prolonged litigation." And Lee De Forest, who by now had nearly forty patents in wireless, also went under when he was convicted with officers of his company in a stock fraud.[5] Concerning Lowenstein, Tesla was backing his protégé's attempts to install equipment on U.S. Navy ships using equipment based on some of his fundamental designs. "He is much abler than the rest of the wireless men," he wrote Scherff, "[so] this has given me great pleasure."[6] The one edge Lowenstein had over Marconi was the Italian's insistence on an all-or-nothing deal with the military. Either all ships were hooked into his system, or none were. The U.S. government, however, was loath to relinquish their upper hand to a private concern, so Marconi had great difficulty integrating his system into the American marketplace.

Nevertheless, Marconi was still the only major competitor, so Tesla set out to reestablish his legal right. Conferring with his lawyers, the pioneer began a strategy to sue the pirate in every country he could.

In England, Tesla had allowed an important patent to lapse, so progress there was halted. Oliver Lodge, on the other hand, was able to prevail and received 1,000 pounds per year for seven years from the Marconi concern there.[7] In the United States, where Tesla was applying for a renewal of his most fundamental patent, he had yet to formalize the suit, but in "the highest court in France" the inventor achieved a resounding success. Sending his written testimony to Judge M. Bonjean in Paris, Tesla explained his work in 1895, when he "erected a large wireless terminal above the building...[and] employed damped and undamped oscillations." He also enclosed two patents from 1897 and specifications for his telautomaton, which confirmed that he had displayed it before G. D. Seely, the examiner-in-chief of the U.S. Patent Office in Washington, D.C. in 1898. Concerning Marconi's June 2, 1896, patent, Tesla testified that the patent was "but a mass of imperfection and error.... If anything, it has been the means of misleading many experts and retarding progress in the right direction....It gives no hint as to the length of the transmitting and receiving conductors and the arrangements illustrated preclude the possibility of accurate tuning....[Marconi replaced] the old-fashioned Rumhkorff coil [with]...the Tesla coil."[8]

Speaking in support of Tesla's cause for Popoff, Ducretet & Rochefort, the French company that had initiated the action, was electrical engineer M. E. Girardeau, who described in detail the technical accomplishments of Tesla's invention. "Indeed," Girardeau began, "one finds in the American patent extraordinary clearness and precision, surprising even to physicists of today....What a cruel injustice would it be now to try to stifle the pure glory of Tesla in opposing him scornfully."[9]

Judge Bonjean struck down Marconi's patents and reestablished Tesla's as superseding. Most likely, compensation was also paid to him from the French company that won the case.

With this victory, however, came other defeats. The year 1912 saw legal entanglements against Tesla by Edmund K. Stallo, who was seeking damages of $61,000 for advances tendered in 1906; and also the Westinghouse Corporation, which sued for $23,000 for equipment loaned out in 1907. In the first instance, the Stallo syndicate, which had ties to Standard Oil, had invested only $3,500 and was frivolously suing for the promise of an enormous profit. The inventor's liability was thus minimal. In the second instance, Tesla argued that he was not personally liable, as the machinery was loaned to the company he was organizing. He did offer, however, to return the equipment which was still being guarded at the Long Island laboratory.[10] Although his financial losses were small, the negative publicity undermined his reputation and caused the inventor to become a more invisible presence at the prestigious Waldorf, where he was still residing and piling up debt. And then there was Mrs. Tierstein, who Tesla said, "wanted to shoot me for throwing electricity at her." Tesla "pitied the poor woman" but had her sent, by way of Judge Foster, "to the asylum."[11]

The suit by Westinghouse caused quite a stir among the inner circle, and even prompted Tom Edison to pen a rare missive of commiseration.

February 24, 1912

My dear Mr. Edison,

Acknowledging your kind letter...I wish to reiterate the sentiments and express my great regret that I was unable to transmit them in person.

With assurance of high regard, I remain as ever,

N. Tesla[12]

The Westinghouse Corporation, however, was not a monolithic enterprise, and its legal department, in some ways, was an autonomous entity. Tesla continued to borrow equipment on a regular basis throughout the period (1909–17), and he frequently conferred with various engineers, particularly Charles Scott, whom he began educating in the field of wireless. Tesla also continued to meet with the recently deposed chieftain George Westinghouse, who was working in semiretirement at his offices in New York City.

"I suppose you cannot help feeling disappointed at the ingratitude of some who are now heading the great enterprise which your genius has created," the inventor remarked. "I sincerely hope that you will very shortly be again in the position you have heretofore occupied. I know that the large majority of the public shares my sentiments."

"Thank you for your concern," the descendent of Russian noblemen replied.[13]

THE GERMAN CONNECTION

Marconi's greatest rival in the legal arena may have been Nikola Tesla, but in the battleground of the marketplace it was Telefunken, the German wireless concern. Although Marconi had patents in Germany, the Telefunken syndicate had too many important connections on its home front and easily maintained a monopoly there. Formed through a forced merger, under orders of the kaiser, of the Braun-Siemens-Halske and Arco-Slaby systems, Telefunken had fought the Marconi conglomerate vigorously on every front. It was, without doubt, the number-two competitor in the world. Although Marconi had achieved a recent coup in Spain, Telefunken gained an edge in America when it constructed two enormous transatlantic systems in Tuckerton, New Jersey, and Sayville, New York.

For nationalistic reasons, Tesla had been prohibited from obtaining his rightful royalties in Germany, but Professor Adolf Slaby never hid the fact that he considered Tesla the patriarch in the field. Thus, when Telefunken came to America, Slaby sought out his mentor not only on moral grounds but also for gaining a legal foothold against Marconi and to obtain the inventor's technical expertise.

A meeting was held between Tesla and the principles of Telefunken's American holding company, the innocuous-sounding Atlantic Communication Company, at 111 Broadway, the location of their offices. Present was its director, Dr. Karl George Frank, "one of the best known German [American] electrical experts," and his two managers, Richard Pfund, a frequent visitor to Tesla's lab, who was head of the Sayville plant, and "the monocle," Lt. Emil Meyers, head of operations at Tuckerton.

Tesla asked for an advance of $25,000 and royalties of $2,500 per month but settled for $1,500 per month, with a one month advance.[11] The inventor met with Pfund to discuss the turbine deal with the kaiser and also to fix a transmitter that the Germans were working on at the Manhattan office. Shortly thereafter, he traveled out to the two wireless stations with his plan to institute his latest refinements in order to boost their capabilities.[15] With Professor Jonathan Zenneck out at Sayville, Tesla calculated that they were wasting nearly 25 percent of their energy in electromagnetic radiations. "Those waves will dissipate only a few miles off shore," he told Zenneck, "whereas the energy that will reach Germany will be from your ground connection."[16]

After a grueling search along the imponderable ice field, John Jacob Astor's emaciated body was fished from the sea for a New York funeral. Documents from his estate revealed five hundred shares of Tesla Electric Company.[17] A year later, J. Pierpont Morgan, the sultan of Wall Street, the octopus, was dead. Considering the resentment and disillusionment the inventor felt, he also bore great admiration for the man he called a "towering" historical figure.

Although Morgan's son-in-law, Herbert Satterlee, could only provide a back-row seat at the chapel, nevertheless, even to be present at the solemn occasion was a great honor. The hall was speckled with numerous faces of colleagues and adversaries, many of whom could track a large percentage of their wealth, and even their station, to Tesla's creations. No doubt, the inventor endured some snickering, particularly from the hydra's underlings, but Tesla, ever the nobleman, transcended the petty titters as he walked up to Anne and J. P. Junior.

"Please accept my heartfelt condolences on the death of the great man who was the head of your famous firm," he told them. "When I can feel such a void in my heart and brain at the passing of Morgan, I can appreciate, in a measure, the depth of feeling of those who were his lifelong comrades. All the world knew him as a genius of rare powers, but to me he appears as one of those colossal figures of [the] past which mark epochs in the evolution of human thought."[18]

Two months later, within days of the termination of his relationship with Hammond, Tesla approached the new head of the House of Morgan with a proposal to help fund his bladeless turbines. "Its application to the manufacture of iron and steel, alone will yield $100,000,000 a year through the utilization of the waste heat and other economies, and it will have a similar effect on ship propulsion, railroad, automobile and many other large industries." The monarch looked over the proposal and advanced the inventor $5,000.[19]

It was a time for reflection on the death and transmutation of his nemesis. Conflicting emotions of enmity and adoration poured through his being in torrents as he relived the exhilaration of his wireless odyssey and the anguish of its incompleteness.

On July 7, 1913, three days before his fifty-seventh birthday, Tesla took the train back out to Wardenclyffe. He had much to think about, for his friend Johnson had decided to retire from the *Century*. Overseers of the company were pressing to lower the standards. Johnson offered to create a separate magazine, which would be more flashy, but he wanted to maintain the integrity of the original; but he was outvoted. "It was really pathetic to see the way authors would plead with me for another 'damn,'"[20] he told his friend. Katharine became more persistent in demanding that Tesla stop by for another visit. Now in financial trouble, his friends could even lose their house. Tesla tried to scurry past the magnifying transmitter without looking up, to get to the safe in the laboratory, but he was drawn like a magnet to the stalwart frame. As he grabbed a rung from the bottom tier, a sorrow swept through him that stole his breath. He staggered to the door of Stanny's building and let himself in. "I did not exactly cry when I saw my place after so long an interval," he wrote Scherff, "but came very near doing it."[21]

Tesla began to court the new Morgan, trying not to fall into the trap he had laid for himself the first time around. He sent Jack an articulate proposal outlining his wireless enterprise and explanation of his outstanding debt and arrangement with his father and also his plans in the field of "fluid propulsion," that is, bladeless turbines.

"In either of these fields in which I have the good fortune to be a pioneer the possibilities are immense and I can vouch for the fullest success; but my appeal for your support is on a higher plane.... The proposition which I would respectfully submit is to organize two companies and to turn over to you my entire interests in both, of which you may accord me such a part as you deem best."[22]

"I'm greatly impressed with your offer," Jack responded. "But of course I could not consent to doing as you suggest. I wish to make a counter offer, which is that you should proceed to organize your companies, and, if and as they work out, repay to Mr. Morgan's estate either in securities or in cash, the money Mr. Morgan advanced. It seems to me, that you are entitled to the profits of these companies, but that Morgan would, in strict justice, be entitled to a return of his money if it could be arranged."[23]

Tesla left the meeting with the old itch reactivated and another check for $5,000. He thanked Jack for his encouragement and forwarded an open letter which he had written to His Grace, the Most Reverend Archbishop of Ireland. "The day is not distant when the very planet which gave him birth, will tremble at the sound of his voice: he will... harness the inexhaustible... intense energy of microcosmic movement, cause atoms to combine in predetermined forms; he will draw the mighty ocean from its bed, transport it through the air and create lakes and rivers at will; he will command the wild elements; he will push on from great to greater deeds until with his intelligence and force he will reach out to spheres beyond the terrestrial."[24]

"It is now clear to me," Tesla told the new J.P., "that you are moved by the same great spirit of generosity which has animated your father and I am more than ever desirous of enlisting your interest and support. Destiny has placed you in a position of great power and influence and here is a wonderful opportunity.

"As for myself," the modest inventor continued, "I contemplated more than financial success.... A great monument will no doubt be built to Mr. Morgan, but none in marble or bronze could be as lasting as the achievement which I have proposed to link with his memory."[25]

To reassert his dominance as the preeminent inventor in wireless, the inventor forwarded a transcript of the entire French litigation proceedings, wherein Marconi's work was overturned in favor of his. If Jack could help fight the legal battle in the United States, the wireless enterprise, which they contractually shared, could be revived.

Jack, however, was not smitten by the same vision of destiny and graciously declined to become involved in Wardenclyffe in any way. He had not ruled out the turbines, however, and told the inventor to keep him informed of any progress.

Tesla returned to the Edison Waterside Station with a new transfusion of 23 Wall Street blood. To reflect the reactivation of the resurrected alliance, the inventor set out to search for more fashionable chambers. Within a few months he took up residence in the brand-new Woolworth Building. Decorated with a gold-leafed emerald-colored mosaic ceiling in the lobby and located by city hall, near Wall Street on Park Row, the gothic-styled Woolworth soared above the city to the dizzying height of eight hundred feet, eclipsing the Metropolitan Towers as the loftiest skyscraper in the world. Tesla took the Johnsons along to the gala opening. The banquet began with the illumination of the building's eighty thousand lightbulbs by President Woodrow Wilson, who pressed a button in Washington, D.C. Tesla met with the mayor, Mr. "Dime Store" Woolworth, and other dignitaries, and then Katharine lured her two escorts into one of the twenty-four high-speed elevators to the roof, where they could gaze out over the sprawling megalopolis.

"Do not worry about finances, Luka," Tesla said confidently. "Remember, while you sleep, I work and am solving your problems." Johnson brought up the old AC polyphase debacle, and Tesla replied that there were "billions invested [in it] now. I won every suit without exception and had it not been for a 'scrap of paper,' I would have received in royalties Rockefeller's fortune, but just the same, I feel I am safe to invite you to dinner."

Tesla's wit and latest maneuver once again brought a needed smile to the oft-brooding Mrs. Filipov. As usual, when the wizard returned to her orb, she seemed to step back from the veil. Johnson, however, reiterated his concern that without a job, 327 Lexington Avenue could go on the market.

"Please take my words seriously," Tesla insisted. "Do not worry, and write your splendid poetry in perfect serenity. I will do away with all difficulties which confront you. Your talent cannot be turned into money, but mine is one which…can be transformed into car-loads of gold. This is what I am doing now."[26] During this period, Tesla continued to pay back monies owed the Johnsons as he labored with his new engines.

Throughout the latter half of 1913, the inventor prepared a careful marketing plan to exploit his new device. Not only would he show Hammond that he had made a costly mistake, he would also set up an industry that would ensure the finances necessary to return to his beloved Long Island tower. His best leads came from the Ford Motor Company in the United States and the Bergmann Works in Germany.

Tesla had known Sigmund Bergmann since his first days in America. Bergmann had emigrated from Germany even before Tesla and became a valued employee/private partner and manufacturer for Thomas Edison. Maintaining a separate company from the other Edison works, Bergmann became highly successful. He returned to Germany in the early 1900s, becoming one of the leading manufacturers for the kaiser,[27] a man who had also attempted to woo the Serb, especially before the turn of the century, when he was demonstrating his fantastic inventions in Europe and America.

In September, Tesla forwarded photographs of his turbines and invited Morgan to the Waterside Station to see them in person; however, Morgan was sailing for Europe. "Perhaps when I return some time in December," Morgan replied, "it will be possible for me to go into the question."[28] While the new Wall Street czar sailed for the Old World, Tesla set up meetings with emissaries from the most consequential markets. In a letter sent to Morgan upon his return, Tesla outlined his numerous strategies for achieving financial success. His list included:

> 1. Sale of exclusive license for Belgium through the advisor of the King [for] $10,000 cash and fair royalty.
>
> 2. Concessions for Italy through an associate at Crispi [for] $20,000 and royalty. Not yet consummated.
>
> 3. Exclusive license for the U.S. to the Wing Manufacturing Company for turbo draught blowers.
>
> 4. Exclusive license for train lighting, Dressel Railway Lamp Works.
>
> 5. [Manufacture of] automobile lighting with engine exhaust gas.
>
> 6. Use of my wireless system [with Lowenstein] on several battleships.
>
> 7. Agreement with L. C. Tiffany Co. on new invention.
>
> 8. Prospective agreement with Mr. N. E. Brady of the Edison Company in regard to the manufacture of turbines. Very good chance for big business.

Manufacturing had begun at some of these places, and royalty contracts were negotiated with most. "As you can see, Mr. Morgan, here are decidedly valuable results very gratifying to me but, on the other hand, I am almost despairing at the present state of things. I need money badly and I cannot get it in these dreadful times. You are about the only man to whom I can look for help. I have stated my case."[29]

Morgan agreed to defer interest payments on the amount, which was now up to $20,000, but decided not to increase the loan. Tesla, however, simply required the funds and followed up the letter with a testimonial from Excellenz von Tirpitz, minister of marine, "who has been requested to the German Emperor relative to the Tesla Turbine who is greatly interested in this invention." Von Tirpitz had "promised his Excellency that the machine will certainly be here on exhibition about the middle of January, so you know what that means." Tesla also informed Morgan that if the deal was consummated, Bergmann would come through with royalties on the turbine of $100,000 per year.[30]

Considering Jack's keen antipathy for the Germans, their connection to the Jewish banking houses (he was notoriously anti-Semitic), and the long-standing policy of the House of Morgan to shun financial arrangements with Germany after they had double-crossed Pierpont many years before, it would seem unlikely that Jack would reverse his decision. However, unlike his father, the son was able to compromise and allow his heart to rule at times. He graciously changed his mind and forwarded the additional funds.[31]

As Tesla awaited news from the Bergmann Works, he labored to perfect a new speedometer he had invented. The device, much simpler than the one then being used, would cost half as much to produce and had a market in the hundreds of thousands. Considering that his selling price would be around twenty-five dollars each, the potential for large profits was great, and Tesla offered the deal to Morgan. Jack declined and asked again for the interest payment due on the loan.

"In the turbine proposition," Tesla replied, "I have received a painful setback. I installed the machine at the Edison Plant and made some very gratifying tests, but soon discovered that the castings of the bearings were full of small holes which allowed the water to enter and made it hazardous to run." Tesla had to renew the parts, "but spent considerably more money than expected." He also had lawyers fees to contend with because of the upcoming suit with Marconi, so he requested that Morgan either be patient or help with the bills to complete the turbine project or protect their other common interest.[32]

Besides the deaths of Astor and Pierpont Morgan, this period also marked the departure of two other lions from Tesla's den, George Westinghouse and naturalist John Muir; both men died in 1914. Westinghouse's death was cushioned by some weeks of declining health, but Muir's passage took Tesla by surprise. "He seemed so vigorous in mind and body when I saw him not long ago," the technological conservationist told the Johnsons.[33] Only a few years before, the sky had been blanketed with hundreds of thousands of passenger pigeons, and the beauty of nature had been rekindled with Muir's inspirational writings. As Tesla strolled to his favorite spot by the Forty-second Street library to spread seed among the bird's domestic cousins and consider a new tack for instituting his wireless scheme, the last passenger pigeon disappeared from the earth. Jack Morgan spent time with one of the few Jews he liked, Daniel Guggenheim, organizing Kennecott Copper, "America's biggest copper producer."[34]

Tesla sent his tribute to Westinghouse to *Electrical World.* It was published with the comments of other colleagues, such as William Stanley, Lewis Stillwell, and Frank Sprague.

> I like to think of George Westinghouse as he appeared to me in 1888, when I saw him for the first time. The tremendous potential energy of the man had only in part taken kinetic form, but even to a superficial observer the latent force was manifest.... An athlete in ordinary life, he was transformed into a giant when confronted with difficulties which seemed unsurmountable. He enjoyed the struggle and never lost confidence.

When others would give up in despair, he triumphed. Had he been transferred to another planet with everything against him he would have worked out his salvation. . . . His was a wonderful career filled with remarkable achievements. . . . He was a great pioneer and builder whose work was of far-reaching effect on his time and whose name will live long in the memory of men.[35]

40

FIFTH COLUMN (1914–16)

Navy Department
Washington, D.C.

September 14, 1916

Sir:

In the files of the Bureau of Steam Engineering is found a copy of a letter by Nikola Tesla to the Light House Board, under date of September 27, 1899, from the Experimental Station, Colorado Springs. This letter is evidently an answer to a communication from the Light House Board, requesting information as to Tesla's ability to supply wireless telegraph apparatus.

This letter may be made use of in forthcoming litigation in which the Government is involved...as the discovery of suitable proof of priority of certain wireless usages by other than Marconi might prove of great aid to the Government.

Sincerely Yours,
Franklin D. Roosevelt
Acting Secretary of the Navy[1]

Within two weeks of the beginning of World War I, Germany's transatlantic cable was severed by the British. The only reasonable alternative for communicating with the outside world was through Telefunken's wireless system. Suddenly, the Tuckerton and Sayville plants became of paramount concern. The Germans obviously wanted to maintain the stations to keep the kaiser abreast of President Woodrow Wilson's intentions, but the British wanted them shut down.

In March 1914, Marconi was made a *senatore* in Italy, a distinguished man of science, and he spoke before the royal couple. In July in Great Britain, the land of his mother, he was decorated by the king at Buckingham Palace. Now the fight against Telefunken would be fought on military as well as commercial grounds as it became clear that the Germans were using their plants to help coordinate submarine and battleship movements. The wireless lines also marked the burgeoning alliance forming between Italy and the British Empire.[2]

As a pacifist, Wilson maintained a strict policy of neutrality, a position bolstered by war hero and former president Teddy Roosevelt, himself a contender for the upcoming 1916 election. Although officially neutral, the sentiments of the majority of the American population was with England, particularly after Germany stormed through the peaceful kingdom of Belgium. Nevertheless, fully one-tenth of the population was of German stock, and their sentiments were with the other side. George Sylvester Viereck, the country's leading poet, colleague of Johnson, and Tesla friend, began to sense the growing shift away from neutrality, especially after the U.S. Navy appropriated the Tuckerton plant to send its own "radio coded messages abroad."

Having just returned from Berlin and the midst of war, Viereck courted Teddy Roosevelt and emissaries of President Wilson. Simultaneously, he began a new publication with other leading German Americans. Initially welcomed by the press, the *Fatherland* soon achieved a subscription base of 100,000.[3]

Ignoring Viereck's plea for neutrality, Wilson prepared a presidential decree "declaring that all radio stations within the jurisdiction of the United States of America were [to be] prohibited from transmitting or receiving...messages of an unneutral nature....By virtue of authority vested in me by the Radio Act," the president continued, "one or more of the high powered radio stations within the jurisdiction of the United States...shall be taken over by the Government."[4]

Throughout the beginning of the war, Tesla stepped up his legal campaign against Marconi and continued to advise and receive compensation from Telefunken. Since the country was officially neutral (America would not enter the war for another three years), the arrangement was entirely aboveboard. Nevertheless, few people knew about the German-Tesla link, although the inventor made no secret of it to Jack Morgan.

February 19, 1915

Dear Mr. Morgan,

I am expecting to embody in their plant at Sayville some features of my own which will make it practicable to communicate with Berlin by wireless telephone and then royalties will be very considerable. We have already drawn papers.[5]

Camouflaged by the smoke screen of the American-sounding Atlantic Communication Company, Telefunken swiftly moved to increase the power of its remaining station at Sayville. Located near the town of Patchogue out on the flats of Long Island, just a few miles from Wardenclyffe, the Sayville complex encompassed a hundred acres and employed many German workers. With its main offices in Manhattan and its German director, Dr. Karl George Frank, an American citizen, Telefunken was legally covered, for no foreigner could own a license to operate a wireless station in the country. (Thus, Marconi also had an American affiliate.) It was an easy matter for Tesla to confer with Atlantic in the city and also go out to the site of the plant.

Within two months of Tesla's letter to Morgan, the plant at Sayville *tripled* its output by erecting two more pyramid-shaped transmission towers, five hundred feet tall. Utilizing Tesla's theories on the importance of ground transmission, resonating accoutrements spread out over the land for thousands of more feet. Thus, by shifting emphasis away from aerial transmission, Telefunken's output was boosted from 35 kilowatts to over 100, catapulting Germany into the number-one spot in the wireless race. The *New York Times* reported on their front page, "Few persons outside radio officials knew that Sayville was becoming one of the most powerful transatlantic communicating stations in this part of the world."[6]

Tesla Sues Marconi on Wireless Plant
Alleges That Important Apparatus Infringes Prior Rights[7]

Calling wireless "the greatest of all inventions," Tesla made an additional appeal for legal assistance to Morgan. "Can you put yourself for a moment in my place?" he wrote the financier. "Surely, you are too big a man to permit such an outrage and historical crime to be perpetuated as is now being done by cunning promoters." Expecting to "receive satisfaction from the Government," since they had installed "$10,000,000 of [his] apparatus," the inventor also revealed that "the Marconi people approached me to join forces, but only in stock and this is not acceptable."[8]

Again Morgan declined assistance in protecting their patents held in common. The Wall Street mogul, however, had not at all abandoned the field, for he was funding a college radio station near Boston at Tufts University.[9]

The years preceding America's entrance into World War I contained an overwhelming quagmire of litigation involving most countries and virtually every major inventor in the wireless field. At about the time of Tesla's breakup with Hammond, Fritz Lowenstein, who was paying royalties to both men (and to Morgan, through Tesla), began installing wireless apparatus aboard navy ships. Although the equipment was also being used by Hammond to test the guided missiles, this work was classified, and Hammond's patents became immune from litigation.[10]

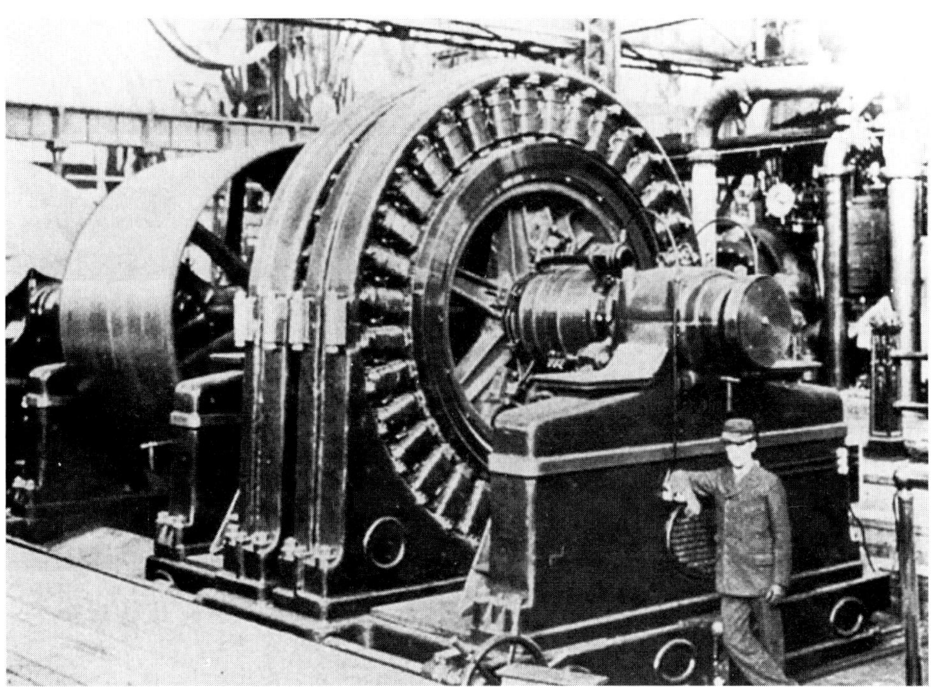

Tesla polyphase generator used by Westinghouse Electric and Manufacturing Company to electrify the Chicago's World Fair of 1893.

Tesla at his Houston Street laboratory in 1898 sending 500,000 volts through his body to light a wireless fluorescent light in a multiple-exposure photograph. (MetaScience Foundation

The wizard at his Colorado Springs laboratory sitting among sixty-foot electrical sparks in this illustrious multiple-exposure photograph. (Nikola Tesla Museum)

Wardenclyffe, circa 1903. (MetaScience Foundation)

Jack Hammond *(center)* with some of his friends, including Leopold Stokowski, far right. Hammond had formed a partnership with Tesla, circa 1912, to perfect remote-controlled torpedoes and guidance systems for the U.S. Navy. (Hammond Castle)

A dinner of the Institute of Radio Engineers in 1915 honoring John Stone Stone, president of the society. Standing along the back wall from the left are Karl F. Braun, winner with Marconi of the Nobel Prize in physics in 1909; John Stone Stone; Jonathan Zeaneck, of the Sayville wireless plant; Lee DeForest, radio pioneer; Nikola Tesla; Fritz Lowenstein, Tesla's longtime associate; Rudolf Goldschmidt, a physicist who worked with Emil Meyer, third from the left, seated in center row, who ran the German wireless plant at Tuckerton, New Jersey. Seated in back, at far left in front of Braun, is David Sarnoff, later head of RCA and NBC-TV. (Smithsonian Institution)

(Opposite) How the Wardenclyffe tower would have looked when completed. (Drawing by sci-fi artist Frank R. Paul; Smithsonian Institution)

(Right) Nikola Tesla illuminated by his wireless cold lamp. (Smithsonian Institution)

(Below) Nikola Tesla, circa 1925 (Nikola Tesla Museum)

(Top) A statue of Nikola Tesla located in the town square of Gospić, Croatia. Designed by Franco Krsinic, this particular statue was purposely destroyed by a bomb during the recent war between Serbs, Croats, and Muslims in the former Yugoslavia. An exact replica sits calmly on Goat Island beside Niagara Falls. (Marc J. Seifer)

(Bottom) Tesla, shortly before his death in 1942, meeting with King Peter of Serbia. Tesla's nephew, Sava Kosanovic, the ambassador from Yugoslavia, is on the left. (Smithsonion Institution)

(Opposite) *Time* celebrated Tesla's seventy–fifth birthday with a cover. (© 1931 Time Inc. Reprinted by permission)

FIFTEEN CENTS July 20, 1931

TIME

The Weekly Newsmagazine

Keystone

NIKOLA TESLA*

All the world's his power house.

(See SCIENCE)

*From a portrait by Princess Lwoff-Parlaghy.

Volume XVIII Number 3

Circulation Office, 350 *East 22nd Street, Chicago* (Reg. U. S. Pat. Off.) Editorial and Advertising Offices, 205 *East 42nd Street, New York.*

1990s promotional mailer. (Westinghouse Corporation)

Aside from Tesla's priority battle, Telefunken was also suing Marconi, who, in turn, was suing the U.S. Navy as well as Fritz Lowenstein for patent infringement.

During the following spring, Marconi was subpoenaed by Telefunken. Due to the importance of the case, he sailed off for America on the *Lusitania,* arriving in April 1915 to testify. "We sighted a German submarine periscope," he told astonished reporters and his friends at dockside.[11] As three merchant ships had already been torpedoed without warning by the German U-boats the month before, Marconi's inflammatory assertion was not taken lightly.

The *Brooklyn Eagle* reported that this suit brought "some of the world's greatest inventors on hand to testify."[12] Declared a victor in the Lowenstein proceedings by a Brooklyn district-court judge, Marconi clearly had the press behind him. Nevertheless, he was beaten by the navy in his first go-around with them,[13] so this case against Telefunken, with all the heavyweights in town, promised to be portentous. Once and for all, it appeared, the true legal rights would be established in America.

Aside from Marconi, there was, for the defense, Columbia professor Michael Pupin, whose testimony was even quoted in papers in California. With braggadocio, Pupin declared, "I invented wireless before Marconi or Tesla, and it was I who gave it unreservedly to those who followed!"[14] "Nevertheless," Pupin continued, "it was Marconi's genius who gave the idea to the world, and he taught the world how to build a telegraphic practice upon the basis of this idea. [As I did not take out patents on my experiments], in my opinion, the first claim for wireless telegraphy belongs to Mr. Marconi absolutely, and to nobody else."[15] Watching his fellow Serb upon the stand, Tesla's jaw dropped so hard, it almost cracked upon the floor.

When Tesla took the stand for Atlantic, he came with his attorney, Drury W. Cooper, of Kerr, Page & Cooper. Unlike Pupin, who could only state abstractly that he was the original inventor, Tesla proceeded to explain in clear fashion all of his work from the years 1891–99. He documented his assertions with transcripts from published articles, from the Martin text, and from public lectures, such as his well-known wireless demonstration which he had presented to the public in St. Louis in 1893. The inventor also brought along copies of his various requisite patents which he had created while working at his Houston Street lab during the years 1896–99.

COURT: What were the [greatest] distances between the transmitting and receiving stations?

TESLA: From the Houston laboratory to West Point, that is, I think, a distance of about thirty miles.

COURT: Was that prior to 1901?
TESLA: Yes, it was prior to 1897....
COURT: Was there anything hidden about [the uses of your equipment], or were they open so that anyone could use them?
TESLA: There were thousands of people, distinguished men of all kinds, from kings and greatest artists and scientists in the world down to old chums of mine, mechanics, to whom my laboratory was always open. I showed it to everybody; I talked freely about it.[16]

As virtually no one knew about the West Point experiment, this statement was somewhat deceptive, although it was true that thousands of people had witnessed Tesla's other wireless experiments, such as in St. Louis in 1893. Referring explicitly to Marconi's system, having brought the Italian's patent along with his own, the inventor concluded:

TESLA: If you [examine these two diagrams]...you will find that absolutely *not a vestige* of that apparatus of Marconi remains, and that in all the present systems there is nothing but my four-tuned circuits.[17]

Another jolt to Marconi came from John Stone Stone (his mother's maiden name, by coincidence, was also Stone). Having traveled with his father, a general in the Union army, throughout Egypt and the Mediterranean as a boy, Stone was educated as a physicist at Columbia University and Johns Hopkins University, where he graduated in 1890. A research scientist for Bell Labs in Boston for many years, Stone had set up his own wireless concern in 1899. The following year, he filed for a fundamental patent on tuning, which was allowed by the U.S. Patent Office over a year before Marconi's.[18] Stone, who never considered himself the original inventor of the radio, as president of the Institute of Radio Engineers and owner of a wireless enterprise, put together a dossier of inventor priorities in "continuous-wave radio frequency apparatus." He wanted to determine for himself the etiology of the invention. Adorned in a formal suit, silk ascot, high starched collar, and pince-nez attached by a ribbon to his neck, the worldly aristocrat took the stand:

Marconi, receiving his inspiration from Hertz and Righi...[was] impressed with the *electric radiation aspect of the subject*...and it was a long time before he seemed to appreciate the real role of the earth..., though he early recognized that the connection of his oscillator to the earth was very material value....Tesla's electric earth waves explanation was the more serviceable in that it explained [how]...the waves were enabled to travel over and

> around hills and were not obstructed by the sphericity of the earth's surface, while Marconi's view led many to place an altogether too limited scope to the possible range of transmission.... With the removal of the spark gap from the antenna, the development of earthed antenna, and the gradual enlargement of the size of stations... greater range could be obtained with larger power used at lower frequencies, [and] the art returned to the state to which Tesla developed it.

Attributing the opposition, and alas, even himself, to having been afflicted with "intellectual myopia," Stone concluded that although he had been designing wireless equipment and running wireless companies since the turn of the century, it wasn't until he "commenced with this study" that he really understood Tesla's "trail blazing" contribution to the development of the field. "I think we all misunderstood Tesla," Stone concluded. "He was so far ahead of his time that the best of us mistook him for a dreamer."[19]

Another case which did not receive much publicity but which became vital to the Supreme Court's 1943 ruling in Tesla's favor, was *Marconi v. U.S. Navy,* brought July 29, 1916, two years after their first go-around. The Italian was seeking $43,000 in damages, suing for infringement of fundamental wireless patent no. 763,772, which had been allowed in June 1904.

E. F. Sweet, acting secretary of the navy, and also Assistant Secretary Franklin D. Roosevelt began a correspondence in September to review Tesla's 1899 file to the Light House Board.[20] The history of Marconi's patent applications to the U.S. Patent Office provided additional ammunition. In 1900, John Seymour, the commissioner of patents, who had protected Tesla against the demands of Michael Pupin for an AC claim at this same time, disqualified Marconi's first attempts at achieving a patent because of prior claims of Lodge and Braun and particularly Tesla. "Marconi's pretended ignorance of the nature of a 'Tesla oscillator' [is] little short of absurd," wrote the commissioner. "Ever since Tesla's famous [1891–93 lectures]... widely published in all languages, the term 'Tesla oscillator' has become a household word on both continents." The patent office also cited quotations from Marconi himself admitting use of a Tesla oscillator.

Two years later, in 1902, Stone was granted his patent on tuning which the government cited as anticipating Marconi, and two years after that, after Seymour retired, Marconi was granted his infamous 1904 patent.[21]

EDWIN ARMSTRONG

"I have had a lot of fun at Columbia," Armstrong said. The lecturer in physics that semester loftily disparaged the experiments of Nikola Tesla.

"He even went so far as to say that there was very little originality about Tesla." Ever the audacious student, Armstrong used the professor's ignorance of Tesla's teachings to cause the man to receive a hearty shock from some electrical equipment. "He couldn't let go, and pulled most of the apparatus off the table before the current was turned off."[22]

Just after graduation, Armstrong invented a feedback amplifier, which was, in essence, a refinement and further development of De Forest's audion tube. Influenced by the "Edison effect" or flow of electrons studied by Tesla in the early 1890s in his "brush" vacuum tube, Armstrong had discovered a way to take the De Forest audion and amplify its sensitivity and boost its power to another magnitude by connecting a second circuit, or wing circuit, to the grid inside the tube and feeding it back into the grid. The end result was that with this new invention the young whippersnapper could pick up wireless messages from Nova Scotia, Ireland, Germany, San Francisco, and even Honolulu.

As Armstrong was one of Professor Pupin's star pupils, Pupin was able to set up meetings with Lee De Forest, David Sarnoff, representing Marconi Wireless, and Dr. Karl Frank, head of the Atlantic Communication Company. Since De Forest's audion lay at the center of the new device, De Forest claimed the "ultra-audion" as his own invention, so Marconi Wireless stepped aside to wait for the dust to settle. Frank, on the other hand, had Armstrong put the equipment into the wireless plant at Sayville and paid him a royalty of $100 per month.[23] Destined to invent AM and FM radio and a nonconformist by nature, this new wizard on the block had come upon his 1912 discovery because he had rejected the inferior Marconi spark-gap apparatus that most of his wireless buddies were still playing with and, like Stone, plunged himself into the continuous-wave technology developed by Tesla.

With Professor Pupin training many of the new breed of electrical engineers, it is no wonder that few of them would have the gumption of an Armstrong to see that Marconi's success was based on the work of another and, furthermore, that Marconi succeeded in spite of the fact that he only partially understood what Tesla had attempted. Blinded by the Hertzian spark-gap research, Marconi spread his "myopic vision" through Pupin to the vast hordes of researchers in the field of wireless; and this policy has continued to this day. With Marconi's highly visible early success, large-scale wireless enterprise, and Nobel Prize on his side of the balance, it became much easier to credit him with the discovery. The ongoing Great War served to further cloud the issue, as the important legal battle between Atlantic Cable Company (Telefunken) and Marconi Wireless was abandoned before it was resolved.

Due to the dangers that existed on the high seas, and the rumors that the Germans were out for Marconi's head, the *senatore* did not sail back on the *Lusitania;* he returned on the *St. Paul* with a disguised identity and under an assumed name.

Marconi set sail as the new head of the AIEE, John S. Stone, was honored at a dinner attended by a potpourri of leaders in the industry. Guests included Lee De Forest, who was about to receive a quarter of a million dollars for sale of his patents to AT&T, J. A. White, editor of *Wireless* magazine; David Sarnoff, on the verge of launching his radio empire; Rudolf Goldschmidt, the force behind the Tuckerton plant; A. E. Kennelly; Fritz Lowenstein, who was about to earn $150,000 from AT&T for one of his inventions; and Nikola Tesla, who stood between De Forest and Lowenstein for an official photograph.[24]

A fortnight later, in May 1915, a German submarine torpedoed the *Lusitania,* killing 1,134 persons. The sinking, in lieu of the alternative procedure of boarding unarmed passenger ships to check cargo, was unheard of. Quite possibly, Marconi could have been a target; however, the Germans used as their reason the cargo of armaments onboard headed for Great Britain. With only 750 survivors, this "turkey shoot" took almost as many lives as the sinking of the *Titanic.* According to Lloyd Scott of the Navy Consulting Board, "press reports stated that the Germans seemed to revel in this crime, and that various celebrations were held in Germany on account of it. Medals were being struck to commemorate the sinking, and holidays were given to school children."[25] No longer neutral, Teddy Roosevelt hailed the event as "murder on the open seas."

The huge loss of life, however, did not stop George Sylvester Viereck from supporting the German position. Having traveled by zeppelin above Berlin during the war, Viereck stated in the *New York Times* that had the weapons made it to England, "more Germans would have been killed than died in the [boat] attack." Viereck's callous argument inflamed the populace against him. The former renowned poet was now hailed as "a venom-bloated toad of treason."[26]

The enemy seemed within. German spies were everywhere. Reports started filtering in that the Germans were creating a secret submarine base around islands off the coast of Maine. It was also alleged that the broadcasting station out at Sayville was not merely sending neutral dispatches to Berlin but also coded messages to battleships and submarines.

As Tesla, just a few months earlier, had boasted to Morgan that he was working for the Germans and with the *Times* reporting on their front page that "Grand Admiral von Tirpitz [was] contemplat[ing] a more vigorous campaign against freight ships…[and planning] a secret base on this side of the Atlantic,"[27] it is quite possible that the inventor became tainted with a smattering of "venom-bloated toad's blood."

On July 2, 1915, the senate chambers in Washington were rocked by a terrorist bomb. The following day, the fanatic who planted it, Frank Holt, a teacher of German from Cornell University, walked into Jack Morgan's Long Island home toting a six-gun in each hand. With his wife and daughter leaping at the assailant, Morgan charged forward. Shot twice in the groin, Morgan was able, with the help of his fearsome wife, to wrestle the guns from the man and get him arrested. Recuperating at the hospital, the hero received a get-well letter from Nikola Tesla.[28]

When questioned, Holt claimed he had not planned to kill the Wall Street monarch. He only wanted the financier to stop the flow of arms to Europe. A few days later, as Morgan recovered quickly and fired all German and Austrian workers from his office, the self-righteous pacifist committed suicide in his jail cell. His secret had been unveiled. Holt's real name was Dr. Erich Muenter; he was a former German teacher from Harvard who had disappeared after murdering his wife with poison in 1906.[29]

A week later, on Tesla's fifty-ninth birthday, the *Times* reported that not only were the Germans dropping bombs over London from zeppelins; they were also "controlling air torpedoes" by means of radiodynamics. Fired from zeppelins, the supposed "German aerial torpedo[es] can theoretically remain in the air three hours, and can be controlled from a distance of two miles.... Undoubtedly, this is the secret invention of which we have heard so many whispers that the Germans have held in reserve for the British fleet."[30] Although it seemed as if Tesla's devil automata had come into being, as the wizard had predicted a decade before, Tesla himself announced to the press that "the news of these magic bombs cannot be accepted as true, [though] they reveal just so many startling possibilities."

"Aghast at the pernicious existing regime of the Germans," Tesla accused Germany of being an "unfeeling automaton, a diabolic contrivance for scientific, pitiless, wholesale destruction the like of which was not dreamed of before.... Such is the formidable engine Germany has perfected for the protection of her Kultur and conquest of the globe." Predicting the ultimate defeat of the fatherland, the Serb, whose former countrymen were fighting for their own survival against the kaiser, no doubt stopped doing business with von Tirpitz, although he probably continued his relationship with Professor Slaby, who may have been morally opposed to the war.

Tesla's solution to war was twofold, a better defense, through an electronic Star Wars type of shield he was working on, and "the eradication from our hearts of nationalism." If blind patriotism could be replaced with "love of nature and scientific ideal... permanent peace [could] be established."[31]

The period from 1915 to the date of the United States entry into the war, in 1917, was marked by numerous reports of espionage. Spies had infiltrated the Brooklyn Navy Yard to use the station to send secret coded messages to Berlin; through Richard Pfund, head of the Sayville plant, they had also installed equipment on the roof of 111 Broadway, the building that housed Telefunken's offices.[32] Shortly after America's entrance into the war, Tesla informed Scherff that Lt. Emil Meyers, "who ran the Tuckerton operation...[had been placed] in a Detention Camp in Georgia," suspected of spying.[33] Thus, the monthly stipend from Telefunken had come to an abrupt end.[34]

The secretary of the navy, whose job it became to take over all wireless stations, was Josephus Daniels; his assistant was Franklin Delano Roosevelt. In the summer of 1915, Daniels, who was actively monitoring Jack Hammond's work, had read a recent interview with Thomas Alva Edison. Impressed with the contents, the secretary called Edison and set up a meeting with the idea of creating an advisory board of inventors. The hope was that should the country go to war, a civilian think tank, much like the one created in Great Britain, could be created. (Great Britain's consulting board included J. J. Thomson, W. H. Bragg, Sir William Crookes, Sir Oliver Lodge, and Ernest Rutherford.) Edison, who himself had received death threats from the fifth column, became president of this "Naval Consultant Board." Working with Franklin Roosevelt, Edison appointed numerous inventors to various positions, including Gano Dunne, Reginald Fessenden, Benjamine Lamme, Irving Langmuir, R. A. Millikan, Michael Pupin, Charles S. Scott, Elmer Sperry, Frank Sprague, and Elihu Thomson. The journalist Waldemar Kaempffert was also included because of his writing prowess.[35]

It is possible that Tesla's link to Telefunken was the reason his name was not on the list, although many other inventors were also excluded, for example, Hammond, Stone, and De Forest. Tesla also was not one to work again for Thomas Edison. Tesla's work, however, was obviously vital to the government. With President Wilson allowing his adviser, Col. William House, to set up a secret fund for Hammond to advance the Tesla inventions, Tesla himself slipped into a more surreptitious realm.[36]

41

The Invisible Audience (1915–21)

Dear Tesla,

When that Nobel Prize comes, remember that I am holding on to my house by the skin of my teeth and desperately in need of cash!

No apology for mentioning the matter.

Yours faithfully,
RUJ[1]

On November 6, 1915, the *New York Times* published on its front page that Tesla and Edison were to share the Nobel Prize in physics that year. The source for the report was "the Copenhagen correspondent of the [London] *Daily Telegraph*." Although Tesla himself forwarded to J. P. Morgan Jr. original copies of the announcement (which were also carried in a number of other journals),[2] neither Tesla nor Edison ever received the Nobel Prize.

In trying to ascertain what happened, Tesla biographers Inez Hunt and Wanetta Draper wrote in the early 1960s to Dr. Rudberg of the Royal Academy of Science of Sweden. Rudberg, referring to an event which took place a half century before, replied, "Any rumor that a person has not been given a Nobel Prize because he has made known his intention to refuse the reward is ridiculous." Thus, they concluded that the affair was "a sardonic joke."[3]

Curiously, this same *Times* article listed four other people for Nobel Prizes in literature and chemistry who also did not receive the award that year, although three of them eventually obtained it. The fourth, Troeln Lund, like Tesla and Edison, never received the honor.[4]

Although the announcement came in November 1915, the nomination process actually was concluded nine months earlier. There were nineteen scientists on the physics committee, each allowed two bids. Out of the thirty-eight possible bids, two were made for inventors in wireless, E. Branly and A. Righi; two were for the quantum physicist Max Planck; Tom Edison received one bid; and the Braggs took four. According to the Royal Academy's records, Nikola Tesla was *not* nominated that year. (However, two bids, nos. 33 and 34, were missing from their files.) A week after the *Times* announcement, on November 14, Stockholm announced that Prof. William H. Bragg and his son would share the award in physics.

The man who nominated Edison, Henry Fairfield Osborn, president of Columbia University (who, twenty years before, had awarded Tesla an honorary doctorate), apologized to the committee for offering up Edison's name. "Although somewhat out of the line of previous nominations," Osborn wrote, qualifying his decision," I would [like to] suggest the name of Mr. Thomas A. Edison...who is through his inventions, one of the great benefactors of mankind." Tesla would not be nominated until 1937 (by F. Ehrenhaft of Wien, who had previously nominated Albert Einstein).[5]

Certainly, both Tesla and Edison deserved such an award, and it is nothing short of astounding that (1) neither of them ever received it and (2) no one at the time discovered the reason behind this curious quirk of history.

O'Neill, having interviewed Tesla on the subject, stated that Tesla "made a definite distinction between the inventor, who refined preexisting technology, and the discoverer who created new principles....Tesla declared himself a discoverer and Edison an inventor; and he held the view that placing the two in the same category would completely destroy all sense of the relative value of the two accomplishments."[6]

Support for this interpretation can be found in a letter Tesla wrote to the Light House Board in Washington from his Colorado Springs Experimental Station in 1899. The navy had written Tesla that they would "prefer" to give their impending wireless contract to an American rather than to Marconi.

"Gentleman," Tesla responded curtly. "Much as I value your advances I am compelled to say, in justice to myself, that I would never accept a preference on any ground...as I would be competing against some of those who are following in my path....Any pecuniary advantage which I might derive by availing myself of the privilege, is a matter of the most absolute indifference to me."[7] If no one else would recognize his genius, Tesla certainly did. He would not think twice about giving up mere cash when faced with the prospect of being *compared,* in this case to Marconi.

The following letter to Johnson, which the inventor took the time to rewrite in a careful hand, was penned just four days after the announcement and four days *before* Sweden's decision to give the award to the Braggs.

My dear Luka,

Thank you for your congratulations.... To a man of your consuming ambition such a distinction means much. In a thousand years there will be many thousand recipients of the Nobel Prize. But I have not less than *four dozen of my creations identified with my name* in technical literature. These are honors real and permanent which are bestowed not by a few who are apt to err, but by the whole world which seldom makes a mistake, and for any of these I would give all the Nobel Prizes which will be distributed during the next thousand years.[8]

This passage was contained in its entirety in the Hunt and Draper text; however, they incorrectly concluded that this "sober" message was overshadowed by "jubilation" because of the announcement.[9] Johnson also incorrectly understood the full implication of the letter, because in March 1916 he refers to the award, fully expecting Tesla to receive it.[10]

In the *New York Times* interview on the day following their announcement, Tesla stated that Edison was "worthy of a dozen Nobel Prizes." The various Tesla biographers assumed that this was a public statement congratulating Edison when, in fact, it was a piquant snub to the Nobel committee. Tesla was stating between the lines that the Nobel committee recognized only small accomplishments rather than truly original conceptualizations.

"A man puts in here [in my Tesla coil] a kind of gap—he gets a Nobel Prize for doing it.... I cannot stop it."[11] Thus, Edison's many "better mousetraps" could all be honored, but none of them, in Tesla's opinion, concerned the *creation of new principles.* They were simply refinements of existing apparatus.

Edison would probably have agreed with Tesla on this point, for most of his inventions were actually further developments of other people's work. However, Edison did have a number of original discoveries and creations. In his own opinion, his most important contribution was the phonograph, which certainly was the work of genius, even by Tesla's criteria, and deserving of a Nobel Prize. Furthermore, Edison's unparalleled success in bringing promising creations to fruition was exactly Tesla's failure, and that, too, was a gift placing Edison in a category all by himself.

Quite possibly a letter much like the one sent to Johnson or the Light House Board could also have been wired to the Nobel committee. If this hypothesis is correct, a prejudice would have persisted against Tesla and Edison, and this would explain the indefensible position of the Swedish Royal Academy in never honoring either of these two great scientists.

Wizard Swamped by Debts

Inventor Testifies He Owes the Waldorf Hasn't a Cent in Bank[12]

As 1915 was drawing to a close, Tesla began to find himself in deeper and deeper financial straits. Although an efficient water fountain which he designed that year was received favorably,[13] his overhead was still too high. Expenses included outlays for the turbine work at the Edison Station, his office space at the Woolworth Building, salaries to his assistants and Mrs. Skerritt, his new secretary, past debts to such people as the Johnsons and George Scherff, maintenance costs for Wardenclyffe, legal expenses on wireless litigation, and his accommodations at the Waldorf-Astoria.

Some of the costs were deferred, particularly by the hotel, but Mr. Boldt's patience had reached its limit. Tesla's uncanny elusiveness and noble air had worn thin. Rumors began circulating of peculiar odors and cackling sounds emanating from the inventor's suite. The maids were complaining that there was an inordinate amount of pigeon excrement on the windowsills. Boldt sent Tesla a bill for the total rent due, nearly $19,000. Simultaneously, Tesla was hit with a suit for $935 for taxes still owed on Wardenclyffe.

Tesla signed over the Wardenclyffe property to Boldt just as he was called into the state supreme court. Before Justice Finch, the inventor revealed that "he possessed no real estate or stocks and that his belongings, all told, were negligible." Under oath, Tesla revealed that he lived at the prestigious Waldorf "mostly on credit," that his company "had no assets but is receiving enough royalties on patents to pay expenses," and that most of his patents were sold or assigned to other companies. When asked if he owned an automobile or horses, the inventor responded no.

"Well, haven't you got any jewelry?"

"No jewelry; I abhor it."[14]

This embarrassing article was published for all to see in the *World.* Yet, as was his custom with any article about himself, the inventor had his secretary paste the *mea culpa* in the latest volume of press clippings. Looking much like a multivolume encyclopedia, this text, along with his other records and correspondence, would provide for posterity an accurate account of the inventor's rich and complicated life. The inventor had chosen his words carefully when speaking under oath to the judge. As much as he loathed being in a debtor's position, he wanted the Morgans, Marconis, Franklin Roosevelts, and Woodrow Wilsons to know of his plight, for in the final analysis this shame would be theirs as much as his. Even T. C. Martin had turned against him, writing petty letters to Elihu Thomson at this time, complaining of how Tesla chiseled money out of him for the opus he had created of the inventor's collected works a generation ago.[15]

Attempting to raise funds in a variety of ways, Tesla continued to try to market his speedometer, push to get monies from American firms for the bladeless turbines, and collect royalty payments from Lowenstein and Telefunken for the Tuckerton and Sayville plants. The elder statesman of invention also continued to write newspaper articles for the *World* and the *Sun* for ready cash, and he also moved to exploit other creations, such as his electrotherapeutic machines, with Dr. Morrell. Tesla wrote Scherff that he expected the medical market to be \$3–\$4 million.[16]

The publication of his wretched state in the public forum and the transfer of Wardenclyffe to another party produced in Tesla a deep sense of anger and corresponding shame; for now the world had officially branded him a dud. If success is measured in a material way, it was clear that Tesla was the ultimate failure.

On the exterior, the inventor kept up appearances, but this event would mark the turning point in his life. Now he began the slow but steady turning away from society. Simultaneously, he traveled to live in other states, in part to conduct business in a fresh atmosphere and in part to remove himself from a hostile environment. He wrote a letter to Henry Ford in Detroit, hoping, finally, that the auto magnate would recognize the great advantages of his steam engine.

"I can tell, any day, that Ford is going to contact me, and take me out of all my worries," Tesla confidently predicted to Julius Czito, Coleman's son, who was now working for him. "Sure enough, one fine morning a body of engineers from the Ford Motor Company presented themselves with the request of discussing with me an important project," Tesla revealed a few years later.

"Didn't I tell you," the prophet remarked triumphantly.

"You are amazing, Mr. Tesla," Julius responded. "Everything is coming out just as you predicted."

"As soon as these hardheaded men were seated," Tesla continued, "I, of course, immediately began to extol the wonderful features of my turbine, when the spokesmen interrupted me and said, 'We know all about this, but we are on a special errand. We have formed a psychological society for the investigation of psychic phenomena and we want you to join us in this undertaking.'" Flabbergasted, Tesla contained his indignation long enough to escort the wayward explorers to the street.[17]

A MEETING WITH A PRINCESS

Suffering from an attack of the grippe throughout the first month of 1916, Tesla made the newspapers by posing for a portrait for the provocative painter Princess Vilna Lwoff-Parlaghy. Daughter of Baroness von Zollerndorff and married and divorced from Prince Lwoff of Russia, Vilna had painted the portraits of such greats as Field Marshall von Moltke, Bismarck, the Baroness Rothschild, Andrew Carnegie, Thomas Edison, and Teddy Roosevelt. Reluctant at first to sit because of superstitious feelings of foreboding, Tesla soon acquiesced and found a comfortable chair among Her Highness's various pets, which included, at one time or another, "two dogs, an Angora cat, a bear, lion cub, alligator, ibis and two falcons." Recently kicked out of the Plaza for unpaid bills totaling $12,000, perhaps the princess shared a good laugh with the Serbian nobleman, who himself was in a similar predicament. The painting was reproduced in *Electrical Experimenter* in 1919 and, once again, on the cover of *Time* for Tesla's seventy-fifth birthday, in 1931.[18]

This time period also saw the arrival of Tesla's sister's son, Nicholas Trbojevich, himself an inventor who wanted to work as Tesla's assistant. Apparently, Tesla was unable to spend much time with his nephew. Feeling rebuked, Trbojevich turned to the local Serbian community, where he found a willing Prof. Michael Pupin, who took the lad "under his wing" and on a tour of the city. Trbojevich endeared himself to the great professor and they became friends. Trbojevich would come to develop, in the 1920s and 1930s the hypoid gear and several sophisticated improvements in steering for the automotive industry. Working with mathematical principles, this inventor designed an elegant way to lower the driveshaft, running from the motor to the rear axle, nearly a foot. This advance enabled running boards to be eliminated, thereby allowing the car to become more streamlined. Simultaneously, it earned Trbojevich a tidy sum. Moving to Detroit in the late 1920s, Trbojevich continued to correspond with Uncle Nikki, who also came to visit around the time of the Great Depression.[19]

In February, Tesla received a letter from ardent admirer John (Jack) O'Neill, who was now working as a news correspondent for a Long Island daily and about to transfer jobs to the *Herald Tribune.* The young man reminded him of their 1907 encounter in the subway and enclosed the following poem, "To Nikola Tesla," as an "infinitesimal tribute to [the inventor's] greatness":

Most glorious man of all ages
Thou wert born to forecast greater days
Where the wonders thy magic presages
Shall alter our archaic ways.

Your coils with their juice oscillating
Sent electrical surges through the earth
Sent great energies reverberating
From the center to the outermost girth.

Is thy mind a power omnipresent
That fathoms the depths of all space
That speaks to an adolescent
The future triumphs of the race?[20]

Tesla sent the youngster a letter in return "thanking him heartily," although "your opinion of me is immensely exaggerated." Enigmatically, he also suggested that O'Neill write a poem for J. Pierpont Morgan, "one man today on whom the world is depending more than any other." Should O'Neill do this, "it might be instrumental in putting [him] in possession of a check."[21] Considering that Pierpont was dead, this was a rather peculiar recommendation.

The Edison Medal

> Were we to seize and to eliminate from our industrial world the results of Mr. Tesla's work, the wheels of industry would cease to turn, our towns would be dark, our mills would be dead and idle. Ye[s], so far reaching is this work, that it has become the warp and woof of industry.
>
> B. A. Behrend, 1917[22]

To those with eyes for the truth, Tesla's state of crisis cut deep. One engineer in particular, Bernard A. Behrend, the Swiss émigré who had refused to testify against him during the malevolent AC-patent litigation days, felt the urgency to act. Clearly, it was Behrend's goal to help restore the reputation of his spiritual benefactor. Having devoted a large measure of his life to refining Tesla's invention of the induction motor, Behrend informed his mentor that he, Tesla, had been nominated to receive the Edison Medal. In fact, it had been Behrend who had proposed the idea to the committee. Winners from the past had included Alexander Graham Bell, Elihu Thomson, and George Westinghouse.

That Tesla would be nominated by an organization dwelling under the banner of the Edison name was shocking enough to the brooding Serb. Edison himself must have allowed the presentation to be made. It does not appear that Edison, having just turned seventy, was plagued by the reciprocal feeling of animosity that Tesla exhibited. It is more likely that the thought of giving Tesla the medal brought a broad smirk to the Menlo Park Wizard's visage.

Tesla's first reaction was abhorrence, and he flatly rejected the offer, but Behrend persisted. Here was an opportunity to recognize a worthy recipient alone for his singular contributions. "Who do you want remembered as the author of your power system?" Behrend inquired. "Ferraris, Shallenberger, Stillwell, or Steinmetz?" Tesla reluctantly capitulated.

The presentation of the Edison Medal was made on May 18, 1917, just two months before Tesla found out by telephone that vandals had broken into his Wardenclyffe laboratory and wrecked equipment valued at $68,000 and that "the Tower [was] to be destroyed by dynamite."[23] Many familiar faces dotted the crowd. The Johnsons and Miss Merrington attended, as did Charles Scott and Edward Dean Adams, the man most responsible for recommending Tesla for the Niagara Falls enterprise.

The opening speech was delivered by A. E. Kennelly, former Edison crony, who was now teaching at Harvard. Long a Tesla adversary, having been active in executing animals with AC current during the heated Battle of the Currents in the early 1890s, Kennelly spoke for fifteen minutes. During this time, the good professor managed to not mention Tesla's name even once.

"Many people," Professor Kennelly began, "suppose that the Edison Medal is presented by Mister Edison, but that is a mistake. In fact, Tom Edison has been so busy during his life receiving medals that he has not time to dispense any." The speaker droned on, making Tesla more nervous with each obsequious sentence. "Every time a worthy recipient is honored with this Medal, Thomas Edison is also honored. In fact," the Edison man continued, "We may look forward to a time, say a thousand years hence, when like this evening the one thousand and seventh recipient will receive the Edison Medal, and once again Edison's achievements will be honored."[24]

As legend has it, Tesla disappeared from the room. Panic-stricken, Behrend ran out of the building to look for him, while Charles Terry, a prominent executive from the Westinghouse Corporation, reviewed Tesla's great accomplishments. According to the story, Behrend found the lonely inventor across the street by the library, feeding his precious pigeons.[25]

During Behrend's introduction, he stated, perhaps to counter Kennelly's opening speech, "The name of Tesla runs no more risk of oblivion than does that of Faraday or Edison. What can a man desire more than this. It occurs to me to paraphrase Pope describing Newton, 'Nature and Nature's laws lay hid in night. God said, "Let Tesla be," and all was light.'"

"Ladies and gentleman," Tesla began, "I wish to thank you heartily for your kind appreciation. I am not deceiving myself in the fact of which you must be aware that the speakers have greatly magnified my modest achievements. Inspired with the hope and conviction that this is just a beginning, a forerunner of still greater accomplishments, I am determined to continue developing my plans and undertake new endeavors.

"I am deeply religious at heart, and give myself to the constant enjoyment of believing that the greatest mysteries of our being are still to be fathomed. Evidence to the contrary notwithstanding, death itself may not be the termination of the wonderful metamorphosis we witness. In this way I manage to maintain an undisturbed peace of mind, to make myself proof against adversity, and to achieve contentment and happiness to a point of extracting some satisfaction even from the darker side of life, the trials and tribulations of existence."

The electrical savant would go on to review much of his life—an anecdote from his childhood about a gander who almost pulled his umbilical cord out, his early meetings with Edison and work with Westinghouse, lectures in Europe, success at Niagara, and future plans in wireless.

"I have fame and untold wealth, more than this," the inventor concluded, "and yet—how many articles have been written in which I was declared to be an impractical unsuccessful man, and how many poor, struggling writers have called me a visionary. Such is the folly and shortsightedness of the world!"[26]

Tesla was aghast that Boldt had not protected Wardenclyffe adequately, for it was valued at a minimum of at least $150,000. Even though he had signed it over to the hotel, he had done so, according to his understanding, to honor his debt "until [his] plans matured." As the property, when completed, would yield $20,000 or $30,000 a day, Tesla was simply flabbergasted that Boldt would move to destroy the place. Boldt or "the Hotel Management" saw Wardenclyffe now as theirs, free and clear, even though Tesla offered as proof "a chattel mortgage" on the machinery that the inventor had placed at his own expense. The hotel's insurance was only $5,000, whereas Tesla's coverage for the machinery was valued at $68,000. Why would Tesla independently seek to protect the property if he didn't still have an interest in it? Tesla saw the contract as "a security pledge," but the paper he signed did not specify any such contingency. According to the Hotel's lawyer, Frank Hutchins of Baldwin & Hutchins, "it was bill of sale with the deed duly recorded two years ago. We fail to see what interest you have," Hutchins callously concluded.[27]

Storming into their offices on Pine Street, Tesla demanded to find out firsthand what was to happen.

"You will have to ask Smiley Steel Company. They are the ones in charge of salvage operations."

J. B. Smiley informed Tesla that indeed the tower was to be taken down, its parts sold to cover outstanding debts. "A great wrong has been done," the inventor wrote in reply, "but I am confident that justice will prevail."[28]

"Pay no attention to Tesla whatsoever, but proceed immediately with wrecking as contracted," Smiley told his wrecking crew after conferring with Hutchins.[29]

Waldorf-Astoria Hotel Company

July 12, 1917

Gentlemen:

I have received reports which have completely dumbfounded me all the more so as I am now doing important work for the Government with a view of putting the plant to a special use of great moment....

I trust that you will appreciate the seriousness of the situation and will see that the property is taken good care of and that all apparatus is carefully preserved.

Very truly yours,
N. Tesla[30]

The wizard decided that the only way to save Wardenclyffe was to extol its virtues as a potential defensive weapon for the protection of the country. Capitalizing on the excellent Nobel Prize publicity, the inventor once again strained the reader's credulity with another startling vision.

Tesla's New Device Like Bolts of Thor

He Seeks to Patent Wireless Engine for Destroying Navies by Pulling a Lever

To Shatter Armies Also

Nikola Tesla, the inventor, winner of the 1915 Nobel Physics Prize, has filed patent applications on the essential parts of a machine the possibilities of which test a layman's imagination and promise a parallel to Thor's shooting thunderbolts from the sky to punish those who had angered the gods. Dr. Tesla insists there is nothing sensational about it....

"It is perfectly practicable to transmit electrical energy without wires and produce destructive effects at a distance. I have already constructed a wireless transmitter which makes this possible."

"Ten miles or a thousand miles, it will be all the same to the machine," the inventor says. Straight to the point, on land or on sea, it will be able to go with precision, delivering a blow that will paralyze or kill, as it is desired. A man in a tower on Long Island could shield New York against ships or army by working a lever, if the inventor's anticipations become realizations.[31]

Tesla would not draw up an official paper on the particle-beam weapon, or "death ray," for another twenty years yet it is clear that he had conceived the machine by this time, probably creating prototypes as far back as 1896, when he was bombarding targets with Roentgen rays.

In "a serious plight," with nowhere else to turn, the inventor contacted Morgan once again to ask for assistance. This was his last chance to protect his wireless patents and save the tower. "Words cannot express how much I have deplored the cruel necessity which compelled me to appeal to you again," the inventor explained, but it was to no avail.[32] He still owed Jack $25,000 plus interest; the financier ignored the entreaty and quietly placed Tesla's account in a bad-debt file.

In February 1917, the United States broke off all relations with Germany and seized the wireless plant at Sayville. "Thirty German employees of the German-owned station were suddenly forced to leave, and enlisted men of the American Navy have filled their places."[33] Guards were placed around the plant as the high command decided what to do with the remaining broadcasting stations lying along the coast. Articles began springing up like early crocus to announce the potential "existence of [yet another] concealed wireless station [able] to supply information to German submarines regarding the movements of ships."[34]

19 More Taken as German Spies

Dr. Karl George Frank, Former Head of Sayville Wireless Among Those Detained[35]

On April 6, 1917, President Wilson issued a proclamation "seiz[ing] all radio stations. Enforcement of the order was delegated to Secretary Daniels.... It is understood that all plants for which no place can be found in the navy's wireless system, including amateur apparatus, for which close search will be made, are to be put out of commission immediately."[36] Clearly, an overt decision had to be made about the fate of Wardenclyffe.

Tesla's expertise was well known to Secretary Daniels and Assistant Secretary Franklin Roosevelt, as they were actively using the inventor's scientific legacy as ammunition against Marconi in the patent suit. Coupled with the inventor's astonishing proclamation that his tower could provide an electronic aegis against potential invasions, Wardenclyffe must have been placed in a special category. However, there were two glaring strikes against it. The first was that Tesla had already turned over the property to Mr. Boldt to cover his debt at the Waldorf; and the second was the transmitter's record of accomplishment: nonexistent. What better indication of the folly of Tesla's dream could there be then the tower's own perpetual state of repose. To many, Wardenclyffe was merely a torpid monument to the bombastic prognostications of a not very original mind gone astray. From the point of view of the navy, Tesla may have been the original inventor of the radio, but he was clearly not the one who made the apparatus work.

A HISTORY OF NAVY INVOLVEMENT

In 1899 the U.S. Navy, via Rear Admiral Francis J. Higginson, requested Tesla to place "a system of wireless telegraphy upon Light-Vessel No. 66 [on] Nantucket Shoals, Mass, which lies 60 miles south of Nantucket Island."[37] Tesla was on his way to Colorado and was unable to comply. Moreover, the navy did not want to pay for the equipment, but rather wanted Tesla to lay out the funds himself. Considering the great wealth of the country, Tesla feigned astonishment at the penurious position of John D. Long, secretary of the navy, via Commander Perry, who brazenly forwarded the financial disclaimer on U.S. Treasury Department stationery.

Upon Tesla's return to New York in 1900, he wrote again of his interest in placing the equipment aboard their ships. Rear Admiral Higginson, chairman of the Light House Board, wrote back that his committee would meet in October to discuss with Congress "the estimates of cost."[38] Higginson, who had visited Tesla in his lab in the late 1890s, wanted to help, but he had been placed in the embarrassing position of withdrawing his offer of financial remuneration because of various levels of bureaucratic inanity. Tesla spent the time to go down to Washington to confer face-to-face with the high command—Hobson also negotiated on his friends behalf—but Telsa was essentially ignored and returned to New York empty handed and disgusted with the way he was treated.

From the point of view of the navy, wireless telegraphy was an entirely new field, and they were unsure what to do. Furthermore, they may have been turned off by Tesla's haughty manner, particularly when it came to being "compared" to Marconi, which had always enraged Tesla. (Keep in mind, however, that the navy took over ten years to recompense Hammond for his work on radio-guided missiles, and even then they almost didn't come through. Tesla was by no means the only one to get the runaround from the military, and Hammond had the best connections possible through his influential father.)

In 1902 the Office of Naval Intelligence called Comdr. F. M. Barber, who had been in retirement in France, back to the States and put him in charge of the acquisition of wireless apparatus for testing. Although still taking a frugal position, the navy came up with approximately $12,000 for the purchase of wireless sets from different European companies. Orders were placed with Slaby-Arco and Braun-Siemans-Halske of Germany and Popoff, Ducretet and Rochefort of France. Bids were also requested from De Forest, Fessenden, and Tesla in America and Lodge-Muirhead in

England. Marconi was excluded because he arrogantly coveted an all-or-nothing deal.[39]

Fessenden was angry with the navy for obtaining equipment outside the United States and so did not submit a bid. Tesla was probably too upset with his treatment from the past and too involved with Wardenclyffe, which was under active construction at that time, to get involved, and so the navy purchased additional sets from De Forest and Lodge-Muirhead.

In 1903 a mock battle with the North Atlantic fleet was held five hundred miles off the coast of Cape Cod. With the "White Squadron" commanded by Rear Adm. J. H. Sands and the "Blue Squadron" by Rear Admiral Higginson, Tesla's ally, the use of wireless played a key role in determining the victor. Commander Higginson, who won the maneuver, commented, "To me, the great lesson of the search we ended today is the absolute need of wireless in the ships of the Navy. Do you know we are three years behind the times in the adoption of wireless?"[40]

Based on comparison testing, it was determined that the Slaby-Arco system outperformed all others, and the navy ordered twenty more sets. Simultaneously, they purchased an eleven-year lease on the Marconi patents.[41]

With the onset of World War I, the use of wireless became a necessity for organizing troop movements, surveillance, and intercontinental communication. While the country was still neutral, the navy was able to continue their use of the German equipment—until sentiments began to shift irreversibly to the British side. Via the British navy, Marconi had his transmitters positioned in Canada, Bermuda, Jamaica, Columbia, the Falkland Islands, North and South Africa, Ceylon, Australia, Singapore, and Hong Kong. His was a mighty operation. In the United States, the American Marconi division, under the directorship of the politically powerful John Griggs, former governor of New Jersey and attorney general under President McKinley, had transmitters located in New York, Massachusetts, and Illinois.[42] One key problem, however, was that the Marconi equipment was still using the outmoded spark-gap method.

In April 1917, the U.S. Navy completed the seizure of all wireless stations, including those of their allies, the British. At the same time, Marconi was in the process of purchasing the Alexanderson alternator, which was, in essence, a refinement of the Tesla oscillator. Simultaneously, the Armstrong feedback circuit was becoming an obvious necessity for any wireless instrumentation. However, the Armstrong invention created a judicial nightmare, not only because it used as its core the De Forest audion but also because De Forest's invention was overturned in the courts in favor of an electronic tube developed by Fessenden. Never mind that Tesla, as far back as 1902, had beaten Fessenden in the courts for this development. With the Fessenden patent now under the control of Marconi, the courts would come to rule that no one could use the Armstrong feedback circuit without the permission of the other players.

The most important ruling, concerning the true identity of the inventor of the radio, became neatly sidestepped by the War Powers Act of President Wilson, calling for the suspension of all patent litigation during the time of the war. France had already recognized Tesla's priority by their high court, and Germany recognized him by Slaby's affirmations and Telefunken's decision to pay royalties; but in America, the land of Tesla's home, the government backed off and literally prevented the courts from sustaining a decision. The Marconi syndicate, in touch with kings from two countries, with equipment instituted on six continents, was simply too powerful.

With the suspension of all patent litigation and the country in the midst of a world war, Franklin Roosevelt, assistant secretary of the navy, penned the famous Farragut letter. This document allowed such major companies as AT&T, Westinghouse, and American Marconi the right to pool together to produce each other's equipment without concern for compensating rightful inventors. Furthermore, it "assured contractors that the Government would assume liability in infringement suits."[43]

On July 1, 1918, Congress passed a law making the United States financially responsible for any use of "an invention described in and covered by a patent of the United States." By 1921, the U.S. government had spent $40 million on wireless equipment, a far cry from Secretary Long's policy of refusing to pay a few thousand dollars for Tesla's equipment eighteen years before. Thus, the Interdepartmental Radio Board met to decide various claims against it. Nearly $3 million in claims were paid out. The big winners were Marconi Wireless, which received $1.2 million for equipment and installations taken over (but not for their patents). International Radio Telegraph received $700,000; AT&T, $600,000; and Edwin Armstrong, $89,000. Tesla received a minuscule compensation through Lowenstein, who was awarded $23,000.[44]

In 1921 the navy published a list of all the inventors in wireless who received compensation from them. The list contained only patents granted after 1902. Inventors included Blockmen, Braun, Blondel, De Forest, Fuller, Hahnemann, Logwood, Meissner, Randahl, Poulsen, Schiessler, von Arco, and Watkins. Note that both Tesla's and Marconi's names are missing.[45] Marconi's could be missing either because his patents had lapsed or, more likely, because they were viewed as invalid from the point of view of the government. In the case of Tesla, all of his twelve key radio patents had "expired and [were] now common property."[46] However, Tesla had renewed one fundamental patent in 1914,[47] and this should have been on the list, as should have Armstrong's feedback patent.

RADIO CORPORATION OF AMERICA

The U.S. government, through Franklin Roosevelt, *knew* that Marconi had infringed upon Tesla's fundamental patents. They knew the details of Tesla's rightful claims through their own files and through the record at the patent office. In point of fact, it was Tesla's proven declaration which was the basis and central argument that the government had against Marconi when Marconi sued in the first place, and it was this same claim, and the same Navy Light House Board files, that would eventually be used by the U.S. Supreme Court to vindicate Tesla three months after he died, nearly twenty-five years later, in 1943.

Rather than deal with the truth and with a difficult genius whose present work appeared to be in a realm above and beyond the operation of simple radio telephones and wireless transmitters, Roosevelt, Daniels, President Wilson, and the U.S. Navy, in the midst of war, took no interest in protecting Tesla's tower.

In July 1917, Tesla packed his bags and said goodbye to the Waldorf-Astoria. Having lived there for nearly twenty years, he talked George Boldt Jr. into allowing him to keep a large part of his personal effects in the basement of the hotel until he found a suitable place for transferring them. "I was sorry to hear about your father," Tesla told the new manager, George Boldt Sr. having died just a few months before.

Preparing to move to Chicago to work on his bladeless turbines, Tesla was invited to the Johnsons for a farewell dinner. Robert was now directing the affairs of the American Academy of Arts and Letters, an organization which counted among its ranks Daniel Chester French, Charles Dana Gibson, Winslow Homer, Henry James and his brother William, Charles McKim, Henry Cabot Lodge, Teddy Roosevelt, and Woodrow Wilson. Katharine had been in bed for over a week with the grippe, but this evening was too important, and she dragged herself out of bed and put on her best gown.

Dressed in straw hat, cane, white gloves, and his favorite green suede high-tops, Tesla arrived with a large bouquet of flowers and a check for Johnson.

"Kate's been ill," Robert managed to say before the lady of the house appeared.

Taking center stage, as she always tried to do when "He" was around, Kate radiated an intense glow of amorous pride as she held back the flood of tears while she chatted on an on about "how crazy [she was] about all of her grandchildren."[18]

Taking a weekend train to Chicago, Tesla moved into the Blackstone Hotel, alongside the University of Chicago. On Monday morning the inventor hired a limousine to drop him at the headquarters of Pyle

National Corporation. Having already shipped prototypes to give them a head start, he would now work at an intense pace in an entirely new setting, his goal being the perfection of his revolutionary bladeless turbines.[49]

At night he liked to walk down the street from his hotel to the Museum of Arts and Sciences, the only building remaining from the World's Fair of 1893. There he could stand by the great columns and think back to a time when, daily, hundreds of thousands would stream into a magical city powered by his vision. One Saturday, in the heat of summer, he took the mile walk along Lake Michigan, past the Midway, to a series of small lakes and a park which was once the Court of Honor. There, at the entranceway, to a place that once was, he found, to his delight, the Statue of the Republic still standing, its gold plating all worn away. With him was a letter from George Scherff.

> August 20, 1917
>
> Dear Mr. Tesla,
>
> I was deeply grieved and shocked when I read the enclosed, but I have the supreme confidence that more glorious work will arise from the ruins.
>
> I trust that your work in Chicago is progressing to your satisfaction.
>
> Yours respectfully,
> George Scherff[50]

At the height of the world conflagration, the Smiley Steel Company's explosives expert had circled the gargantuan transmitter to place a charge around each major strut and nail the coffin shut on Tesla's dream. With the Associated Press recording the event and military personnel apparently present, the magnifying transmitter was leveled, the explosion alarming many of the Shoreham residents.

And with the death of the World Telegraphy Center came the birth of the Radio Broadcasting Corporation, a unique conglomerate of private concerns under the auspices of the U.S. government. Meetings were held behind closed doors in Washington between President Wilson, who wanted America to gain "radio supremacy,"[51] Navy Secretary Daniels, his assistant Franklin Roosevelt, and representatives from GE, American Marconi, AT&T, and the Westinghouse Corporation. With J. P. Morgan & Company on the board of directors and the Marconi patents as the backbone of the organization, RCA was formed. It would combine resources from these megacorporations, all of which had cross-licensing agreements with each other and co-owned the company.[52] (Cross-licensing agreements also existed with the government, which also owned some wireless patents.) Here was another *entente cordiale* reminiscent of the AC polyphase days, which was not so for the originator of the invention. It was a second major time

Tesla would be carved from his creation,[53] a secret deal probably concocted which absolved the government from paying any licensing fee to Marconi in lieu of their burying their Tesla archives. David Sarnoff, as managing director, would soon take over the reins of the entire operation.

The *New York Sun* inaccurately reported:

U.S. Blows Up Tesla Radio Tower

> Suspecting that German spies were using the big wireless tower erected at Shoreham, L.I., about twenty years ago by Nikola Tesla, the Federal Government ordered the tower destroyed and it was recently demolished with dynamite. During the past month several strangers had been seen lurking about the place.[54]
>
> The destruction of Nikola Tesla's famous tower...shows forcibly the great precautions being taken at this time to prevent any news of military importance of getting to the enemy.[55]

At the end of the war President Wilson returned all remaining confiscated radio stations to their rightful owners. American Marconi, now RCA, of course, was the big beneficiary.[56]

In 1920 the Westinghouse Corporation was granted the right to "manufacture, use and sell apparatus covered by the [Marconi] patents."[57] Westinghouse also formed an independent radio station which became as prominent as RCA. At the end of the year, Tesla wrote a letter to E. M. Herr, president of the company, offering his wireless expertise and equipment.

> November 16, 1920
>
> Dear Mr. Tesla,
>
> I regret that under the present circumstances we cannot proceed further with any developments of your activities.[58]

A few months later, Westinghouse requested that Tesla "speak to our 'invisible audience' some Thursday night in the near future [over our...] radiotelephone broadcasting station."[59]

> November 30, 1921
>
> Gentleman,
>
> Twenty-one years ago I promised a friend, the late J. Pierpont Morgan, that my world-system, then under construction...would enable the voice of a telephone subscriber to be transmitted to any point of the globe....
>
> I prefer to wait until my project is completed before addressing an invisible audience and beg you to excuse me.
>
> Very truly yours,
> N. Tesla[60]

42

TRANSMUTATION (1918–21)

I come from a very wiry and long-lived race. Some of my ancestors have been centenarians, and one of them lived 129 years. I am determined to keep up the record and please myself with prospects of great promise. Then again, nature has given me a vivid imagination. . . .

NIKOLA TESLA[1]

Tesla's lifework was his World Telegraphy Center. Partially materialized on the physical plane as Wardenclyffe, this was the inventor's Holy Grail, the key to anointment. In 1917 the project was demolished, and in that sense, so was the inventor. Capable of recognizing the absurdities of life and drawing from transcendent energies, the mystic sought regeneration by consummating his grand plan in fantasy form and by seeking a new philosopher's stone.

One year earlier, when Tesla's project was at its bleakest, he had formed an alliance with one of his most ardent admirers, Hugo Gernsback, editor of *Electrical Experimenter*. Gernsback had first heard about Tesla when he was a child growing up in Luxembourg in the late 1890s. It was at this time that the ten-year-old came across the fantastic picture of the emblazoned electrician sending hundreds of thousands of volts through his body and the declaration in the accompanying article that he was the grandest wizard of the age. Considered by most futurologists to be the "founder and father of science fiction," Gernsback studied electronics at Bingen Technicum in Europe, before immigrating to America, at the age of nineteen, in 1903.[2]

With his mind totally captivated by the fantastic union of science and fantasy, the exuberant youth wrote a spectacular tale which took place in the year 2660 called *RALPH 124C41+*, which he serialized in his new magazine *Modern Electronics*. Simultaneously, he also opened up Hugo

Gernsback's Electro Importing Company, an all-purpose electronics shop located under the "el" at Fulton Street. There the new breed of amateur ham operators could buy whatever they wanted and browse through "the biggest bunch of junk you ever saw."[3]

Gernsback's first meeting with Tesla was in 1908, when he stopped at the inventor's lab to view the new turbine.[4]

Gernsback wrote, "The door opens, and out steps a tall figure—over six feet high—gaunt but erect. It approaches slowly, stately. You become conscious at once that you are face to face with a personality of a high order. Nikola Tesla advances and shakes your hand with a powerful grip, surprising for a man over sixty. A winning smile from piercing light blue-gray eyes, set in extraordinarily deep sockets, fascinates you and makes you feel at once at home.

"You are guided into an office immaculate in its orderliness. Not a speck of dust is to be seen. No papers litter the desk, everything just so. It reflects the man himself, immaculate in attire, orderly and precise in his every movement. Drest [*sic*] in a dark frock coat, he is entirely devoid of all jewelry. No ring, stickpin or even watch-chain can be seen."[5]

In 1916 the inventor edited a consequential article for Gernsback on the magnifying transmitter. The inventor also promised to think more seriously about putting his life story down on paper; in fact, he wrote a short first draft for *Scientific American* which he embellished for the Edison Medal acceptance speech.[6]

By this time, Gernsback had also secured the talents of the gifted illustrator Frank R. Paul. Destined to be the most influential science-fiction artist of the twentieth century, Paul was able to "render the possible development of any invention [from]...a raw idea into a picture fantasy." With a penchant for drawing futuristic scenarios such as Goliath-sized insects, spaceships orbiting planets, and a variety of humanoidian mad scientists conquering galactic empires, Paul advanced to become the premier cover artist for *Electrical Experimenter*, and later *Amazing Stories* and *Science Wonder Stories*.[7] He was assigned the role of completing Tesla's tower in picture form. The drawing, replete with fully functioning Wardenclyffe transmitters and Tesla wingless airfoils beaming down death rays to incoming ships, not only became a fantastic cover for *Electrical Experimenter;* it also became the centerpiece of the wizard's new letterhead.

As alchemist, Tesla transformed the ruins of his station into a fantastic Gernsbackian World Telegraphy Center, as he also transformed himself, leaving New York City to begin anew with his next major creation.

Before he left, in June 1917, the inventor wrote Jack Morgan, hoping, optimistically, because of new developments, to pay off his debt to the financier "in about four months.... My big ship is still to come in, but I have now a marvelous opportunity having perfected an invention which will astound the whole world." Cryptically, Tesla said that the invention would "afford an effective means for meeting the menace of the submarine." Whether he was talking about a long-range radar system, a remote-controlled torpedo, or some other invention is uncertain.[8]

The following month, Tesla moved to Chicago, and he stayed there through November 1918, working with Pyle National on the perfection of his turbines. Here, during the day, with the slate clean, the gangly mechanic could continue to battle the demons by plunging himself into a brand-new endeavor. At night, as creative author, the cognoscente sketched out the first draft of his expanded autobiography.

Most of the time, he drew from his own capital for fear of causing difficulties with the new partners.[9] He knew he would eventually receive compensation because the Chicago company had signed an agreement promising "cash payments and guarantees" with the expiration of their option, but carrying costs were becoming a problem.[10]

To handle expenses in the interim, the inventor requested that Scherff step up the pressure on receiving royalties from the various wireless companies. His greatest source of income was probably the Waltham Watch Company, which was now in the active stage of marketing his speedometer. Even though the war was still going on, the inventor expected to receive compensation from Telefunken "after the hostilities cease," even though he would have to "apply to the War Trade Board under the Trading with the Enemy act for a license to receive payment."[11]

Progress on the turbines was hampered by numerous obstacles. Nevertheless, the inventor was delighted with the "extraordinarily efficient personnel" and overall organization of the Chicago firm. As the disks could rotate at speeds ranging from 10,000 to 35,000 rpm, the centrifugal force tended to elongate them. Thus, they were subject to fatigue and ran the risk of cracking after performing for long periods of time. Perceived by skeptical engineers as fatal flaws, Tesla endeavored to hammer home the point that stress was a factor in all engines.[12] Thus, much of the time in Chicago was spent experimenting with different alloys and inventing means for instantaneously regulating orthorotational speed and centrifugal pressure to minimize the stress factor. "For instance, suppose that the steam pressure of the locomotive would vary from say 50 to 200 lbs, no matter how rapidly, this would not have the slightest effect on the... performance of the turbine."[13]

In January 1918, the U.S. Machine Manufacturing Company inquired about placing one of Tesla's turbines inside an airplane, and a few months later, the Chicago Pneumatic Tool Company also expressed interest. Tesla was writing Scherff, expecting the invention to yield $25 million per year. However, there was still the difficulty of perfecting it, and Tesla was still not free of the numerous other problems of his life, such as the past debts and the continuing quagmire of litigation. During the summer, the inventor twisted his back and was laid up for several weeks.[11]

During Tesla's time in Chicago, he calculated his operating expenses at $17,600, with revenues of $12,500. Pyle National tried to get out of their debt by sending a check for $1,500, but Tesla returned the token payment and threatened suit. Meanwhile, back at home, the sheriff took possession of the Woolworth office, so Tesla had to wrestle some capital from Pyle National to release his company. In New York, George Scherff continued to handle all of the details.

Concerning his relationship with the government (as stated in chapter 41), most of Tesla's wireless patents had expired, and his 1914 patent was complicated by its clash with the Marconi claim. However, he was negotiating with the government on an engine for a plane, writing to the Bureau of Steam Engineering at this time.[15] In litigation, Tesla won a few thousand dollars from Lowenstein, lost a $67,000 case against a Mr. DeLaVergne, in part, because he refused to travel back to New York to testify, and had to pay out $1,600 to A. M. Foster, for nonpayment of services rendered.[16]

Before returning to Manhattan for the last months of 1918, the inventor traveled to Milwaukee to visit the Allis Chalmers people. There he was met by the astute but pedantic head engineer Hans Dahlstrand. After providing various articles and records from his work at the Edison station and Pyle National, a contract was drawn up for Tesla to return to Milwaukee and develop the engine with Dahlstrand. Skeptical from the start, the learned head engineer reluctantly agreed to defer to Tesla's wishes and begin a preparatory investigation of the turbine before he arrived.

Throughout the period 1917–1926, the inventor spent most of his time outside New York City. In the years 1917–1918, he was in Chicago with Pyle National; in 1919–22 he was in Milwaukee with Allis Chalmers; for the last month of 1922 he was in Boston with the Waltham Watch Company; and in the years 1925–26 he was in Philadelphia working on the gasoline turbine at Budd Manufacturing Company.[17]

Tesla also sold a motor which was used in motion-picture equipment to Wisconsin Electric in 1918 and a valvular conduit, or "unidirectional fluid flow tube," to an unspecified oil company.[18] This last invention, which can also be called a fluid diode, could not only be used to pump oil from the ground but also be attached to the bladeless turbine to turn it into a combustion engine. According to Tesla expert Leland Anderson, this invention "is the only valving patent without moving parts. It has been used in attempts to develop micro-miniature radiation hardened logic circuits and simple fluid computers."[19]

Waltham Speedometers & Automobile Clocks

> Every progressive automobile manufacturer is adding improvements to his car. This is why the only Air-Friction Speedometer in the world, invented by Nikola Tesla, perfected and developed by Waltham...has won the unqualified approval of the world's great automotive engineers. You will find this...instrument upon such cars as the Cunningham, Lafayette, Leach-Biltwell, Lincoln, Packard, Pierce-Arrow, Renault, Rolls-Royce, Stevens-Duryea, Wills-Sainte Claire and others.
>
> The Speedometer of Instantaneous Accuracy[20]

The inventor arrived at the Copley Plaza in Boston to negotiate with Mr. May, the manager of the factory, the advance and royalty schedule.[21] Concerning revenues, Tesla received $5,000 from Waltham, assigning them three of his patents in 1922 for a speedometer and tachometer. This agreement included royalties which he received until at least 1929. Pyle National eventually paid him $15,000, and maybe $30,000 in 1925; from Budd National, he received $30,000 for the turbines, and probably a similar amount from Allis Chalmers, from whom he was expecting profits on the order of a quarter of a million dollars per year.[22] George Scherff received 5 percent from most of these contracts.

Tesla arrived back home at the tail end of 1918 in time for Christmas dinner with the Johnsons. He stayed for a brief time at the Waldorf and then moved to the Hotel St. Regis, where he lived off and on for the next few years. The great influenza epidemic was just in its beginning stages, and Katharine was one of the first to display signs of its ravages. In the next year, over a billion people were infected and 20 million died worldwide. She was lucky to survive. Her health deteriorated throughout the year, and by the following Christmas she experienced episodes during which she lost consciousness three times within a single day.[23] Perhaps heightened by the severity of the situation and with new income from Waltham, Tesla paid Robert checks totaling at least $1,500 during this period.

Throughout 1919, Tesla's autobiography appeared in serialized form in Gernsback's *Electrical Experimenter*. Paired with photographs and a series of spectacular Frank Paul drawings, the story began as an unusual tale of a wizard-child growing up in another era in a faraway land. The account of the early years of Tesla's life oozed charm and wit, with its numerous Mark Twainian depictions of amusing anecdotes, harrowing experiences, life with his inventive mother, preacher father, prodigal brother, and three doting sisters. Digging deep into his past, Tesla explored the tragedy of his brother's death, how it impacted his career decision, the traumatic move away from the idyllic farm to the clutter of Gospić, his college years, engineering training in Europe before coming to America, and his early meetings with Edison, Westinghouse, and members of the Royal Society of London. Also included was an uncommon description of his peculiar powers of eidetic imagery, out-of-body experiences, childhood illnesses, phobias, and idiosyncrasies. In month after month of fascinating reading, the pundit detailed the development of his ideas, his physical breakdown and "opening up of the third eye" experience and accompanying revelation which led to the development of the rotating magnetic field, his creation of the telautomaton, work in Colorado Springs, and the grand Wardenclyffe world-wireless design.

This liaison with Gernsback supplied the inventor with a steady income and helped *Electrical Experimenter* boost its circulation to around 100,000. Simultaneously, *My Inventions* also provided the world with a notable autobiographical testimony of one of the most singular and controversial personalities of the age.

The year also saw numerous articles about Marconi's recent experiences intercepting impulses possibly emanating from extraterrestrials. With Professor Pickering writing Elihu Thomson that he might have detected vegetation on the moon[21] and a resurgence of interest in the "Canals of Mars" scenario, the press jumped at the Italian's far-out declaration and grilled him for additional details.

Stealing Tesla's thunder even on this front, Marconi proclaimed that he "had often received strong signals out of the ether which seemed to come from some place outside the earth and which might conceivably have proceeded from the stars." As to the language problem of communicating with the Martians, Marconi said, "It is an obstacle, but I don't think it is insurmountable. You see, one might get through some such message as 2 plus 2 equals 4, and go on repeating it until an answer came back signifying 'Yes.'... Mathematics must be the same throughout the physical universe."[25]

Seeking redress in a variety of ways, Tesla sought publicity in *Electrical World,* where he attributed the Italian's signals to an undertone metronome effect emanating from other wireless operators. Anticipating the possibility that a critic might ascribe the same mechanism to his own extraterrestrial encounter of 1899, Tesla added: "At the time I carried on those investigations there existed no wireless plant [capable of]... produc[ing] a disturbance perceptible in a radius of more than a few miles."[26] This, of course, was a false premise, as Marconi at that time was already sending messages hundreds of miles.

Johnson wrote Tesla that "When Marconi repeats [your] idea, it is no longer laughed at," but in some circles, this did not seem to be the case.

Celesial Movies

Mr. Tesla has small confidence in the Marconian idea of getting into communication by way of mathematics. He would prefer to

> send pictures by wireless: the human face, for example. But suppose Mars does not like your face. That would be a regrettable rebuff to scientific investigation. If civilization on Mars is as old as we are asked to believe, the Martians have no doubt acquired their own taste in faces.[27]

Although the Christmas dinner of 1919 was marred by Katharine's ill health, it was overshadowed by good news: President Wilson had appointed Robert ambassador to Italy! With mixed emotions and Katharine apparently recovering her health, Tesla's friends left for Europe, where they stayed throughout the following year.

Now, really alone, the wizard continued his slide from public scrutiny. Copied, mocked at, and ultimately abandoned by the world he helped create, Tesla tried to keep his life in perspective and contain his anger by doing his best to transform it; but over time the irony of it all took its toll and caused an already eccentric individual to exaggerate already strange ways. Tesla would become more fanatical about cleanliness and spend more time walking the streets after hours, circling his block three times before entering the St. Regis and avoiding stepping on cracks on the sidewalks. Some said he peeked in windows and liked to watch others in voyeuristic ways. Practicing "gastronomical frugality,"[28] the celibate slowly turned away from the meat and potatoes of life and eventually from eating solids altogether. Now he would rarely write in pen, preferring the less definite pencil. He would spend more time by himself, feeding the pigeons at midnight by the Forty-second Street Library or stealing away via the Staten Island ferry to a quiet farm where he could block out the city and search once again for his fountainhead.[29] With the Johnsons' departure, he left for Milwaukee to consummate his relationship with Allis Chalmers.

Much of his time in Wisconsin was invested in trying to perfect the turbine. However, he had reached an impasse, what Sartre calls a "counterfinality," or unforeseen event which opposes the goal intended, with Hans Dahlstrand, head engineer. Thwarted, Tesla had no recourse but to return to New York. So upset was he that he refused to talk about it when his biographer, Jack O'Neill, questioned him on the Milwaukee experience.[30]

Allis Chalmers had issued Dahlstrand's detailed report describing a long list of serious problems, as he saw it, in the manufacture of the turbine. Aside from the fatigue and cracking of the disks, Dahlstrand also cited additional impediments, including only a 38 percent efficiency performance, a decrease of mechanical efficiency as steam pressure increased, a problem in designing attachment gears needed to join the turbine to other units, and a high cost of production. Another factor was that the present-day motors, such as the Parsons turbine which was being developed by Westinghouse, or the Curtis motor, being developed by GE, were operating satisfactorily.[31]

This question of the failure of the Tesla turbine was posed to a number of Tesla experts. Leland Anderson found that manufacturers interested in the Tesla turbine "all say it is a fine concept and an excellent machine, but there [are] too [many]...support systems...to be replaced for a machine *not that much better in performance.* And that is the point—the Tesla turbine is good, but not that much better."[32]

C. R. Possell, president and chief engineer of the American Development & Manufacturing Company, one of the only existing organizations working on manufacturing Tesla bladeless turbines and pumps, offered a somewhat different explanation. Mr. Possell, who initially worked on the Tesla "boundary layer drag turbine" during the Korean War for the military and who has been actively trying to perfect the turbine for thirty-five years, stated that the main problem had simply to do with the high cost of research and development.

According to Possell, "Tesla was about twenty-five to thirty years ahead of his time. Metallurgy was not what it is today. Magnetic bearings are a whole new science. He didn't have the right materials. Instrumentation [for measuring performance] was in its infancy, and it was hard to demonstrate the turbine adequately. Somewhere between the first prototype and the first use of it, you are going to have hundreds and hundreds of man hours, and the turbine didn't get that." Possell gave as just one example (and there are numerous others), the "millions of man hours" required to get a plane to fly at Mach One.

At present, the Tesla pump, based on the same technology, has been used by Jerry LaBine as a replacement for the motor in the jet ski recreational vehicle, and also, it has been further developed by Max Gurth, inventor of the "Discflo pump." Utilizing Tesla's basic idea and principles associated with the structure of a vortex (responsible for such events as whirlpools and tornadoes) and laminar flow, (i.e., the natural, gentle movement through fluids), Gurth has been able to increase the space between the disks. Thus, he has improved its ability to move such difficult products as solid waste and petrochemicals. Whereas a normal pump would have their blades pitted and corroded by coming into contact with the assorted troublesome products, the boundary layer drag pump has no blades and therefore avoids that entire problem![33]

Possell not only sees a day when the pump will be used inside the human body, as, for instance, a heart valve, but also a day when the turbine is perfected. One of the great advantages of a bladeless Tesla engine is its ability to withstand extremely high temperatures. "Bladed turbines are about at their maximum," Possell said, meaning that they can run at about 2,000 degrees Fahrenheit, "although GE is experimenting with turbines that can run at 2,200 degrees. If you could boost the temperature an additional 350 degrees, you would double its Horse Power output." Possell is convinced that the bladeless turbine built with new ceramic components could run at about 2,700 degrees, which would effectively "triple the Horse Power performance." Thus, Possell is also working to design an engine to compete with the Pegasus engine found in the VTOL (vertical takeoff and landing) Harrier jet. This VTOL of the future has been named the Phalanx. The vehicle will not come about, however, without large funding and commitment from the highest levels of industry and government.[34]

The waiter was surprised to see an elegant gentleman sitting at the breakfast counter before the restaurant was officially opened. "Aren't you Dr. Tesla," the fellow inquired, amazed to see such an important man back in town after so many years.

Having received permission from the owner to eat as early as possible, Tesla replied in the affirmative. He had journeyed to Colorado Springs from Milwaukee, retracing his past and looking toward a possible future when he might erect another wireless station. With a key from Dean Evans of the local engineering school, the inventor was able to utilize the lab to work on some technical calculations. Enjoying the much-needed respite, and perhaps a quick jaunt in a hot spring, the inventor had returned to his beloved retreat. There the spry mountaineer could perch himself like a phoenix on a cliff, to sit and contemplate Thor's design, and watch the lightning storms that crackled along the jagged horizon.[35]

43

THE ROARING TWENTIES (1918–27)

I have been feeding pigeons, thousands of them for years. But there was one, a beautiful bird, pure white with light grey tips on its wings; that one was different. It was a female. I had only to wish and call her and she would come flying to me.

I loved that pigeon as a man loves a women, and she loved me. As long as I had her, there was purpose to my life.

NIKOLA TESLA[1]

In November 1918, Germany signed the armistice ending the Great War. Shortly thereafter, Kaiser Wilhelm II abdicated his throne and fled to Holland; his country had incurred a debt of $33 billion to the Allies. The new heroes of the age were aeronauts, like Eddie Rickenbacker, hailed as the top ace with twenty-six downed Messerschmitts. Humans were leaping continents the following year, with the British propelling the sturdy Dirigible R-34 from Edinburgh to Roosevelt Field and back to London in seven days. This first-ever round-trip transatlantic airship journey was commanded by Maj. G. H. Scott of the Royal Air Force, complete with his thirty-man crew and Willy Ballantyne, a twenty-three-year-old stowaway. The same year, with Tesla, Thomson, Marconi, and Pickering bickering about Martian signals and lunar plant life, Robert Goddard, military rocket expert and physics professor from Clark University, proposed a seemingly outrageous trajectory for sending a man to the moon. Even Tesla thought the scheme far-fetched, for the known fuels of the day did not have sufficient "explosive power," and even if they did, he doubted that a "rocket...would operate at 459 degrees below zero—the temperature of interplanetary space."[2]

In 1920, William Jennings Bryan led the campaign to institute Prohibition; Anne Morgan and her suffragettes gained the right to vote for women; and four motion-picture celebrities, Charlie Chaplin, D. W. Griffith, Mary Pickford, and her new husband, Douglas Fairbanks, formed United Artists. As the war faded, sports figures became the new heroes, the young Red Sox pitcher Babe Ruth making the papers after being sold to the Yankees for a whopping $125,000.

Hugo Gernsback tried to put Tesla on the masthead of yet another futuristic *Electrical Experimenter* spin-off, but his financial offer was, in Tesla's eyes, puny, and he rejected it. Feeling that he had been underpaid for his autobiography, Tesla replied, "I appreciated your unusual intelligence and enterprise, but the trouble with you seems to be that you are thinking only of H. Gernsback first of all, once more, and then again."[3] Gernsback, however, never wavered in his praise of Tesla and continued to feature Teslaic articles and drawings in his various periodicals. On the topic of thought transference, as a materialist, Tesla completely rejected any concept related to ESP; however, he did think that it was possible to read out the thoughts of another person's brain by attaching TV equipment to the rods and cones of the retina, which was, in his view, the arena of cognitive processing.[4] This invention, called the "thought recorder," provided the basis for a number of Frank Paul spectaculars, such as his October 1929 *Amazing Stories* cover depicting two humans wearing thought-reading helmets.

REVISITING WARDENCLYFFE

The 1920s marked a period of turmoil and revolution. Homeostasis had yet to settle in. With the Johnsons still in Europe, Tesla was forced to face the painful Wardenclyffe fiasco once again without the solace of his close friends. With his Manhattan attorney, William Rasquin Jr., Tesla took the train out to the Supreme Court of Suffolk County to battle agents for the George C. Boldt estate and the Waldorf-Astoria, who were trying, once again, to recoup approximately $20,000 in unpaid rent. The referee was the Honorable Rowland Miles.

The case dragged on for months and covered over three hundred pages of testimony. Tesla testified that in March 1915 he had put up Wardenclyffe as collateral against past monies owed to Francis S. Hutchins, personal counsel for George C. Boldt and the Waldorf-Astoria. Hutchins and the hotel interpreted the transaction as an outright transference of the deed. Since the hotel now thought that they owned the property, they felt that it was their right to resell the land and take down the tower to sell the lumber and other parts for salvage.

When Tesla took the stand, he was asked if he remembered the day he delivered the deed.

"I distinctly remember [telling] Mr. Hutchins that the plant had cost an enormous amount of money in comparison with which this indebtedness was a trifle, and that I expected great realizations from the plant, $30,000 a day, if the plant had been completed." Tesla assumed that if he paid the $20,000 owed, he would have gotten back the plant. He further assumed that the Waldorf-Astoria would take good care of the property because of its enormous value. They did not take good care, however. Vandals broke in and stole equipment, such as expensive lathes.

"Can you describe the structures and any other equipment that was in the laboratory," Tesla's counsel asked. The plaintiff's attorney tried to block the testimony, but the judge allowed Tesla to begin.

The inventor sat back, removed his white gloves, placed them on the podium, and proceeded. "The building formed a square about one hundred feet by one hundred. It was divided into four compartments, with an office and a machine shop and two very large areas." "The engines were located on one side, and the boilers on the other side, and in the center, the chimney rose."

When asked how big the boilers were, Tesla said that there were two 300-horsepower boilers surrounded by two 16,000 gallon water tanks that utilized the ambient heat for hot water. "To the right of the boiler plant were the engines. One was a 400-horsepower Westinghouse engine, and a 35-kilowatt outfit which, with the engine, drove the dynamo for lighting and furnished other conveniences." There were high- and low-pressure compressors, various kinds of water pumps, and a main switchboard for operating everything.

"Towards the road, on the railroad side, was the machine shop. That compartment was one hundred by thirty-five feet with a door in the middle and it contained I think eight lathes. Then there was a milling machine, a planer and shaper, a spliner, three drills, four motors, a grinder and a Blacksmith's forge.

"Now, in the compartment opposite, which was the same size as the machine shop, there was contained the real expensive apparatus. There were two special glass cases where I kept historical apparatus which was exhibited and described in my lectures and scientific articles. There were at least a thousand bulbs and tubes each of which represented a certain phase of scientific development. Then there was also five large tanks, four of which contained special transformers created so as to transform the energy for the plant. They were about, I should say, seven feet high and about five by five feet each, and were filled with special oil which we call transformer oil, to stand an electric tension of 60,000 volts. Then there was a fifth similar tank for special purposes. And then there were my electric generating apparatus. That apparatus was precious, because it could flash a message across the Atlantic, and yet it was built in 1894 or 1895."

The court sat humbled. The opposing attorney tried to block Tesla's further testimony, but the judge allowed the inventor to continue.

"Beyond the door of this compartment," Tesla continued, "there were to be the condensers, what we call electric condensers, which would store the energy and then discharge and make it go around the world. Some of these condensers were in an advanced state of construction, and others were not. Then there was a very expensive piece of apparatus that the Westinghouse Company furnished me, only two of this kind have ever been constructed. It was developed by myself with their engineers. That was a steel tank which contained a very elaborate assemblage of coils, an elaborate regulating apparatus, and it was intended to give every imaginable regulation that I wanted in my measurements and control of energy."

Tesla also described "a special 100-horsepower motor equipped with elaborate devices for rectifying the alternating currents and sending them into the condensers. On this apparatus alone I spent thousands of dollars. Then along the center of the room I had a very precious piece of apparatus." It was Tesla's remote-controlled boat.

"Was that all there was, generally speaking?"

"Oh, no, nowhere near," Tesla replied. The inventor then proceeded to describe a series of closets that housed numerous other appliances, "each representing a different phase" of his work. There was the testing room, which included precious instruments given to him by Lord Kelvin, a breach, and other instruments such as voltmeters, wattmeters, ampere meters. In that small space there was a fortune.

The opposing attorney asked that the statement "there was a fortune" be taken out.

"Yes, strike it out," said the judge.

Tesla then went on to discuss the tower. After describing the structure above the ground, he described the shaft. "You see," Tesla said, "the underground work was one of the most expensive parts of the tower." He was referring particularly to special apparatus he invented for "gripping the earth."

"The shaft, your Honor, "was first covered with timber and the inside with steel. In the center of this there was a winding stairs going down and in the center of the stairs there was a big shaft again through which the current was to pass, and this shaft was so figured in order to tell exactly where the nodal point is, so that I could calculate exactly the size of the earth or the diameter of the earth and measure it exactly within four feet with that machine.

"And then the real expensive work was to connect that central part with the earth, and there I had special machines rigged up which would push the iron pipes, one length after another, and I pushed, I think sixteen of them, three hundred feet. The current through these pipes [was to] take hold of the earth. Now that was a very expensive part of the work, but it does not show on the tower, but it belongs to the tower.

"The primary purpose of the tower, your Honor, was to telephone, to send the human voice and likeness around the globe. That was my discovery, that I announced in 1893, and now all the wireless plants are doing that. There is no other system being used. Then, the idea was to reproduce this apparatus and connect it just with a central station and telephone office, so that you may pick up your telephone and if you wanted to talk to a telephone subscriber in Australia you would simply call up that plant and that plant would connect you immediately. And I had contemplated to have press messages, stock quotations, pictures for the press and the reproductions of signatures, checks and everything transmitted from there, but...

"And then I was going to interest people in a larger project and the Niagara people had given me 10,000-horse power..."

"Did you have any conversation with Mr. Hutchins or anybody representing the plaintiffs concerning the taking down of the tower or anything like that?" asked the judge.

"No, sir. It came like a bolt from the blue sky."

As the deed had been transferred in a legal manner with Tesla's full compliance, Judge Miles ruled in favor of the hotel. The inventor's lawyer countered, arguing that the Waldorf-Astoria sold equipment which they did not account for and destroyed a property worth $350,000 to try and recoup the $20,000 owed. "The property despoiled exceeded the value of the mortgage, and therefore the plaintiffs [the hotel management] should have been held to account to defendant Tesla." Precedent cases were cited.

The Waldorf-Astoria, however, had the last word. "As a solace to the wild hopes of this dreamy inventor," their lawyer wrote, "if prior to that time he should grasp in his fingers any one of the castles in Spain which always were floating about in his dreams, and had he paid the board bills which he owed, this wild scrubby woodland, including the Tower of Babel thereon, would cheerfully have been reconveyed to him. By no fair inference or construction can [Tesla's counterclaim make void this judgment]. It was merely a sop to the vanity of a brilliant but unpractical mind. The judgement should be affirmed with costs."[5]

In the summer of 1922, Robert Johnson and his ailing "ambassadress"[6] returned to the States from Italy. They arrived in time to attend Paderewski's comeback piano concerto with their elusive friend at Carnegie Hall in November.

Robert's autobiography, *Remembered Yesterdays,* which was just completed, highlighted not only a memorable meeting between Tesla and Paderewski in the late 1890s but also the virtuoso's 1919 stint as president of Poland. As the pianist had held office for only ten months, Tesla was moved to jest that it was "just long enough to gain publicity for his next tour."

"That's a terrible thing to say, Mr. Tesla," Kate sparkled as they stepped into the limousine that was to take the trio to the opening. Dressed in black capes, canes, and silk high hats, the tall "angular" gentlemen struck a smart pair as they accompanied the suddenly recovering and radiant Mrs. Filipov.

"Seeing Paderewski again is like falling in love all over," she said between her men. Tesla looked down and noticed the sorrow that lay hidden beneath her brow. His was apparent to her as well. Robert's upper lip held them steady.

The Bolsheviks were taking over in Russia; Communist and anarchist uprisings reverberated throughout the world. In the United States there were race riots in Chicago, Negro lynchings in Minnesota, a suspicious explosion outside the J. P. Morgan Building, in New York, killing thirty people and wounding three hundred others, and forty thousand Klansmen marching on Washington. It was time to do something to stop the tide, so Attorney General A. M. Palmer rounded up three hundred Communists and sixty-seven anarchists in thirty-three cities. The last group arrested faced deportation for bombing out the windows and homes of Palmer and also Assistant Secretary of the Navy Franklin Roosevelt. Eugene V. Debs, still in prison for violating the Espionage Act, was nominated once again to run for president by the Socialist party; Woodrow Wilson received the Nobel Peace Prize.

The election of 1920 was the first to be broadcast by radio to a national audience; Lee De Forest announced the wrong winner four years earlier to a much smaller crowd. With his running mate Calvin Coolidge, Warren Harding trounced Democratic contender James Cox and vice presidential hopeful Franklin Roosevelt.

By this time, RCA was a megacorporation, writing million-dollar checks to John Hays Hammond Jr. and Edwin Armstrong. Having uncovered a great new market, RCA had increased its radio audience in 1924 to 5 million listeners. Profits were made not only in selling air space to advertisers but also in selling the radios themselves. By 1928 national broadcasts would link all forty-eight states, and soon after, regular shows featuring Buck and Will Rogers, Amos 'n' Andy, Burns and Allen, the Shadow, Stoopnagle and Budd, and Jack Benny would become daily fare. Such advertisers as Lucky Strike, Maxwell House, Canada Dry, Chesterfield, and Pontiac would soon insinuate themselves into the mass psyche. Tesla would say that he cared not to listen to the radio because he found it "too distracting."

Other milestones during this period included the anointment of the "Manassa Mauler," Jack Dempsey, as world heavyweight champion, a soaring stock market, and a number of key trials, most notably, Sacco and Vanzetti's, alleged anarchists accused of murder, the Scopes monkey trial, and the $500 fine and ten-day incarceration of Mae West for lewd improvisations during her hit Broadway play *Sex*. Newest crazes included speakeasies, Al Capone, flapper dresses, and such dances as the Charleston, waltzing till you dropped, and the Shimmy. Although the radio was czar at home, for a night on the town, silent movies were king. Deaths during the Roaring Twenties included T. C. Martin, Jacob Schiff, Henry Clay Frick, Andrew Carnegie, Enrico Caruso, William Roentgen, Alexander Graham Bell, Woodrow Wilson, Warren Harding, and the thirty-one-year-old heartthrob Rudolf Valentino, who, along with Harry Houdini, died of a ruptured appendix, Vladimir Lenin, Sarah Bernhardt, Princess Lwoff-Parlaghy, and Katharine Johnson, who passed on during autumn 1925.[7]

October 15, 1925

Dear Tesla,

It was Mrs. Johnson's injunction that last night of her life that I should keep in touch with Tesla. This is a pretty hard thing to do, but it will not be my fault if it is not done.

Yours faithfully,
Luka[8]

THE MYSTERIOUS MR. BETTINI

Throughout the world, wireless inventors were becoming a precious commodity. In Italy, Mussolini "adroitly" redirected the Italian senate's Fascist salute to Guglielmo Marconi for having established a national broadcasting system.[9] A few years later, *Il Duce* approached Jack Hammond to institute a "foolproof secret radio system," which, to Jack's later revulsion, became a tool for killing anti-Fascists.[10]

In the Soviet Union, Lenin contacted Tesla to ask him to come over to his country to institute his AC polyphase and "regional power-distribution stations."[11] Sending emissaries to lure the Serbian nobleman. Tesla became enmeshed in a shady organization known as the Friends of Soviet Russia. With over 5 million people dying there of famine in 1922, the celebrated inventor had been approached by Ivan Mashevkief, of the Russian Workers Club of Manhattan, and by Elsie Blanc, a Communist leader from Massachusetts, to speak at "monster mass-meeting" at the Grange Hall in Springfield in June 1922. The purpose of the conclave, which was co-organized by a group of "Italian radicals," was ostensibly to raise money for clothing and food for the starving and dying people of Russia. Since a

Russian "bomb factory" was discovered in a warehouse in Manhattan at that time, no doubt some of the funds were also siphoned off for more nefarious activities.

Traveling to Springfield with Mashevkief, who described "with considerable imagination the manufacturing industries of Russia," Tesla heard the first speaker announce "that the only solution to the economic problem [in Europe] was in the hands of the working class...[which] will have charge of all means of production. 'They will do this for humanity's sake and not for profit.' The speaker prophesied that an economic collapse of the entire industrial structure of Europe will come and when it does, the working class will then secure full control of affairs. The speaker emphatically stated that the famine in Russia to-day is caused by counter-revolutionary forces backed by world capitalists and not because of the alleged poor rule of the bolshevists."

According to Adrian Potter, the FBI agent who monitored the event, " 'Nicolo Tesla' was addressed by several Italians as 'Bettini'.... Tesla or Bettini prophesied that Italy was soon to adopt a communist form of Government."[12]

Clearly, Tesla was, in some sense, a revolutionary and on the side of the worker, but more for the purpose of transforming and uplifting their station. Tesla's inventions were purposely constructed so as to reduce consumer cost, preserve natural resources, and relieve humanity of unnecessary manual labor. The Serb believed in the profit motive and strove all his life to become what Lenin loathed, so the reader should read this FBI report with caution, as Tesla's supposed statement and motive for attending the meeting are not totally clear. Most likely he was concerned with the plight of the starving people in Russia (the U.S. government would send a reported $60 million in aid to feed the Soviets over the next decade),[13] and he was also looking to sell his inventions to this new regime, with an accompanying vast potential market.

Where the Soviet leadership sought Tesla out, the aging gnome-wizard and odd-combination capitalist-socialist Charles Steinmetz initiated his own contact, writing the Soviet premier a letter in February 1922. "Wishing Lenin success," Steinmetz "express[ed] confidence that he would complete the astonishing work of social and industrial construction which Russia had undertaken under difficult circumstances."

Joining a variety of Soviet organizations, Steinmetz also publicized his correspondence with Lenin and "published two papers in the *Electrical World* that described Russia's electrification plans." Scratching his gnarled arthritic fingernails on GE's capitalist blackboard, the self-righteous $100,000-per-year supposed academic implacably "called for American capital to support the project."[14]

Although Lenin's correspondence with Tesla has not been located, his response to Steinmetz is well known. "Lenin replied...that 'to my shame' he had heard the name of Steinmetz only several months ago...thanked Steinmetz for his help, but suggested that the absence of diplomatic relations between the United States and Soviet Russia would impede its implementation." Lenin, however, would publish the note from the prominent engineer and send Steinmetz an autographed photograph of himself, which he received a few months later.[15]

Just one year later, Charles Proteus Steinmetz, the four-foot giant of electrical engineering, bon vivant, and family man, was dead. He was fifty-eight.[16]

Tesla was living on Fifth Avenue, two blocks from Central Park, in the Hotel St. Regis, room 1607, for the years 1920–1923.[17] Commuting to Milwaukee and paying an exorbitant fifteen dollars a day in rent, the inventor neglected to compensate the hotel for a seven-month period and was promptly sued for the balance, over $3,000.[18] Forced to find other premises, he moved into the Hotel Marguery on Park Avenue and Forty-eighth Street, just a few blocks from his favorite stomping grounds: Bryant Park, behind the New York Public Library, and the great commuter's hall at Grand Central Station. After hours, in the dead of night, the inventor would grab his coat, cane, white gloves, and derby and prance out for a tour of the park by the library, where he cogitated and fed his precious pigeons. Rumors began spreading about the gaunt eccentric who fed the pigeons, as Tesla purposely kept his identity concealed. "Midnight is the hour he chooses for his visits....Tall, well dressed, of dignified bearing [the man] whistles several times, a signal for the pigeons on the ledges of the building to flutter down about his feet. With a generous hand, the man scatters peanuts on the lawn from a bag. A proud man, yet humble in his charities—Nikola Tesla."[19]

According to several researchers, Tesla was a homosexual, and it was supposedly there, in the Hotel Marguery, that he liked to meet "his special friends." More likely a celibate, the inventor did have one homosexual admirer, the young journalist Kenneth Swezey.[20]

Born in 1905 and raised in a Brooklyn apartment, where he stayed his whole life, Swezey had constructed his first radio at the age of thirteen, during the height of World War I. Shortly thereafter, forsaking secondary school, he began to write science articles for a number of the local newspapers and magazines and eventually a textbook on chemistry. Able to reduce complex ideas to a level understandable by the masses, Swezey was later congratulated by Albert Einstein for explaining Archimedes' principle.[21]

Having sifted through the data on the wireless, Swezey came to realize that Tesla was the unsung author of the invention and sought the hermit out for an interview.

With a round, boyish face, glasses, and a quick and perceptive mind, Swezey quickly endeared himself to Tesla, who expressed surprise at the writer's youth. Only nineteen at the time, Swezey and Tesla began a special friendship that would last until the end of the inventor's life. Often they would meet in Tesla's apartment to go over some articles Swezey was writing or to discuss aspects of Tesla's work. Afterward, the youngster might join the inventor for dinner, or Tesla would walk the boy back to the gate of the subway.[22] As the friendship grew, the aging sage also came to rely on Swezey when he needed assistance, and by the time he was in his late seventies, their friendship became so familiar that, according to Swezey, Tesla sometimes greeted him at the door stark naked. As the years progressed, Tesla's new publicist became virtually one of the family, befriending Agnes Holden, Robert Johnson's daughter, and also Sava Kosanovic, Tesla's nephew, who often traveled to New York from the newly formed Yugoslavia as its first ambassador.

Swezey, who himself described Tesla as "an absolute celibate," began to compile a large holding of articles, letters, and original manuscripts as he raced, unwittingly, Jack O'Neill, Tesla's other journalist-compadre, to write the quintessential biography. Concerning Tesla's habits, Swezey confirmed that the inventor rarely slept. Tesla claimed he slept less than two hours per night. The inventor, however, did admit to "dozing" from time to time "to recharge his batteries." For exercise, the inventor would walk "8-10 miles per day" and also loosen up in the bathtub (although he also touted a waterless bath which involved charging his body with electricity in such a way as to repel all foreign particles). Later, Tesla would add to his repertoire the squishing and unsquishing of his toes one hundred times for each foot every night. He claimed the practice stimulated his brain cells. "And how this man worked! I will tell you about a little episode.... I was sleeping in my room like one dead. It was three after midnight. Suddenly the telephone ring awakened me. Through my sleep I heard his voice, "Swezey, how are you, what are you doing?" This was one of many conversations in which I did not succeed in participating. He spoke animatedly, with pauses, [as he]...work[ed] out a problem, comparing one theory to another, commenting; and when he felt he had arrived at the solution, he suddenly closed the telephone."[23]

In 1926, shortly after moving to the Hotel Pennsylvania, the inventor agreed to an interview for *Colliers* magazine. The sixty-eight-year-old philosopher chose as his topic of the evening the female of the species. Viewing the woman's movement as "one of the most profound portents of the future," the "tall, thin, ascetic man" told the interviewer, "This struggle of the human female toward sex equality will end in a new sex order, with the female as superior."[24] Happy with the article, Tesla forwarded a copy to Anne Morgan, with whom he still kept in touch and Anne wrote back to review her own twenty-year odyssey as an advocate in the women's movement.[25]

During this same time period Tesla divulged in the *World* his unbridled attachment to the city's pigeons. "Sometimes I feel that by not marrying I made too great a sacrifice to my work," he told the reporter, "so I have decided to lavish all the affection of a man no longer young on the feathery tribe. I am satisfied if anything I do will live for posterity. But to care for those homeless, hungry or sick birds is the delight of my life. It is my only means of playing."

In the same article, Tesla poignantly reveals a fondness for one particular pigeon that had a broken wing and leg. "Using all my mechanical knowledge, I invented a device by which I supported its body in comfort in order to let the bones heal." Carrying the bird up to his suite, Tesla calculated that "it cost me more than $2,000 to cure [her]." It took over a year and one-half of daily care, and afterward Tesla hand-carried the bird to one of his favorite farms, where "it is now one of the finest and prettiest birds I have ever seen."[26]

Concerning his potential affinity for males, Tesla certainly displayed an affection for muscle men, often, in the later years, inviting boxers such as Henry Doherty, Jimmy Adamick, and the Yugoslav welterweight champion Fritzie Zivic out to dinner or to his apartment.[27] Having "made a study" of the 1892 championship match between Gentleman Jim Corbett and John L. Sullivan (which had been held in New Orleans), Tesla made the sports headlines in 1927 by predicting the outcome of the rematch between Gene Tunney and Jack Dempsey, the "Manassa Mauler" having been deposed the year before in a ten-round decision.

Dr. Tesla Picks Tunney on Basis of Mechanics

> Sitting in this suite at the Hotel Pennsylvania, the 71-year-old inventor...did not hedge or pussyfoot, but declared that Tunney was "at least a ten to one favorite....Tunney will hit Dempsey continuously and at will....[In addition], he is single, and other things being equal, the single man can always excel the married man."
>
> Dr. Tesla smiled significantly. He is a lifelong bachelor.[28]

With Katharine's death came a new level of intimacy between Tesla and Luka. Often they would meet for dinner or at the cinema, and when Spanish-American War hero Richmond P. Hobson returned with his wife to live in the city, he would join them. According to Mrs. Hobson, "these two dear friends [Tesla and Hobson], about once a month or sometimes oftener, would meet and go to a movie and then sit in the Park, and talk till after midnight! Richmond always came home with enthusiasm over some new invention of Tesla's and well I recall the night he told Richmond, 'I can shake the world of its orbit, but I won't do it, Hobson!'"[29]

Certainly Tesla's relationship with Katharine had been, at least at one time, provocative, and no doubt he reviewed details with Hobson, but apparently, according to his own words, he had "never touched a woman."[30] One could make a case for a germ phobia, although his relationship with the pigeons would seem to dispel that myth, magnified greatly by the O'Neill biography;[31] or Tesla's aversion to sexual intimacy with women could be evaluated from a psychoanalytic perspective. In 1924, Tesla wrote in a condolence note to Jack Morgan, "The mother's loss grips one's head more powerfully than any other sad experience in life."[32]

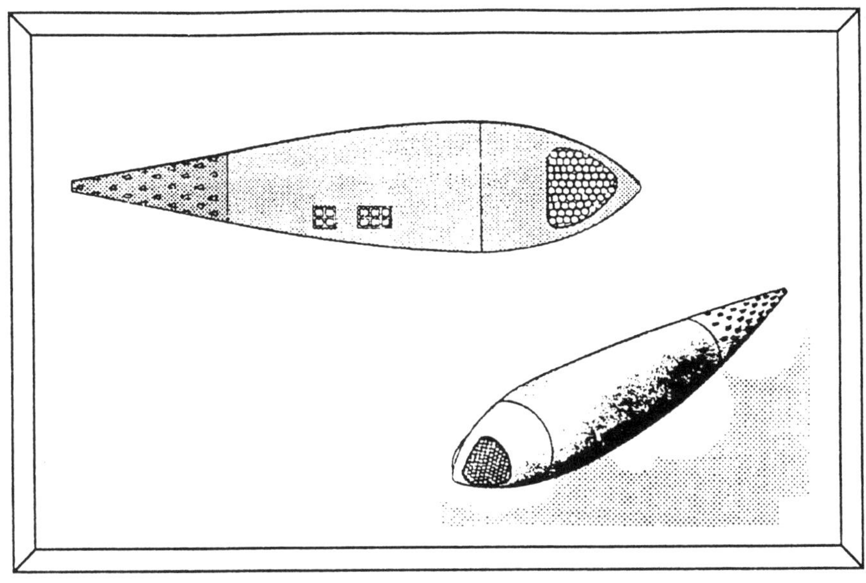

Tesla's reactive jet dirigible, circa 1909. This model is a forerunner of various flying wing prototypes such as the new Lockheed Martin X-33, which has been designed to replace the Space Shuttle. (MetaScience Foundation)

44

FASTER THAN THE SPEED OF LIGHT (1927–40)

June, 1931
Potsdam, Germany

Dear Mr. Tesla!

I'm happy to hear that you are celebrating your 75th birthday, and that, as a successful pioneer in the field of high frequency currents, you have been able to witness the wonderful development of this field of technology.

I congratulate you on the magnificent success of your life's work.

Albert Einstein[1]

For the balance of the wizard's life, he would continue to speak cryptically about a number of entirely new and revolutionary inventions. These included (a) a machine for harnessing cosmic rays; (b) a means for transmitting mechanical energy; (c) a particle-beam weapon; and (d) a mechanism for communicating with other planets. In addition, Tesla also continued to refer to (e) his Wardenclyffe idea. The identification of each separate invention became a somewhat confusing task for journalists and researchers because each of these ideas involves the transmission of energy to distant places; and the third invention, the so-called death ray apparently, in its final form, comprised features from some, if not all, of the other inventions.

Throughout Tesla's seventies, that is, from the mid-1920s until about 1934, Tesla continued his practice of traveling to industrial centers throughout the Northeast and Midwest in his quest to market his wares. While commuting to Philadelphia during the years 1924–25 to work on his gasoline turbine (he had worked on the steam turbine in Chicago and

Milwaukee), Tesla met with John B. Flowers, inspector of airplanes and engines at the local naval aircraft factory. He had known the inspector since 1917.[2] As it became more apparent that the bladeless turbine was stuck in the endless cycle of research and development, Tesla returned to his first love, wireless transmission of power, and began a publicity campaign to espouse its merits. By implementing a series of central stations to pump energy into the ground and surrounding medium, the ultimate conservationist-pragmatist theorized that airplanes and automobiles, equipped with specially designed receiving devices, could operate without fuel onboard; they would simply derive their power from his towers.

On October 10, 1925, Flowers traveled to New York City to confer with the wizard in his suite at the Hotel Pennsylvania. There they drafted out the entire scheme so that it could be presented to physicist J. H. Dillinger, head of the Radio Laboratory, Bureau of Standards, in Washington, D.C.

In a carefully worded ten-page document, complete with schematic drawings of the earth imbued with Tesla-created standing waves, Flowers unveiled a plan for operating cars and planes powered by electromagnetism. "Dr. Tesla said that the Wireless Power System would supply power to airplanes at any point around the earth," Flowers told Dillinger. "In addition," Flowers continued, "Dr. Tesla has already developed the oscillator to provide the power and is willing to furnish the U.S. Government his plans if they agree to build the plant." Flowers also set up a meeting in Washington to go over the proposal.

In the interim, Dillinger referred the proposal to H. L. Curtis, a fellow expert. After canny consideration, Curtis rejected the plan, his main objection being that "as he understood it, Tesla's scheme was to create standing electrical waves around the earth as a sphere. There would then be considerable concentration of energy at the nodes and it was at the nodal points Tesla expected to develop his energy. The system proposed by Mr. Flowers does not have this feature. He proposes to collect energy at any point....[Thus] some means would have to be devised for concentrating this energy and making it available. No such method is proposed, and I do not think of any that appears feasible....[Furthermore] I do not know of any wireless apparatus...of sufficient magnitude to warrant the expectation that power can be economically transmitted by radio methods."[3]

The basic criticism that the energy would not be available at any point of the globe, but only at the nodal points, was countered on numerous occasions by Tesla (although, apparently, the towers, which were not near power sources, would have to be placed at nodal points). One of Tesla's favorite analogies was to view electricity as a kind of fluid and his magnifying transmitters as a series of pumps. Just as with a hydraulic system the fluid would be present at all points in equal pressures, so, too, would Tesla's electrical oscillations. And just as electrical energy is present at every connected electrical outlet in the world but is not used until an appliance is plugged in, so, too, was Tesla's electricity available, but not used until the receiver was turned on.

In a comprehensive article published in *Telegraph & Telephone Age* in October 1927, which was probably written as a rebuttal to Curtis and Dillinger, Tesla also explains that the oscillations would spread from the magnifying transmitter

> with a theoretically infinite speed, slowing down first very quickly and afterward at a lesser rate until the distance is about six thousand miles, when it proceeds with the speed of light. From there on it again increases in speed, slowly at first, and then more rapidly, reaching the antipode with approximately infinite velocity. The law of motion can be expressed by stating that the waves on the terrestrial surface sweep in equal intervals of time over equal areas, but it must be understood that the current penetrates deep into the earth and the effects produced on the receivers are the same as if the whole flow was confined to the earth's axis joining the transmitter with the antipode. The mean surface speed is thus about 471,200 kilometers per second—57% greater than that of the so-called Hertz waves.[4]

Tesla likened the effect to the moon's shadow spreading over the earth during an eclipse. Here was the first of a number of instances in which Tesla disagreed with the findings of Einstein's theory of relativity, as the so-called Tesla wave supposedly traveled faster than light.[5]

In 1928, Tesla traveled to Philadelphia to attempt construction of his helicopter-airplane, probably with John Flowers, and to Detroit to try to market it as a "flying automobile" to GM. On a more practical level, he also peddled his speedometer to the Ford Motor Company.

One of the problems in the speedometer was cost, the Tesla invention having become a premium item only found in the more expensive vehicles. He also visited with his nephew Nicholas Trbojevich, who was helping fund the helicopter, and who was on the verge of becoming very wealthy from his various automobile inventions associated with economizing the transmission and steering. Trbojevich, like his uncle, was somewhat of a workaholic, and Tesla cautioned his wife to give her husband "unceasing care [and love]," as in the long run "your husband is sure to acquire great wealth and when his battle is won, you will have everything to your heart's desire."[6]

Shortly after, Tesla returned to Detroit and met Trbojevich for a late snack at the Book-Cadillac Hotel, the city's "finest." According to William Terbo, "the maitre d' suggested they wait five minutes, when the five dollar cover charge would be lifted. Tesla would hear none of it and marched in."

As this was during the Great Depression, when a quarter could buy three hot dogs and two cokes, this wasted expense was enormous, and it became a great point for laughter among the Trbojevich family, which tended to view Tesla as simply their old eccentric uncle rather than one of the most important inventors in the world. When the nephew tried tactfully to bring up the matter of the cover charge, Tesla evaded by responding, "I'll never die rich unless the money comes in the door faster than I can shovel it out the window."[7]

During this period (1925–38) Tesla also negotiated with Myron Taylor, CEO of U.S. Steel. Interested in the steel company for a variety of reasons, the ever-prodigious inventor had developed special equipment for purifying ores, "degasification of steel," and also conservation of sulphur during iron processing. In the late 1920s he asked Taylor if he could install equipment to see if the procedure worked. Taylor agreed, and so Tesla traveled up to their Worcester plant in September 1931 to install it. Although he hoped for a successful demonstration, this apparently did not occur, since the archives of U.S. Steel have only one short paragraph referring to Tesla's dealings with the company.[8] Tesla's ultimate plan, which apparently was not tested, was to install his bladeless turbine in the heat exhaust system, with the idea of converting the enormous amount of wasted heat into useful electricity. Ever the conservationist, this was one of Tesla's most elegant ideas.

From Worcester, Tesla moved on to Buffalo for a top-secret experiment, according to Peter Savo, a cousin living in New York. There the inventor reportedly refitted an automobile that, according to the story, ran on electrical power from an outside source.

> The car [was] a standard Pierce Arrow, with the engine removed and certain other components installed instead. The standard clutch, gear box, and drive train remained.... Under the hood, there was a brushless electric motor, connected to [or in place of] the engine.... Tesla would not divulge who made the motor.
>
> Set into the dash was a "power receiver" consisting of a box...containing 12 radio tubes.... A vertical antenna, consisting of a 6 ft. rod, was installed and connected to the power receiver [which was] in turn, connected to the motor by two heavy, conspicuous cables.... Tesla pushed these in before starting and said: "We now have power."[9]

If this tale is to be believed, it would mean that Tesla had also installed one of his powerful oscillators somewhere near Niagara Falls to provide the wireless energy needed to power the vehicle. An alternative possibility was that Tesla was testing one of his gasoline or steam turbines in the automobile, and Savo mistook it for the wireless device. "The aging inventor, a tall, thin, almost spiritual figure in the sort of brown cutaway suit that older men wore before the World War, received interviewers in one of the public rooms in the Hotel New Yorker, where he lives. Before he would speak of his present work, he reviewed his past achievements, which entitle him more than Edison, Steinmetz or any other, to be called the father of the power age...."[10]

There was a new king of the hill. Ever since the 1919 confirmation of his theory of relativity, that space was curved and that light traveled at a constant speed irrespective of the movement of its source, Einstein began to occupy the spot formerly held by such technical wizards as Bell, Edison, the Wright brothers, or Tesla. First postulated in 1905, Einstein's theories not only shifted the prevailing space-time paradigm, that self-assuring Newtonian world that the old guard grew up in; his theories also threatened Tesla's position as premier mastermind. Although the measurement of the starlight bending around the sun during the 1919 eclipse was experimental proof of Einstein's new postulate,[11] for the most part the theoretical physicist was exactly that, a theorist, whereas Tesla, as hands-on creator of new technologies, was able to prove out his assumptions in the everyday world. This was the inventor's advantage, and he used it to attack the new Nobel Prize–winning upstart.

As Einstein's theory abandoned the old nineteenth-century ether, it explained the bending of light rays around large bodies as being caused by the non-Euclidean curving of space-time. This, in essence, became the new and more abstract ether. Mathematical equations accurately predicted the precise amount of bending that occurred. "In general relativity, the gravitational field and the structure or geometry of space are identical....The gravitational field *is* the curved space."[12]

Tesla completely disagreed with the concept of space being curved, saying that it was "self-contradictory." Since "every action is accompanied by an equivalent reaction," it appeared to Tesla's "simple mind...[that] the curved spaces must react on the bodies and, producing the opposite effect, straighten out the curves." To Tesla, the light bent because the large body (e.g., the sun) had a force field which influenced it.[13]

Ironically, Einstein's contemporaries at the Carnegie Institution of Washington were using the Tesla coil in their new 1929 experiments in attempting to split the atom,[14] while Tesla was discussing a more esoteric source of energy, cosmic rays:

> A principle by which power for driving the machinery of the world may be derived from the cosmic energy which operates the universe, has been discovered by Nikola Tesla, noted physicist and inventor....

> This principle, which taps a source of power described as "everywhere present in unlimited quantities" and which may be transmitted by wire or wireless from central plants to any part of the globe, will eliminate the need of coal, oil, gas or any other of the common fuels...."The central source of cosmic energy for the earth is the sun," Dr. Tesla said, but "night will not interrupt the flow of the new power supply."[15]

On July 10, 1931, Tesla turned seventy-five. *Time* honored the senior inventor by placing his portrait on the cover. Tesla's life was briefly reviewed, and his most recent mysterious research on harnessing "an entirely new and unsuspected source [of energy]" was discussed. Unwilling to reveal more about the adjuvant, the venerated iconoclast startled the interviewer by referring obliquely to his most esoteric invention, the "Tesla-scope," a device for signaling the nearby stars: "I think that nothing can be more important than interplanetary communication. It will certainly come some day, and the certitude that there are other human beings in the universe, working, suffering, struggling like ourselves, will produce a magic effect on mankind, and will form the foundation of a universal brotherhood that will last as long as humanity itself."[16] Hugo Gernsback couldn't have said it better.

Simultaneously, Kenneth Swezey wrote a flurry of letters to every notable he could think of, requesting a birthday greeting. Accolades poured in (many quoted throughout this text) from E. F. Alexanderson, B. A. Behrend, W. H. Bragg, Lee De Forest, Gano Dunn, Jack Hammond, A. E. Kennelly, Arthur Korn, Oliver Lodge, Robert Millikan, D. McFarlan Moore, Valdemar Poulsen, Charles F. Scott, Georg Graf von Arco, H. H. Westinghouse, and Albert Einstein. VIPs who wrote back to decline included Guglielmo Marconi and Michael Pupin.[17]

By October, Thomas Alva Edison was dead; the lights in the city were dimmed in honor of the great man's passing. Perhaps it was the death of his nemesis or the new round of adulation, or Tesla's advanced age that prompted him to alter his style of avoiding publicity. Whatever the reason, from 1931 on the inventor made it an annual practice on his birthday to invite the press to his flat and announce his latest discoveries. With the talent of a mystery writer, the electrician stretched out the secrets of his various creations, revealing just a little more each year.

By 1935, on his seventy-ninth birthday, Tesla, although exceedingly gaunt, was still exuberant and expected to live past 110. His mind ever evolving, the sorcerer utilized this occasion to lay out in considerable detail the various particulars of a number of his more exotic creations. With movie cameras rolling and the inventor "treating the press, about 30 in number, to a gourmet's luncheon... Mr. Tesla sat at the head of the table." Eating little more than bread and warm milk, which he heated up in a chafing dish at the table, the wizard talked while the reporters "feasted."[18]

Inventor, 81, Talks of Key to Interstellar Transmission

& Tube to Produce Radium Copiously and Cheaply

> Reports of discoveries by which it will be possible to communicate with the planets and to produce radium in unlimited quantities for $1 a pound were announced by Dr. Nikola Tesla yesterday on his 81st birthday at which he was honored with high orders from the Yugoslav and Czechoslovak Governments....
>
> "I am expecting to put before the Institute of France an accurate description of the devices with data and calculations and claim the Pierre Guzman prize of 100,000 francs for means of communication with other worlds, feeling perfectly sure that it will be awarded to me. The money, of course, is a trifling consideration, but for the great historical honor of being the first to achieve this miracle I would be almost willing to give my life.
>
> "I am just as sure that the prize will be awarded to me as if I already had it in my pocket. They have got to do it. It means it will be possible to convey several thousand units of horsepower to other planets, regardless of the distance. This discovery of mine will be remembered when everything else I have done, is covered with dust."[19]

In discussing this invention, one runs into murky waters, for it appears that Tesla tied the concept of the radium-producing tube to the interplanetary communicator. These, however, may be two unrelated creations. Another problem was that the inventor was also discussing at this time the idea of capturing cosmic rays that travel at velocities fifty times greater than that of light. If this invention utilized cosmic rays, it would imply that Tesla planned to transcend the speed of light and communicate with other stars.

In reading the text carefully, it appears that Tesla does not mention other stars but, rather, the planets, which are relatively close to the earth; furthermore, he does not really discuss communicating with extraterrestrials so much as transmitting energy. It is known that as early as 1918, while working with Coleman Czito's son, Julian, the inventor was bouncing laserlike pulses off the moon and testing some type of "scope."[20] Therefore, it is possible that he was working on more than one device to send energy into outer space.

Verification for Tesla that there existed particles that traveled faster than the speed of light were purportedly discovered in the late 1890s when he invented a device to capture radiant energy. The machine, patented on November 5, 1901, comprised, in essence, an insulated plate, resembling a fly swatter, made out of "the best quality of mica as dielectric." This was attached to a condenser. Stemming from his work with radiant energy, X rays, and Lenard tubes, the device could also capture what he called cosmic rays.[21]

> I made some progress in solving the mystery until in 1899 I obtained mathematical and experimental proofs that the sun and other heavenly bodies similarly conditioned emit rays of great energy which consist of inconceivably small particles animated by velocities vastly exceeding that of light. So great is the penetrative power of these rays that they can traverse thousands of miles of solid matter with but slight diminution of velocity. In passing through space, which is filled with cosmic dust, they generate a secondary radiation of constant intensity, day or night, and pouring upon the earth equally from all directions.[22]

Since Victor Hess's discovery in 1911 and Robert Millikan's confirmation, there have been many scientists who have measured cosmic rays. We now know that emitted uncharged elementary particles known as neutrinos possess the penetrative powers suggested by Tesla, but no researcher, to my knowledge, has discovered rays that transcend the speed of light. This supposed finding of Tesla also violated relativity.

Tesla was insistent that such particles did exist; he saw them as a source which could be converted into electrical power. During the summer of 1932 he told Jack O'Neill that he had "harnessed the cosmic rays and caused them to operate a motive device.... The attractive features of the cosmic rays is their constancy. They shower down on us through the whole 24 hours, and if a plant is developed to use their power, it will not require devices for storing energy, as would be necessary with devices using wind, tide or sunlight." When pressed for more details, Tesla revealed that he would tell O'Neill "in a general way [their modus operandi....] The cosmic rays ionize the air, setting free many charges—ions and electrons. These charges are captured in a condenser which is made to discharge through the circuit of the motor." Tesla also told O'Neill that he "had hopes of building such a motor on a large scale."[23]

FREE ENERGY?

As the years rolled on, it became a challenge for reporters to wrestle more details about each invention from the wizened prestidigitator, since Tesla continued to maintain his perpetual reticence about revealing particulars. Concerning the cosmic-ray accumulator, the reporters were able collectively to pry from the inventor the following: "My power generator will be of the simplest kind—just a big mass of steel, copper and aluminum, comprising a stationary and rotating part, peculiarly assembled.... Such a source of power obtainable everywhere will solve many problems with which the human race is confronted....[The] machinery for harnessing it would last more than 5,000 years."[24]

Cosmic rays, he asserted, are produced by the force of "electrostatic repulsion"; they consist of powerfully charged positive particles which come to us from the sun and other suns in the universe. He determined, "after experimentation," that the sun is charged "with an electric potential of approximately 215,000,000,000 volts."[25]

> Owing to its immense charge, the sun imparts to minute positively electrified particles prodigious velocities which are governed only by the ratio between the quantity of free electricity carried by the particles and their mass, some attaining a speed exceeding fifty times that of light....
>
> At great altitudes, the intensity of the rays is more than 10,000% greater than at sea level.... The energy of the cosmic radiations impinging upon the earth from all sides is stupendous, such that if all of it were converted into heat the globe quickly would be melted and volatilized... Rising air currents... partially neutralize [their intensity].... Those who are still doubting that our sun emits powerful cosmic rays evidently overlook that the solar disk, in whatever position it may be in the heavens, cuts off the radiations from beyond, replacing them by its own.[26]

Coupled with his view that all bodies in the universe obtain their energy from external sources, and possibly influenced by Walter Russell, an artist, philosopher, and longtime friend of Tesla's who hypothesized that the periodic table of elements was constructed in a hierarchical spiral of octaves, Tesla was "led to the inescapable conclusion that such bodies as the sun are taking on mass much more rapidly than they are dissipating it by the dissipation of energy in heat and light."[27] Similarly, radioactive decay was not caused by the disintegration of the nucleus of the atom; rather, it was a "secondary effect of external rays and two-fold—one part coming from the energy stored, the other from that continuously supplied."[28] In other words, radioactive material was, to Tesla, apparently a kind of conduit for the ever-present primary substance "Akasa," which was being *absorbed* in such a way as to cause the emission of the radioactive material.

These Tesla offerings would appear as evidence of a great mind gone astray, for the various discoveries and suggestions inherent in Tesla's theory violate not only such accepted theories as relativity and quantum physics but also, on the surface, common sense. The idea that the inventor could construct a simple device made up basically of a receiving plate and a condenser to provide electricity to run motors from cosmic rays harks the reader back to the inane days of the Keely motor and the obtuse concepts of perpetual motion and free energy. Yet underneath the veneer of the theory is an exciting notion that the sun is somehow absorbing energy from the universe and that there does exist some form of it which transcends the limiting factor of the speed of light. Called by other researchers tachyons (i.e., particles that travel faster than the speed of light) and linked to other such concepts as black holes, worm holes, string theory nonlocality, the implicate order, hyperspace, gravitons, and Mach's principle, Tesla's theories, when viewed within the matrix of the bizarre new physics, may not be so far-out.

Another of Tesla's discoveries involved the transmission of mechanical energy to distant places. By strategically placing one of his mechanical oscillators on, for example, solid bedrock, a mechanical impulse could be sent into the ground to accomplish "at least four practical possibilities. It would give the world a new means of unfailing communication; it would provide a new safe means for guiding ships at sea into port; furnish a kind of divining rod for locating ore deposits...; and finally, it would provide scientists with a means for laying bare the physical conditions of the earth."[29] The essential principle behind this invention is, of course, used today in sonar for ships and by geophysicists in studying the interior of the earth, mapping fault lines, studying the core, and so on.

Tesla, at 78, Bares New 'Death-Beam'

> Dr. Tesla...has perfected a method and apparatus... which will send concentrated beams of particles through the free air, of such tremendous energy that they will bring down a fleet of 10,000 enemy airplanes at a distance of 250 miles from a defending nation's border and will cause armies of millions to drop dead in their tracks.[30]

Tesla's discovery of a death ray stems all the way back to his work in the early 1890s with his creation of a button lamp that could bounce electrons off of a central filament made of almost any substance (e.g., carbon, diamonds, zirconia, rubies) onto the interior of a self-reflective bulb and then bounce back to the source. This device would not only produce an extraordinarily brilliant light, it could also "vaporize" the button. As stated previously, it was only a short step from this machine to the invention of the ruby laser. For instance, if there was a scratch or imperfection in the coating of the glass, the energy would stream out through this opening in laserlike fashion.

In the late 1890s, Tesla was bombarding targets with X rays at distances in excess of forty feet, and by 1915 he had announced in the *New York Times* a type of electronic defensive shield which today corresponds to what has been called SDI, or the Strategic Defense Initiative.

HARRY GRINDELL-MATHEWS

During World War I, another Teslarian, Harry Grindell-Mathews, was provided with 25,000 pounds by the British government for the creation of a searchlight beam which he said could control aircraft. A wireless electrician and veteran of the British army, wounded at the turn of the century during the Boer War, Grindell-Mathews eventually refined this invention and changed it into a "diabolical ray." This new electronic beam, he said, could not only destroy zeppelins and airplanes, but also immobilize marching armies and nautical fleets. Although he would not divulge the specifics of his creation, he made no secret of his admiration for Tesla, whose technologies had "inspired" its groundwork.

In July 1924, Grindell-Mathews traveled to America to see an eye specialist. He probably met with Hugo Gernsback at that time and might also have visited Tesla. Staying at the Hotel Vanderbilt, the British inventor was interviewed by a number of the local dailies. "Let me recall to you the air attacks on London during the [world] war. Searchlights picked up the German raiders and illuminated them while guns fired, hitting some but more often missing them. But suppose instead of a searchlight you direct my ray? So soon as it touches the plane this bursts into flame and crashes to the earth."[31]

Grindell-Mathews was also convinced that the Germans had such a ray. They were using a high-frequency current of 200 kilowatts, which as of yet they were "unable to control."

Working with the French government in Lyons and performing successful tests before members of the British War Office, Grindell-Mathews instituted destructive effects at distances of sixty feet but was hoping to extend the force to a radius of six or seven miles. Asked for specifics, he said that his device utilized two beams, one as a carrier ray and the other as the "destructive current." The first beam would constitute a low frequency and would be projected through a lens; the second, of a higher frequency, would increase conductivity so that destructive power would be more easily transmitted. The motor of an airplane, for instance, could be the "contact point" at which the paralyzing ray would do its handiwork. He admitted, however, that if the object were grounded, it would be protected against such a force.[32]

Hugo Gernsback, along with Dr. W. Severinghouse, a physicist from Columbia University, tried unsuccessfully to duplicate the effects using heat beams, X rays, and ultraviolet rays. Doubting Grindell-Mathews's claims, Gernsback nevertheless featured the "diabolical ray" with characteristic Frank Paulian panache on the cover of his magazine and with a series of exposés.[33]

Leaders from other countries were less critical than Gernsback, and many proclaimed that their scientists also had such diabolical rays. Herr Wulle, a member of the German Reichstag, announced that "three German scientists have perfected apparatus that can bring down airplanes, halt tanks and spread a curtain of death like gas clouds of the recent war." Not to be outdone, Leon Trotsky stated that the Soviets had also invented such a device. Warning all nations, Trotsky proclaimed, "I know the potency of Grammachikoff's ray, so let Russia alone!"[34]

This theme of all-powerful efficient weaponry reappeared during the 1930s as the seeds of World War II were sown. At this time, Tesla began to reveal more and more about his own diabolical ray as he criticized the Grindell-Mathews scheme.

"It is impossible to develop such a ray," [Dr. Tesla says]. "I worked on that idea for many years before my ignorance was dispelled and I became convinced that it could not be realized. This new beam of mine consists of minute bullets moving at a terrific speed, and any amount of power desired can be transmitted by them. The whole plant is just a gun, but one which is incomparably superior to the present." The inventor further claimed that the new weapon, which was to be used for defense only, comprised "four new inventions": (1) an apparatus for producing the rays; (2) a process for producing immense electrical power; (3) a method for amplifying the power; and (4) a tremendous electrical repelling force.[35]

Working in two undisclosed locations, including a secret laboratory under the Fifty-ninth Street Bridge, near Second Avenue,[36] Tesla perfected his particle-beam weapon, as he conspired with unabashed anarchist and architect Titus deBobula, to design the all-purpose power plant that could generate high voltages or capture cosmic rays and convert them into his defensive electronic shield.[37] Believing that entire countries could be protected by such plants, the inventor clandestinely approached the war departments of each of the Allies with his scheme.

45

LIVING ON CREDIT (1925–40)

If you mean the man who really invented, in other words, originated and discovered—not merely improved what had already been invented by others, then without a shade of doubt, Nikola Tesla is the world's greatest inventor, not only at present, but in all history.

HUGO GERNSBACK[1]

Tesla was leading a double life in his later years, one as the elegant author of the electrical power system and father of the wireless and another, more labyrinthe existence as the quintessential mad scientist whose ultimate creations would rule not only the earth but other worlds as well.

In 1935, with the help of newsreel photographers, Tesla designed and produced an electrified extravaganza which he offered to Paramount Pictures. "Paramount said the film came out unusually good, both in respect to pictorial and sound effects," he told George Scherff, "but they feel that the subject was too technical."[2] Nevertheless, Teslaic themes continued to make their way into the mass consciousness. The most important movie person to implement the inventor's wizardry was revolutionary movie producer Carl Laemmle and his special-effects expert Kenneth Strickfadden. Together, they unleashed one of Tesla's unforgettable coils in the Boris Karloff classic *Frankenstein.* (Strickfadden also resurrected the same paraphernalia forty years later when Mel Brooks recreated the set for the spoof *Young Frankenstein.*)[3] Tesla was partial to Laemmle, whom he referred to as a genius, because Laemmle, too, successfully fought the powerful Edison clique a generation earlier when Edison held a monopoly on key movie making patents and would not allow competitors to use them. Routing his product through Europe, Laemmle was able to withstand more than 200 legal actions against him to create Universal Pictures and thereby defeat Edison.[4]

Hugo Gernsback, of course, also continued to espouse Teslaic motifs in *Science Wonder Stories*[5] with such fresh galactic escapades as "The Mightiest Machine," "Interplanetary Bridges," and "A City on Neptune."

Other apostles, such as John Hays Hammond Jr. and Edwin Armstrong, were riding out the Great Depression living like kings. Hammond's castle, throughout the 1930s, became a haven for Hollywood stars, corporate giants, and virtuosos; the retreat also doubled as a top-secret military think tank.

As Tesla continued to work at his hideaway by the Fifty-ninth Street Bridge, Edwin Armstrong continued to engage in a never-ending court battle with Lee De Forest over the invention of the heterodyne. Holding eighty thousand shares of RCA stock, Armstrong was powerful enough to ride out the litigation as he continued to design a variety of new patents. Starting up his own radio station, Armstrong unveiled his newest invention, FM radio, a novel system which reduced such problems as static caused by ground interference so often encountered with AM. Little did Armstrong know that the litigation caused by this latest creation would make the De Forest trial seem like a kindergarten spat.[6]

Getting on in years, Tesla decided to hire a few Western Union boys to feed the pigeons for him. Dressed in their official caps and snappy uniforms, the lads could be seen like clockwork at 9:00 A.M. and 4:00 P.M. at three different locations around the city: in front of the New York Public Library, in Bryant Park, at the library's rear, and at St. Patrick's Cathedral.[7] The inventor had constructed special wooden cages complete with a birdbath for taking care of wounded as well as healthy feathered friends, and he befriended other dove fanciers to whom he could deliver the birds.

In 1925, Tesla's office was moved from 8 West Fortieth Street, near the New York Public Library, to fashionable quarters at 350 Madison Avenue. His secretaries, Dorothy Skerrett and Muriel Arbus, during the later years, shared duties with Slavic professor Paul (Rado) Radosavljevich, from New York University, who edited Tesla's articles and screened incoming visitors.[8] By 1928, however, the upkeep on the office had become too burdensome, and Tesla closed it for good. All of his holdings, consisting of thirty trunks, including his priceless correspondence, theoretical papers, and prototype inventions, were carted to the basement of the Hotel Pennsylvania, and that is where they remained until November 21, 1934, when he transferred them to the Manhattan Storage Warehouse, located at Fifty-second Street and Seventh Avenue.[9]

A contented member of the cognoscenti who had lived a life full of triumph was merely the persona, for underneath the appearance Tesla was often bitter, seeking an essentially solitary existence, displacing his anger in editorials that lashed out at Edison and Marconi and on unfortunate hotel managers who had to contend with the thought of throwing the grand master out on his coattails for not paying his rent. In 1930, Tesla was escorted out of the Hotel Pennsylvania after residents complained about the interminable droppings from his "flying rats," and because he was "$2000 behind in his rent."[10] As B. A. Behrend quietly reimbursed the hotel to the best of his abilities, the inventor hired a crew to cart his beloved avian companions to George Scherff's home north of the city. Escaping their confinement, the pigeons returned to Manhattan just in time to move in with Tesla at his new abode, the Hotel Governor Clinton.[11]

Working on a variety of new fronts, Tesla entered a furtive realm which would put him in touch with a series of nefarious agents and heads of many governments. Naturally, he required funds, for he began to lag behind yet again in his rent.

When he stopped by Hugo Gernsback's office for another twenty dollars,[12] the science-fiction editor showed Tesla an article on Westinghouse's new radio machinery. Realizing that this company was essentially pirating his wireless patents, Tesla strode into their offices and demanded royalty payments. He met with Victor Beam, assistant to the vice president.

"It would be painful to me to resort to legal proceedings against a great corporation whose business is largely founded on my inventions," the inventor stated matter-of-factly, "and I trust that you will recognize the advantage of an amicable understanding."

"Which statement do you claim to be an infringement?" Beam replied, feigning naïevté.

"Statement!" Tesla shot back. "Surely you must admit that my claim is too palpably evident to be denied."

Beam inquired as to a price for purchasing Tesla's wireless patent no. 1,119,732, but it was really a stall tactic, as no genuine offer was made. In exasperation, Tesla went home to draft a technical letter spelling out each and every infringement of his fundamental work and concluded: "We [Charles Scott and Tesla] have offered this revolutionary invention repeatedly virtually on your own terms and you did not want it. You have preferred to take it by force. You have robbed me of the credit that is due and injured me seriously in business. Instead of showing a willingness to adjust the matter in an equitable way, you say you want to fight. You may think you can secure an advantage by such matters, but we doubt it, and certainly they will not meet with public approval when all the facts are published."[13]

One of the problems Tesla had to deal with was the continuing legacy of resentment toward him by some members of the corporation. Unfortunately, one of the key antagonists was Andrew W. Robertson, a Westinghouse official who would soon become chairman of the company. Just a few years later, while Tesla was still alive, Robertson came to write a small treatise on the AC polyphase system for the 1939 World's Fair. In it, he neatly sidestepped any clear mention of Tesla's role in the development of the system, suggesting, rather, that William Stanley was the inventor. Robertson even had the audacity to write the following:

> In George Westinghouse's time, an inventor was recognized as owner of his ideas and was given a patent to protect him in that ownership. Now we are told that patents are evil monopolies, used to prevent people from getting the full benefits of an individual's work. If we are thinking clearly, we must draw the conclusion that these signs all point *to a common hostility against the... great inventor*.... If this hostility continues, it cannot but result in an environment certain to interfere with the growth and development of individual research and inventiveness.[14]

Here is a classic case of the Freudian defense mechanism known as projection, whereby one's real feelings are attributed to others: Robertson suggests that common people resent the great inventor when in fact it is really he who harbors this resentment. Tesla had first offered his wireless patents to the company in the early 1920s; a long time elapsed before the issue was resolved.

As Tesla continued to make headlines for his invention of a diabolical ray, he also was becoming more and more adept at slipping past the manager of the Governor Clinton. If he had to wait for Westinghouse to come through, so would the hotel.

Tesla was now working with the notorious architect and arms merchant Titus deBobula, whose offices were located at 10 East Forty-third Street; deBobula was hired to design the tower, power plant, and housing for the inventor's "impenetrable shield between nations."

"We can project destructive energy in thread-like beams as far as a telescope can discern an object," said the seventy-eight-year-old inventor. "Dr. Tesla's death ray can annihilate an army 200 miles away. It can penetrate all but the thickest armor plate, and a country's whole frontier can be protected [with plants] producing these beams every 200 miles." Dr. Tesla concluded: "The plane is thus absolutely eliminated as a weapon; it is confined to commerce."[15]

Born in 1878 in Hungary, with ties to Tesla probably through the Puskas brothers, deBobula had emigrated to the United States during the Gay Nineties. At that time, Tesla "took the youth under his protection" and aided him in obtaining passage for a boat trip back to his homeland.[16] Short and stocky in stature, with a brush mustache and ruddy complexion, deBobula returned to the States a few years later to study architecture. Having borrowed the money from Tesla ostensibly because he needed medical help in Budapest, deBobula had actually returned home to help his father out in his moving business and because he wanted to complete other studies at the local "polytechnic." Never reimbursing the inventor for his aid and having lied about his real intentions, deBobula apologized and appealed for funds once more, in 1901. Writing from Marietta, Ohio, where he was attempting to design a church and school for a parish, he requested "$70 or $80." Perhaps in lieu of repayment, the new architect offered to draw up plans for the laboratory at Wardenclyffe, but this assignment had already been undertaken by Stanford White.[17]

At about 1908, deBobula moved to Pittsburgh, where he met and married Eurana Mock, niece of Bethlehem Steel czar Charles Schwab. Shortly thereafter, he designed and built Schwab's new mansion; he also secured loans from the steel magnate to finance a series of real estate ventures.

By 1910, deBobula was back in New York, earning his way by designing churches and constructing large apartment buildings in Manhattan and the Bronx. Playing fast and loose with Schwab's capital, deBobula returned to Ohio, where he crossed the border into West Virginia and Kentucky to purchase eleven thousand acres.[18] Now well connected, the Hungarian offered to set up a syndicate of wealthy English steel men to help refinance Wardenclyffe, promising to raise a million pounds "without going to very much trouble, providing, of course, that we could demonstrate things to them satisfactorily." But Tesla declined the offer and "resolved to fight my own battles."[19]

Known as a "racketeer," having never paid taxes on the land deal, deBobula also reneged on a series of other loans. Naturally, Schwab became angry, particularly because he had lent deBobula additional monies to keep him out of debtor's prison. Paradoxically, deBobula became interested in workers' rights at this time and in the growing anarchist movement.

Seen as a gold digger and bully by Schwab, the wealthy financier was reported to have said that "Bob is dishonest, and I would give a million dollars if he would jump out this window right now." This event unleashed a powder keg of animosity between the two, deBobula suing Schwab for $100,000 for defamation of character, Schwab severing connections between his niece and deBobula and the family.[20]

Attracted to violent activities, deBobula continued to associate with a variety of radical and paramilitary groups; he also caught the eye of the Secret Service. In 1923, during a bizarre interlude, deBobula returned to his home in Budapest and aligned himself with a pro-Hitler group. There he authored a paper which attacked Jewish physics and espoused the developing new world order. Charged with conspiring to overthrow the Hungarian government, he escaped back to America.

Throughout this entire period, deBobula would regularly correspond with Tesla to discuss various ideas he had, such as how to perfect a projector bomb, which he had a patent pending on, and he also discussed his latest encounter with a cabal of international warlords. The following letter suggests the possibility that Tesla's attacking Einstein may have been prompted by anti-Semitic sentiments as much as by philosophical differences:

MUNITIONS INC.
295 Madison Ave.
New York, NY

My dear Mr. Tesla,

I hugely enjoyed your comments on relativity, which we, way back in 1921, attacked in my Budapest paper, as a theory of the fundamentals of which, if logically developed, would inevitably lead to an anthropomorphic Jehovah with all the evil trimmings in philosophy and social order.

Yours sincerely,
Titus deBobula[21]

In full support of Tesla's desire to resurrect a new Wardenclyffe, deBobula carefully drafted for Tesla blueprints of a 120-foot-tall teleforce powerhouse and transmission tower; at the same time, he set up a factory for his munitions company in New Jersey with Capt. Hans Tauscher. The tower, set up somewhat like a high-tech Van deGraaff generator, with Van deGraaff's primitive cardboard belt replaced by a vacuum stream of ionized air, had at its bulbous apex a turret-rotating particle-beam cannon, able to move about to attack planes and airships with information provided by an earth-current radar system the wizard had also designed during his days at Wardenclyffe.

Tauscher, a German American, was closely linked with the Fatherland through his daughter, who still lived there. Through this and other channels, Tesla could be introduced to potential weapons buyers. Simultaneously, he could also be supplied with small amounts of dynamite, which he used to test telegeodynamic equipment that he was trying to sell to such companies as Texaco Oil Company for geophysical exploration.[22]

Titus deBobula was now making his living selling rifle grenades, fragmentation and gas bombs, and other armaments to both domestic police departments and foreign governments in Europe and South America. Unfortunately for the Hungarian, unwished-for newspaper publicity concerning a lawsuit with Tauscher drew the attention of the Internal Revenue Service (IRS). Having hid his relationship with the munitions factory, deBobula not only attracted the tax collectors but also the seemingly omniscient J. Edgar Hoover, who monitored his activities for the next ten years. He was accused of being a German agent and illegal alien involved in subversive activities, and Hoover "recommended that [de-Bobula] be considered for custodial detention in the event of a national emergency."

Although they maintained their relationship during this stormy period, Tesla's antennae were raised, especially when deBobula requested loans from him "in amounts you would not miss" just at the time he declared bankruptcy: DeBobula was $750,000 in debt. Hoping to make Tesla a partner in the company, deBobula raised Tesla's ire when he used his name as a reference when working an arms deal with the minister of Paraguay. Tesla called deBobula and stated emphatically that he was not to use the inventor's name.[23]

Unable to maintain his Manhattan apartment, deBobula moved to the Bronx, where he was eventually arrested. A search of his apartment revealed an arsenal of hand grenades, dynamite, and tear-gas bombs. The architect claimed that they were simply part of the inventory from his company, and he was interrogated and released. Presumably an anti-Semite who apparently had ties to Henry Ford, deBobula disavowed any relationship to either the Communist party or the German-American Bund. Although he was monitored throughout the World War II years, the Federal Bureau of Investigation was unable to prove that he had violated any federal statute, and deBobula was allowed to move to Washington, D.C., where he set up another munitions arsenal right in the heart of the city! In 1949 he wrote directly to J. Edgar Hoover to "get them off his back," and apparently Hoover complied, as articles and references to him faded away at that time.[24]

As the Great Depression wore on, Tesla's expenses continued to mount. He was now $400 in debt to the hotel, which, during those dreary days, was a considerable sum. Cornered by the management, the sly conceptualist fell back on a strategem that had worked so well when he was at the Waldorf-Astoria: he would live on credit. He thereupon offered as collateral "a working model" of his death ray, which he told the establishment was worth $10,000. They asked for a receipt in writing, and he agreed. The treacherous device and his note were carefully placed in locker no. 103 in the hotel's back-room vault.[25]

Although he had solved the problem of meeting the rent, Tesla's finances were still in a precarious state. He was even having trouble coming up with the fifteen dollars per month storage costs at the Manhattan Storage Warehouse. In 1934, with a bitter taste in his mouth, he drafted a long letter to J. P. Morgan Jr. In it the inventor revealed that he has made overtures to sell his "Chinese wall of defense" to war departments in America and England. "The Russians are very anxious to render their border safe against Japanese invasion, and I have made them a proposal which is [also] being seriously considered." The inventor went on to acknowledge his outstanding debt to Morgan from the turbine affair and his willingness to pay back the sum, about $40,000, and then made an appeal. "If I had now $25,000 to secure my property and make convincing demonstrations, I could acquire in a short time colossal wealth." Tesla ended the letter by lambasting FDR's "New Deal," which "poured out billions in public money [just so] it can remain in power indefinitely." This "perpetual motion scheme," in Tesla's view, was antidemocratic, "destructive to established industries, and decidedly socialistic."[26]

Needless to say, Morgan did not lend Tesla any money. However, what is interesting in this passage is Tesla's reference to property he needed to secure in order to make convincing demonstrations. This suggestion that he had, in fact, erected a working death ray would become important later, when the question was raised whether a functioning model had actually been constructed.

According to Hugo Gernsback, through to his prompting, the Westinghouse Corporation agreed to help their former champion out. Gernsback stated that he had called up the Westinghouse officials "in the late 30's... to discuss what could be done. I apprised [them] of the fact that Tesla was a very proud man who, under no circumstances, would consider charity. I suggested that perhaps he could be put on the staff in an honorary capacity as a consultant. This was agreed upon and from that time until his death... Tesla received a modest pension."[27]

Gernsback may certainly have called the company. However, his interpretation of what occurred conflicts with the timing of the contract and the reasons for the deal. Tesla had settled with Westinghouse about four years before the science-fiction publisher's gracious call. On legal footing, the inventor continued to press his case. Finally, on January 2, 1934, the president, F. A. Merrick, acquiesced and agreed to pay Tesla "to act as a consulting engineer for $125/month for such a period as may be mutually agreeable."[28] To solve the aging patriarch's psychological problem of refusing to pay his rent, Merrick also agreed to cover Tesla's rent. The debt owed to the Hotel Governor was never settled; rather, Tesla moved, after signing of the agreement to the Hotel New Yorker; and there he would live, so far as he was concerned, rent free to the end of his days.[29]

46

LOOSE ENDS (1931–43)

Then there appeared the vision of a man from a strange, new world. A tall, thin man, whose eyes blazed with an unearthly light, entered the room so quietly that one was hardly aware of his presence. He bowed himself to his seat....He beamed upon [Viereck and his wife] paternally. He greeted the guests with a kindly nod. Before he could be introduced, Tewson blurted out, "Nikola Tesla!"

ELMER GERTZ[1]

The first-time ambassador Stanko Stoilkovic saw the celebrated Serb he was standing in front of the library with "two white pigeons on his arm...pecking seeds from his palm." That was in 1918, and they met only briefly. A decade later, Stoilkovic returned to America as an emissary of the Yugoslavian consulate, and they became close friends for the next ten years. At the age of ninety, Stoilkovic would recall their visits.

As with many other Serbs, Stoilkovic was unhappy about the split between Tesla and Pupin, and like others before him, he tried to bring the two men together. Perceiving Pupin as ungrateful and deeply hurt by his association with "that donkey" Marconi, Tesla wanted nothing to do with his fellow Serb. The feeling, of course, was mutual.

May 29, 1931

My dear Mr. Swezey:

I have not seen Mr. Tesla for nearly 20 years. In the beginning of the World War a difference of opinion created a split between Mr. Tesla and myself. Neither he nor I have ever had, since that time, an opportunity to cure that split. In 1915 I offered through a mutual friend, to forgive and forget, but somehow the offer was not accepted. I regret, therefore,

> that...I could not transmit to Mr. Tesla a letter of greeting or congratulation on his seventy fifth birthday.
>
> Yours very sincerely,
> M. J. Pupin[2]

In Dunlap's classic text *Radio's 100 Men of Science*, the author writes, "Pupin was a man of complete intellectual honesty; if he made an error in an equation on the blackboard, he would quickly admit the mistake, rub out the blunder and begin again."[3]

We have discussed in the past Pupin's insistence that many of Tesla's inventions were his own and how Pupin removed Tesla's name from the discussion of the history of the AC polyphase system and of the wireless in his four-hundred-page Pulitzer Prize–winning autobiography, which he had dedicated to "the rise of idealism as a qualified witness whose testimony had competence and weight."[4] This tack also carried over to his legendary courses at Columbia University in which he camouflaged Tesla's role in the etiology of a variety of inventions. "When Marconi came to New York, in 1927 to lecture...Dr. Pupin presided at the Institute of Radio Engineers....'Marconi, we love you,' said Dr. Pupin; 'we have come not so much to hear what you have to say, but to see your boyish smile.'"[5]

When Pupin became ill in 1935, he asked his secretary to go see Stoilkovic and "plead with him to get Tesla to visit Pupin in the hospital. He wanted to make peace before he passed on."

Greeting Stoilkovic at the door in his favorite lounging outfit, a red robe with blue slippers, Tesla was taken aback by the request. He said he needed to sleep on the matter. The following day, he called his friend and said he would go if Stoilkovic would accompany him.

> In Pupin's room there were a few doctors alongside his bed. The meeting was most touching. Tesla approached the sick man and held his hand out and said, "How are you, my old friend?"
>
> Pupin was speechless from emotion. He cried and tears were coming down his face. We all went out of the room and left the two men alone. Tesla was able to talk with Pupin eye to eye....In parting Tesla had mentioned that they would meet again in the Science Clubrooms and converse as before....Immediately after Tesla's visit, Pupin passed away. Tesla attended the funeral.[6]

Four years Tesla's senior, Robert Underwood Johnson left for a tour of England as a widower in 1927 and went again to France and Italy in 1928. Tesla lent his friend $500 for the trip and $800 for the mortgage on his home. Upon his return in 1929, Johnson joined Tesla and Richmond P. Hobson for a movie and an evening out on the town. Hobson, who was living with his wife, Grizelda, at the Hotel Weylin on Fifty-fourth Street, also had a residence in the nation's capital. Tesla was seeing his friends on a daily basis at this time; Johnson was particularly lonely, even though he had children and grandchildren who frequently came to visit. The following year, now an octogenarian, the old poet set off again for Europe. On this trip he was hoping to interview Madame Marie Curie.

Although Tesla was able to spend much of the 1930s with Johnson, Hobson left the city after only a short stay to purchase a cattle ranch with his son in Vancouver. In April 1937, Tesla forwarded a new biography of his life that had been translated from Serbo-Croatian into English to Johnson, who passed it on to the editor in chief of the *New York Times*. At eighty-five years of age, Johnson was too weak to write a thank-you letter in his own hand, but he was able to sign it "R. U. Johnson—Luka Filipov." Both Johnson and Hobson died shortly thereafter; the famous lieutenant, barely sixty years old, was buried in Arlington Cemetery. With the help of Agnes Holden, Robert's daughter, Tesla sent a "gorgeous flowering azalea" to Grizelda, who appreciated greatly the kind thoughts of this dear friend.[7]

Another character who stayed friendly with Tesla in the later years was George Sylvester Viereck, wunderkind poet. Viereck was a sensualist, cynic, and German propagandist during World War I and Nazi spokesman during the 1930s and World War II. Tesla's link to Viereck extended back thirty years, to when Gilder and Johnson introduced such provocative Viereck poems as "The Haunted House" in the *Century* in 1906, which was a poem about an enchanting partner whose body was, alas, "a haunted place.... When I did yield to passion's swift demand," Viereck penned, "one of your lovers touched me with his hand. And in the pang of amorous delight, I hear strange voices calling through the night."[8]

Viereck, who was probably an illegitimate grandson of Wilhelm II, abdicated kaiser of Germany, was a multifaceted character and self-proclaimed genius. Having interviewed many of the greatest minds of the epoch, the German-American intellectual had rattled the souls of such individuals as Theodore Roosevelt, George Bernard Shaw, occultist Aleister Crowley, H. G. Wells, Sigmund Freud, Albert Einstein, Kaiser Wilhelm, and Adolf Hitler. A man in intimate contact with his irrational side, Viereck had passed many a day with Sigmund Freud, explicating the great theoretician's libido theory, influencing Freud's writings, and applying the master's psychology to modern life. Freud, Viereck wrote, was "motivated not only by a desire to complete his own world view, but by a belief that each individual represented a special expression of the world spirit."[9]

Not an anti-Semite, having, for instance, cowritten a series of books with a Jewish professor, Viereck, a lifelong German apologist, was nevertheless somehow able to rationalize the Nazi spiel, and he became an American spokesman for Adolf Hitler. Although Freud saw in this journalistic "lion hunter" a great mind, he said that Viereck suffered from "narcissism, had delusions of persecution and a fixation on the Fatherland." Once Viereck began to rationalize Hitler's rhetoric, Freud saw the journalist as "debasing himself" and would not correspond with him anymore.[10]

During one of a series of interviews Viereck conducted with Tesla, he revealed that Tesla "was not a believer [in God] in the orthodox sense.... To me, the universe is simply a great machine which never came into being and never will end. The human being is no exception to the natural order.... What we call the 'soul' or 'spirit' is nothing more than the sum of the functionings of the body. When this function ceases, the 'soul' or 'spirit' ceases likewise." According to John O'Neill, this "meat machine" theory was really a ruse used to hide the numerous mystical experiences that Tesla had.

On his vision of the twenty-first century, Tesla foresaw a world in which eugenics would be "universally established." Perhaps spurred by Viereck's discussion of the Aryan vision or the iniquitous American practice of sterilizing criminals and some mentally retarded individuals, Tesla supported the idea of "sterilizing the unfit and deliberately guiding the mating instinct. A century from now," the celibate concluded, "it will no more occur to a normal person to mate with a person eugenically unfit than to marry a habitual criminal."

On diet, the slim epicurean revealed that he had given up meat altogether. Tesla believed that in the future inexpensive and healthy food would be derived from milk, honey, and wheat. As the 1930s progressed, the inventor would continue to live on an ever-receding subsistence diet, moving from meat to fish to vegetables and finally to warm milk, bread, and something he called "factor actus." Eliminating solid food altogether, the thinning wizard had concocted a health potion made up of a dozen vegetables, including white leeks, cabbage hearts, flower of cauliflower, white turnips, and lettuce hearts.[11] Although he still maintained that he might live to 140, a psychoanalytic view of this regimen of meager victuals could only construe it as an anorexic, unconscious plan for self-extinguishment.

"Long before the next century dawns," the oracle predicted, "systematic reforestation and the scientific management of natural resources will have made an end of all devastating droughts, forest fires and floods. Long distance transmission of electrical power by harnessing waterfalls will dispense with the necessity of burning fuel; robots and thinking machines will replace humans, and the trend of spending more on war and less on education will be reversed." A major reason for this would be Tesla's latest discovery of his defensive shield between nations.

"If no country can be attacked successfully, there can be no purpose in war. My discovery ends the menace of [war]. I do not say that there may not be several destructive wars before the world accepts my gift. I may not live to see its acceptance."[12]

In the habit of meeting with Viereck and his family at their home on Riverside Drive on a somewhat regular basis, Tesla accepted an invitation to attend a dinner party. Present was Viereck's son Peter. Today, as a Pulitzer Prize–winning poet and English professor, Peter remembers Tesla as almost an uncle.[13] Also present was the youthful Elmer Gertz, who, at the time, was preparing a biography of Viereck. A friend of Carl Sandburg and also biographer of voluptuary Frank Harris, Gertz would later come to defend such illustrious individuals as Nathan Leopold (and Loeb), pornographic writer Henry Miller, and assassin Jack Ruby (in an appeal of the death sentence). At a spry eighty-five years of age, Gertz remembered the meeting, which took place fifty-seven years before, when he was twenty-nine.

"In a communicative mood, [at the dinner party] Tesla told his life story unostentatiously, simply, with quiet eloquence. He told of his platonic affairs of the heart...explained the inventions that have made the world his debtor...[and] told of his plans, of his credo, of his foibles. It was a tale of wonders, told with guileless simplicity."

Struck by the fact that they were in "the same house that had seen Einstein, Sinclair Lewis, and countless others," Gertz noted that "Viereck was comparatively silent much of the night, but he was subtly responsible for the intellectual thrills of the evening."[14]

Pressed for other details, Gertz revealed that "Tesla knew all of Viereck's poems by heart." Tesla also discussed his platonic affair with Sarah Bernhardt, whom he had met while in Paris at the 1889 exposition. Differing with O'Neill's oft-quoted suggestion that Tesla avoided her eyes when he picked up her handkerchief, Gertz said that Tesla had met with her on several occasions, perhaps also in New York. So taken with her was he that "Tesla had saved and preserved her scarf without ever washing it" and still had it to that day.[15]

Viereck provides a fascinating link between Tesla and Sigmund Freud: self-denial and eros. Stoilkovic told a story of how he was invited to Tesla's New Yorker apartment and a valet "brought in a bottle of wine in a dish of ice," but Tesla never allowed it to be opened. When this event was replayed another evening, Tesla revealed that he kept the bottle there to prove that he could prevent himself from drinking it. A man of rigid habits, Tesla denied himself certain pleasures as a way to supposedly establish total control over himself. And yet Tesla was a complete slave to his idiosyncrasies and to a cauldron of phobias.

Going out of his way to avoid handshakes, displacing his amorous affections onto birds, keeping hotel servants at a distance of at least three feet, throwing out collars and gloves after one use, Tesla had other rigid requirements as well. The hotel was requested to keep one table permanently reserved for himself. No one else was allowed to eat there. If a fly landed on the table, it had to be reset and a new plate of food brought forth. When it came to money matters, Tesla also showed little ability to restrain himself, and as we have seen, he habitually refused to honor rent payments. From the Freudian point of view, Tesla was an anal-compulsive personality, fixated in the latent stage of sexual repression, with a displacement of his energies into scientific endeavors. By denying his libido, his censor had converted his primal sexual energy into an odd mixture of prelogical behavior patterns that tended to diffuse, redirect, and sublimate the highly cathected complexes the inventor wished to deny.

It appears quite possible that Viereck strove to psychoanalyze Tesla, to have the inventor dig deep into his childhood in an effort to release submerged events that may have blocked or redirected his life in neurotic fashion. Viereck, who was known to partake of "opium tincture," probably entered an altered state when conversing with the ethereal prophet.[16] Nearly eighty years old at the time, exceedingly frail and thin, Tesla wrote Viereck a long letter reviewing the traumas of his childhood.

> It was a dismal night with rain falling in torrents. My brother, a youth of eighteen and intellectual giant, had died. My mother came to my room, took me in her arms and whispered, almost inaudibly, "Come and kiss [Dane]." I pressed my mouth against the ice cold lips of my brother knowing only that something dreadful had happened. My mother put me again to bed and lingering a little said with tears streaming, "God gave me one at midnight and at midnight took away the other one."[17]

Viereck not only wanted Tesla to reactivate the hidden complexes associated with the death of Dane but also to reflect on the very process of how his fertile mind gave birth to ideas. One can only guess as to whether or not Viereck trod into the realm of Oedipus and Narcissus, dead-brother kissing rituals, and Tesla's monastic life of quirks, self-denial, and affection displaced onto feathered friends.

In 1937, on his eighty-first birthday, at a luncheon in his honor, Tesla was "bestowed" both the Order of the White Lion from the minister of Czechoslovakia and the Grand Cordon of the White Eagle, the highest order of Yugoslavia, from Regent Prince Paul by order of King Peter. Belgrade also set up an endowment of $600 per month, which they paid to him for the rest of his days. Looking much like an eagle himself, with his long, hawk-shaped proboscis accentuated by his extreme leanness, the skeletal inventor "followed his annual custom [after the award ceremony] by playing host to a group of newspaper men at his Hotel New Yorker suite." There, dressed in his finest tuxedo, the wizard read a prepared treatise outlining his latest inventions and plans for contacting nearby planets.[18]

Just a few months later, in late autumn, while negotiating with emissaries from the war departments of Yugoslavia, Czechoslovakia, England, the Soviet Union, and the United States, Tesla was run down by a taxicab. Refusing to see a doctor, the inventor managed to limp home. He stayed in bed on and off for six months. He had cracked three ribs. Thus, in May 1938, still in the process of recuperation, he declined an invitation to attend another award ceremony at the National Institute of Immigrant Welfare in honor of him, Felix Frankfurter, of Harvard Law School, and Giovanni Martinelli, of the Metropolitan Opera.

Accepting the award in Tesla's honor was Dr. Paul Radosavljevich, professor of pedagogy at New York University. Rado read a statement from Tesla which apparently verified the famous 1885 Edison story that Edison had laughed off the promised debt to Tesla of $50,000 for redesigning machinery.[19]

In 1939, just as World War II was about to start, George Sylvester Viereck set off on a surreptitious journey to the Fatherland. There, amid the pageantry of swastikas and the Gestapo, he met yet again with Adolf Hitler and received a communiqué signed by the Führer in his own hand. The date was February 26, 1939. Upon his return, Viereck continued his practice of preaching the Nazi line in a variety of publications, writing some articles under an assumed name, interpreting FDR as having a "messianic complex" and Hitler as "a dynamic genius, and a poet of passion...first in war, first in peace, first in the hearts of his country-men." Viereck was arrested and indicted on two counts of seditious conspiracy. It was quickly established that he was on the German payroll supposedly as a journalist but obviously as a paid propagandist. Conceited and self-deluded, the bewildered and obtuse philosopher was sent to jail, where he wrote poetry for the next few years.[20] Just as references to Tesla were deleted from engineering texts, Viereck's name was "dropped from many anthologies, and *Who's Who*."[21] Thus, both individuals disappeared from history books but for totally different reasons.

As World War II began, Tesla was becoming more feeble, lapsing in and out of states of coherence. During one of his more lucid moments, he wrote (with the help of his nephew Sava Kosanovic) the foreword to Vice President Henry Wallace's address on "The Future of the Common Man" for the Serbo-Croatian edition. The essay not only portrays a prophet who envisions a better world in the future; it also betrays the conflict and humiliation he himself suffered in his dealings with the greedy industrialists who capitalized on his inventions with little regard for his well-being, let alone the welfare of mankind as a whole: "Out of this war, the greatest since the beginning of history, a new world must be born that would justify the sacrifices offered by humanity, where there will be no humiliation of the poor by the violence of the rich; where the products of intellect, science and art will serve society for the betterment and beautification of life, and not the individuals for achieving wealth. This new world shall be a world of free men and free nations, equal in dignity and respect."[22]

The old man gazed out the window as his nimble fingers unconsciously preened the ruffled feathers of his beloved white pigeon with the brown-tipped wings. Although it was January, a lightning storm rumbled in the distance. "I've done better than that," the wizard mumbled as the sun peeked through the clouds to reveal the iridescent purples, violets, greens, and reds of the neck feathers of another of the more hearty birds that came to visit. Tesla thought back affectionately to his days as a boy on the farm, rolling down the hill carefree with Mačak, his pet cat. And then his mind swirled with thoughts of violent arguments with Morgan, his unfinished fifteen-story transmission tower, and his friend Mark Twain, who was now in financial trouble. Requesting funds from Kosanovic, Tesla handed the money to a messenger to deliver to Twain, giving as the address his old laboratory on South Fifth Avenue, a street that no longer existed. Unable to locate the deceased writer, the boy returned to Tesla, but his explanation was ignored. The old man told the boy to keep the money if he could not deliver it.[23]

Having neglected to pay the rent on his belongings held at Manhattan Storage, the wizard had managed somehow to mail out a check for $500 to a Serbian church fund-raising event held in Gary, Indiana.[24] G. J. Weilage, manager of Manhattan Storage, threatened to put the lot up for auction. The outstanding bill was $297. Perhaps too disillusioned to care, Tesla ignored the last warning, and Weilage made good on his promise, placing an announcement in the local papers. Noticing the advertisement, Jack O'Neill rushed to contact Tesla's nephew Sava Kasanovic, who was now the Yugoslavian ambassador stationed in New York. Kasanovic covered the debt and continued to make the carrying charge (fifteen dollars per month), saving this invaluable legacy from tragic dismantlement.

"One night," Tesla wrote, "as I was lying in bed in the dark, solving problems as usual, [my beloved pigeon] flew through the open window and stood on my desk. As I looked at her I knew she wanted to tell me—she was dying. And then, as I got her message, there came a light from her eyes—powerful beams of light. When that pigeon died, something went out of my life. I knew that my life's work was finished."[25]

Lingering at the abyss throughout the beginning years of World War II, Tesla continued to lead his dual life, meeting with friends and dignitaries whenever possible and lending out his secret papers to mysterious men. A few months later, he was dead. The date was January 7, 1943; he was eighty-six years old.

2000 Are Present at Tesla's Funeral

Great in Science Attend

> The President and I are deeply sorry to hear of the death of Mr. Nikola Tesla. We are grateful for his contribution to science and industry and to this country.
>
> Eleanor Roosevelt

The funeral service was held in Serbian in the Cathedral of St. John the Divine. With the coffin open, the Sermon for the Dead was conducted by the venerable Reverend Dushan Shoukletovich, rector of the Serb Orthodox Church of St. Sava. Over the radio, New York mayor Fiorello La Guardia read a moving eulogy written by Croatian author Louis Adamić, as a long line of mourners filed past. The list of honorary pallbearers included Dr. Ernest Alexanderson of GE, who gained his wealth and fame from inventing a powerful high-frequency transmitter; Dr. Harvey Rentschler, director of the research laboratories of Westinghouse; Edwin Armstrong, father of FM radio; Consul General D. M. Stanoyevitch of Yugoslavia; William Barton, curator of the Hayden Planetarium, where Tesla often went to meditate; and Gano Dunn, president of J. G. White Engineering and Tesla's assistant a half century before, during his paradigm-shifting experiments delivered just a few blocks away at Columbia University.[26]

"We cannot know, but it may be that a long time from now, when patterns are changed, the critics will take a view of history," Hugo Gernsback wrote propitiously in his magazine. "They will bracket Tesla with Da Vinci, or with our own Mr. Franklin....One thing is sure," Gernsback concluded. "The world, as we run it today, did not appreciate his peculiar greatness."

Col. David Sarnoff, president of RCA, also took the soapbox in Tesla's behalf: "Nikola Tesla's achievements in electrical science are monuments that symbolize America as a land of freedom and opportunity.... His novel ideas of getting the ether in vibration put him on the frontier of wireless. Tesla's mind was a human dynamo that whirled to benefit mankind."

Edwin Armstrong, who was about to sue Sarnoff and RCA for infringing his FM patents, helped place Tesla in the proper historical perspective when he said, "Who today can read a copy of *The Inventions, Researches and Writings of Nikola Tesla,* published before the turn-of-the-century, without being fascinated by the beauty of the experiments described and struck with admiration for Tesla's extraordinary insight into the nature of the phenomena with which he was dealing? Who now can realize the difficulties he must have had to overcome in those early days? But one can imagine the inspirational effect of the book forty years ago on a boy about to decide to study the electrical art. Its effect was both profound and decisive."[27]

On September 25, 1943, just nine months after Tesla's death, the Bethlehem Fairfield Shipyards near Baltimore launched the USS *Nikola Tesla,* a ten-thousand-ton Liberty ship. Sponsors at the ceremony included a number of Croats, such as Louis Adamić and violinist Zlatko Balokovic, and also Serbs, such as Tesla's nephews Sava Kosonavic and Nicholas Trbojevich.[28]

The *New York Sun* editorialized:

> Mr. Tesla was eighty-six years old when he died. He died alone. He was an eccentric, whatever that means. A nonconformist, possibly. At any rate, he would leave his experiments and go for a time to feed the silly and inconsequential pigeons in Herald Square. He delighted in talking nonsense; or was it? Granting that he was a difficult man to deal with, and that sometimes his predictions would affront the ordinary human's intelligence, here, still, was an extraordinary man of genius. He must have been. He was seeing a glimpse into that confused and mysterious frontier which divides the known and the unknown.... But today we do know that Tesla, the ostensibly foolish old gentleman at times was trying with superb intelligence to find the answers. His guesses were right so often that he would be frightening. Probably we shall appreciate him better a few million years from now.[29]

47

The FBI and the Tesla Papers (1943–56)

War Department
Military Intelligence Service

22 January 1946

Alien Property Custodian
Dear Sir:

This office is in receipt of a communication from Headquarters, Air Technical Service Command, Wright Field, requesting that we ascertain the whereabouts of the files of the late scientist, Dr. Nichola Tesla, which may contain data of great value to the above Headquarters.... In view of the extreme importance of those files... we would like to request also that we be advised of any attempt by any other agency to obtain them.

Sincerely yours,
Col. Ralph E. Doty
Chief, Washington Branch[1]

After Tesla's death, the FBI, the Office of Alien Property (OAP), and factions of the War Department conspired to impound and protect the Tesla secret-weaponry papers. The United States was in the midst of a world war, and Tesla's ties to arms merchants, Communists, through his Yugoslavian nephew, and a notorious German propagandist all helped prompt the covert agencies to safeguard this material until it was properly analyzed. A half century later, they have yet to release it.

Raised to the level of a national hero within the Slavic countries, Tesla was considered practically of royal stock. Thus, his nephew, Sava Kosanovic, rose to become a representative of the newly forming

Yugoslavia republic at the "Eastern European Planning Board," which met in Czechoslovakia.[2] Kosanovic, like Tesla, wanted a unified country, but their orientations were different.

In 1941 the Nazis continued their policy of intimidation and deception by trying to force a treaty with King Peter of Yugoslavia. Backed by the people, Peter refused to agree to an alliance and therefore suffered a fatal blow when Germany masterminded a brutal invasion involving troops from Bulgaria, Italy, and Hungary and also three hundred Luftwaffe bombers.[3]

Prof. Michael Markovitch of Long Island University, a Serb living in Croatia during World War II, has said that ninety thousand Serbs were killed by the Croats; and when I inquired why, he said, "Because the Croats were fascists," that is, because they had aligned themselves with the Nazis. As a youth in the midst of the war, Markovitch had watched the bodies float down the river. When asked why he survived, his reply was "sheer luck."

Concerning the Tesla mythology, Markovitch said that ever since he was a child, he was aware that Tesla was considered a great national hero. Years later, as Hitler's invasion became imminent, Markovitch and his countrymen had expected Tesla himself to return to Belgrade and shield it from the Nazis by harnessing his impenetrable death shield! Unfortunately, Tesla never came.[4]

Kosanovic was not as romantic a figure, and although a Serb, he abandoned the exiled king in order to back the rising Croat leader, Joseph Tito (Josip Broz), and his Communist doctrine. Tito was a solid choice; although he was an ally of the Soviets, he was able to maintain autonomy. He also sought to unify the warring factions; his marriage to a Serbian woman was a powerful symbol in advancing this goal.[5]

Since the Soviet Union was an ally, Kosanovic, as a Yugoslavian ambassador, was able to travel freely to America to discuss various diplomatic tactics with the new leadership. Thus, during the course of World War II, he was able to attend to his ailing uncle in New York and also try to finalize plans to set up a museum in Belgrade in order to honor the great inventor.

In 1942, Tesla became more seriously ill and suffered from palpitations and fainting spells. Although Tesla's true commitment was to the young exiled king Peter, Tesla's nephew was able to coax the inventor into sending a message to Tito which preached unification between Serbs and Croats. Kosanovic also admitted shielding Tesla from factions of the Serbian royalty at the same time; however, when Peter arrived in New York City, he actively helped arrange a meeting between Tesla and the king.

Having unsuccessfully talked with Churchill in England and Roosevelt in Washington, both of whom were reluctantly backing Tito, King Peter was at least consoled by Eleanor Roosevelt, who attended a large party in his honor at the Colony Club in New York. Organized by the American Friends of Yugoslavia, the king's mother, Queen Marie, and also Anne Morgan, Pierpont's daughter, attended, but Tesla was too ill to come.

Therefore, King Peter (with Kosanovic) took a cab to the Hotel New Yorker to confer with the virtual patriarch of his country. Shocked by Tesla's cadaverous condition and upset by the terrible chain of events in his country, Peter told the inventor that he had hoped that he could have returned to Yugoslavia to free it from the Nazis. In his diaries, Peter also revealed that he and Tesla wept together "for all the sorrows that had torn apart [their] homeland."[6]

A few months later, Tesla was dead. A maid discovered his body on January 8, 1943. While Hugo Gernsback rushed to make a death mask, Kenneth Swezey, Sava Kosanovic, and George Clark, director of a museum and laboratory at RCA, entered the apartment. With a locksmith and the hotel management present, they removed various documents from the inventor's safe. Although the FBI alleged that "valuable papers, electrical formulas, designs, etc., were taken," the hotel management confirmed that Kosanovic removed only three pictures and Swezey took the 1931 testimonial autograph book created for the commemoration of Tesla's seventy-fifth birthday.

These events were monitored by the ubiquitous surveillance mastermind J. Edgar Hoover, hard-line anti-Communist and protector of American interests. Hoover wrote in a memorandum under the heading "Espionage," that he feared that Kosanovic, as heir to the Tesla estate, "might make certain material available to the enemy." Kosanovic had been identified as a member of the Eastern European Planning Board, but because of the complicated condition of the Balkan states, there was essentially no way for Hoover to ascertain exactly where Kosanovic's alliances rested. He could have been affiliated with King Peter, the Communist Tito, Fascist factions associated with Mussolini, Hitler, the Soviet Union, or none of the above.[7]

Consistent with his suspicious nature, Hoover also questioned Tesla's sympathies, even though the inventor was friends with Vice President Henry Wallace and Franklin Roosevelt through letters to his wife, Eleanor. One of the main reasons for Hoover's concerns was Tesla's address before the Friends of Soviet Russia which he gave at the Grange Hall in Springfield, Massachusetts, in 1922.[8]

On January 8, Abraham N. Spanel, the forty-two-year-old president of the International Latex Corporation of Dover, Delaware (now Playtex), who was residing in New York City, had called FBI agent Fredrich Cornels to discuss Tesla's death-ray experiments. As the inventor had just died, Spanel feared that Kosanovic would obtain the pertinent papers and pass them to the Soviets.

Spanel had already begun to make a name for himself in media and military circles by having invented floating pontoon stretchers for soldiers wounded in amphibious landings and by turning back the million-dollar profits to the government for the war effort. Born in Odessa in 1901, Spanel would later became a vociferous anti-Communist who spent upward of $8 million throughout the 1940s and 1950s "buying space in the United States press to reprint articles that would contribute to an understanding of world problems." Having fled to France in 1905 to escape the anti-Semitic pogroms of Russia with his family as a child, Spanel, at the age of seven, came to the United States in 1908. A graduate of the University of Rochester, Spanel had invented electrical appliances and pneumatic products in the early 1920s before starting the International Latex Corporation in 1929.[9] Realizing the potential importance of the Tesla invention in the "democratic" fight for world supremacy, Spanel had contacted Dr. D. Lozado, adviser to Vice President Wallace, and a Mr. Bopkin of the Department of Justice. Bopkin agreed to contact J. Edgar Hoover regarding the affair, and Lozado conferred with Wallace and perhaps even FDR, calling Spanel back shortly after their conversation to convey that the government was "vitally interested in Tesla's papers."[10]

Spanel had also contacted one Bloyce Fitzgerald, whom the FBI had pegged as "an electrical engineer who was a protégé of Tesla's," who had also called Cornels. Having met Fitzgerald at an engineering meeting a few years earlier, Spanel became highly interested in the Tesla weapon, possibly hoping to become involved in a profitable business developing the death-beam device for the U.S. military.

Fitzgerald, still in his twenties, had been in postal communication with Tesla since the late 1930s. He had called the inventor on his birthday in 1938 to congratulate him and continued this practice for the next four years. In 1939, Fitzgerald tried to meet Tesla, but it appears that he did not do so at that time. Just two weeks prior to Tesla's death, Fitzgerald proposed another meeting; it may have taken place. He was working at the Massachusetts Institute of Technology with Professors Keenan, Woodruff, and Kay "in the solution of certain problems regarding the dissipation of energy from rapid fire weapons" and desired to discuss his "radiation problem" with the elderly inventor.[11]

It is plausible that at this delicate time Fitzgerald was able to borrow the various papers in which he was interested. Coincidentally, Tesla had declared that "efforts had been made to steal the invention. [My] room had been entered and [my] papers examined, but the...spies left empty handed."[12] Fitzgerald, who also worked for the Ordinance Department of the U.S. Army, later told Cornels that he "knows [that] the complete plans, specifications and explanations of the basic theories of these things are some place in the personal effects of Tesla...[and] that there is a working model [of the death ray]...which cost more than $10,000 to build in a safety deposit box of Tesla's at the Governor Clinton Hotel."[13]

To corroborate this story, another acquaintance of Tesla's, Charles Hausler, a hired hand who took care of the inventor's pigeons, later said that Tesla "had a large box or container in his room near the pigeon cages. He told me to be very careful not to disturb the box as it contained something that could destroy an airplane in the sky and he had hopes for presenting it to the world." Hausler also added that the device was later stored in the basement of a hotel.[14]

Fitzgerald reported that Tesla had claimed that he had eighty trunks in different locations in the city containing inventions, manuscripts, and plans of his various work. The young engineer reiterated the need for the government to obtain the Tesla papers "for use in war." He was also worried about the "loyalty and patriotism to the Allied Nations" of Sava Kosanovic and another nephew, Nicholas Trbojevich.

During these same days, Cornels's overseer, D. E. Foxworth, an assistant director of the FBI, assured those concerned that "this matter would be properly handled," that Tesla's "nephew, who is his heir," would not be able to send the papers to "the Axis Powers."[15] On the eleventh, yet another FBI agent, T. J. Donegan, brought up the possibility that the New York district attorney could have Kosanovic and Swezey picked up "discretely on a burglary charge." However, this was not done. Three days later, Donegan notified Hoover that the situation "was being handled as an enemy custodian matter and therefore we should take no further action."[16]

OFFICE OF ALIEN PROPERTY

What happened appears to be this: The FBI attempted to remove themselves from the responsibility in the Tesla case, thereby allowing the OAP to take charge. Nevertheless, because of the FBI's initial involvement, numerous people contacted them through the years in attempts to gain access to the Tesla estate. The OAP questioned the legality of their own jurisdiction, since Tesla was a naturalized citizen. However, as Kosanovic was probably legally entitled to his uncle's estate, the OAP had justification in considering the material alien property. According to Irving Jurow, who was the attorney assigned to the Tesla case at the time of his death, "the activities of the OAP were not only not 'illegal,' it was the only government agency with statutory power to seize 'enemy assets' without court order."[17] Because of this unique jurisdiction, it was the OAP and only the OAP that maintained legal control over the papers until they were released ten years later. Naturally, both real and imagined concerns involving the war situation were the key factors influencing what to the uninformed appeared to be an illegal action. The Germans still controlled a large part of Europe, and the outcome of the war was by no means determined in January 1943. Rumors that the enemy was also developing an ultimate weapon were also well founded.[18]

Walter Gorsuch, alien property custodian, thereupon ordered all of Tesla's belongings, including the safe from his apartment and other holdings in the basement of the New Yorker, shipped to where Tesla's other possessions were, Manhattan Storage. Gorsuch, however, was out of the office the day Tesla's body was discovered, and it was the young attorney Irving Jurow who handled the case.

"At about noon, on Saturday, January 9th," Jurow recalled, fifty years later, "I was ordered by telephone from the Washington office not to close up shop, but to wait for further instructions. I was informed that a Nikola Tesla had just died, that he was reputed to have invented, and in the possession of a 'death ray,' a significant military device capable of destroying incoming war planes [presumably Japanese on the West Coast] by 'projecting' a beam into the skies and creating a 'field of energy' which would cause the planes to 'disintegrate.' Moreover, it was suspected that German [enemy] agents were in 'hot pursuit' to locate the device or the plans for its production."

With orders to impound all of Tesla's belongings, Jurow was also instructed to "visit other hotels where Tesla had resided and to take similar action." Jurow was accompanied by four individuals from the Office of Naval Intelligence, army intelligence, and the FBI. Arriving at the Hotel New Yorker, "We learned that Tesla had been found dead by the service maid. We were told that he was laid out in his bed...with only a pair of stockings on." The officials were also told that Kosanovic had been in the room and had removed three photographs.

The military officers were "concerned about the death ray model, but I was the only one who had authority." Taking a taxi, Jurow and the others visited each of the hotels, which included the St. Regis, the Waldorf-Astoria, and the Governor Clinton, and they also visited Manhattan Storage. Tesla's possessions were impounded there, and the safe-deposit box at the Governor Clinton was also impounded.

With Walter Gorsuch, Jurow went to visit Ambassador Kosanovic at his hotel on Central Park South, where they apparently also met Nikola Trbojevich, Tesla's other nephew, and an elderly lady who did not speak English. Gorsuch and Jurow saw the photos on a table and left. "I was told later," Jurow recalled, "probably through the staff of the OAP, that Tesla's trunks contained mostly newspapers and bird seed, and that the safe deposit box contained a model of some type of device, whether the 'death ray' or not, is not clear. It was also rumored that the Soviet Union had offered Tesla $50 million to come to the USSR and work on his 'death ray' but he refused."

As Jurow had never heard of Tesla before January 8, 1943, he saw the inventor as a "deadbeat" because he did not pay his hotel bills. "He may have been 'disturbed' because he spent so much time feeding pigeons," Jurow said. But the story was too strange and too incomplete for him, so he called the Westinghouse people to try to verify who Tesla was. "They were ecstatic," Jurow recalled. "They said that without Tesla there would have been no Westinghouse."[19]

The Manhattan Storage inventory did not mention the birdseed, which played so prominently in Jurow's memory. Possessions listed included "12 locked metal boxes, 1 steel cabinet, 35 metal cans, 5 barrels and 8 trunks." Gorsuch also ordered the "large hotel box at the Hotel Governor Clinton held for over 10 years as security for unpaid bills sealed."[20] Jack O'Neill's papers were also confiscated,[21] although they were probably returned to him, for he was able to publish his extensive biography a year later.

Although Kosanovic assured O'Neill that "there was no reason to worry" and that the OAP "conveyed full rights" to the Tesla papers to him; in fact, Kosanovic was highly concerned. He hired Philip Wittenberg, from Wittenberg, Carrington & Farnsworth, to protect his interest. Although the lawyer pleaded the case, the government countered with advice from the War Policies Unit of the Department of Justice. They ruled that Kosanovic could not touch the estate. This edict was maintained throughout the 1940s. Tesla's secret weaponry papers were scrutinized by various divisions of the military, although the nephew was given the combination to the safe and took care of the fifteen dollars-per-month rent for the storage of the property during the entire period.

Within a week of Tesla's death, Walter Gorsuch met with his Washington representative, Joseph King, and together with H. B. Ritchen of the Antitrust Division of the Department of Justice, they called in Colonel Parrott of Military Intelligence and "Bloyce Fitzgerald of the U.S. Army," whom they considered "a former employee of Tesla's." One key problem discussed was that Tesla was "supposed to have been working for, and in the pay of the Yugoslav government-in-exile." Fitzgerald also discussed the supposed Tesla model held in a vault at the Governor Clinton Hotel.[22]

It was determined that before Tesla's estate could be released to Kosanovic, a thorough probe of its contents should be undertaken. Prof. John O. Trump, director and founder of MIT's High Voltage Research Laboratory and secretary of the Microwave Committee at the National Defense Research Committee (NDRC) of the Office of Scientific Research and Development, was commissioned to go to the warehouse and conduct the investigation of the contents of the eighty-eight–odd trunks. They were held in rooms 5J and 5L. Trump set aside two days for the task. He was aided in the search by an inventory of the Tesla holdings compiled by Mr. O'Sullivan, one of the guards at Manhattan Storage.

Trump was accompanied by five individuals: two members of the OAP—John Newinton, from the New York office, and Charles Hedetneimi, chief investigator from Washington—and three from Naval Intelligence—Willis George, a civilian agent; John Corbett, who served as stenographer; and Edward Palmer, who took photographs and probably microfilm copies. Both Corbett and Palmer were also listed as chief yeomen of the U.S. Marine Reserves.[23]

As the only qualified scientist able to comprehend the work, Trump spent little more than half this time actually perusing the wizard's cache. "The second day was somewhat cursory in character," Hedetneimi reported reluctantly, "since Dr. Trump was confident that nothing valuable would be found. He was entirely convinced that it would be useless to look in the 29-odd trunks which...had...been stored since 1933."[24]

The Trump papers, which included a synopsis of about one dozen articles by or about Tesla, began with an opening letter. The professor acknowledged that he and his colleagues investigated the Tesla trunks at Manhattan Storage on January 26 and 27, summarizing first that (1) "no investigation of the Tesla trunks held for 10 years in the basement of the Hotel New Yorker was conducted"; (2) "no scientific notes, descriptions of hitherto unrevealed methods or devices or actual apparatus...of scientific value to this country or which would constitute a hazard in unfriendly hands [was found]....I can therefore see no technical or military reason why further custody of the property should be retained." Nevertheless, Trump "removed...a file of various written materials which covers typically and fairly completely the ideas which he [Tesla] was concerned [with] during the later years" and forwarded it or copies of it to Mr. Gorsuch of the OAP.

Trump concluded in his report that the last fifteen years of Tesla's life were "primarily of a speculative, philosophical and somewhat promotional character."

On his return to Washington, Trump met with Homer Jones, chief of the Division of Investigation and Research. "Sir," the MIT professor smugly concluded, "upon the basis of my examination, it is my opinion that the Tesla papers contain nothing of value for the war effort, and nothing which would be helpful to the enemy if it fell into enemy hands."

"Are you quite certain in this conclusion, Dr. Trump?"

"I am willing to stake my professional reputation on it."[25]

Satisfied, Jones sent the report and Trump's recommendations to

Lawrence M. C. Smith, chief, Special War Policies Unit of the War Division of the Department of Justice, and that, for one faction of the government, ended the matter.[26]

Trump drew up a report which described a number of articles by the inventor, interviews, and scientific treatises. Exhibits D, F, and Q refer to a highly technical and heretofore underground Tesla treatise written in 1937 entitled *The New Art of Projecting Concentrated Non-dispersive Energy through the Natural Media.* This article, in contradiction to Trump's statement, contained explicit information which had never been published describing the actual workings of a particle-beam weapon for destroying tanks and planes and for igniting explosives. Novel features included (1) an open-ended vacuum tube sealed with a gas jet "while at the same time, permitting and facilitating the exit of the particles"; (2) a way to generate many millions of volts for charging minute particles; (3) a method of creating and directing a nondispersive stream of such particles with a trajectory of many miles.

Written virtually as a patent application, the Tesla article presents in clear and straightforward terms the mathematical equations and schematics of his death ray. Aside from the unpublished drawing and mathematical analysis of its capability, it employed three most unusual features. The first was its mechanism for creating a nondispersive beam of particles. "I perfected means for increasing enormously the intensity of the effects, but was baffled in all my efforts to materially reduce dispersion and became fully convinced that this handicap could only be overcome by conveying the power through the medium of small particles projected, at prodigious velocity, from the transmitter. Electrostatic repulsion was the only means to this end.... Since the cross section of the carriers might be reduced to almost microscopic dimensions, an immense concentration of energy, irrespective of distance, could be attained."

The second feature involved the creation of an open-ended vacuum tube by replacing the walled enclosure or glass window with a "gaseous jet of high velocity"; and the third outstanding feature was the means for generating large voltages. Having studied the precursors in the Van de Graaff electrostatic generator (a device which Tesla said was all but useless for generating usable amounts of energy), Tesla replaced the circulating cardboard belt that transferred the charge with an ionized stream of air hermetically sealed in a 220-foot-long circular vacuum chamber. Analogous to the way a shock can be created and transferred by rubbing one's shoes along a carpet on a dry day, the new fluid airstream belt achieved the same end but to a degree "many times greater than a belt generator." This charge, which apparently could be as much as 60 million volts, was in turn transferred to the myriad small bulbs at the top of the tower, their round shape and internal structure constructed to augment the accumulation of energy.

Atop this domed citadel, which was planned to be over a hundred feet in height, was the particle-beam weapon. Nestled in a turret as a supergun, the weapon was set up so that tungsten wire could be fed into its high-vacuum firing chamber. There minute "droplets" of this metal would be sheared off and repelled out the long barrel at velocities exceeding 400,000 feet per second.[27] The entire apparatus apparently was also constructed for nonmilitary purposes, such as for transmitting streams of electrical energy to distant places, much like microwave wireless telephone trunk lines do today.

Although Trump downplayed the importance of this paper, it is, to the present day, classified top secret by the U.S. military, with copies at the time going to naval intelligence, the FBI, the OAP, the NDRC, Wright-Patterson Air Force Base, MIT, and most likely, the White House.

AMTORG TRADING CORPORATION

Exhibits D, F, and Q state explicitly that Tesla sold the plans for the construction of his particle-beam weapon to A. Bartanian, a Soviet agent of the Amtorg Trading Corporation! These exhibits also specify that Tesla offered the device to the U.S. military, Great Britain, and Yugoslavia.[28]

Surprisingly, the FBI did not exploit this blatant Soviet connection, even though this was just the kind of thing that J. Edgar Hoover thrived on. One possible reason was that the Soviet Union was an ally at that time. Furthermore, a number of major corporations, such as Bethlehem Steel, RCA, and Westinghouse, were selling equipment to the Soviets via Amtorg, a company that did over a billion dollars of business in America by the time of World War II. FDR, for instance, in 1933, approved $4 million in credit to Amtorg to purchase cotton from American suppliers. Amtorg, in return, supplied the country with furs, caviar, oil, and precious metals. Still operating in America today, Amtorg was unable "to find any mention of Mr. Tesla [in their] records."[29]

If Tesla really did receive $25,000 from Amtorg in 1935, which the communiqué with the Soviets implied, why wouldn't he have paid off his debts to the Hotels Pennsylvania and Governor Clinton and retrieved his secret device held as collateral? An amount as large as $25,000 at the height of the Great Depression was worth roughly twenty times that figure today, yet there is no indication that Tesla obtained great wealth during that period, although he may have received this amount and used it to pay off other debts and purchase other equipment.

A few days after viewing the estate, Trump went to the Governor Clinton to view the actual death ray held in their vault. Charles Hedetneimi of the OAP, reported that "officers of the hotel showed us the handwritten letter in which Tesla stated that he was leaving the equipment as security and *that it was worth $10,000.*" Trump later recalled the incident: "Tesla had warned the management that this "device" was a secret weapon, and it would detonate if opened by an unauthorized person. Upon opening the vault...the hotel manager and employees promptly left the scene."

The Trump letter went on to describe his reluctance to remove the brown paper covering and that before summoning his courage, he looked outside and noticed that the day was pleasant. "Inside was a handsome wooden chest bound with brass...[containing] a multidecade resistance box of the type used for a Wheatstone bridge resistance measurements—a common standard item found in every electric laboratory before the turn of the century!"[30]

"At this point," Hedetneimi concluded, "Dr. Trump indicated that he had no further interest in the case."

An FBI report written just two weeks prior to Trump's visit to the hotel described the managers' assessment of the inventor somewhat differently, and it does not appear that they took him as seriously as Trump alleges. "The Hotel managers report he [Tesla] was very eccentric, if not mentally deranged during the past ten years and it is doubtful if he has created anything of value during that time, although prior to that he probably was a very brilliant inventor."[31]

DID TESLA ACTUALLY BUILD A PROTOTYPE OF THE DEATH RAY?

The inventor appears to have told both Hausler, his pigeon caretaker, and Fitzgerald, the army engineer, that he did build a working model; and in an interview I had with Mrs. Czito, whose husband's father and grandfather were both trusted Tesla employees, she recalls that her father-in-law used to recount stories of Tesla bouncing electronic beams off the moon. This is not a death ray, but it certainly supports the hypothesis that the inventor created working models along these lines.

The well-known columnist Joseph Alsop, who interviewed Tesla at the Hotel New Yorker and who was one of the first to fully report Tesla's work with particle-beam weapons, described an experience Tesla had when experimenting with cathode-ray tubes, "Sometimes a particle larger than an electron, but still very tiny, would break off from the cathode, pass out of the tube and hit him. He said he could feel a sharp, stinging pain where it entered his body, and again at the place where it passed out. The particles in the beam of force, [i.e., the] ammunition...will travel far faster."[32]

Other evidence has been reported by Corum and Corum, who assert that this invention was an outcropping of Tesla's work on X-ray machines patented in the mid-1890s. Thus, we can trace the Tesla death ray to at least three earlier inventions, his Tesla coil and work in high-frequency currents from the early 1890s, his work in bombarding targets with Roentgen rays in 1896, and also his 1901 ideas associated with transmitting energy by means of wireless by beaming up an ionizing ray from his magnifying transmitter and using it is a conduit to reach the ionosphere. With this mechanism, Tesla planned not only to circle the globe with information but also to illuminate shipping lanes over the oceans and control the weather.

The deBobula schematic, the Corums write, having studied the plans in Belgrade, was also taken to Alcoa Aluminum, which said that it was ready to supply the materials as soon as Tesla raised the necessary capital.[33] Alcoa, however, was unable to locate any reference to Tesla in their corporate records.[34]

At the age of eighty-one, Tesla stated at a luncheon attended by ministers of Yugoslavia and Czechoslovakia that he had constructed a number of beam-transmission devices, including the death ray, for protecting a country from incoming invasions and a laserlike machine that could send impulses to the moon and other planets (in Exhibit I). He also said that he was going to take the death ray to a Geneva conference for world peace. When pressed by the columnists to "give a full description... Dr. Tesla said, 'But it is not an experiment.... I have built, demonstrated and used it. Only a little time will pass before I can give it to the world.'"[35]

Considering that Tesla had two secret laboratories throughout the 1930s in which no reporter ever set foot,[36] we are left with a mystery. Did Tesla really "scam" the hotel by frightening the management into accepting a bogus invention in lieu of $400 rent? Feeling that the world owed him a place in which to live, perhaps he purposely chose not to pay for his housing. This certainly was a compulsive and self-destructive pattern during the latter quarter of his life. Lesser mortals often fore the brunt of his repressed rage. He was known to be abrasive to maids and office secretaries. Having made it a habit to live on credit, Tesla may have derived great pleasure during those lugubrious nights when he was forced to face his failures in thinking back to the chaps at the Hotel Governor Clinton trembling at the terrible weapon they kept protected in their midst.

THE CONSPIRACY SCENARIO

Secret agents break into Tesla's New Yorker Hotel safe without Kasanovic knowing, remove keys to his Hotel Governor Clinton vault, and steal the death-ray prototype, substituting the equipment Trump found a week or two later. This potential incident would have had to have taken place between January 9 and January 29, 1943, the dates of Tesla's death, and the end of Trump's investigations. The agents who performed this task, if it occurred, could have been Bloyce Fitzgerald and Ralph E. Doty. The evidence is as follows:

Sava Kosavonic's secretary at this time was Charlotte Muzar. She reports that she saw Tesla during his last days for the purpose of delivering funds that he required and that she was also present at the opening of the safe after Tesla died. Present were Kenneth Swezey, Sava Kosanovic, and George Clark. According to her story and official reports, a locksmith was called in to change the combination of the safe and give the new combination to Kosanovic, who was the only one who had it. In the safe, before he locked it again, was a set of keys and the 1917 Edison Medal. About ten years later, when the estate was finally shipped to Belgrade and the safe opened, the Edison Medal and the keys were found to be missing. The medal was never recovered, but the keys were found outside the safe "in one of the numerous cases of documents."[37]

A January 12, 1943, OAP memorandum states that Charles McNamara, assistant manager of the Hotel Governor Clinton, "permitted [the OAP] to seal safety deposit box #103, which contained the $10,000 machine.... Box #103 is *not* a specially constructed box as Fitzgerald said. It is, however, one of the largest boxes of the lower tiers." The day before this, at the Hotel New Yorker, Tesla's bedside safe was opened. Present, according to another OAP memorandum, aside from Swezey, Clark, and Kosanovic, were two hotel personnel from the New Yorker, Mr. L. O. Doty, credit manager, and Mr. L. A. Fitzgerald, assistant credit manager.

It strikes this researcher as a rather odd coincidence that these two individuals carried the same last names as a colonel from military intelligence and the ever-present Bloyce Fitzgerald.

If, in fact, these two supposed credit managers were really government agents, it would have been a fairly easy task for them to retrieve the key (or make a copy) to box no. 103 and swap the device.

A further investigation of the FBI files reveals that on October 17, 1945, E. E. Conroy from the New York office, sent two copies of the Trump report to J. Edgar Hoover and reviewed with him, once again, the roles played by Fitzgerald, Spanel, a censored person "X" and Kosanovic. Conroy said that "X" (probably another FBI agent) suspected that Spanel was "definitely pro-Russian in attitude" and that Spanel was spreading pro-Communist propaganda in full-page ads in various newspapers and yet also suing these newspapers for libel. Conroy also reiterated that Spanel had ties to Vice President Henry Wallace, so caution was advised.

It appears that Spanel had met Fitzgerald (a friend of "X") at an engineering meeting in November 1942. Fitzgerald at the time was an army private at Wright Field, Dayton, Ohio. He was described in the FBI report as "a brilliant 20-year-old scientist who spent endless hours with Tesla before his death.... Fitzgerald had developed some sort of anti-tank gun." Spanel tried to form a partnership with Fitzgerald in order to sell this weapon to the Remington Arms Company, but for some reason "Spanel blocked the final sale" and then tried to arrange a more lucrative deal with Eiogens Ship Building Company of New Orleans.

In November 1943, Eiogens fired Fitzgerald, and a year later the young engineer returned to the army. "Today, [1945, Fitzgerald is]...engaged in a highly secret experimental project at Wright Field...In spite of his rank of private, Fitzgerald is actually director of this research and is working with many top young scientists...on perfection of Tesla's 'death ray' which in Fitzgerald's opinion is the only defense against offensive use by another nation of the atom bomb."[38]

Conroy suggested cooperating with Fitzgerald in order to secure "legal possession of Tesla's effects." The goal, of course, was to obtain and protect the details of the weapons system, yet also to set up a "memorial foundation...for the preservation of the inventor's memory." Fitzgerald purportedly also interested Henry Ford in the project.

On October 19, 1945, Brig. Gen. L. C. Craigee, Chief Engineering Division Control Equipment Branch, Wright Field, writing at the request of Bloyce Fitzgerald, David Pratt, Herbert Schutt, and P. E. Houle, all engineers working at Wright Field, contacted Harvey Ross of the FBI in New York in order to officially "request...in the interest of National Defense, access...to the effects of Dr. Nicola Tesla held in Manhattan Storage Warehouse." Col. Ralph Doty, from military intelligence in Washington, followed up the inquiry by working as a liaison officer between the War Department, the OAP, and the FBI.[39]

Since the FBI had no jurisdiction over the Tesla estate, Fitzgerald, Conroy, and Craigee were referred to the OAP.

On September 5, 1945, Lloyd Shaulis, of the OAP, mailed off two copies of the Trump report to Colonel Holliday of the Equipment Laboratory, Propulsions and Accessories Subdivision, who no doubt forwarded them to Fitzgerald. "These were the full photostatic copies, not merely the abstracts."[40] Two years later, Colonel Duffy of Air Material Command, Wright Field, wrote to the OAP that the papers were still being evaluated.

By 1950, Kosanovic was still barred from the warehouse.[41] He was now officially the Yugoslavian ambassador to both the United States and the United Nations, and his patience had reached its limit. Kosanovic wanted his uncle's effects in Belgrade, where they would be rightly honored. In March he went to the warehouse to inform them of the inventor's wishes to send the estate to the Tesla Museum. It was at this time that the ambassador was informed that the FBI had microfilmed the entire contents. He called J. Edgar Hoover to request a copy of the microfilm, but Hoover said that they did not have such a copy. Kosanovic had probably been misinformed; one of the warehouse people may have mistaken the Trump people for the FBI; or another group microfilmed the papers (now held in the Library of Congress) at a different time.

Finally, in 1952, arrangements were made, and the entire eighty trunks were shipped to Belgrade. Included were many valuable original papers completely unknown to O'Neill, such as Tesla's 1899 Colorado diary, various photographs, tens of thousands of letters, and most of his inventions, including the remote-controlled boat, wireless fluorescent lamp, motors, turbines, plans for his vertical takeoff flivver plane, and a model of the magnifying transmitter. Tesla's ashes were sent at a later date.

Spanel would be questioned by Joseph MaCarthy, and the copies of the death ray papers moved further underground.

THE PROBABLE TRAIL OF THE SECRET PAPER

In 1984, Andrija Puharich presented Tesla's secret death-ray paper to the International Tesla Centennial Symposium, held in Colorado Springs. It was published in the proceedings. Puharich told this author that the original source of the report was Ralph Bergstresser, author of a notable Tesla article published in 1957. Puharich thought that Bergstresser, along with an associate (who may have been Bloyce Fitzgerald), was linked to the FBI and thus obtained the paper in that capacity. In the 1940s, Bergstresser worked for RCA and "Marconi's boys. I always believed that Marconi was the inventor of the wireless.... Then I found out it was all a lie." Bergstresser, who was about eighty years old when I spoke to him, recalled that Marconi had lived on a ship on the high seas to avoid subpoenas Tesla tried to give him.

At the outbreak of World War II, Bergstresser began to work on "behalf of the war effort... [when] he [Tesla] turned over to me [his various papers]. I would take and read them and return them to him." According to one source, Bergstresser was under the command of the new secret organization the Office of Strategic Services (OSS), (later the Central Intelligence Agency [CIA]), and in that capacity he was analyzing the papers for their military significance.

Bergstresser recounted that he had only known Tesla during the last six months of his life. "He was skinny, tall, stooped, emaciated—didn't eat right." In further questioning, he also said that he had known Jack O'Neill and his colleague William Lawrence, author of the 1940 *New York Times* death-ray article, which sparked such interest by the FBI and factions of the armed forces, and that he and Bergstresser had attended Tesla's funeral together in 1943.

During the conversation, I asked Bergstresser if he had any proof to support his contention that Tesla's papers had been systematically removed from libraries. He said that the conspiracy was massive and extremely complex, going back to J. Pierpont Morgan and his wish to suppress Tesla's wireless power distribution inventions because they threatened to provide cheap or free power for the masses. He was upset that Tesla's entire estate was "hauled away...behind the Iron Curtain" and partially blamed Lawrence, who he said was later found to be a Communist. He agreed that Bloyce Fitzgerald had probably taken the papers to Wright-Patterson Air Force Base, but he declined to comment on any relationship to Fitzgerald or the trail of the secret papers.

Puharich had said that the particle beam article had been passed from Bergstresser to Bob Beck of the U.S. Psychotronics Society around 1981 and from there they were passed to him.[42]

INFLUENCE OF THE McCARTHY PERIOD

It was true that by the time the FBI and federal government became actively involved in suppressing Tesla's scientific papers, the inventor had already slipped from the mass consciousness. Nevertheless, an air of secrecy, supported by the taint of ties to a German fifth column or the Communists via his Yugoslav heritage, served to submerge even further the inventor's work. This was augmented by the transfer of Tesla's papers to the remote city of Belgrade.

STAR WARS

Tesla's numerous inventions could be applied in a variety of ways for military purposes, for example, particle-beam weapons, worldwide radar, earthquake contrivances, brain-wave manipulation. One or more magnifying transmitters could supposedly send destructive impulses through the earth to any location. Thus, a well-placed jolt of many millions of volts could theoretically destroy the communications network of any major city. Recent discourses on potential future warfare technologies stemming mainly from war games analyst Lt. Col. Tom Bearden and parapsychologist Andrija Puharich, M.D., suggest that the Soviets had harnessed various Tesla weapons, including apparatus for seismic, weather, and mind control.[43]

According to Bearden, the Tesla magnifying transmitter produced a fundamental one-point gravity vector (or electrostatic scalar wave) that disturbs the very fabric of the space/time grid itself and therefore is not bound by the speed of light. Thus, a theoretical instantaneous Tesla wave, emitted from the magnifying transmitter, could potentially affect the planet's geomagnetic pulse and thereby be directed to any number of targets on any continent.[44]

This research is highly controversial and speculative, and should be read with caution. Nevertheless, in 1977, in its May 2 issue, *Aviation Week* published a seven-thousand-word article on Soviet particle-beam weapons. The exposé, which "shook Washington," was also abstracted in *Science.* It contained a schematic drawing of a particle beam weapon which bears a remarkable resemblance to Tesla's then unpublished drawings made four decades earlier. Coupled with the realization that the Soviets were well advanced in this area, this is strong evidence in support of the claim that Tesla did, in fact, sell the schematics of such a device to them in the mid-1930's.

Aviation Week also described the use of "young geniuses under [the age of] 29...located at the Wright-Patterson Air Force Base in Dayton Ohio" who were trying to conceive of a breakthrough in the technology; and also, surprisingly, that "the President [Jimmy Carter] was screened from vital technical developments by the bureaucracy of the CIA and Defense Intelligence Agency." The source was General George Keegen, former head of Air Force Intelligence.[45]

Three intriguing points emerge: (1) the concept of intense secrecy of particle-beam weapons; (2) the mention of Wright-Patterson Air Force Base; and (3) the policy of utilizing bright young geniuses. All of these variables are evident in the FBI files on Tesla reviewed earlier. Great support is lent to the conspiracy hypothesis that Tesla's work and papers were systematically hidden from public view in order to protect the trail of this top-secret research, which today is known as Star Wars.

Particle-beam weaponry is still more a dream than a reality, at this time in 1996. Therefore, if indeed the inner sanctum of one or both superpowers has access to Tesla's plans, why is it that no death ray has ever been constructed? Perhaps there are prototypes, but it seems to me that they should have been tried out in the field during such wars as Vietnam, Afghanistan, or Kuwait/Iraq. This part of the story must remain a mystery.

48

THE WIZARD'S LEGACY

When a man looks back over the events associated with his work...he begins to realize how minor is that part which he himself has played in shaping the events of his career, how overpowering the part played by circumstances utterly beyond his control.

EDWIN ARMSTRONG ACCEPTING THE EDISON MEDAL[1]

As an echo from the Tesla-Morgan affair, Edwin Armstrong had to battle the giant RCA to get FM accepted over AM radio, even though it was a superior means for transmitting music. Col. David Sarnoff, "Napoleonic" head of RCA and NBC, had simply pirated his friend's FM invention *because he required it for use in television*. With the help of CBS, Sarnoff also manipulated the Federal Communications Commission (FCC) to create a stipulation that severely restricted the power of FM transmitters "to one tenth their intended level...and shifted the FM band to a less desirable frequency range." That is why AM became the dominant band for long-distance radio transmission.

Required to rebuild his own radio station because of the FCC ruling and forced to battle the media Goliath in the courts for patent infringement, Armstrong cashed in his remaining block of RCA stock to finance the enormous expenditure that would be required. Lawyers for NBC were able to persuade the high court in New York that it was their engineers that invented FM and not Armstrong! Although the Institute of Radio Engineers as a group filed a formal protest (an unprecedented move in such situations), which eventually caused the ruling to be reversed by the Supreme Court, it was too late for Armstrong. With his marriage in shambles, his key patents expired, and his fortune drained by litigation, he jumped out a thirteen-story window in the dead of winter of 1954, spiritually trammeled by the very people who profited from his creative endeavors. That Tesla survived such a fate is a testament to his fortitude and transcendent nature.[2]

In 1956 there was a celebration of the centenary of Tesla's birth. To highlight the event, with Nobel laureate Niels Bohr as a speaker, a centennial congress was held in Tesla's honor. Simultaneously, the Yugoslav Postal Service issued a commemorative Tesla stamp, and the Yugoslav government placed Tesla on the 100 dinar note (equivalent to a U.S. dollar). Statues of the inventor were placed at museums in Zagreb and Vienna, a school was named after him in Illinois, a Tesla day was proclaimed in Chicago, and in Munich the Institute Electrotechnical Committee agreed to adopt the name "tesla" as the unit of magnetic flux density. Now Tesla could take his rightful place beside such other luminaries as Ampère, Faraday, Volta, and Watt.

Twenty years later, as a gift from the Yugoslavian people, in 1976 a statue of Tesla by sculptor Franco Krsinic, was placed at Niagara Falls and an identical companion statue erected on the village square in Gospić, Croatia, where Tesla was raised as a boy. (Unfortunately, this Gospić statue was purposely blown apart during the war in 1993.) President Tito gave a speech in Tesla's honor in Smiljan before thousands of Croats, Serbs, and Bosnians (who were separated by armed guards and demarcation fences), and the celebration also continued in America. Plaques were posted at the site of his residence at the Hotel Gerlach in New York City and at the Wardenclyffe laboratory, which still stands at Shoreham, Long Island.

In 1983 the U.S. Post Office honored Tesla, along with Charles Steinmetz, Philo Farnsworth, and, alas, Edwin Armstrong, in a block-of-four commemorative stamp. In Boston a giant Tesla coil can be found at the Museum of Science, and Tesla's picture can be found hanging in the Smithsonian Institution in Washington, D.C. There are also two major organizations in his honor, the Tesla Memorial Society, in Lackawanna, New York, and the International Tesla Society, which has held conferences every two years since 1984 at the society's headquarters in Colorado Springs.

William Whyte wrote in *The Organization Man* that as beneficial as the large corporations can be, they are also static, delusory, and self-destructive. During the inevitable conflict between the individual and society, the organization man is caught in a bind because the company provides a livelihood, but at the expense of the worker's individuality. This is what Whyte calls a "mutual deception": "It is obvious to fight tyranny; it is not easy to fight benevolence, and few things are more calculated to rob the individual of his defenses than the idea that his interests and those of society can be wholly compatible.... One who lets the Organization be the judge ultimately sacrifices himself."[3]

The corporate view becomes the very embodiment of rationality itself, structuring, restructuring, and thereby controlling consciousness.[1] As was the case with the Tesla situation, the more the corporate world rejected the Wardenclyffe idea, the more unworkable it appeared to the engineers working in the industry, because they, as products and extensions of the corporations, had their consciousness shaped by its policies. Ultimately, Tesla's worldview became a threat, and it was easier to dismiss him as an eccentric than consider that his plans may have been viable.

A modern example of a famous innovator who became somewhat of a nonperson is Steven Jobs, cofounder of Apple Computer. Realizing, in the mid-1980s, that the Motorola microprocessor was superior for graphics capabilities to the one in the Apple II computer and the one being used by Bill Gates/Microsoft for IBM, Jobs produced the Macintosh. First-generation Macintoshes, which were admittedly unattractive for business use and constructed so that they were not expandable, were not immediate successes in the marketplace. The inferior IBM disk operating system (DOS) was, of course, the standard. Undeterred, Jobs wanted to scrap the obsolete but highly profitable Apple II to go solely into production of Macs. Even though he was the wunderkind cocreator of the billion-dollar enterprise, Jobs became a threat to its financial stability. In a stunning move, Jobs was not only deposed; he was literally barred from working at Apple, even though he was still the largest shareholder! A decade later, the Bill Gates/IBM-compatible (Intel/Pentium) chip, although endowed with graphics (Windows) capabilities, is still based on an inferior design as compared to the Power-Mac. Nevertheless, the Gates/Intel chip is by far the leading system in the country, even though the Power-Mac by necessity is the standard in such fields as graphic design. Coyly, Gates said it this way: "People were coming out with completely new operating systems, but we had already captured the volume, so we could price it low and keep selling....[And] believe me, it would have been a lot easier to write Windows so it didn't run DOS applications. But we knew we couldn't make the transition without that compatibility."[5]

PSYCHOANALYSIS

Being aware of the criticisms of the Freudian paradigm and the problem of oversimplification, I argued in my doctoral dissertation that in order to explain Tesla's unusual personality, self-proclaimed celibacy, and alleged homosexuality, it is possible that he may have suffered repressed guilt feelings associated with the untimely death of his older brother Dane when Tesla was five years old. In the throes of the Oedipal complex and admittedly overattached to his mother, young Niko experienced great trauma not only because Dane was Djouka's favorite son but also because Niko was at that age of gaining his sexual identity and learning to transfer love bestowed upon himself to others. Possibly the mother rejected Tesla after Dane's death, and thus Tesla bestowed the love back upon himself and became narcissistic. Wishing to gain back the perceived loss of love from his mother and lost brother, later in life Tesla would unconsciously seek out figures that would combine the dynamics of "older brother/mother surrogates," for example, strong, nurturing authority figures, such as Westinghouse and Pierpont Morgan. This combination of brother and mother might also explain the confusion associated with Tesla's sexual identity.

To recapture their lost love and, in the symbolic sense, to bring the brother back from the dead (for that would be the only way to repair the damage), Tesla would have to form a sacrifice as penance. In Westinghouse's case, he ripped up the royalty clause, and it cost him millions of dollars (he could have set up a deferred payment schedule);[6] and in Morgan's case, the inventor insisted that the financier take the larger share of the partnership, that is, 51 percent, even when Morgan suggested fifty-fifty.

Due, however, to a multitude of personality factors, including egomania, overambition, and his impatient wish to crush the competition, Tesla breached his contract with Morgan by deciding to build a larger tower than was agreed upon. This was a self-destructive move (although it might have succeeded and was also, on another level, simply a calculated risk). From the psychoanalytic standpoint, unconsciously he was hoping that the financier would forgive his sins (show his surrogate son that he still loved him) by providing him with the additional funds to complete the tower. When Morgan said no, the unconscious could not face such a harsh rejection, so Tesla tried obdurately to turn the man around; and even when it was clear that Morgan would never come through, the self-perceived surrogate son continued to try.

Turning to Tesla's persistent wish to contact extraterrestrials, from a psychoanalytic perspective these outer space entities may have symbolized beings existing in the afterworld. Certainly the need to believe in extraterrestrials is a powerful and popular one. It explains why so many people accepted Percival Lowell's "Canals of Mars" hypothesis and, in today's world, the extreme popularity of such movies as *Star Wars, Star Trek,* and *ET.* In Tesla's case the extraterrestrials may have prelogically stood for his dead brother and mother. The insistence that he had probably been contacted by Martians became an unconscious safety valve which allowed him to hypercathect (release) much of the anxiety associated with the death of his older brother, as the brother would still be, in a sense, alive. Thus, if Dane still existed, the trauma of his death would be diminished, and Niko's mother would love *him* again.

This form of regressive behavior could also explain Tesla's obsession with caring for pigeons. After Dane died, the family was wrested from their idyllic farm in Smiljan to the bustling city of Gospić. The pigeons were not only substitute mistresses for Tesla; they also symbolized a return to the utopia of his early and untroubled childhood in Smiljan.

As a counterhypothesis to the psychoanalytic paradigm (and as no solid evidence of homosexuality has been discovered by this researcher), one must place Tesla within his time period. Like William James and other intellects of that day, the idea of dedicating oneself to science at the expense of marriage was not a unique occurrence. Tesla was keenly aware that the responsibilities of marriage would have interfered too greatly with his inventive élan. Purposely, and admittedly through self-denial, he transformed his instincts in the alchemical sense to raise them to a higher level. This view, however, does not completely explain the natural proclivity to partake of the passions, especially when one considers that at his height, in the Gay Nineties, the rising star could probably have had his pick of any number of willing females, for example, Marguerite Merrington. Nor does it explain his overattachment to the city pigeons.

Nevertheless, the Freudian paradigm falls short in its attempts to explain the nature of Tesla's wizardry in that it tends to see this ability as a sublimation rather than an end in itself. Tesla's emphasis on ritual and such obsessions as cleanliness and self-denial were just as much linked to his childhood bouts with cholera, caused by impure drinking water, as to his wish to change his state of consciousness through a set routine so that he could prepare his mind to do its work. Unlike most inventors, Tesla's creativity did not just lie in one plane. He adapted his mental faculties to a number of separate fields, designing fundamental inventions in lighting, electrical power distribution, mechanical contrivances, particle-beam weaponry, aerodynamics, and artificial intelligence. This great versatility of achievement places the inventor in a category all his own. Ultimately, Tesla was a journeyman searching for the Holy Grail. His goal was nothing short of altering the very direction of the human species through extensions of *his* effort.

CULT FIGURE

> It was the coil that I noticed first—because I had seen drawings like it years ago.... "Hank, do you understand? Those men, long ago tried to invent a motor that would draw static electricity from the atmosphere, convert it and create its own power as it went along. They couldn't do it. They gave it up." She pointed at the broken shape. "But there is it."
>
> *Atlas Shrugged* by Ayn Rand[7]

At 8:00 P.M. on June 20, 1957, in the ballroom of the Hotel Diplomat in New York City, the *Interplanetary Sessions Newsletter* announced a meeting to coordinate an expected visit by the "Space People" to the planet Earth. The event was planned by three individuals: George Van Tassel, author of *I Rode a Flying Saucer;* George King, purported telepathic contactee with extraterrestrials, and Margaret Storm, author of the occult Tesla biography *Return of the Dove,* a book whose "transcripts were received on the Tesla set, a radio-type machine invented by Tesla in 1938 for interplanetary communication." By July 1 it was assured that the "Martians" would have "full scale operations" in Washington, D.C., New York, and "general North American areas." It was also revealed that "Tesla was a Venusian, brought to this planet as a baby in 1856, and left in a remote mountain province in what is now Yogoslavia [*sic*]."

In attendance at this meeting was a man who preferred to remain unnoticed. He was an FBI agent assigned to continue the expanding file on the enigmatic Serbian inventor Nikola Tesla. It is quite likely that a copy of this newsletter, which is in the FBI file, was also read by J. Edgar Hoover, a man concerned about the growing interest in the flying-saucer phenomena and the secrecy surrounding various adherents.

Margaret Storm's supposition that Tesla was born on another planet to give us our entire electric power and mass communications systems stemmed from a colorful history of the inventor's ties to the group-fantasy belief that life on Mars was a virtual certainty. Fueled by McCarthyism and the fear of Communist (alien) infiltration and also theosophical literature, Storm proclaimed that Tesla had descended from the sixth-root race, a new species of human that was evolving on the planet. To complicate matters, she was also friends with Arthur Matthews, author of *Wall of Light: Nikola Tesla and the Venusian Spaceship.* A bizarre electrician who had once written to Tesla in the 1930s, Matthews contended that he and his supposed employer, Tesla, had traveled many times to nearby planets aboard Venusian spacecraft and that Tesla, as late as 1970, was still alive, living as an extraterrestrial.[8]

Relegated to occult status for many decades, Tesla has also been fictionalized as one of any number of mad scientists in science-fiction literature, as part of the composite New Age hero John Galt in Ayn Rand's novel *Atlas Shrugged,* as a source for future technology in James Redfield's 1996 bestseller *The Tenth Insight,* and as the extraterrestrial (played by rock star David Bowie) in the Nicholas Roeg movie *The Man Who Fell to Earth.* As a cult figure in the United States, Tesla has also seen a panegyric resurgence with the younger generation because of a rock band which goes by the same name.

In Japan, however, Tesla's cult status is much more complicated. On the one hand, Dr. Yoshiro NakaMats, the world's leading inventor, whose creations include the floppy disk, is a great proponent of Tesla and has created a celebration in Tesla's honor. On the other hand, Tesla's secret weaponry work has also attracted the attention of one of the most dangerous cults of modern times.

Just one month after the January 1995 earthquake in Kobe destroyed the city and killed 5,000 people, followers of the "charismatic psychopath" Shoko Ashara, the man responsible for poisoning the subways with Sarin gas in Tokyo, flew into Belgrade to infiltrate the Tesla Museum with hopes of obtaining the inventor's schematics for his supposed telegeodynamic earthquake machine. Ashara's cult, known as Aum Supreme Truth, had an Internet of supposedly tens of thousands of members in a half dozen countries with access to state-of-the-art high-tech devices, a large data base of military secrets, firearms, and laser apparatus, and other more esoteric New Age weaponry systems. Seeking world domination, their plans were interrupted by the earthquake. Rather than seeing the event as a natural occurrence, Ashara suggested that the Kobe disaster was caused by electromagnetic experiments conducted by one of the Japanese mega-corporations or by American or possibly Russian military tests of a top-secret Tesla telegeodynamic instrument.

Raised in a country that had seen its cities destroyed by the atom bomb, Ashara had been influenced by apocalyptic science fiction stories, by the book *Tesla Superman* written by Japanese author Masaki Shindo, and by such Teslaphiles as Lt. Col. Tom Bearden, whose model for the Tesla magnifying transmitter hypothesized that it could be used as an intercontinental "electrostatic scalar-wave" delivery system. Realizing that Armageddon was at hand, the Kobe earthquake being proof of the prophecy, Ashara now planned to seize the high ground by constructing his own "Tesla howitzer" while at the same time perfecting the death ray.[9]

Christopher Evans, in his text *Cults of Unreason,* suggests that cults appear as "stop-gaps" for people in society—ways for them to deal with life's mysteries and also for the unsettling feeling associated with the rapid pace of our times.[10] According to Evans, cults exist in order to discover the "Holy Grail," the supposed secret behind the universe. Tesla himself called his magnifying transmitter the "philosopher's stone." To him, it was the mechanism by which to transform society and interlink the entire globe. Following a Goethean path, Tesla's weltanschauung suggested a hierarchy of intelligence to the universe. Not only were his creations derived from natural law, through them, humans could attain godlike status and communicate with other interstellar neighbors.

"According to the idea of esotericism, as applied to history," says Russian philosopher P. D. Ouspensky, "no civilization ever begins of itself."[11] Esoteric schools involve other (higher) dimensions, says astrologer Dane Rudyar. Certain individuals, often referred to as avatars, Rudyar suggests, are actually "seed men" who one way or another have within their being knowledge that can lead a culture into transcendence.

Tesla had possession of certain knowledge which for various reasons was rejected by mainstream science and society or suppressed by powers that perceived his contributions as threatening. Yet the essence of Tesla's work is available for the seeker. Described by popular New Age writer Robert Anton Wilson as an "illuminati," Tesla remains a cult hero because of his esoteric status, because his life's work has served as a template for numerous science-fiction characters and cinematic themes, and because he provides answers for those who study his work for its inner meaning.[12]

Unlike so many other esoteric figures, however, Tesla is in a unique position because so many of his inventions *were* incorporated into our modern high-tech world. Had his ultimate world broadcasting plan actually coalesced during his heyday, there is no telling how history might have proceeded and how the quality of our lives might have changed.

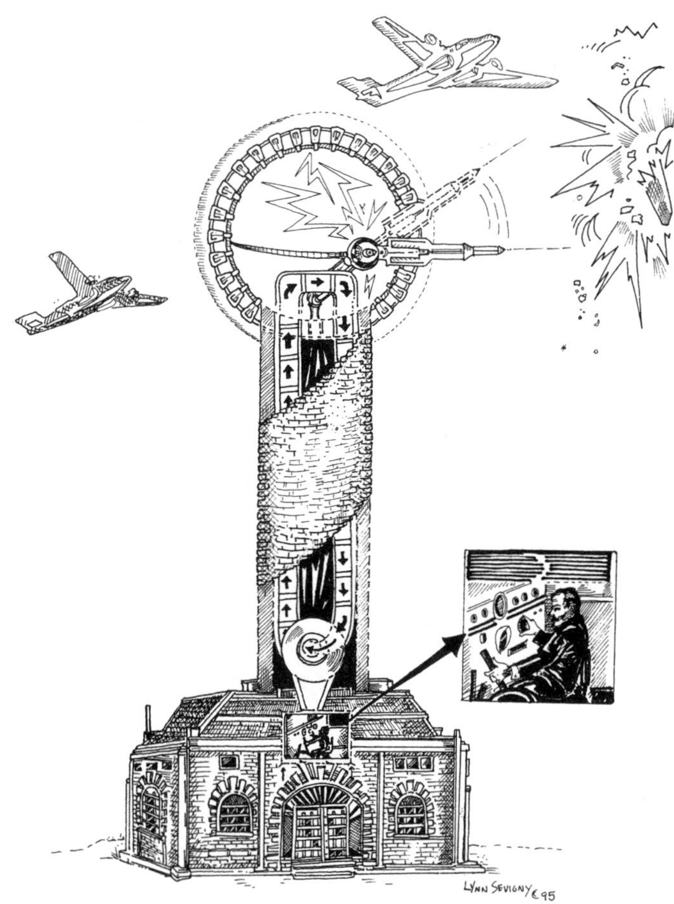

The Tesla particle beam weapon. (© Lynn Sevigny 1995)

APPENDIX A
THE MAGNIFYING TRANSMITTER: A TECHNICAL DISCUSSION

At the 1990 International Tesla Conference held in Colorado Springs, Dr. Alexander Marincic, curator of the Tesla Museum in Belgrade; Robert Golka, the only modern electrical engineer to construct large-scale magnifying transmitters; and I were engaged in a conversation on the viability of Tesla's plan. Both Marincic and Golka concurred that Tesla's ultimate plan, that of sending energy around the earth for industrial purposes, was not practicable.

Leland Anderson, an electrical engineer and Tesla expert for nearly forty years, agreed. Tesla's experiments at Colorado Springs, according to Anderson, were probably a local effect caused by the fortuitous placement of his tower next to the Pikes Peaks range along a great plain. When Tesla detected lightning discharges and standing waves, he made the incorrect assumption that these waves would encircle the globe. In fact, Anderson wrote, the standing waves Tesla detected were probably "unrecognized reinforcement effects" rebounding off Pikes Peak, and his generated waves probably did the same thing. His conclusion was based on measurements that electronic scientist Ralph Johler made of thunderstorms occurring along the peak.[1]

Two experts who have concluded that the Tesla apparatus was viable are Professor James Corum and Eric Dollard, both designers of transmitting equipment based on Tesla's findings. Dollard writes that the invention (circa 1920) of the "multiple loaded flat top antenna" by "Steinmetz's protégé, Ernst F. W. Alexanderson," was really "fashioned after those developed by Tesla." One such plant, located at Bolinas, California, sets up a resonant transformer between two separate "earth plates" and an "elevated plate." This arrangement produces three separate types of wireless frequencies: "atmospheric induction, antenna transmission, and earth induction." Acting as a "virtual ground" the aerial transmitting energy down into the earth sets up standing waves that "continuously bounce back and forth between the earth and the reflecting capacitance at a rate tuned to a natural rate of the earth."[2]

A simple tuning fork experiment can be used to explain the importance of the ground connection. It resonates much more powerfully when the fork is attached to a ground, such as a table. Due to the conductive property of the earth, individualization of impulse transmission is also facilitated. The electrical energy "does not pass through the earth in the ordinary acceptance of the term, it only penetrates to a certain depth according to the frequency."[3]

Corum, who holds a doctorate in physics and is a former professor of electrical engineering at West Virginia University writes, "It has been common in the past to discard Tesla's far-sighted vision as baseless. [I] believe that such depreciation has stemmed from critics that were uninformed as to Tesla's actual technical measurements, and physical observation." Having performed various experiments himself, Corum surmises that Tesla's mathematical results written in his May 16, 1900, patent application "could have only been obtained as a result of authentic terrestrial resonance measurements." In other words, he concludes that Tesla's claims that he (*a*) measured a terrestrial pulse that rebounded off the antipode of the earth and (*b*) calculated the resonant frequency of the earth are essentially correct.[4]

Looking at the wireless project from the technical point of view, most likely, Wardenclyffe was to be set up mainly for the purpose of distributing information and meager amounts of electrical power, just enough to run clocks and stock tickers, but not enough to charge factories. Each tower could act as a sender or a receiver. In a letter to Katherine Johnson, Tesla explains the need for well over thirty such towers.[5]

Further, the magnifying transmitter was conceived to transmit electricity in a variety of ways. Tesla could utilize the carrier waves traveling within the earth (e.g., the Shuman cavity, and/or the earth's geomagnetic pulse), he could transmit frequencies through the air, or he could beam a carrier wave up to the ionosphere and use it for transport:

> I will confess that I was disappointed when I first made tests along this line on a large scale. They did not yield practical results. At the time, I used about 8,000,000 to 12,000,000 volts of electricity. As a source of ionizing rays, I employed a powerful arc reflected up into the sky...trying only to connect a high

> tension current and the upper strata of the air, because my pet scheme for years ha[d] been to light the ocean at night.[6]

Central towers, acting much like today's wireless trunkline microwave transmitters for the phone company, could then be hooked up via conventional wires to numerous households within a given radius.

WORLD BROADCASTING SYSTEM

From what I can ascertain, I believe that Tesla's magnifying transmitter in a completed state would have operated as follows: A transmission tower would have been constructed so that its height and ability to radiate electrical oscillations was in a resonant relationship to the size, electronic and geophysical properties of the earth.

Rather than utilize transverse electromagnetic waves exclusively, Tesla would be utilizing longitudinal waves (such as those found in the impulses transmitted by earthquakes and by sound).[7] The gigantic Tesla coil was also calculated to take into account the wavelength of light. In other words, the length of the wires wound in the transformer were in a harmonic relationship to the distance light would travel in a given time. With the production of standing waves resonant with the planet, "nodal points" on the earth's surface were also plotted out.[8]

A tremendous charge, in excess of 30 million volts and in a harmonic frequency to the electrical and/or geophysical state of the earth, would be driven down the tower, into the ground, and out to sixteen 300-foot- long iron spokes positioned in a spiral down the entire length of the 120-foot well. Thereby gripping the earth, this pulse would generate an electronic disturbance in a harmonic relationship with the naturally occurring geomagnetic pulse that would reach the other side of the globe and in turn would bounce back up the tower. By controlling the period of frequency, this pulse could be modulated and actually increased in intensity in the same way one can make a well-made bell resound in increasing loudness by tapping it at precisely timed faster and faster rates. Also, the energy would be stored at the top of the tower and in specially built condensers by the laboratory. Stationary waves in resonance with known earth currents would thereby be established.[9]

Like a vibrating spring with a weight on it, this device enabled Tesla to determine and manipulate the electrostatic capacity (in analogy, like the pliability of the spring) and the inductance (analogous to the weight on the spring) of the carrier vibrations.[10] Tesla also maintained that the use of liquefied air (−197°F) would greatly augment the production and/or reception of very high frequencies while also reducing impedance caused by friction or heat.[11] By *transforming energy to higher frequencies on the rebound flow,* Tesla increased the efficiency of his towers. Each could act as both a sender and receiver. One tower situated near a waterfall could "jump" energy to another tower situated at another point on the globe.

Just as electricity is available throughout the electrical circuits that run through the transmission lines that circumscribe our planet, electricity would also be available throughout the entire electromagnetic grid of the earth itself. In the same way electricity is not utilized by conventional means until a plug is placed in a socket and a switch turned on, electricity would also not be utilized in the Tesla system until it too was connected up to a wireless instrument and that instrument was turned on. Electricity by the Tesla system would not be wasted by being diffused, no more so than electricity is wasted by present means, such as with wireless car telephones or by being made available through transformers and high-tension wires that run from transmission pole to transmission pole.[12]

It appears that the tower could at this point serve in a variety of ways. For instance, intelligible signals (wireless telephone) could be transmitted to any region of the globe. Power also could be provided by the same mechanism, probably within a confined region of each tower, to thousands of specific machines after they sent a coded request impulse or simply to another tower not located by a power source. And this second tower, situated in a remote area, could be connected to home appliances and telephones by way of conventional wires or by wireless. If two transmitters were utilized and separated by many miles, vector waves could more easily place impulses in desired locations.[13]

Referring to Figure 1: A power source (such as coal or water) would generate energy into a transformer comprising both a secondary (tuned to the wavelength of light) and primary coil. The secondary coil in the transmitting tower would be the inside thinner one, which is longer and has more turns. The generated frequency would be lowered when induced into the thicker primary, which has fewer turns and is shorter. The transmitter would then pump the energy into the natural medium, broadcasting it via earth or air (i.e., two different ways). According to Tesla:

> At the receiving station, a transformer of similar construction is employed; but in this case, the longer coil [of many turns]...constitutes the primary, and the shorter coil [of fewer turns]...the secondary.... It is to be noted that the phenomenon here involved in the transmission of electrical energy is one of true conduction and not to be confounded with the phenomena of electrical radiation.[14]

An Additional Criticism

E. Kornhauser, a professor of electrical engineering at Brown University, in reviewing this section, is doubtful that this form of power transmis-

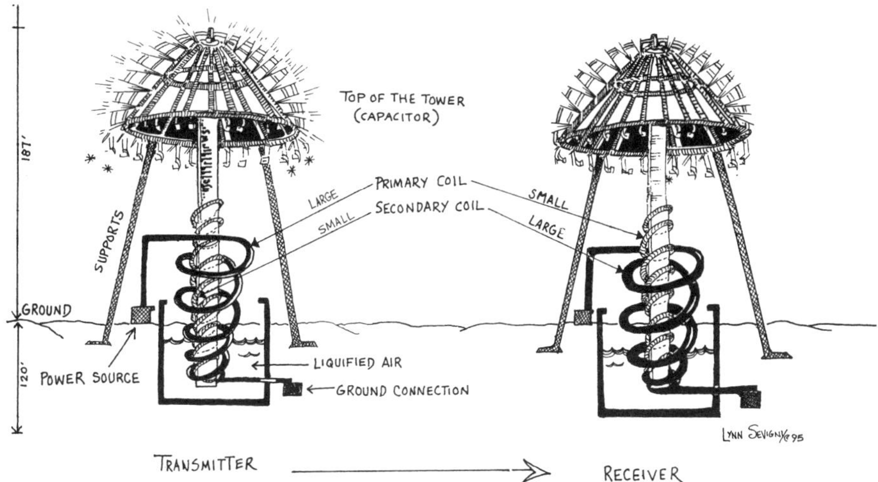

1. TRANSFORMER comprising:
   a. **Thick Coil** of short length and few turns, which acts as the primary in the transmitter and as the secondary in the receiver.
   b. **Thin Coil** of longer length and many turns, which acts as the secondary in the transmitter and as the primary in the receiver. This coil would be 50 miles in length or one fourth the wavelength of a lightwave whose circuit was 185,000 miles long.[15]
   c. **Magnetic Core** attached to the earth and elevated terminal.
2. POWER SOURCE deriving energy from coal or a waterfall.
3. GROUND CONNECTION.
4. CONTAINER OF LIQUID AIR (– 197°F) which causes "an extraordinary magnification of oscillation in the resonating-circuit[s]."[16]
5. ELEVATED TERMINAL or bulbous top for accumulating stored charge. To obtain highest possible frequency, a terminal of small capacity (like a taught spring) and high pressure is employed.

Figure 1. The sending and receiving magnifying transmitters are built essentially the same way. The length and size of the tower and transformer is in a harmonic relationship to the electromagnetic properties of the earth. It has a multipurpose function. Standing waves generated in resonant relationship to known Earth currents could be used as carrier frequencies for transmitting electrical power.

sion could be effectively achieved because the earth is not an efficient conductor (e.g., as compared to a copper wire). Concerning the possibility of creating wireless communication that could circumscribe the planet, Kornhauser conceded it was possible. He stated that the navy had unsuccessfully tried to institute a world radar system utilizing extremely low frequencies. Project Seafarer, as it was named, purportedly could have set up communication even with submarines deep underwater at any point of the globe. However, the plan was scrapped, it appears, mainly because of the potential to markedly disturb existing radio and television frequencies and fear of damage in the environment.

The efficiency of Tesla's radio receiving tubes was also questioned by Kornhauser, who thought it was doubtful that they would have been efficient enough as it would take another fifteen years before radio tubes of any merit came into being. Kornhauser did say, however, that the modern AM radio broadcasting stations use the earth as their primary means of transmitting their impulses. FM and television also use the earth, but the atmosphere in these instances is the more important medium for impulse transmission.

APPENDIX B

THE TUNGUSKA INCIDENT

A question often asked is whether Tesla had anything to do with the massive explosion which occurred in Tunguska, Siberia in June of 1908. As no meteor or crater was found, a rumor stemming from Oliver Nicholson, picked up by Tad Wise in his Tesla novel suggested that Tesla used Wardenclyffe to deliver the charge. Since the tower became disoperational in 1903, I saw no reason to include the incident in the first edition of *Wizard.* However, because the story was repeated on TV, rumors have persisted. Roy Gallant estimates in his book *The Day the Sky Split Apart* that the Tunguska explosion devastated a forty square mile area, and released energy 2,000 times greater than the atom bomb dropped on Hiroshima! Tesla expert James Corum allowed that if Tesla had the capability of releasing just 1% of the earth's magnetic charge, *in theory,* he could have produced comparable results. However, both Corum and the author are in agreement that Tesla not only did not do this, but further, Wardenclyffe simply had nowhere near that kind of capability. As Gallant suggests, the Tunguska explosion was probably caused by a comet or asteroid, which barely missed the earth by skipping along the atmosphere two or three miles above the site.

BIBLIOGRAPHY

Frequently Used Abbreviations

NT	Nikola Tesla
FOIA	Freedom of Information Act
CSN	Colorado Springs Notes
ITS	International Tesla Society, Colorado Springs, Colo.
TMS	Tesla Memorial Society, Lackawanna, N.Y.

Research Facilities

LA	Leland Anderson, personal archives, Denver, Colo.
	American Friends of the Hebrew University, New York, N.Y.
SWP	Avery Library, Manuscript Division, Columbia University, New York, N.Y. (Stanford White papers)
BLCA	Bancroft Library, Manuscript Division, University of California, Berkeley, Calif.
	Brown University Library, Providence, R.I.
BLCU	Butler Library, Manuscript Division, Columbia University, New York, N.Y. (Robert U. Johnson, George Scherff, and Michael Pupin papers)
WBP	Cornell University Library, Manuscripts Division, Ithaca, N.Y. (William Broughton papers)
TAE	Edison National Historic Site, West Orange, N.J. (Thomas Alva Edison, Charles Batchelor, and Nikola Tesla papers)
	Engineering Societies Library, New York, N.Y.
FBI	Federal Bureau of Investigation, Washington, D.C.
	George Arents Research Library. Syracuse University, Syracuse, N.Y.
GP	Gernsback Productions, Farmingdale, N.Y.
HC	Hammond Castle, Gloucester, Mass.
	Health Research Publishers, Mokelumne, Calif.
HL	Houghton Library, Harvard University, Cambridge, Mass.
LC	Library of Congress, Washington, D.C. (Tesla correspondence on microfilm)

JPM	J. Pierpont Morgan Library, New York, N.Y.
	Lloyd's of London, England
MSF	MetaScience Foundation Library, Kingston, R.I.
	National Academy of Science, Stockholm, Sweden
NAR	National Archives, Washington, D.C.
NYPL	New York Public Library, New York, N.Y.
NYHS	New York Historical Society, New York, N.Y.
OAP	Office of Alien Property, Washington, D.C.
	Port Jefferson Library, Port Jefferson, N.Y.
	St. Louis Public Library, St. Louis, Mo.
KSP	Smithsonian Institution, Washington, D.C. (Kenneth Swezey Papers)
NTM	Tesla Museum, Belgrade, Yugoslavia
	University of Prague Library, Archiv Univerzity Karlovy, Prague, Czechoslovakia
	University of Rhode Island Library and Interlibrary Loan, Kingston, R.I.
	USX Corporation, Pittsburgh, Penn.
GWA	Westinghouse Corporation Archives, Pittsburgh, Penn.
YL	Yale University Library, New Haven, Conn.

Frequently Cited Periodicals

BE	*Brooklyn Eagle*
CL	*Current Literature*
EE	*Electrical Engineer*
EEX	*Electrical Experimenter*
ER	*Electrical Review*
EW	*Electrical World*
EW & E	*Electrical World & Engeineer*
NYHT	*New York Herald Tribune*
NYS	*New York Sun*
NYT	*New York Times*
NYW	*New York World*
PACE	*Planetary Association for Clean Energy*
R of R	*Review of Reviews*

Tesla Correspondence

JJA	John Jacob Astor
TdB	Titus deBobula
RFL	Reginald Fessenden Litigation
JHH Jr	John Hays Hammond Jr.
JH	Julian Hawthorne
AH	Admiral Higginson
KJ	Katharine Johnson
RUJ	Robert Underwood Johnson
TCM	Thomas Commerford Martin
JPM	J. Pierpont Morgan
JPM Jr	J. Pierpont Morgan Jr.

GS George Scherff
NT Nikola Tesla
ET Elihu Thomson
GSV George Sylvester Viereck
SW Stanford White
GW George Westinghouse
GWC George Westinghouse Corporation

For example: NT/JHHjr January 3, 1911: Tesla wrote Hammond on that date. GS/NT April 7, 1902: Scherff wrote Tesla on that date.

Frequently Cited Sources by or About Nikola Tesla

NT 1894 *The Inventions, Researches and Writings of Nikola Tesla.* T. C. Martin, ed. New York: Electrical Engineer.

NT 6/1900 "The Problem of Increasing Human Energy." *Century,* June 1900, pp. 175–211.

NT 1916 *Nikola Tesla: On His Work With Alternating Currents and their Application to Wireless Telegraphy, Telephone, and Transmission of Power.* L. Anderson (ed.), Denver Colo.: Sun, 1992.

NT 1919 *My Inventions: The Autobiography of Nikola Tesla.* Ben Johnston, ed., Williston, Vt.: Hart Brothers, 1981.

NT 1937 "The New Art of Projecting Concentrated Non-dispersive Energy Through the Natural Medium." In Elizabeth Raucher and Toby Grotz, eds. *Tesla: 1984: Proceedings of the Tesla Centennial Symposium,* Colorado Springs, Colo.: International Tesla Society, 1984, pp. 144–50.

NT 1956 *Nikola Tesla: Lectures, Patents, Articles.* Belgrade: Nikola Tesla Museum.

NT 1961 *Tribute to Nikola Tesla: Letters, Articles, Documents.* Belgrade: Nikola Tesla Museum, 1961.

NT 1979 *Colorado Springs Notes and Commentary.* Alexander Marincic, ed. Belgrade: Nikola Tesla Museum, 1979.

NT 1981 *Solutions to Tesla's Secrets.* J. Ratzlaff, ed. Milbrae, Calif.: Tesla Book Co., 1984.

NT 1984 *Tesla Said.* J. Ratzlaff, ed. Millbrae, Calif.: Tesla Book Co., 1984.

R & A. Ratzlaff, J., and Anderson, L. *Dr. Nikola Tesla Bibliography 1884–1978.* Palo Alto, Calif.: Ragusen Press, 1979.

Bibliography of Major Sources

Abraham, J., and R. Savin. *Elihu Thomson Correspondence.* New York: Academic Press, 1971.

Adams, E. D. *Niagara Power: 1886–1918.* Niagara Falls, N.Y.: Niagara Falls Power Co., 1927.

Anderson, F., ed. *Mark Twain's Notebooks & Journals.* iii, 1883–91. Berkeley, Calif.: University of California Press, 1979.

Anderson, Leland, ed. "John Stone Stone on Nikola Tesla's Priority in Radio." *Antique Wireless Review,* 1 (1986).

———. "Priority in the Invention of the Radio: Tesla v. Marconi." *Tesla Journal* 2, no. 3 (1982-83): pp. 17–20.

______. *Nikola Tesla; Lecture Before the New York Academy of Sciences.* Breckenridge, Colo.: 21st Century Books.

Asimov, Isaac. *Asimov's Biographical Encyclopedia of Science and Technology.* Garden City, N.Y.: Doubleday, 1964.

Baker, Paul. *Stanny: The Gilded Life of Stanfored White.* New York: Free Press, 1989.

Baker, E. C. *Sir William Preece: Victorial Engineer Extraordinary.* London: Hutchinson, 1976.

Barnow, Erik. *A Tower in Babel: A History of Broadcast USA.* New York: Oxford University Press, 1966.

Barrett, W., ed. *The Smithsonian Book of Invention.* New York: Norton, 1978.

Bearden, Tom. "Solutions to Tesla's Secrets and the Soviet Tesla Weapon." In *NT 1981,* pp. 1–45.

______. "Tesla's Electromagnetics and Soviet Weaponization." In Raucher and Grotz eds. *NT 1984,* pp. 119–138.

Beckhard, A. *Electrical Genius, Nikola Tesla.* New York: Messner, 1959.

Behrend, B. A. *The Induction Motor.* New York: McGraw Hill, 1921.

Birmingham, Stephen. *Our Crowd.* New York: Pocket Books, 1977.

Blackmore, J. *Ernst Mach, His Work, Life and Influence.* Berkeley, Calif.: University of California Press, 1972.

Bulwer-Lytton, Edward. *The Coming Race.* London: Routledge, 1871.

Cameron, W. *World's Columbian Exposition.* New Haven, Conn.: James Brennan, 1893.

Cantril, H. *The Invasion of Mars: A Study in the Psychology of Panic.* New York: Harper, 1940/1966.

Carlson, Oliver. *Brisbane: A Candid Biography.* New York: Stackpole, 1937.

Cases Adjudged in the Supreme Court. United States v. Marconi. 373, pp. 1–80, April 9, 1943.

Cheney, M. *Tesla: Man Out of Time.* Englewood Cliffs, N.J.: Prentice-Hall, 1981.

Chernow, Ron. *The House of Morgan.* New York: Atlantic Monthly Press, 1990.

Clark, Roland. *Einstein: The Life and Times.* New York: World, 1971.

Crockett, Albert Stevens. *Peacocks on Parade.* New York: Sears, 1931.

Crookes, William. "Some Possibilities of Electricity." *Forthnightly Review* (February 1892): vol. no. 3, pp. 173–81.

Coit, Margaret. *Mr. Baruch.* New York: Houghton Mifflin, 1957.

Conot, R. *Streak of Luck: The Life Story of Edison.* New York: Bantam Books, 1980.

Corum, James, and Ken Corum. "A Physical Interpretation of the Colorado Springs Data." In Raucher and Grotz, *NT ITS 1984 Proceedings.*

Corum, James. "100 Years of Resonator Development." *NT ITS 1990 Proceedings,* pp. 2-1–2-18.

Cowles, Virginia. *The Astors.* New York: Knopf, 1979.

David, C., ed. *Chronicle of the 20th Century.* Mt. Kisco, N.Y.: Chronicle, 1987.

Davis, John. *The Guggenheims: An American Epic.* New York: Morrow, 1978.

De Forest, Lee. *Father of Radio: An Autobiography.* Chicago: Wilcox & Follett, 1950.

DelRay, L. *Science Fiction Art.* New York: Ballantine, 1975.

Dickson, W., and A. Dickson. *The Life & Inventions of T. A. Edison.* New York: Crowell, 1892.

Djilan, Milovan. *Land Without Justice.* New York: Harcourt Brace, 1958.

Edison, Thomas. "Pearl Street." *ER* (January 12, 1901): pp. 60–62.

Einstein, A., B. Podolsky, and N. Rosen. "Can Quantum Mechanical Description of Physical Reality be Considered Complete?" *Physical Review* 47 (1935): p. 777.

Elswick, Steven. *1986 Proceedings of the International Tesla Society. Colorado Springs, Colo: ITS Press.*

______. *NT ITS 1988 Proceedings.*

______. *NT ITS 1990 Proceedings.*

______. *NT ITS 1992 Proceedings.* In press.

______. *NT ITS 1994 Proceedings.* In press.

Encyclopedia of Science & Technology. New York: McGraw-Hill, 1987.

Evans, Christopher. *Cults of Unreason.* New York: Dell, 1973.

Fabin, Sky. *The Zenith Factor* (video). Santa Fe, N.M.: Southwestern College of Life Sciences, 1985.

Gates, John D. *The Astor Family.* Garden City, N.Y.: Doubleday, 1981.

Gardiner, P., ed. *Theories of History.* Glencoe, Ill.: Free Press, 1959.

Gernsback, Hugo. "Nikola Tesla: Father of Radio." *Radio Craft* February 1943, pp. 263–65; 307–10.

Gillifan, S. *The Sociology of Invention.* Cambridge, Mass.: MIT Press, 1935.

Gertz, Elmer. *Odyssey of a Barbarian: The Biography of G. S. Viereck.* Buffalo, N.Y.: Prometheus, 1978.

Goethe. *Faust.* C. Brooks, ed./trans. Boston: Ticknor & Fields, 1856.

Gorowitz, B., ed. *The Steinmetz Era.* Schenectady, N.Y.: Elfun Hall, 1977.

Hammond, John Hays Sr. *Autobiography of John Hays Hammond.* New York: Farrar & Rhinehart, 1935.

Hammond, John Hays Jr. "The Future in Wireless." *National Press Reporter,* 15, no. 110 (May 1912).

Harding, Robert. *George H. Clark Radiona Collection.* Washington, D.C.: Smithsonian Institution, 1990.

Hawkings, L. *William Stanley, His Life and Times.* New York: Newcomen Society, 1939.

Hayes, Jeffrey. *Boundary—Layer Breakthrough.* Security, Colo.: High Energy Enterprises, 1990.

Held, David. *Introduction to Critical Theory.* Berkeley, Calif.: University of California Press, 1980.

Howeth, L. S. *History of Communications-Electronics in U.S. Navy.* Washington, D.C.: U.S. Government Printing Office, 1963.

Hoyt, Edwin. *The House of Morgan.* New York: Dodd, Mead, 1966.

______. *The Guggenheims and the American Dream.* New York: Funk & Wagnalls, 1967.

Huart, M. "The Genius of Destruction." *ER* (December 7, 1898): p. 365.

Hughes, T. *Networks of Power.* Baltimore, Md.: Johns Hopkins University Press, 1983.

Hunt, Inez, and Wanetta Draper. *Lightning in His Hands: The Life Story of Tesla.* Hawthorne, Calif.: Omni, 1964/1977.

Jobs, Steven. Interview. *Playboy,* February 1985, 49–50, 54, 58, 70, 174–80.

Johnson, Niel. *George Sylvester Viereck: German/American Propagandist.* Chicago: University of Illinois Press, 1972.

Johnson, R. U. *Songs of Liberty.* New York: Century, 1897.

______. *Remembered Yesterdays*. Boston: Little, Brown, 1923.

Jolly, W. *Marconi*. New York: Stein & Day, 1972.

Josephson, M. *The Robber Barons*. New York: J. J. Little, 1934.

______. *Thomas Alva Edison*. New York: McGraw-Hill, 1959.

Jovanovic, Branimir. *Tesla I Svet Vazduhoplovstva*. Belgrade: Nikola Tesla Museum Press, 1988.

Kuhn, T. *The Structure of Scientific Revolutions*. Chicago: University of Chicago Press, 1970.

Leonard, J. *Loki: The Life of Charles Proteus Steinmetz*. Garden City, N.Y.: Doubleday, 1928.

Lessing, Lawrence. *Man of High Fidelity: Edwin Howard Armstrong*. Philadelphia: Lippincott, 1956.

Lowell, Perceival. *Mars and its Canals*. New York: Macmillan, 1906.

Lyons, E. *David Sarnoff, A Biography*. New York: Pyramid, 1970.

Mannheim, K. *The Sociology of Knowledge*. London: Routledge, 1952.

Marconi, D. *My Father Marconi*. New York: McGraw-Hill, 1962.

Marconi, Guglielmo. "Wireless Telegraphy." *ER* 38 (June 15, 1901): pp. 754–56; 781–86.

Marincic, Alexander. "Research on Nikola Tesla in Long Island Laboratory." *Tesla Journal*, 6, no. 7, (1988/89): pp. 25–28.

Martin, T. C. "Nikola Tesla." *Century*, February 2, 1894, pp. 582–85.

Mooney, Michael. *Evelyn Nesbit & Stanford White*. New York: Morrow, 1976.

O'Hara, J. G., and W. Pricha. *Hertz & the Maxwellians*. London: Peregrinus, 1987.

O'Neill, John. *Prodigal Genius: The Life Story of N. Tesla*. New York: Ives Washburn, 1944. Reprinted by David McKay, circa 1972.

Ouspensky, P. D. *New Model of the Universe*. New York: Vintage, 1971.

Passer, Harold. *The Electrical Manufacturers: 1875–1900*. Cambridge, Mass.: Harvard University Press, 1953.

Petkovich, Dragislav. "A Visit to Nikola Tesla." *Politika*, March 27, 1927.

Petrovich, Michael Boro. *A History of Modern Serbia*. New York: Harcourt Brace, 1976.

Prout, Henry. *George Westinghouse: An Intimate Portrait*. New York: Wiley, 1939.

Pupin, M. *From Immigrant to Inventor*. New York: Scribner, 1925.

Rand, Ayn. *Atlas Shrugged*. New York: 1957.

Ratzlaff, J., and F. Jost, eds. *Tesla/Scherff Correspondence*. Milbrae, Calif.: Tesla Book Co., 1979.

Raucher, Elizabeth, and Toby Grotz, eds. *Tesla: 1984: Proceedings of the Tesla Centennial Symposium*. Colorado Springs, Colo.: International Tesla Society, 1984.

Robertson, Andrew. *About George Westinghouse and the Polyphase Currents*. New York: Newcomen Society, 1939.

Roman, Klara. *Handwriting: A Key to Personality*. New York: Free Press, 1971.

Rubin, C., and K. Strehlo. "Why So Many Computers Look Like 'IBM Standard.'" *Personal Computing* March 1984, pp. 52–65, 182–89.

Rubin, Nancy. *John Hays Hammond, Jr.: A Rennaissance Man in the 20th Century*. Gloucester, Mass.: Hammond Museum Press, 1987.

Rudhyar, Dane. *Occult Preparations for a New Age*. Wheaton, Ill.: Quest, 1975.

Sartre, J. *Search for a Method*. New York: Knopf, 1963.

Satterlee, H. *J. Pierpont Morgan: An Intimate Portrait.* New York: Macmillan, 1939.
Scott, Lloyd. *Naval Consulting Board of the U.S.* Washington, D.C.: U.S. Government Printing Office, 1920.
Siegel, Mark. *Hugo Gernsback: Father of Modern Science Fiction.* San Bernadino, Calif.: Borgo Press, 1988.
Silverberg, R. *Light for the World.* Princeton, N.J.: Van Nostrand, 1967.
Smith, Page. *The Rise of Industrial America,* vol. 6. New York: McGraw-Hill, 1984.
Sobel, Robert. *RCA.* New York: Stein & Day, 1986.
Steinmetz, Charles. *AC Phenomena.* New York: McGraw-Hill, 1900.
______. *Theoretical Elements of Electrical Engineering.* New York: McGraw-Hill, 1902.
Stockbridge, Frank Parker. "Tesla's New Monarch of Mechanics." *NYHT,* October 18, 1911, 1.
Stoilkovic, Stanko. "Portrait of a Person, a Creator and a Friend." *Tesla Journal,* 4, no. 5 (1986/87): pp. 26–29.
Storm, Margaret. *Return of the Dove.* Baltimore, Md.: Margaret Storm, 1956.
Tate, Alfred O. *Edison's Open Door.* New York: Dutton, 1938.
Telford, H. *Applied Geophysics.* New York: Cambrdidge University Press, 1976.
Thompson, Silvanus. *Polyphase Electric Currents.* New York: American Technical Book Co., 1897.
Toomey, J. *Location of Sources of Elf Noises.* Kingston, R.I.: University of Rhode Island Press, 1970; Philadelphia: Lippincott, 1963.
Wheeler, G. *Pierpont Morgan and Friends: Anatomy of a Myth.* Englewood Cliffs, N.J.: Prentice-Hall, 1973.
White, W. *The Organization Man.* Garden City, N.Y.: Doubleday, 1956.
Wise, Tad. *Tesla: A Biographical Novel.* Atlanta, Ga.: Turner, 1994.
Wolff, Robert Lee. *The Balkans in Our Time.* Cambridge, Mass.: Harvard University Press, 1956.
Wolff, Werner. *Diagrams of the Unconscious.* New York: Grune & Stratton, 1948.
Woodbury, David. *Beloved Scientist: Elihu Thomson.* New York: McGraw-Hill, 1944.

Works About Tesla by the Author

1977 Nikola Tesla [Harry Imber, pseud.] "The Man Who Fell to Earth." *Ancient Astronauts,* September, pp. 23–26.
1978 (H. Smukler, coauthor) "The Tesla/Matthews Outerspace Connection," as told by Andrija Puharich. *Pyramid Guide,* May, part 1, p. 5; July, part 2, p. 5.
1979 "Forty Years of the Handwriting of Nikola Tesla." Paper presented before the National Society for Graphology, New York, N.Y.
1982 "On Nikola Tesla." Letter to Editor. *Radio Electronics,* June, p. 24.
1983 "Nikola Tesla: The Forgotten Inventor." In J. Dorinson and J. Atlas, eds. *Psychohistory: Persons & Communities.* New York: Long Island University, pp. 209–31.
1984 "The Belief in Life on Mars: A Turn-of-the-century Group Fantasy." In J. Dorinson and J. Atlas eds. *Proceedings: Sixth Annual International Psychohistory Convention,* pp. 101–19.
1984 *Tesla: Mad Scientist of the Gilded Age.* New York: Windsor Total Video, produced in association with Bob Henderson, directed by Marc Seifer, narrated by J. T. Walsh, original score by Marshall Coid. Videocassette.
1985 "Nikola Tesla: The Lost Wizard." In E. Raucher and T. Grotz, eds., *Tesla 1984:*

Proceedings of the Tesla Centennial Symposium. Colorado Springs, Colo.: International Tesla Society, pp. 31–40.

1986 "The Inventor and the Corporation: Case Studies of Nikola Tesla, Steven Jobs and Edwin Armstrong." In S. Elswick, ed., *Proceedings: 1986 Tesla Symposium*. Colorado Springs, Colo.: International Tesla Society, pp. 53–74.

1986 "Nikola Tesla: Psychohistory of a Forgotten Inventor." Ph.D. diss., Saybrook Institute, San Francisco, Calif.

1988 "Tesla: The Interplanetary Communicator?" *Hands on Electronics,* December, pp. 62–66; 102.

1988 "The History of Lasers and Particle Beam Weapons." In S. Elswick, ed., *Proceedings: 1988 Tesla Symposium*. Colorado Springs, Colo.: International Tesla Society.

1988 *The Lost Wizard*. Screenplay, cowritten with Tim Eaton.

1989 "Nikola Tesla: Psychohistory of a Forgotten Inventor." 1990 Dissertation Chapter Abstracts. *Tesla Journal*. Lackawanna, N.Y.: Tesla Memorial Society, pp. 49–57.

1990 Nikola Tesla and John Hays Hammond, Jr.: "Pioneers in Remote Control." In S. Elswick, ed., *Proceedings: 1990 Tesla Symposium*. Colorado Springs, Colo.: International Tesla Society. In press.

1991 *Nikola Tesla: The Man Who Harnessed Niagara Falls*. Kingston, R.I.: MetaScience Publications.

1991 "Nikola Tesla and John Muir: Ecologists." *IV International Tesla Symposium*. Belgrade, Yugoslavia: Serbian Academy of Sciences and Arts and Nikola Tesla Museum, pp. 317–28.

1992 "Nikola Tesla and FDR: The Secret History of Wireless." In S. Elswick, ed., *Proceedings: 1992 Tesla Symposium,* Colorado Springs, Colo.: In press.

1994 Nikola Tesla: The Lost Years. In S. Elswick, ed., *Proceedings: 1994 Tesla Symposium*. Colorado Springs, Colo.: International Tesla Society. In press.

1996 "Taking on Einstein," *Extraordinary Science* 8, no. 1 (Jan., Feb., Mar.), pp. 38–43.

1996 "Wardenclyffe," *Extraordinary Science* 8, no. 2 (Apr., May, June), pp. 5–10.

1996 "John Jacob Astor and Nikola Tesla." *Proceedings: 1996 Tesla Symposium*. Colorado Springs, Colo.: International Tesla Society. In press.

Other Works

Handwriting & Brainwriting (collected works)
Staretz Encounter (novel)
Hail to the Chief (screenplay)
The Steven Rosati Story (true-crime biography)

Correspondence

Alker, Henry, Saybrook Institute, San Francisco, Calif.
†Anderson, Leland, Denver, Colo.
Basura, Nick, Los Angeles, Calif.
*Bearden, Tom, Huntsville, Ala.
‡Bergstresser, Ralph, Phoenix, Ariz.
Bromberg, Joan, Woburn, Mass.

Burg, David, Lexington, Ky.
Call, Terrence, University of Rhode Island, Kingston, R.I.
*Corum, James, Columbus, Ohio
*Corum, Ken, Franconia, N.H.
Clark, Peggy McKinnon, Shoreham, N.Y.
†Czito, Nancy, Washington, D.C.
Eaton, Tim, Industrial Light & Magic, San Rafael, Calif.
Elswick, Steve, International Tesla Society, Colorado Springs, Colo.
Feeley, Terrence, Johnston, R.I.
‡Gant, James, Washington, D.C.
‡Gertz, Elmer, Chicago, Calif.
Gold, Harry, Tesla Coil Builders Association, Glens Falls, N.Y.
*Golka, Robert, Colorado Springs, Colo.
Grotz, Toby, Colorado Springs, Colo.
Hessen, Robert, Stanford University, Calif.
*Hardesty, James, Ithaca, N.Y.
†Jankovitch, M., Paris, France
*Jovanovich, Branimir, Belgrade, Yugoslavia
†*Jurow, Irving, Washington, D.C.
Kasanovich, Nicholas, Tesla Memorial Society, Lackawanna, N.Y.
Kline, Ronald, Cornel University, Ithaca, N.Y.
Kramer, Jurgen, Saybrook Institute, San Francisco, Calif.
Krippner, Stanley, Saybrook Institute, San Francisco, Calif.
McCabe, Bob, Flint, Mich.
McGinnis, J.W., Colorado Springs, Colo.
*Marincic, Alexander, Belgrade, Yugoslavia
*Markovitch, Michael, Brooklyn, N.Y.
Neuschatz, Sanford, Shannock, R.I.
Parry, F., Washington, D.C.
*Possell, Jake, Colorado Springs, Colo.
†Puharich, Andrija, Colorado Springs, Colo.
*Ratzlaff, John, Milbrae, Calif.
Romero, Sid, Salt Lake, Utah
Seifer, Stanley, West Hempstead, N.Y.
Seifer, Thelma, West Hempstead, N.Y.
Shriftman, Elliott, Manhassett, N.Y.
Smukler, Howard, Berkeley, Calif.
‡Terbo, William, New York, N.Y.
Vagermeerch, Richard, University of Rhode Island
‡Viereck, Peter, Mt. Holyoke, Mass.
*Vujovic, Ljubo, Brooklyn, N.Y.
Walsh, J.T., Studio City, Calif.
White, Debra, Saybrook Institute, San Francisco, Calif.
White, John, Cheshire, Conn.

*Interviewed because of special knowledge about the subject.
†Personally knew associates of Tesla.
‡Personally met with Tesla.

NOTES

Preface, pp. xiii–xvi

1. Margaret Storm, *Return of the Dove* (Baltimore: M. Storm Productions, 1956).

2. Nikola Tesla, *My Inventions: The Autobiography of Nikola Tesla* (Williston, Vt.: Hart Brothers, 1982) [Originally published, 1919].

Chapter 1: Heritage, pp. 1–4

1. Nikola Tesla, "Zmai Ivan Ivanovich, the Chief Servian Poet of To-day." In R. U. Johnson, ed., *Songs of Liberty and Other Poems* (New York: Century Company, 1897).

2. Personal trip to Yugoslavia, 1986.

3. Robert Lee Wolff, *The Balkans in Our Time* (Cambridge, Mass.: Harvard University Press, 1956).

4. Michael Markovitch, personal interview, 1988.

5. Louis Adamic, *My Native Land* (New York: Harper & Brothers, 1943)

6. Michael Boro Petrovich, *The History of Nineteenth-Century Serbia*, 2 vols. (New York: Harcourt Brace Jovanovich, 1976), p. 5.

7. Petrovich, *History,* p. xii.

8. Markovitch, interview, 1988.

9. Markovitch, interview, 1988.

10. Adamic, *My Native Land,* p. 270.

11. Ibid; Petrovich, *History,* pp. 142–43, 350–51.

Chapter 2: Childhood, pp. 5–14

1. NT, "A Story of Youth" (1939). In John Ratzlaff, ed., *Tesla Said* (Milbrae, Calif.: Tesla Book Co., 1984), pp. 283–84. Articles included in this volume will be referred to hereafter by two dates—of original composition and the date of the Ratzlaff anthology.

2. John O'Neill, *Prodigal Genius: The Life Story of Nikola Tesla.* (New York: Ives Washburn, 1944), p. 12.

3. Ibid., p. 13.

4. Nikola Pribic, personal correspondence, April 19, 1988.

5. V. Popovic, *Nikola Tesla.* (Belgrade: Tecnicka Knjiga, 1951).

6. O'Neill, *Prodigal Genius,* p. 11.

7. Nikola Tesla, "Scientists Honor Nikola Tesla." Unidentified newspaper article, 1894. Displayed at the Edison Archives, Menlo Park, N.J.

8. Nikola Tesla, *My Inventions: The Autobiography of Nikola Tesla* (Williston, Vt: Hart, 1982), p. 29; originally published in *Electrical Experimenter,* in six monthly installments, February–July, 1919.

9. William Terbo, interview, 1988.

10. Birth charts from M. Markovitch archives.

11. Tesla, *My Inventions,* p. 30.
12. Terbo interview, 1988.
13. T. C. Martin, "Nikola Tesla," *Century,* February 1894, pp. 582–85; Tesla, April 22, 1893. Branimira Valic, ed., *My Inventions* (Zagreb: Moji Pronalasci; Skolska Kanjiga, 1977).
14. O'Neill, *Prodigal Genius,* p. 10.
15. NT, *My Inventions,* p. 31.
16. O'Neill, *Prodigal Genius,* p. 12.
17. Tesla, 1939/1984.
18. T. C. Martin, "Nikola Tesla," *Electrical World* 15, no. 7 (1890), p. 106.
19. NT, *My Inventions,* p. 45.
20. NT, 1939/1984, p. 285.
21. Ibid., pp. 284–85 (condensed).
22. NT, *My Inventions,* p. 29.
23. Ibid., p. 30.
24. NT, "Nikola Tesla and His Wonderful Discoveries," *N.Y. Herald,* April 23, 1893, p. 31.
25. D. Budisavljevic, "A Relative of Tesla's Comments on the Tesla Library," in Nikola Kasanovich, ed., *Tesla Memorial Society Newsletter,* Spring 1989, p. 3.
26. NT, *My Inventions,* p. 30; NT, 1939/1984, p. 283.
27. The date may have been 1861.
28. NT, *My Inventions,* p. 28. Rumors suggesting Niko pushed his brother down a flight of stairs stem from A. Beckhard, *Electrical Genius, Nikola Tesla* (New York: Julian Messner, 1959). Beckhard's book, clearly written for young adults, utilized only one referenced source, the O'Neill work. An imaginative writer, Beckhard made up the names of the townspeople from Tesla's childhood as well. In *Tesla: Man Out of Time* (Englewood Cliffs, N.J.: Prentice-Hall, 1981) Margaret Cheney repeats the rumor without referencing it. Leland Anderson, who helped on the research, stated that Cheney heard the story at the Tesla Museum. The original source was probably still Beckhard, as the book is prominently referred to by V. Popovic, professor at Belgrade University and vice president (in 1976) of the Tesla Society in Belgrade in his article "Nikola Tesla—True Founder of Radio Communications," in *Tesla: Life and Work of a Genius* (Belgrade: Nikola Tesla Society, 1976).
29. Ibid., p. 47.
30. Phillip Callahan, "Tesla the Naturalist," in Steven Elswick, ed., *Tesla Proceedings,* 1986, pp. 1–27.
31. Michael Markovitch personal archives, New York City.
32. NT, *My Inventions,* p. 46.
33. Ibid., p. 32, 36 (condensed).
34. Ibid., pp. 32–33 (condensed).
35. Ibid.
36. Ibid., pp. 36–37.
37. Marc Seifer, *Nikola Tesla: Psychohistory of a Forgotten Inventor* (San Francisco: Saybrook Institute, 1986). (Doctoral dissertation.)
38. NT, *My Inventions,* p. 53.
39. Ibid., pp. 35–36 (condensed).
40. TCM "Nikola Tesla," p. 106.
41. NT, *My Inventions,* p. 47 (condensed).
42. NT, April 23, 1893.
43. NT, *My Inventions,* p. 53.
44. Nikola Pribic, "Nikola Tesla: A Yugoslav Perspective," *Tesla Journal* 6&7 (1989/1990), pp. 59–61 (condensed).
45. NT, *My Inventions,* p. 54; Valic, 1977, p. 101.
46. Ibid., p. 53.
47. Ibid., p. 54.
48. Ibid., p. 55.
49. Ibid.
50. O'Neill, *Prodigal Genius,* p. 29.
51. NT to R. U. Johnson, April 5, 1900 [BLCU].

Chapter 3: College Years, pp. 15–26

1. NT, *Electrical Engineer,* September 24, 1890.

2. NT, *My Inventions,* pp. 56–57.
3. Franz Pichler, "Tesla's Studies in College," ITS Conference, Colorado Springs, 1994.
4. Kosta Kulishich, "Tesla Nearly Missed His Career as Inventor: College Roommate Tells," *Newark News,* August 27, 1931.
5. NT, *My Inventions,* p. 56.
6. Pichler, "Tesla's Studies in College."
7. K. Kulishich, "Tesla Nearly Missed..."
8. NT, *My Inventions,* p. 37.
9. Thomas Edison, "A Long Chat With the Most Interesting Man in the World," *Morning Journal,* July 26, 1891, p. 17 [TAE].
10. T.C. Martin, *The Inventions, Researches and Writings of Nikola Tesla* (New York: Electrical World Publishing, 1894), p. 3.
11. NT, reconstructed from: "A New Alternating Current Motor," *The Electrician,* June 15, 1888, p. 173; NT, *My Inventions,* p. 57.
12. NT, *My Inventions,* p. 37.
13. Ibid.
14. Alfred O. Tate, *Edison's Open Door* (New York: Dutton, 1938), p. 149.
15. Timothy Eaton, to author, quoting William Terbo, 1988.
16. W. Terbo, "Remarks at Washington, D.C., premiere of *The Secret Life of Nikola Tesla." (Yugoslavia: Zagreb Films, 1983).*
17. Kulishich, "Tesla Nearly Missed..."
18. Nikola Pribic, personal discussion with the author, Zagreb, 1986.
19. Dragislav Petkovich, "A Visit to Nikola Tesla," *Politika,* April 27, 1927, p. 4 [LA].
20. Blackmore, John T. *Ernst Mach: His Life and Work.* (Berkeley, Calif.: University of California Press, 1972), pp. 38–39.
21. Karel Litsch, director, Archiv Univerzity Karlovy, Prague, Czech Republic, to author, September 28, 1989.
22. William James, quoted in J. Blackmore, *Ernst Mach,* p. 76.
23. Ibid.
24. Robert Watson, *The Great Psychologists: Aristotle to Freud* (New York: Lippincott, 1963), pp. 198–200.
25. NT, "How Cosmic Forces Shape Our Destiny," (February 7, 1915), in *Lectures, Patents, Articles,* 11956, p. A-173.
26. Blackmore, *Ernst Mach,* pp. 41–43.
27. C. C. Gillispie, ed, "Ernst Mach," in *Dictionary of Scientific Biography* (New York: Scribners, 1977).
28. Ibid.
29. NT, *My Inventions,* p. 59.
30. Velac, p. 102.
31. Batchelor to Edison, October 24, 1881; Batchelor to Mr. Bailey, April 11, 1882 [TAE].
32. Inez Hunt and Waneta Draper, *Lightning in His Hands: The Life Story of Nikola Tesla* (Hawthorne, Calif.: Omni Publications, 1977), p. 33; originally published, 1964.
33. Petkovich, "A Visit to Nikola Tesla."
34. T. C. Martin, "Nikola Tesla," *Century,* February 1894, p. 583.
35. Anthony Szigeti, Deposition to the State of New York (February, 1889), in *Tribute to Nikola Tesla* (Belgrade: Tesla Museum, 1961), p. A-398.
36. NT, *My Inventions,* pp. 60–61.
37. For a full discussion of this event from both a neurological and metaphysical point of view, see author's doctoral dissertation, chapter 52 on "Creativity, Originality and Genius."
38. NT, *My Inventions,* pp. 59–60 (condensed).
39. P. Lansky, "Neurochemistry and the Awakening of Kundalini," in J. White, ed., *Kundalini, Evolution & Enlightenment* (Garden City, N.Y.: Doubleday, 1979), pp. 295–97.
40. NT, *My Inventions,* p. 61.
41. NT, "A New System of Alternate Current Motors and Transformers" (1888), in T. C. Martin (ed.), *The Inventions, Researches, and Writings of Nikola Tesla* (New York: Electrical Review, 1894), pp. 11–16.
42. NT, *My Inventions,* pp. 62–63.
43. Ibid., p. 65.
44. Ibid., p. 66.

45. Walter Baily, "A Mode of Producing Arago's Rotation," *Philosophical Magazine,* 1879, pp. 286–90.

46. Silvanus R. Thompson, *Polyphase Electrical Currents* (New York: American Technical Book Co., 1899), p. 86.*See also* Kline, 1987, p. 287.

47. Ronald Kline, "Science & Engineering Theory in the Invention and Development of the Induction Motor," *Technology & Culture* (April 1987), pp. 283–313.

48. "Marcel Deprez Gets Publicity for Efficient Power Transmission," *New York Times,* November 2, 1881.

49. Henry Prout, *George Westinghouse: An Intimate Portrait* (New York: Wiley, 1939), p. 102.

50. Ibid., p. 100.

51. Galileo Ferraris, "Electromagnetic Rotations With an Alternating Current," *Electrician,* vol. 36 (1895), pp. 360–75; C. E. L. Brown, "A Personal Conversation With G. Ferraris." *Electrical World,* February 6, 1892; O'Neill, *Prodigal Genius,* p. 115; Thompson, *Polyphased Electric Currents,* 1897, p. 88.

52. T. Hughes, *Networks of Power* (Baltimore: Johns Hopkins University Press, 1983), p. 118.

53. EW. Silvanus P. Thompson, portrait, January 9, 1892, p. 20; Thompson, 1897, pp. 93–96.

54. Tesla stated that the invention began to take form while he was attending the University of Prague. A review of the holdings of the university by the director of archives, K. Litsch, reveals that *Philosophical Magazine* was not subscribed to at that time.

55. Ibid., Tesla quoted in Thomson, 1897, pp. 96–97.

56. "Sweeping Decision of the Tesla Patents," *Electrical Review,* September 19, 1900, pp. 288–91.

57. "Westinghouse Sues General Electric on the Tesla Patents," *Electrical Review,* March 22, 1899, p. 183; 9/19/1900, pp. 288–91; "Tesla Split-Phase Patents," *Electrical World,* April 26, 1902, p. 734; "Tesla Patent Decision," May 17, 1902, p. 871; September 30, 1903,p. 470.

58. "Tesla Split-Phase Patents," *Electrical Review,* p. 291.

Chapter 4: Tesla Meets the Wizard of Menlo Park, pp. 27–39

1. Thomas Edison, quoted in "Wizard Edison Here. 'Sage of Orange' Tells About Tesla's Enormous Appetite as a Youth," *Buffalo New York News,* August 30, 1896 [TAE].

2. Charles Batchelor to T. Edison, November 21, 1881 [TAEs].

3. Alfred O. Tate, *Edison's Open Door* (New York: Dutton, 1938), p. 148; *New York Evening Sun,* December 22, 1884.

4. Batchelor papers, Edison Archives; *New York Evening Sun,* ibid.

5. Szigeti, in NT, 1961.

6. Charles Batchelor to T. Edison, January 2, 1881 [TAE].

7. Ibid., November 26, 1881.

8. Ibid., October 22, 1883.

9. NT, *My Inventions,* p. 66.

10. Ibid.

11. TCM, "Nikola Tesla," *Century,* 1894, p. 4.

12. Branimir Jovanovich interview, Belgrade, 1986.

13. NT, *My Inventions,* p. 34.

14. Ibid., pp. 34–35.

15. The timing is taken from NT, *My Inventions,* and Batchelor to T. Edison, September 24, 1882, and November 22, 1882 [TAE].

16. Charles Batchelor Papers, 1883 [TAE].

17. NT, *My Inventions,* p. 67.

18. Batchelor to T. Edison, January 23, 1883 [TAE].

19. Batchelor to T. Edison, January 9, 1882 [TAE].

20. Ibid., October 28, 1883 [TAE].

21. NT, *My Inventions,* p. 67.

22. Ibid.

23. Szigeti, in NT, *Tribute to Nikola Tesla* (1961), pp. A399–400.

24. NT, *My Inventions,* p. 67.

25. Ibid., p. 70.

26. O'Neill, *Prodigal Genius,* p. 60.

27. Batchelor to T. Edison, March 1884 [TAE]. It is possible that Batchelor returned to Paris before Tesla's arrival between March and late spring of 1884.

28. Edison, "Wizard Edison Here...," August 30, 1896.

29. O'Neill, *Prodigal Genius,* p. 58.

30. Thomas Edison, quoted in "An Interview With the Most Interesting Man in the World," *New York Journal,* July 26, 1891.

31. Batchelor to T. Edison, October 23, 1883 [TAE].

32. NT to RUJ, April 5, 1900 [BLCU].

33. Nicholas Kosanovich, ed. and trans., *Nikola Tesla: Correspondence with Relatives* (Lackawanna, New York: Tesla Memorial Society and the Nikola Tesla Museum, 1995), p. iv.

34. NT, *My Inventions,* p. 70.

35. NT, *Tesla Said,* Letter to the National Institute of Immigrant Welfare (May 11, 1938), in John Ratzlaff, ed. (Milbrae, Calif.: Tesla Book Co., 1984), p. 280.

36. M. Josephson, *Thomas Alva Edison* (New York: McGraw-Hill, 1959), p. 178.

37. Ibid., p. 184; H. Passer, *The Electrical Manufacturers: 1875–1900* (Cambridge, Mass.: Harvard University Press, 1953), pp. 144–45, 178–79.

38. Josephson, *Thomas Alva Edison,* pp. 194–99.

39. Robert Conot, *Streak of Luck* (New York: Bantam Books, 1979), pp. 151–52.

40. R. Silverberg, *Light for the World* (Princeton, N.J.: Van Nostrand, 1967), pp. 134–35.

41. Herbert Satterlee, *J. Pierpont Morgan* (New York: Macmillan, 1939), p. 207.

42. TCM, "Nikola Tesla," *Electrical World,* 1890, p. 106.

43. NT, "Letter to National Institute..." (May 11, 1938) in *Tesla Said,* p. 280.

44. Ibid.

45. NT, quoted in "Tesla Has Plan to Signal Mars," *New York Sun,* July 12, 1937.

46. NT, "Some Personal Recollections," *Scientific American,* June 5, 1915, p. 537, 576–77.

47. W. Dickson and A. Dickson, *The Life and Inventions of T. A. Edison* (New York: Thomas Crowell, 1892), p. 236.

48. NT, "Letter to National Institute..." (May 11, 1938), in *Tesla Said,* p. 208.

49. TCM, "Nikola Tesla," *Century,* February 1894, pp. 582–85.

50. NT, quoted in "Tesla Says Edison Was an Empiricist," *New York Times,* October 19, 1931, p. 25.

51. TCM, "Nikola Tesla," *Century,* February 1894, p. 583.

52. NT, *My Inventions,* p. 71.

53. NT, October 19, 1931, quoted in "Tesla Says Edison Was..." October 19, 1931.

54. NT, *My Inventions,* p. 72.

55. NT, October 19, 1931, quoted in "Tesla Says Edison Was..." October 19, 1931.

56. Josephson, *Thomas Alva Edison,* p. 9.

57. Harold Passer, *The Electrical Manufacturers: 1875–1900* (Cambridge, Mass.: Harvard University Press, 1953), p. 180.

58. Carole Klein, *Gramercy Park: An American Bloomsbury* (Boston: Houghton Mifflin, 1987).

59. T. Edison, "Pearl Street," *Electrical Review,* January 12, 1901, pp. 60–62 [condensed].

60. R. Conot, *Streak of Luck: The Life Story of Edison* (New York: Bantam, 1981), p. 305.

61. Ibid., p. 259.

62. David Woodbury, *Beloved Scientist: Elihu Thomson* (New York: McGraw-Hill, 1944), pp. 155–57.

63. F. Dyer and T. C. Martin, *Edison: His Life and Inventions* (New York: Harper Bros., 1910), p. 391.

64. Josephson, *Thomas Alva Edison,* pp. 230–32.

65. Edison, "Pearl Street," January 12, 1901.

66. Dickson and Dickson, *Life and Inventions,* p. 236.

67. NT, *My Inventions,* p. 72.

68. Kenneth Swezey, "Nikola Tesla," *Science,* May 16, 1958, pp. 1147–58; NT, *My Inventions,* p. 72.

69. Ibid., p. 41.

70. Alfred Tate, *Edison's Open Door* (New York: Dutton, 1938), p. 147.

71. Batchelor correspondence, July 14, 1884 [TAE].

72. Tate, *Edison's Open Door,* 146–47.

73. NT, "Some Personal Recollections," *Scientific American,* June 5, 1915, pp. 537, 576–77.

74. NT, Letter to National Institute... May 11, 1938, in *Tesla Said*, p. 280.

75. Kenneth Swezey, archival material, Smithsonian Institution, Washington, D.C.

76. Conot, *Streak of Luck*, pp. 272x73 [condensed].

Chapter 5: Liberty Street, pp. 40–50

1. NT, "Tesla Has Plan to Signal Mars," *New York Sun*, July 12, 1937, p. 6.

2. Leland Anderson, ed., *Nikola Tesla: On His Work With Alternating Currents and Their Application to Wireless Telegraphy, Telephony and Transmission of Power* (Denver, Colo.: Sun Publishing, 1992). This work contains Tesla's original testimony before his patent attorneys on the origins of the invention of the wireless in 1916.

3. NT, March 18, 1891/1980, p. 15; NT, 1959, p. P-199; R. Conot, *Streak of Luck: The Life Story of Edison* (New York: Bantam, 1981), p. 597.

4. "Tesla Electric Co." (advertisement), *Electrical Review*, September 14, 1886, p. 14.

5. NT, *My Inventions: The Autobiography of Nikola Tesla*, Ben Johnston, ed., p. 72; Anderson, *Nikola Tesla*, p. 12.

6. "Tesla Electric Co.," September 14, 1886, p. 14.

7. Kulishich, "Tesla Nearly Missed His Career," 1931.

8. NT, Letter to the National Institute of Immigrant Welfare (May 11, 1938), in *Tesla Said*, 1984, p. 280.

9. NT, *My Inventions*, p. 72.

10. NT, Letter to the National Institute of Immigrant Welfare (May 11, 1938), in *Tesla Said*, 1984, p. 280.

11. Ibid.

12. Alfred S. Brown, "Arc Lamp Patents," *Electrician and Electrical Engineer*, 1886.

13. NT. 12/1931, p. 78.

14. Hugo Gernsback, "Tesla's Egg of Columbus," *Electrical Experimenter*, March 19, 1919, p. 775 [paraphrased].

15. NT. *Nikola Tesla: Lectures, Patents, Articles* (Belgrade: Nikola Tesla Museum, 1956).

16. O'Neill, *Prodigal Genius*, p. 67.

17. TCM, *Nikola Tesla*, 1890, p. 106.

18. "Thomas Commerford Martin Dies," *Electrical World*, May 24, 1924, p. 1100.

19. Ibid.; *Who's Who of Electrical Engineers*, 1924 ed.

20. W. J. Johnston, "Mr. Martin's Lawsuit: Why and How It Failed," *Electrical World*, Part I, September 30, 1893, pp. 253–54; Part VII, November 11, 1893, pp. 382–87.

21. "Thomas Commerford Martin Dies," *Electrical World*, May 24, 1924.

22. Ibid., p. 5.

23. M. Josephson, *Thomas Alva Edison* (New York: McGraw-Hill, 1959), p. 356.

24. H. Byllesby to GW, May 21, 1888 [GWA].

25. Leonard Curtis in Henry Prout, *George Westinghouse: An Intimate Portrait* (New York: Wiley, 1939), p. 101.

26. Prout, pp. 101–4.

27. Charles F. Scott, "Early Days in the Westinghouse Shop," *Electrical World*, September 20, 1924, p. 586.

28. T. Hughes, *Network of Power* (Baltimore: Johns Hopkins University Press, 1983), pp. 101–3.

29. Prout, *George Westinghouse*, p. 95.

30. Scott, "Early Days."

31. Robert Silverberg, *Light for the World* (Princeton, N.J.: Van Nostrand, 1967), p. 233.

32. Alfred O. Tate, *Edison's Open Door* (New York: Dutton, 1938), p. 148.

33. Laurence Hawkins, *William Stanley: His Life and Times* (New York: Newcomen Society, 1939).

34. George Westinghouse, "No Special Danger," *New York Times*, December 13, 1888, 5:3.

35. Josephson, *Thomas Alva Edison*, p. 346.

36. David Woodbury, *Beloved Scientist: Elihu Thomson* (New York: McGraw-Hill, 1944), pp. 169, 179.

37. Josephson, *Thomas Alva Edison*, p. 346.

38. N. Tesla, "A New Alternating Current Motor," *Electrician*, June 15, 1888, p. 173.

39. Leland Anderson, Nikola Tesla (slide presentation) (Colorado Springs, Colo.: International Tesla Society, 1988) symposium. August 1988.

40. William Anthony, quoted in NT, "A New System of Alternate Current Motors and Transformers," (May 16, 1888), in *Lectures, Patents and Articles* (1956), p. L11.

41. Elihu Thomson, quoted in NT, i bid., p. L12.

42. Ibid., p. L12.

43. H. Byllesby to GW, May 21, 1888.[GWA].

44. H. Byllesby to GW, May 21, 1888 [GWA].

45. Ibid.; see also Harold Passer, *The Electrical Manufacturers: 1875–1900* (Cambridge, Mass.: Harvard University Press, 1953), p. 175.

46. H. Byllesby to GW, December 13, 1888.

47. C. C. Chesney and Charles F. Scott, "Early History of the AC System in America," *Electrical Engineering,* March 1936, pp. 228–35.

48. NT. "Mr. Tesla on Alternating Current Motors," letter to the editor, *Electrical World,* May 25, 1888, pp. 297–98; NT, *Tesla Said,* (1984), p. 4.

49. Henry Carhart, "Professor Galileo Ferraris," *Electrical World,* February 1887, p. 284, "as I understand it, there is a gigantic step from Ferraris' whirling pool to Tesla's whirling magnetic field," Pupin to Tesla, December 19, 1891 [NTM].

50. Passer, *Electrical Manufacturers,* p. 177.

51. G. Westinghouse, internal memorandum. July 5, 1888 [GWA].

52. Ibid.; see also Passer, *Electrical Manufacturers,* pp. 277–78.

Chapter 6: Induction at Pittsburgh, pp. 51–60

1. N. Tesla, "Death of Westinghouse," *Electrical World,* March 21, 1914, p. 637.

2. Charles F. Scott, "Early Days in the Westinghouse Shops," *Electrical World,* September 20, 1924, pp. 585–87.

3. Ibid., p. 586.

4. NT, "Tribute to George Westinghouse," *Electrical World & Engineer,* March 21, 1914, p. 637.

5. H. Passer, *The Electrical Manufacturers: 1875–1900* (Cambridge, Mass.: Harvard University Press, 1953), p. 279.

6. G. Westinghouse, memorandum, July 11, 1888 [GWA].

7. Undated memorandum [GWA]; Passer, *Electrical Manufacturers,* said that the author was Byllesby, July 7, 1888.

8. NT to GW, January 2, 1900 [LC].

9. NT to GW, September 12, 1892; November 29, 1898 [LC].

10. NT to JJA. January 6, 1899 [NTM].

11. Westinghouse Co. annual report, *Electrical Review,* June 30, 1897, p. 313.

12. The figure most often noted is $1 million, and the source is O'Neill. This same amount was mentioned by R. U. Johnson in his chapter on Tesla in his autobiography, "This to the man who had sold the inventions used at Niagara to the Westinghouse Company for a million dollars and lived to rue the bargain!" (*Remembered Yesterdays* [Boston: Little Brown, 1923], 401). As Johnson was Tesla's closest confidant, the figure must have originally come from Tesla.

13. Letter to Westinghouse Corporation, February 6, 1898 [LC]; Tesla may have also been influenced by the consensus concerning the noble profession of scientist. For instance, Louis Pasteur also refused to seek financial compensation for his discoveries. To do so, Pasteur said, a scientist would "lower himself.... A man of pure science would complicate his life and risk paralyzing his inventive faculties" (quoted in M. Josephson, *Thomas Alva Edison* [New York: McGraw-Hill, 1959], p. 336).

14. Leland Anderson, ed., *Nikola Tesla: On His Work With Alternating Currents...* (1916), pp. 64–65.

15. P. Callahan, "Tesla Stationary Obtained from Tesla Museum."

16. Scott, September 20, 1924.

17. Charles F. Scott, to NT, July 10, 1931 [BCU].

18. Ibid.

19. L. Hawkings, *William Stanley: His Life and Times* (New York: Newcomen Society, 1939), p. 32; Stanley advertisement, "The S.K.C. Two Phase System," *Electrical Review,* January 16, 1895, p. vii.

20. Charles F. Scott, *George Westinghouse Commemoration* (New York: American Society of Mechanical Engineers, 1936, 1985), p. 21.

21. Henry Prout, *W. Westinghouse: An Intimate Portrait* (New York: Wiley, 1939), p. 129.
22. NT, *My Inventions*, p. 23.
23. "Brown Executes Dogs," *New York Times*, July 31, 1888, 4:7.
24. "A Humane Method of Capital Punishment," *Electrical Review*, December 24, 1887; "One Dead Dog," ibid., July 20, 1889, p. 2.
25. "Edison and Capital Punishment," *Electrical Review*, June 30, 1888, p. 1; "Edison Says It Will Kill," *New York Sun*, July 4, 1889.
26. "Electricity on Animals," *New York Times*, December 13, 1888, p. 2.
27. George Westinghouse, "No Special Danger," *New York Times*, December 13, 1888, p. 5.
28. Harold P. Brown, "Electric Currents," *New York Times*, December 18, 1888, p. 5.
29. "Cockran Debates McKinley at Madison Square Garden," *New York Press*, August 19, 1896, pp. 1–2.
30. "Electricity as a Means of Execution," *Elecrical Review*, August 3, 1889; "Edison Says It Will Kill," *New York Sun*, July 24, 1889.
31. "Electricity as a Means," *Electrical Review*, August 3, 1889.
32. "Electrical Execution a Failure," *Electrical Review*, August 16, 1890, pp. 1–2.
33. "Kemmler Dies in Electric Chair," *New York Times*, August 6, 1890, p. 1.
34. B. Lamme, *An Autobiography* (New York: Putnam's, 1926), p. 60.
35. O'Neill, *Prodigal Genius*, p. 83.
36. B. Lamme, *Autobiography*, p. 60.
37. Ibid., p. v.
38. Francis Jehl, *Menlo Park Reminiscences* (Dearborn, Mich.: Edison Institute, 1939), p. 336.
39. Charles F. Scott, "Nikola Tesla's Achievements in the Electrical Art," *AIEE Transactions*, 1943, p. 3.

Chapter 7: Bogus Inventors pp. 61–65

1. "Who Is the Greatest Genius of Our Age?" *Review of Reviews*, July 1890, p. 45.
2. Nikola Tesla, "The True Wireless," *Electrical Experimenter*, May 1919, p. 28, in NT, *Solutions to Tesla's Secrets*, J. Ratzlaff, ed. (1981), p. 62.
3. John O'Neill, *Prodigal Genius* (New York: Ives Washburn, 1944), p. 77.
4. NT to JPM, December 10, 1900 [LC]
5. NT, "On the Dissipation of the Electrical Energy of the Hertz Resonator," *Elecrical Engineer*, December 21, 1892; in NT, *Tesla Said*, J. Ratzlaff, ed. (Milbrae, Calif.: Tesla Book Co., 1984), p. 22.
6. J. G. O'Hara and W. Pricha, *Hertz and the Maxwellians* (London: Peter Peregrinus Ltd. in assoc. with the Science Museum, 1987), p. 42.
7. NT, December 21, 1892; "New Radio Theories," *New York Herald Tribune*, Sepember 22, 1929, in NT, *Tesla Said*, pp. 225–26.
8. Nikola Pribic, "Nikola Tesla: The Human Side of a Scientist," *Tesla Journal* no. 2/3, 1982–83, p. 25.
9. 1889 newspaper clipping, Edison Archives, Menlo Park, N.J.
10. R. Conot. *Streak of Luck* (New York: Bantam Books, 1981), pp. 344–46; M. Josephson, *Thomas Alva Edison* (New York: McGraw-Hill, 1959), pp. 335–37.
11. Ambrose Fleming, "Nikola Tesla," in NT, *Tribute to Nikola Tesla: Letters, Articles* (1961), p. A-222.
12. Louis Hamon, *My Life With the Occult* (Garden City, N.Y.: Doubleday, 1933, 1972), p. 243.
13. *Review of Reviews*, July 1890, p. 45.
14. "Was Keely a Charlatan?" *Public Opinion*, December 1, 1898, p. 684.
15. T. Carpenter Smith, "Our View of the Keely Motor," *Engineering Magazine*, vol. 2, 1891–92, pp. 14–19.
16. "Keely Not Yet in Jail," *New York Times*, September 19, 1888, p. 1.
17. "Keely's Latest Move," *New York Times*, August 24, 1888, p. 5.
18. "Keely in Contempt," *New York Times*, November 11, 1888, p. 6.
19. "Inventor Keely in Jail," *New York Times*, November 18, 1888, p. 3.
20. Francis Lynde Stetson, quoted in *William Birch Rankine*, deLancy Rankine, ed. (Niagara Falls, N.Y.: Power City Press, 1926), p. 30.
21. "Science and Sensationalism," *Public Opinion*, December 1, 1898, pp. 684–85.

22. W. Barrett, "John W. Keely," in R. Bourne, ed., *The Smithsonian Book of Invention* (New York: Norton, 1978), pp. 120–21.
23. Ibid.
24. NT to RUJ, June 12, 1900 [BLCU].

Chapter 8: South Fifth Avenue pp. 66–72

1. Joseph Wetzler, "Electric Lamps Fed From Space, and Flames That Do Not Consume," *Harper's Weekly,* July 11, 1891, p. 524.
2. NT to Petar Mandic, August 18, 1890, in Nicholas Kosanovich, ed. and trans., *Nikola Tesla: Correspondence with Relatives* (1995), p. 15.
3. Ibid., May 17, 1894.
4. Ibid. Angelina Trbojevic to NT, January 2, 1897, p. 65.
5. Ibid. Jovo Trbojevic to Nikola Tesla, February 27, 1890; Milutin Tesla (a cousin) to Nikola Kosanovic, November 10, 1892.
6. Ibid. NT to Petar Mandic, December 8, 1893, p. 41.
7. Ibid. NT to Pajo Mandic, January 23, 1894, p. 42.
8. Ibid. Milkin Radivoj to NT, September 24, 1895, p. 51.
9. Karl Marx, "The Materialist Conception of History;" in P. Gardiner, ed., *Theories of History* (Glencoe, Ill.: Free Press, 1959), p. 134.
10. NT, "Problem of Increasing Human Energy," *Century,* June 1900, pp. 178–79.
11. T. C. Martin to NT, August 5, 1890 [NTM].
12. William Anthony, "A Review of Modern Electrical Theories," *AIEE Transactions,* February 1890, pp. 33–42. See also J. Ratzlaff and L. Anderson, *Dr. Nikola Tesla Bibliography, 1884–1978* (Palo Alto, Calif.: Ragusen Press, 1979), p. 6.
13. M. Pupin, *From Immigrant to Inventor* (New York: Scribners, 1923), p. 144.
14. Oscar May, "The High-Pressure Transmission of Power Experiments at Oerlikon," *Electrical World,* April 18, 1891, p. 291.
15. Louis Duncan, "Portrait," *Electrical World,* April 5, 1890, p. 236; "Alternating Current Motors, Part 2," June 16, 1891, pp. 357–58; Ratzlaff and Anderson, *Bibliography,* p. 7.
16. Pupin, *From Immigrant to Inventor*, pp. 283–84.
17. Elihu Thomson, "Phenomena of Alternating Currents of Very High Frequency," *Electrical World,* April 4, 1891, p. 254.For previous aspects of the debate, see also E. Thomson, "Notes on Alternating Currents of Very High Frequency, *Electrical World,* March 14, 1891, pp. 204–5; "Phenomena of Alternating Currents of Very High Frequency," *Electrical World,* April 11, 1891, pp. 223–24.
18. NT, "High Frequency Experiments," *Electrical World,* February 21, 1891, pp. 128–30.
19. *Electrical World,* February 21, 1891, pp. 128–30.
20. Wetzler, "Electric Lamps Fed From Space," *Harper's Weekly,* July 11, 1891, p. 524.
21. Ibid.
22. E. Raverot, "Tesla's Experiments in High Frequency," *Electrical World,* March 26, 1892.
23. Gano Dunn to NT, June 1931, in NT, *Tribute to Nikola Tesla: Letters, Articles* (1961), LS-54.
24. It was the term "without effort" which I believe has been misinterpreted. From Tesla's point of view, energy was not truly available without effort. Machines instead of humans could be constructed that would extract this "free energy." Solar, wind, and water power are all ways to extract "free energy" without the exertion of human effort.
25. Sperry's gyroscope, of course, is based upon the principles inherent in the Tesla rotating egg, and Tesla should therefore be considered ahead of Sperry in this invention.
26. *Electrical World,* May 20, 1891, p. 288.
27. Robert Millikan to NT, 1931, in NT, *Tribute to Tesla,* p. LS-30.
28. Petkovich, p. 3.
29. Michael Pupin to NT, December 19, 1891, in NT, *Tribute to Tesla,* p. LS-11.

Chapter 9: Revising the Past, pp. 73–82

1. Charles Steinmetz, *Alternating Current Phenomena* (New York: McGraw-Hill, 1900), pp. i–ii [condensed].

2. Oscar May, "The High-Pressure Transmission of Power Experiments at Oerlikon," *Electrical World,* April 18, 1891, p. 291.

3. T. Hughes, *Networks of Power* (Baltimore: Johns Hopkins University Press, 1983), pp. 131–33.

4. Ibid.

5. Dragislov Petkovich, "A Visit to Nikola Tesla," *Politika,* April 27, 1927, p. 3.

6. Hughes, *Network of Power.*

7. "C. E. L. Brown Portrait," *Electrical World,* October 12, 1891, p. 284.

8. M. Dobrowolsky, "Electrical Transmission of Power by Alternating Currents," *Electrical World,* September 14, 1891, p. 268.

9. Carl Hering, "Comments on Mr. Brown's Letter," *Electrical World,* November 7, 1891, p. 346.

10. Jonathan Leonard, *Loki: The Life of Charles Proteus Steinmetz* (Garden City, N.Y.: Doubleday, 1932), p. 109.

11. John Winthrop Hammond, *Charles Proteus Steinmetz* (New York: Century Co., 1924).

12. "Charles Steinmetz," in M. Pupin, "Pupin on Polyphasal Generators," *AIEE Transactions,* December 16, 1891, pp. 591–92.

13. Harold Passer, *The Electrical Manufcturers: 1875–1900* (Cambridge, Mass.: Harvard University Press, 1953).

14. NT to Villard, October 10, 1892 [Houghton Library, Harvard University].

15. M. Josephson, *Edison: A Biography* (New York: McGraw-Hill, 1959), p. 361.

16. Ibid., p. 392.

17. Ibid.

18. J. Leonard, *Loki: The Life of Charles Proteus Steinmetz* (Garden City, N.Y.: Doubleday, 1928), p. 202.

19. H. Prout, *George Westinghouse: An Intimate Portrait* (New York: Wiley, 1939), p. 125.

20. *Electrical World,* September 16, 1893, p. 208, cited in Passer, *Electrical Manufacturers,* p. 292.

21. Charles Steinmetz, *Theoretical Elements of Electrical Engineering* (New York: McGraw-Hill, 1902), pp. iii–iv.

22. Pupin, *From Immigrant to Inventor,* pp. 285–86.

23. Ibid., p. 289.

24. Gisbert Kapp to NT, in NT, *Tribute to Nikola Tesla,* p. LS-6.

25. B. A. Behrend, *The Induction Motor* (New York: McGraw-Hill, 1921), p. 1.

26. C. E. L. Brown, "Reasons for the Use of the Three-Phase Current in the Lauffen-Frankfort Transmission," *Electrical World,* November 7, 1891, p. 346.

27. Carl Hering, "Comments on Mr. Brown's Letter," in ibid., p. 346.

28. W. H. Johnston, "Mr. Tesla and the Drehstrom Systems," *Electrical World,* February 6, 1892, p. 83.

29. Carl Hering, "Mr. Tesla and the Drehstrom System," *Electrical World,* February 6, 1892, p. 84.

30. Behrend, *Induction Motor,* pp. xiii–xiv.

31. Ibid., p. 261.

Chapter 10: The Royal Society: pp. 83–97

1. "Mr. Tesla Before the Royal Institution, London," *Electrical Review,* March 19, 1892, p. 57.

2. The Tesla oscillator conceived at this time became the basis for all of his later transmitters, such as at Colorado Springs and also Wardenclyffe (see especially, patent nos. 462,418—November 13, 1891; 514,168—February 6, 1894; and 568,178—September 22, 1896).

3. NT, "Electric Oscillators," *Electrical Experimentation* (July 7, 1919), in NT, *Nikola Tesla: Lectures,* 1956, p. A-78–93.

4. NT, "The Problem of Increasing Human Energy," *Century,* June 1900, p. 203.

5. T. C. Martin, "Tesla's Oscillator and Other Inventions," *Century,* April 1895.In NT, *Tribute to Nikola Tesla,* p. A-16.

6. "NT and J.J. Thomson" (1891), in NT, *Nikola Tesla: Lectures,* 1956, pp. A-16–21.

7. NT, "High Frequency Oscillators for Electro-Therapeutic and Other Purposes," *Electrical Engineer,* November 17, 1898, pp. 477–81.

8. T. C. Martin, J. Wetzler, and G. Sheep to Tesla, January 8, 1892 [NTM].

9. William Preece to NT, January 16, 1892 [NTM].

10. M. Josephson, *Thomas Alva Edison* (New York: McGraw-Hill, 1959), pp. 275–77; E. C. Baker, *Sir William Preece: Victorial Engineer Extaordinary* (London: Hutchinson, 1976), pp. 185–86.

11. "Mr. Tesla Before the Royal Institution, London," *Electrical Review,* March 19, 1892, p. 57; NT, *The Inventions, Researches and Writings of Nikola Tesla,* T. C. Martin, ed. (New York: Electrical Review Publishing Company, republished, Mokelumne Hill, Calif.: Health Research, 1970), p. 200.

12. Most of the titles of these distinguished scientists were obtained later in their career; for example, Dewar became knighted in 1904; Fleming in 1924.William Thomson became Baron or Lord Kelvin a few months after Tesla's lecture.

13. Ibid., p. 198 [paraphrased].

14. Ibid., p. 200.

15. Ibid., p. 186.

16. Ibid.

17. NT, *Inventions, Researches,* pp. 130–131; 228–229 [paraphrased in part].

18. Ibid., pp. 287–88 [paraphrased].

19. Ibid., p. 235.

20. W. Kock, *Engineering Applications of Lasers and Holography* (New York: Plenum Press, 1975), pp. 28–35. I Hunt and W. Draper, *Lightning in His Hands: The Life Story of Tesla* (Hawthorne, Calif.: Omni Publications, 1964), were the first to suggest that Tesla invented the laser.

21. NT, "On Electrical Resonance," *Electrical Engineer,* June 21, 1893, pp. 603–5.

22. NT, "On Light and High Frequency Phenomena," *Electrical Engineer,* March 8, 1893, pp. 248–49.

23. NT, 1916/1992, on his work with alternating currents, p. 62.

24. NT, "Mr. Tesla Before the Royal Institution," pp. 247–49.

25. Ibid., pp. 250–52.

26. Ibid., p. 292.

27. Ibid.

28. Ibid. [paraphrased in part].

29. Isaac Asimov, *Asimov's Biographical Encyclopedia of Science and Technology* (Garden City, N.Y.: Doubleday, 1964), p. 347.

30. NT, *My Inventions,* p. 82.

31. Leland Anderson, Slide presentation and lecture before the International Tesla Society, Colorado Springs, Colo., August 1988.

32. NT, *My Inventions,* p. 82 [condensed].

33. J. A. Fleming to NT, February 5, 1892, in NT, *Tribute to Nikola Tesla* 1961, p. LS-13.

34. Asimov, *Asimov's Biographical Encyclopedia,* p. 364.

35. William Crookes to NT, March 5, 1892, in NT, *Tribute to Nikola Tesla,* p. LS-12.

26. William Crookes, "Some Possibilities of Electricity," *Fortnightly Review,* February 1892, pp. 173–81.

37. Crookes became president of the Society of Psychical Research in 1896; Lodge, in 1901; and Rayleigh, in 1919.J. J. Thomson was a vice president. See A. Koestler, *Roots of Coincidence* (New York: Vintage, 1972), pp. 32–34.

38. William Crookes, "D.D. Home," *Quarterly Journal of Science,* January 1874 [condensed]. See also C. J. Ducasse, "The Philosophical Importance of Psychic Phenomena," in J. Ludwig, ed., *Philosophy and Parapsychology* (Buffalo, N.Y.: Prometheus Books, 1978), p. 138.

39. Crookes to NT, March 5, 1892.

40. NT, "Elliott Cresson Gold Medal Presentation," in *Tribute to Nikola Tesla,* p. D-4.

41. NT, "Mechanical Therapy" (undated), in *Tesla Said,* p. 286.

42. Robert O. Becker, "Direct Current Neural Systems," *Psychoenergetic Systems* 2 (1976), pp. 190–91.

43. "Tesla's Experiments," *Electrical Review,* April 9, 1892, p. 1.

44. NT to GW, September 12, 1892 [LC].

45. NT, *Tribute to Nikola Tesla,* p. LS-69; see also B. A. Behrend, *The Induction Motor* (New York: McGraw-Hill, 1921), pp. 6–7.

46. NT, *My Inventions,* pp. 94–95.

47. Ibid., p. 95.
48. Ibid., p. 104.
49. Ibid.
50. Ibid., pp. 104–5.
51. William Broad to author, 1986.
52. "Honors to Nikola Tesla from King Alexander I," in *Electrical Engineer,* February 1, 1893, p. 125.
53. N. Pribic, "Nikola Tesla: The Human Side of a Scientist," *Tesla Journal* November 2 and 3, 1982/1983, p. 25.
54. Ambrose Fleming, "Nikola Tesla," *Journal of Institution of Electrical Engineers,* London, 91, February 1944, in *Tribute to Nikola Tesla,* p. A-215.
55. J. G. O'Hara and W. Pricha, *Hertz and the Maxwellians* (London: Peter Peregrinus, 1987), p. 5.
56. Hertz's decision to eliminate scalar potentials was also a puzzlement to Oliver Heaviside, who corresponded frequently with the German scientist during this same period. "I am quite of your opinion, that you have gone further on than Maxwell," Heaviside wrote in 1889, "[but] electrostatical (scalar) potential and magnetical (scalar) potential ought to remain I think." Heaviside, however, like Hertz, was in agreement with the idea of dispensing with vector potentials.
57. NT, "On the Dissipation of the Electrical Energy of the Hertz Resonator," *Electrical Engineer,* December 21, 1892, p. 587–88, in *Tesla Said,* pp. 22–23.
58. "NT tells of New Radio Theories," *New York Herald Tribune,* September 22, 1929, pp. 1, 29; in NT, *Tesla Said,* pp. 225–26.
59. NT, "The True Wireless," *Electrical Experimenter,* May 1919, p. 28.
60. Tesla researcher Tom Bearden has gone so far as to say that the Hertzian decision to eliminate scalar waves and vector potentials from Maxwell's equations created a flaw in the next theoretical development called quantum mechanics. It was for this reason, Bearden speculates, that Einstein could not create a unified field theory. Bearden suggests bringing back these components along with another abandoned aspect called quaternion theory. He further suggests that by utilizing Tesla transmitters to produce converging powerfully pumped scalar waves, spinners and twisters can be created, that is, local space/time can be curved, and large amounts of power can be transmitted wirelessly over long distances (Tom Bearden, "Scalar Waves and Tesla Technology," paper presented at the International Tesla Society Symposium, Colorado Springs, Colo., August 1988).
61. NT, *My Inventions,* p. 83.

Chapter 11: Father of the Wireless, pp. 98–109

1. NT, *The Inventions, Researches, and Writings of Nikola Tesla,* T. C. Martin, ed. (1893), p. 149.
2. J. Ratzlaff and L. Anderson, *Dr. Nikola Tesla Bibliography, 1884–1978* (Palo Alto, Calif.: Ragusen Press, 1970), p. 21.
3. John O'Neill, *Prodigal Genius: The Life Story of N. Tesla* (New York: Ives Washburn, 1944), p. 101.
4. Moses King, *King's Handbook of New York* (New York: F. A. Ferris & Co., 1894), p. 230
5. Walter Stephenson, "Nikola Tesla and the Electric Light of the Future," *Scientific American Supplement,* March 30, 1895, pp. 16408–09; NT to Simp. Majstorovic, Jan. 2, 1893, in *Correspondence with Relatives,* p. 31.
6. NT, "On the Dissipation of Electrical Energy of the Hertz Resolution," (Dec. 21, 1892), in *Tesla Said,* pp. 22–23.
7. NT, *Inventions, Researches and Writings,* p. 347.
8. NT to Fodor, September 9, 1892; November 27, 1892; January 1, 1893; March 19, 1893 [LC].
9. NT to Petar Mandic, Dec. 8, 1893, in *Correspondence with Relatives,* p. 41.
10. NT to Thurston, November 4, 1892; January 23, 1893; February 21, 1893; October 23, 1893 [WBP].
11. NT to GW, September 27, 1892 [LC].
12. Henry Prout, *George Westinghouse: An Intimate Portrait* (New York: Wiley, 1939), p. 143.
13. Reconstructed from NT to GW, September 12, 1892 [LC].
14. Benjamin Lamme, *An Autobiography* (New York: Putnam's, 1926), p. 66.

15. NT to GW, September 12, 1892 [LC].

16. Page Smith, *The Rise of Industrial America.* vol. 6 (New York: McGraw-Hill, 1984), p. 486–88.

17. NT, "On Light and Other High Frequency Phenomena" (Feb./Mar. 1893), in *Inventions, Researches,* pp. 294–95.

18. Ibid.

19. Ibid., p. 299.

20. Ibid., p. 299.

21. James Coleman, *Relativity for the Layman.* New York: Mentor Books, 1958, p. 44.

22. NT, "Radio Power Will Revolutionize the World," *Modern Mechanix & Invention, 71,* 1934, pp. 40–42, 117–19.

23. T. C. Martin, "The Tesla Lecture in St. Louis," *Electrical Engineer,* March 18, 1893, pp. 248–49.

24. NT, "Experiments with Alternate Currents..." (May 20, 1891), in *Inventions, Researches,* p. 148.

25. "An infinitesimal world, with molecules and their atoms spinning and moving in orbits, in much the same manner as celesial bodies, carrying with them and probably spinning with them ether, or in other words, carrying with them static charges, seems to my mind the most probable view, and one which in a plausible manner, accounts for most of the phenomena observed. The spinning of the molecules and their ether sets up the ether tensions or electrostatic strains; the equalization of ether tensions sets up ether motions or electric currents, and the orbital movements produce the effects of electro and permanent magnetism." NT, "Experiments With Alternate Currents of Very High Frequency and Their Application to Methods of Artificial Illumination," lecture delivered before the American Institute of Electrical Engineers at Columbia College (May 20, 1891). In T. C. Martin, ed., *The Inventions, Researches, and Writings of Nikola Tesla* (New York: Electrical Engineer, 1893), p. 149.

26. Orrin Dunlop, *Radio's 100 Men of Science* (New York: Harper and Bros., 1944), pp. 156–58.

27. NT, "How Cosmic Forces Shape Our Destiny," *New York American,* February 27, 1925, in *Lectures, Patents, Articles,* p. A-172.

28. Ibid.

29. NT, "On Light and Other High Frequency Phenomena," (Feb/March 1893), in *Inventions, Researches,* p. 301.

30. Ibid., p. 347.

31. Ibid., p. 347.

32. William Broughton Jr., "William Broughton Dedication Speech," Schenectady Museum, Schenectady, N.Y., February 6, 1976 [Nick Basura Archives].

33. NT, *Inventions, Researches and Writings,* p. 348.

34. NT, *My Inventions,* p. 29.

35. William Preece, "On the Transmission of Electrical Signals Through Space," *Electrical Engineer,* August 30, 1893, p. 209.

36. O. E. Dunlap, 1944, pp. 58–59; also James Corum lecture, *One Hundred Years of Resonator Development,* ITS Conference, Colorado Springs, Colo., 1992.

37. M. Josephson, *Thomas Alva Edison* (New York: McGraw-Hill, 1959), p. 128.

38. R. Conot, *Streak of Luck* (New York: Bantam, 1981), p. 95.

39. Preece, "On the Transmission of Electrical Signals."

40. A. Slaby, "The New Telegraphy," *Century,* 1897, pp. 867–77.

41. Oliver Lodge, *Talks About Wireless* (New York: Cassell, 1925), p. 32.

42. NT, "The True Wireless," *Electrical Experimenter,* May 1919, pp. 28–30, 61–63, 87; in *Solutions to Tesla's Secrets,* pp. 62–68.

Chapter 12: Electric Sorcerer, pp. 110–121

1. "New Electric Inventions," *New York Recorder,* June 15, 1891.

2. NT, "Nikola Tesla and His Wonderful Discoveries," *Electrical World,* April 29, 1893, pp. 323–24.

3. "Tesla and His Wonderful Discoveries," *New York Herald,* April 23, 1893; NT, "Nikola Tesla and His Wonderful Discoveries," pp. 323–24.

4. [WBP].

5. TCM, "Tesla's Lecture in St. Louis," *Electrcial Engineer,* March 8, 1893, pp. 248–49.

6. [WBP].

7. TCM, "Tesla's Lecture in St. Louis," *Electrical Engineer,* March 8, 1893.

8. NT, "On Light and Other High Frequency Phenomena," *Electrical Engineer,* June 28, 1893, p. 627.

9. NT, "Nikola Tesla & His Wonderful Discoveries," *Electrical World,* April 29, 1893, pp. 323–24.

10. Ibid.

11. NT, "On Phenomena Produced by Electric Force," in *Inventions, Researches and Writings,* February/March 1893, p. 318.

12. Ibid., p. 318–19.

13. TCM, "A New Edison on the Horizon," *Review of Reviews,* March 1894, p. 355.

14. Martin, "Tesla's Lecture in St. Louis," March 8, 1893.

15. NT, *Inventions, Researches and Writings,* p. 349.

16. M. Josephson, *Thomas Alva Edison* (New York: McGraw-Hill, 1954), p. 235.

17. Thomas Edison, "A Long Chat With the Most Interesting Man in the World," *Morning Journal,* July 26, 1891 [TAE].

18. NT, "Nikola Tesla and His Wonderful Discoveries," *Electrical World,* April 29, 1893, from *New York Herald,* April 23, 1893.

19. NT, *My Inventions,* p. 41.

20. Ibid., p. 83.

21. TCM, "Tesla's Oscillator and Other Inventions," *Century,* April 1895, pp. 916–33.

22. Ibid. See also NT, *Nikola Tesla: Lectures* 1956, pp. P-141–145, P-225–231.

23. NT, "On Phenomena Produced by Electrostatic Force," in *Inventions, Researches, and Writings,* February/March, 1893, pp. 319–21.

24. William Cameron. *The World's Fair: A Pictorial History of the Columbian Exposition* (New Haven, Conn.: James Brennan & Co., 1894), pp. 108, 669–70; Stanley Applebaum, *The Chicago World's Fair of 1893: A Pictorial Record* (New York: Dover, 1980), pp. 96–97, 106.

25. W. E. Cameron, *World's Fair,* pp. 641–85.

26. Ibid., p. 316.

27. Ibid., p. 318.

28. J. Barrett, *Electricity at the Columbian Exposition* (Chicago: Donnelley & Sons, 1894), pp. 168–69; "Mr. Tesla's Personal Exhibit at the World's Fair," *Electrical Engineer,* November 29, 1893, pp. 466–68.

29. Cameron, *World's Fair,* p. 325; G. R. Davis, *World's Columbian Exposition, 1893* (Philadelphia: W. Houston & Co., 1893), p. 127; *World's Fair Youth Companion* (Boston: 1893), p. 19.

30. "Electricians Listen in Wonder to the 'Wizard of Physics,'" *Chicago Tribune,* August 26, 1893 (Edison Archives).

31. "Tesla's Egg of Columbus," *Electrical Experimenter,* March 1919, p. 775.

Chapter 13: The Filipovs, pp. 122–131

1. TCM, "Nikola Tesla," *Century,* February 1894, pp. 582–85.

2. "Electricians Listen in Wonder to the 'Wizard of Physics'," August 26, 1893.

3. TCM, "A New on the Horizon," *Review of Reviews,* March 1894, p. 355.

4. Arthur Brisbane, "Our Foremost Electrician, Nikola Tesla," *World,* July 22, 1894.

5. Robert Underwood Johnson, *Remembered Yesterdays* (Boston: Little Brown, 1923).

6. W. T. Stephenson, "Electric Light of the Future," *Outlook* March 9, 1895, pp. 384–356.

7. Ibid. [The experience of this reporter was adapted to the Johnson meeting]

8. Ibid.

9. NT to RUJ, January 8, 1894 [BCU].

10. NT to RUJ, December 7, 1893 [BCU].

11. NT, "Introductory Note on Zmai," in R. U. Johnson, *Songs of Liberty and Other Poems* (New York: Century, 1897), pp. 43–47.

12. KJ to NT [NTM].

13. Ibid., April 3, 1896.

14. Ibid., December 6, 1897.

15. Ibid., June 6, 1898.

16. TCM to KJ, January 8, 1894 [BLCU].

17. TCM to NT, January 22, 1894 [NTM].
18. Johnson, *Remembered Yesterdays,* p. 400.
19. Ibid.
20. Mark Twain to NT, March 4, 1894; RUJ to NT, March 5, 1894; NT to RUJ, April 26, 1894 [BLCU].
21. Mark Twain Papers [BLCU].
22. F. Anderson, ed., *Mark Twain's Notebooks and Journals,* vol. 3, 1883–1891 (Berkeley, Calif.: University of California Press, 1979), p. 431.
23. Ibid.
24. NT, *My Inventions,* p. 53.
25. NT, 1897, pp. 286–87.
26. NT to RUJ, May 2, 1894 [BLCU].
27. NT to KJ, May 2, 1894 [BLCU].
28. TCM to NT, February 17, 1894 [NTM].
29. Nicholas Pribic, "Nikola Tesla: The Human Side of a Scientist," *Tesla Journal,* nos. 2 & 3 (1982–83), p. 25.
30. TCM to NT, February 6, 1894 [NTM].
31. J. Abraham and R. Savin, *Elihu Thomson Correspondence* (New York: Academic Press, 1971), p. 352.
32. TCM to RUJ, February 7, 1894 [BLCU].
33. NT, "Elliott Cresson Gold Medal Award," *Tribute to Nikola Tesla,* p. D-5.
34. RUJ to H. G. Osborn, May 7, 1894 [BLCU].
35. H. G. Osborn to Seth Low, January 30, 1894 [BLCU].

Chapter 14: Niagara Power, pp. 132–137

1. NT, *My Inventions,* p. 48.
2. E. D. Adams, *Niagara Power: 1886–1918* (New York: Niagara Falls Power Co., 1927), pp. 148–49; H. Passer, *The Electrical Manufacturers: 1875–1900* (Cambridge, Mass.: Harvard University Press, 1953), pp. 283–84.
3. Ibid.
4. T. Hughes, *Networks of Power* (Baltimore: Johns Hopkins University Press, 1983), pp. 97–98, 238–39.
5. J. A. Fleming, "Nikola Tesla," in *Tribute to Nikola Tesla* (1961), p. A-222.
6. Hughes, in *Networks of Power,* wrote, "It is difficult to understand why he [Ferranti] and his financial backers took such a great leap beyond the state of existing technology in their Depford project." Hughes, loath to give Tesla unequivocal credit, was therefore unable to make the connection.
7. H. Satterlee, *J. Pierpont Morgan: An Intimate Portrait* (New York: Macmillan, 1939), pp. 194, 221, 228, 269, 300, 307, 325.
8. R. Conot, *Streak of Luck: The Life Story of Edison* (New York: Bantam Books, 1981), p. 340.
9. H. Passer, *Electrical Manufacturers,* p. 285.
10. E. D. Adams, *Niagara Power,* pp. 173, 176, 185.
11. Charles Scott, "Nikola Tesla's Achievements in the Electrical Art," *AIEE Transactions,* 1943 [Archives, Westinghouse Corp.].
12. Ibid.
13. Ibid., pp. 179–87.
14. David Woodbury, *Beloved Scientist: Elihu Thomson* (New York: McGraw-Hill, 1944), p. 214.
15. *Electrical World,* May 25, 1895, p. 603.
16. H. Passer, *Electrical Manufacturers,* p. 292.
17. H. Prout, p. 144.
18. H. Passer, p. 298.
19. Woodbury; Abraham and Savin. Interestingly, Passer, 1953, whose work is a primary source for this event, completely misunderstood Tesla's central role in the Niagara project, even though he had access to the files of G.E. and Westinghouse. Passer could not understand why the contract was given to Westinghouse over G.E.
20. H. Passer, p. 292.

21. F. L. Stetson, in de Lancey Rankine, *Memorabilia of William Birch Rankine,* (Niagara Falls: Power City Press, 1926), p. 28.

22. "Nikola Tesla and His Works," *Review of Reviews,* August 8, 1894, p. 215.

23. "Nikola Tesla and His Work," *New York Times,* September 16, 1894, 20:1–4.

24. "Tesla's Work at Niagara," *New York Times,* July 16, 1895, 10:5.

25. NT to JJA, January 6, 1899 [NTM].

26. "The Nikola Tesla Company," *Electrical Engineering,* February 13, 1895, p. 149.

27. NT to JJA, January 6, 1899 [NTM].

Chapter 15: Effulgent Glory, pp. 138–145

1. D. McFarlan Moore to NT, June 13, 1931.In *Tribute to Tesla* 1961, p. LS-41.

2. TCM to NT, February 6, 1894 (some paraphrasing for readability's sake).

3. TCM to NT, May 7, 1894.

4. T. C. Martin, "Tesla's Oscillator and Other Inventions," *Century,* April 1895, in *Tribute to Nikola Tesla,* 1961, pp. A-11–32.

5. Ibid., July 18, 1894.

6. NT to RUJ, December 4, 1894 [BLCU].

7. Ibid., p. A-20.

8. EE. "American Electr-Therapeutic Association"; "An Evening in Tesla's Laboratory," *Electrical Engineering,* October 3, 1894, pp. 278–79.

9. NT vs. Reginald A. Fessenden, *Interference,* 21:701, April 16, 1902, p. 20 [Scherff papers, BLCU].

10. Herbert Spencer, *The Principles of Biology* (New York: Appleton, 1896).

11. NT, April 16, 1902, p. 19; "The Transmission of Electrical Energy Without Wires As a Means for Furthering Peace," *Electrical World,* January 1905, pp. 21–24.

12. T. C. Martin, op. cit. April 1895, in *Tribute to Nikola Tesla,* 1961, pp. A-31–32.

13. Ibid.

14. "NT and his works," *Review of Reviews,* August 1894, p. 215.

15. "Tesla and Edison," *Watertower Times,* April 24, 1895 [TAE].

16. "Nikola Tesla," *Electrical World,* April 14, 1894, p. 489.

17. F. Jarvis Patten, "Nikola Tesla and His Work," *Electrical World,* April 14, 1894, pp. 496–99; "Tesla and Edison," *Watertower Times,* April 24, 1895 [TAE].

16. Arthur Brisbane, "Our Foremost Electrician," *New York World,* July 22, 1894, Sunday supplement.

17. "Tesla's Triumphs," *St. Louis Daily Globe Democrat,* March 2, 1893, p. 4.

20. NT, "Tuned Lightning," *English Mechanic and World of Science,* March 8, 1907, pp. 107–108.

21. W. T. Stephenson, "Electrical Light of the Future," *Outlook,* March 9, 1895, pp. 384–86.

22. NT vs. Fessenden, April 16, 1902, p. 14.

23. Ibid., Scherff's testimony, p. 89.

24. Patents 454,622 (June 23, 1891); 462, 418 (November 3, 1891); 514,168 (February 6, 1894). In *Lectures, Patents, Articles,* pp. P-221–27.

25. Michael Pupin papers [BLCU].

26. NT, "High Frequency and High Potential Currents," February 1892, in *Inventions, Researches and Writings* (1894), p. 292.

27. Pupin papers, March 28, 1894 [BLCU].

28. Ibid., August 23, 1895.

29. Ibid., May 21, 1895.

30. Ibid., July 25, 1896.

31. NT to RUJ, December 21, 1894 [BLCU].

Chapter 16: Fire at the Lab, pp. 146–151

1. Charles Dana, "The Destruction of Tesla's Workshop," *New York Sun,* March 13, 1895; in *Tribute,* 1961, p. LS-18.

2. D. McFarlan Moore to NT, June 13, 1931; in *Tribute,* 1961, p. LS-41.

3. John O'Neill, *Prodigal Genius,* 1944.

4. J. Ratzlaff and Leland Anderson, eds., *Dr. Nikola Tesla Bibliography* (Palo Alto, CA: Ragusen Press, 1979), p. 34.

5. T. C. Martin, "The Burning of Tesla's Laboratory," *Engineering Magazine*, April 1895, pp. 101–4.

6. "A Calamitous Fire," *Current Literature*, May 1895 [TAE].

7. Michael Boro Petrovich, *A History of Modern Serbia* (New York: Harcourt Brace, 1976), p. 523.

8. T. C. Martin, "The Burning..."

9. J. Ratslaff and L. Anderson, *Tesla Bibliography*, p. 34.

10. J. Abraham and R. Savin, *Elihu Thomson Correspondence* (New York: Academic Press, 1971), p. 352.

11. TCM to NT, May 20, 1895; May 21, 1895; May 28, 1895 [NTM].

12. H. Passer, *Electrical Manufacturers*, p. 297.

13. "Westinghouse Electric. Ad on Tesla polyphase system." *Review of Reviews*, June 1895, p. viii.

14. J. Ratzlaff and L. Anderson, *Tesla Bibliography*, p. 34.

15. John O'Neill, *Prodigal Genius*, p. 123. Concerning Tesla's expenses, including the loss from the fire and the construction of another lab, Tesla wrote, "Before I ever saw Colorado—I think my secretary knows that—I have expended certainly not less than $750,000," NT, *On His Work With A.C.*, p. 172.

16. NT to RUJ, February 14, 1895 [BLCU].

17. Ernest Heinreich to NT, February 13, 1895 [LC].

18. "Tesla in Jersey," *Rochester Express*, April 5, 1895 [TAE].

19. NT to A. Schmid (two letters combined), March 23, 1895; April 3, 1895 [LC].

20. Samuel Bannister to NT, April 8, 1895 [LC].

21. Brisbane, June 22, 1894.

22. "Edison's Rival," *Troy Press*, April 20, 1895 [TAE].

22. J. Ratzlaff and L. Anderson, *Tesla Bibliography*, p. 36.

24. "Tesla Solved the Problem," *Philadelphia Press*, June 24, 1895 [TAE].

25. "The Electric Combinations," *NY Com. Bulletin*, April 18, 1895 [TAE].

26. TCM to NT, May 22, 1895 [NTM].

27. TCM to NT, March 12, 1896 [NTM].

28. "Nikola Tesla and the Electrical Outlook," *Review of Reviews*, September 1895, pp. 293–94.

29. Ibid.

Chapter 17: Martian Fever pp. 152–157

1. Quoted in *New York Sun*, March 25, 1896.

2. John D. Gates, *The Astor Family* (Garden City, N.Y.: Doubleday, 1981), pp. 112–13.

3. Ibid.

4. Camille Flammarion, "Mars and Its Inhabitants," *North American Review* 162 (1896), p. 549.

5. William Pickering, "Pickering's Idea for Signaling Mars," *New York Times*, April 25, 1909, Pt. 5, 1:1–6 [some paraphrasing to improve readability].

6. NT to JJA, February 6, 1895 [NTM].

7. John Jacob Astor, *A Journey in Other Worlds* (New York: D. Appleton, 1894), pp. 115–16.

8. Ibid., p. 161.

9. George DuMaurier, *The Martian* (Boston: Little Brown, 1896).

10. Camille Flammarion, *Stories of Infinity* (Boston: Roberts Brothers, 1873).

11. Carl Jung, *The Portable Jung* (New York: Viking Press, 1961), p. 311.

12. D. Cohan, "Heavenly Hoax," *Air & Space* 4–5 (1986), pp. 86–92.

13. E. Morse, *Mars and Its Mystery* (Boston: Little Brown, 1906), pp. 52–53.

14. Camille Flammarion, "Mars and Its Inhabitants," *North American Review* 162 (1896), pp. 546–57.

15. "Strange Lights on Mars," *Nature*, August 2, 1894; "Mars Inhabited Says Prof. Lowell," *New York Times*, August 30, 1907, 1:7; "Signalling to Mars," *Scientific American*, May 8, 1909.

16. Perceival Lowell, *The Canals of Mars* (New York: Macmillan, 1906), pp. 376–77.

17. W. Von Braun et al., *The Exploration of Mars* (New York: Viking Press, 1956), pp. 84–85.
18. Ibid.; J. Abrahams and R. Savin, *Elihu Thomson Correspondence*.
19. NT, *Tribute*, p. LS-18.
20. "Is Tesla to Signal the Stars?" *Electrical World*, April 4, 1896, p. 369.

Chapter 18: High Society, pp. 158–166

1. Quoted in Paul Baker, *Stanny: The Gilded Life of Stanford White* (New York: Free Press, 1989), p. 137, circa February 25, 1894.
2. Frederick Finch Strong, "Electricity and Life," *Electrical Experimenter*, March 1917, pp. 798, 831.
3. Jennie Melvene Davis, "Great Master Magician Is Nikola Tesla," *Comfort*, May 1896 [NTM].
4. "The Field of Electricity: Edison, Tesla and Moore at Work," untitled newspaper clipping, Omaha, Nebraska, June 14, 1896 [TAE].
5. SW to E. D. Adams, May 14, 1891; August 16, 1892 [ALCU].
6. SW to Adams, December 1891; October 1891 [ALCU].
7. SW to NT, February 25, 1894 [ALCU].
8. SW to NT, February 5, 1895 [ALCU].
9. Ibid.
10. Michael Mooney, *Evelyn Nisbet and Stanford White* (New York: Morrow, 1976), pp. 193–99; Paul Baker, *Stanny*, pp. 249–50.
11. SW to NT, November 30, 1895 [ACU].
12. George Wheeler, *Pierpont Morgan and Friends: Anatomy of a Myth* (Englewood Cliffs, N.J.: Prentice-Hall, 1973), p. 17.
13. G. Scherff, 1902 [BLCU].
14. R. U. Johnson, *Remembered Yesterdays*, pp. 480–81; "True Buddhism, Brooklyn Standard Union," February 4, 1895, in *The Complete Works of Swami Vivekananda*, vol. 2 (Calcutta, India: Advaita Ashram, 1970); Tad Wise to author, April 10, 1996.
15. RUJ to NT, October 25, 1895 [BLCU].
16. NT to KJ, October 23, 1895 [BLCU].
17. Matthew Josephson, *The Robber Barons* (New York: J. J. Little & Ives Co., 1934), pp. 332–34.
18. H. J. Satterlee, *J. Pierpont Morgan*, p. 214.
19. TCM to NT, November 7, 1895; November 17, 1895 [NTM].
20. NT to JJA, December 20, 1895 [NTM].
21. NT to RUJ, December 13, 1895; December 22, 1895 [NTM].
22. NT to SW, January 4, 1896 [NTM].
23. NT to RUJ, January 10, 1896 [BLCU].
24. Swami Vivekananda to W. T. Stead (ed., *Review of Reviews*), in *Letters of Swami Vivekananda* (Pithoragarth, Himalays: Advaita Ashrama, 1981), pp. 281–83; *The Complete Works of Swami Vivekananda* (Calcutta, India: Advaita Ashrama, 1979).
25. JJA to NT, January 18, 1896 [NTM].

Chapter 19: Shadowgraphs, pp. 167–170

1. "Phosphorescent Light," *New York Mail and Express*, May 22, 1896 [TAE].
2. Michael Pupin, *From Immigrant to Inventor* (New York: Scribners, 1925), p. 306.
3. John O'Neill, *Prodigal Genius*, 1944.
4. Søren Kierkegaard, *Either or Or*, 1848, translated by David and Lillian Svenson (New York: Oxford University Press, 1944).
5. NT, "On Roentgen Rays," *Electrical Review*, March 11, 1896; in *Nikola Tesla: Lectures* (1956), p. A-27.
6. Ibid., p. A-29.
7. Ibid., p. A-30.
8. NT, "On Roentgen Radiations," *Electrical Review*, April 8, 1896; in *Nikola Tesla: Lectures* (1956), p. A-43.
9. NT, "Roentgen Rays or Streams," *Electrical Review*, , December 1, 1896; in *Nikola Tesla: Lectures* (1956), p. A-52.

10. Ibid.; "On the Roentgen Streams," *Electrical Review,* December 1, 1896; in *Nikola Tesla: Lectures* (1956), p. A-56.Tesla also associated this idea to Kelvin's "ether vortexes."

11. NT, "On Roentgen Rays: Latest Results," *Electrical Review,* March 18, 1896; in *Nikola Tesla: Lectures* (1956), pp. A-32–38; "On Roentgen Radiations," *Electrical Review,* April 8, 1896; in *Nikola Tesla: Lectures* (1956), p. A-41.

12. NT. "On the Roentgen Streams," *Electrical Review,* December 1, 1896; in *Nikola Tesla: Lectures* (1956), p. A-58.

13. NT. "On the Hurtful Actions of Lenard and Roentgen Tubes," *Electrical Review,* May 5, 1897; in *Nikola Tesla: Lectures* (1956), p. A-65.

14. "Tesla Opposes Edison," *NY Evening Journal,* December 2, 1896 [TAE].

15. "Tesla Says 'Let us hope'." *Philadelphia Press,* November 20, 1896 [TAE].

16. "Scoffs at X rays for the blind," *NY Morning Journal,* December 3, 1896 [TAE].

17. "Combined Devices," *NY Evening Journal,* December 2, 1896; "Triumph of Science: Combination of Tesla and Edison contrivances," *Louisville KY Courier Journal,* November 24, 1896 [TAE].

18. "Edison Caught a Fluke," *NY Morning Journal,* August 10, 1897 [TAE].

Chapter 20: Falls Speech, pp. 171–177

1. Charles Barnard, "Nikola Tesla, the Electrician," *The Chautauguan* 25, (1897), pp. 380–84.

2. "Nikola Tesla: An Interesting Talk With America's Great Electrical Idealist," *Niagara Falls Gazette,* July 20, 1896, 1:1.See also T. Valone, "Tesla's History in Western New York," in S. Elswick, ed., *Tesla Proceedings* (1986), pp. 27–51.

3. William Preece to NT, 1896 [NTM].

4. E. C. Baker, *Sir William Preece: Victorian Engineer Extraordinary* (London: Hutchinson, 1976), pp. 269–70.

5. NT, "Marconi and Preece," *New York World,* April 13, 1930, p. 229, in J. Ratzlaff, *Tesla Said,* p. 229.

6. KJ to NJ, August 6, 1896 [NTM].

7. RUJ to NT, July 28, 1896 [LC].

8. RUJ to NT, November 7, 1896.

9. "History Making Celebration of the Only Electrical Banquet the World Has Ever Seen," *Buffalo Evening News,* January 13, 1897, 1:1–2; 4:2–5.

10. Nikola Tesla, "Niagara Falls Speech," *Electrical World,* February 6, 1897, pp. 210–11. Reprinted in *Nikola Tesla: Lectures* (1956), pp. A101–8.

11. Ibid.

12. Ibid.

13. *Buffalo Evening News,* January 12, 1897; see also D. Dumych, "Nikola Tesla and the Development of Electric Power at Niagara Falls," *Tesla Journal* 6, 7, 1989–90, pp. 4–10.

14. Ibid.

Chapter 21: Luminaries, pp. 178–181

1. R. U. Johnson, *Remembered Yesterdays,* p. 402.

2. Ibid., March 13, 1896.

3. JJA to NT, January 29, 1897 [NTM].

4. SW to NT, January 29, 1897 [ACU].

5. NT to RUJ, March 28, 1896; salutation, March 12, 1896 [BLCU].

6. NT to RUJ, April 8, 1896 [BLCU].

7. R. U. Johnson, *Remembered Yesterdays,* pp. 402–3.

8. Ignace Paderewski and Mary Lawton, *The Paderewski Memoirs* (New York: Scribners, 1938), p. 205–6.

9. NT to KJ, April 8, 1896; April 9, 1896 [BLCU].

10. NT to KJ, April 10, 1896 [BLCU].

11. NT to R. Kipling, April 1, 1901 [BLCU].

12. NT to KJ, March 10, 1899 [BLCU].

13. NT to KJ, March 9, 1899 [BLCU].

14. Peter Browning, ed., *John Muir in His Own Words* (Lafayette, Calif.: Great West Books, 1988).

15. NT to KJ, November 3, 1898 [BLCU].
16. P. Browning, *John Muir in His Own Words,* p. 12.

Chapter 22: Sorcerer's Apprentice, pp. 182–192

1. "Tesla Electrifies the Whole Earth," *New York Journal,* August 4, 1897, 1:1–3.
2. Lee DeForest, *Father of Radio: An Autobiography* (Chicago: Wilcox & Follett, 1950), pp. 76, 81, 85.
3. NT, *Nikola Tesla: Lectures,* (1956). Essential patents for oscillators and transmitters: 454622 June 23, 1891; 462418 November 3, 1891; 514168 August 2, 1893; 568176-180 April 20, 1896–July 9, 1896; remote control: 613809 July 1, 1898; wireless communication: 649621 September 2, 1897; 1119732 January 18, 1902.
4. "Wizard Edison Here," *Buffalo (N.Y.) News,* August 30, 1896 [TAE].
5. "Nikola Tesla on Far Seeing—The Inventor Talks Interestingly on the Transmission of Sight by Wire," *New York Herald,* August 30, 1896. See also Ratzlaff & Anderson, 1979, p. 45.
6. NT, "Developments in Practice and Art of Telephotography," *Electrical Review,* December 11, 1920, in *Lectures* (1956), pp. A-94–97.
7. A. Korn to NT, May 1931, in *Tribute* (1961) pp. 25–27.
8. NT, "Developments in Practice and Art of Telephotography," *Electrical Review,* December 11, 1920, in *Lectures* (1956), p. A-97.
9. Chauncey Montgomery McGovern, "The New Wizard of the West," *Pearson's Magazine,* May 1899, pp. 291–97.
10. NT to Parker W. Page, August 8, 1897 [KSP].
11. Patent no. 649621, September 2, 1897; in *Lectures* (1956) pp. P293–96; C. M. McGovern, "The New Wizard of the West," p. 294.
12. Ibid., pp. P293–96.
13. Preece quoted in E. C. Baker, *Sir William Preece* 1976, p. 270.
14. Vyvyan quoted in D. Marconi, *My Father, Marconi* (New York: McGraw-Hill, 1962), p. 138.
15. W. Jolly, *Marconi* (New York: Stein & Day, 1972), p. 48.
16. D. E. W. Gibb, *Lloyds of London* (London: Lloyds of London Press, 1957), p. 158.
17. Frank Jenkins, "Nikola Tesla: The Man, Engineer, Inventor, Humanist and Innovator," in *Nikola Tesla: Life and Work of a Genius* (Belgrade: Yugoslav Society for the Promotion of Scientific Knowledge, 1976), pp. 10–21. Original source, O'Neill, 1944.
18. NT vs. Marconi, court transcripts, pp. 440–41 [LA].
19. C. M. McGovern, op. cit., 1899, p. 297.
20. "A Crowd to Hear Tesla," *New York Times,* April 7, 1897, 12:2; J. Ratzlaff and L. Anderson, *Tesla Bibliography,* p. 49.
21. For a discussion of this unpublished lecture, see L. Anderson (ed.), *On His Work with Alternating Currents* 1916/1992.
22. "Telegraphy without wires," *Scribners Monthly,* 1897, pp. 527x28.
23. NT vs. Reginald Fessenden litigation, op. cit., 1902 [BLCU].
24. Westinghouse Co. annual report, *Electrical Review,* June 30, 1897, p. 313.
25. Westinghouse memorandum, July 7, 1888.
26. NT to JJA, January 6, 1899 [NTM].
27. NT to E. Heinreich, December 4, 1897 [LC].
28. Marica to NT, March 27, 1891, in A. Marincic, ed., *Tesla's Correspondence with Relatives* (Belgrade: Nikola Tesla Museum) [Zoran Bobic, transl.].
29. "Tesla at 79 Discovers New Message Wave," *Brooklyn Eagle,* July 11, 1935, 1:1, 3:4; see also O'Neill, 1944, pp. 158–64.
30. Allan Benson, "Nikola Tesla: Dreamer," *The World To-Day,* 1915, pp. 1763–67 [Archives, Health Research, Mokelumne Hill, Calif].

Chapter 23: Vril Power, pp. 193–203

1. Edward Bulwer-Lytton, *The Coming Race* (London: Routledge, 1871).
2. NT, "Tesla's Latest Invention: Electrical Circuits and Apparatus of Electrically Controlled Vessels," *Electrical Review,* November 16, 1898, pp. 305–12.
3. NT to RUJ, July 12, 1900 [BCU].

4. This connection between Tesla and Bulwer-Lytton was originally noticed by Desire Stanton, a newspaper columnist in Colorado Springs in 1899. See I. Hunt and W. Draper, *Lightning in His Hands: The Life Story of Nikola Tesla* (Hawthorne, Calif.: Omni Publications, 1964.
5. Bulwer-Lytton, *Coming Race.*
6. NT to JJA, January 27, 1897; July 3, 1897 [NTM].
7. NT to JJA, December 2, 1898 [NTM].
8. Virginia Cowles, *The Astors* (New York: Knopf, 1979), pp. 130–31.
9. NT, *My Inventions*, pp. 107–9.
10. John Oliver Ashton to Lee Anderson, July 17, 1953 [LA].
11. "Tom Edison's Son Explodes Desk by Accident," *New York Times*, May 3, 1898, 7:1.
12. "Tesla's Latest Invention," *Electrical Review*, November 9 and 16, 1898.
13. NT, "Torpedo Boat Without a Crew," *Current Literature*, February 1899, pp. 136–37.
14. "Mr. Tesla and the Czar," *Electrical Engineering*, November 17, 1898, pp. 486–87.
15. NT, "The Problem of Increasing Human Energy," *Century*, June 1900, p. 188.
16. M. Huart, "The Genius of Destruction," *Electrical Review*, December 7, 1898, p. 36.
17. Mark Twain to NT, November 17, 1898 [NTM].
18. "Mr. Tesla and the Czar," *Electrical Engineering*, November 17, 1898, pp. 486–87.
19. "Was Keely a Charlatan?" and "Science and Sensationalism," *Public Opinion*, December 1, 1898, pp. 684–85.
20. NT to RUJ, January 1, 1898 [BLCU].
21. NT to RUJ, January 1, 1898 [BLCU].
22. NT to RUJ, November 28, 1898 [BLCU].
23. NT, "Mr. Tesla's reply," *Electrical Engineer*, November 24, 1898, p. 514.
24. Marc Seifer, *Nikola Tesla: Psychohistory*, 1986, p. 272. Survey derived from Ratzlaff and Anderson, 1979.
25. "His Friends to Mr. Tesla," *Electrical Engineer*, November 24, 1898, p. 514.
26. TCM to Elihu Thomson, January 16, 1917, in H. Abrahams and M. Savin, *Selections from the Scientific Correspondence of Elihu Thomson* (Cambridge, Mass.: MIT Press, 1971), p. 352.
27. T. C. Martin, "The Burning of Tesla's Laboratory," *Engineering*, 11:1, April 1895.
28. NT, "The Problem of Increasing Human Energy," *Century*, June 1900, pp. 175–211.
29. NT, "How Cosmic Forces Shape Our Destiny," 1915, in *Nikola Tesla: Lectures* (1956), p. A-122.
30. NT, "The Problem of Increasing Human Energy," *Century*, June 1900, pp. 173–74.
31. NT, "How Cosmic Forces Shape Our Destiny," 1915/1956, p. A-172.
32. Ibid.
33. NT, "The Problem of Increasing Human Energy," *Century*, June 1900, pp. 184–85.
34. Ibid., pp. 185–86.
35. NT, *My Inventions*. It should also be noted that for many years, in order for a patent to be granted, the inventor *had* to demonstrate his invention.

Chapter 24: Waldorf-Astoria, pp. 204–213

1. NT to RUJ, November 29, 1897 [BLCU].
2. Virginia Cowles, *The Astors* (New York: Knopf, 1979), p. 126.
3. Albin Dearing, *The Elegant Inn* (Secaucus, N.J.: Lyle Stuart, 1986), pp. 75, 78, 87.
4. Ibid., p. 81.
5. NT to U.S. Navy, September 27, 1899 [NAR].
6. P. Delaney, "Telegraphing From a Balloon in War," *Electrical Review*, October 1898, p. 68.
7. NT to JJA, January 3, 1901 [NTM].
8. General Dynamics advertisement, *Smithsonian*, 1990.
9. "Offer of the Holland Owners," *New York Times*, June 4, 1898, 1:4.
10. NT to U.S. Navy, 1899 [NAR].
11. "The Patience of Hobson," *New York Times*, April 20, 1908.
12. "The Merrimac Destroyed?" *New York Times*, June 4, 1898, 1:4.
13. Martha Young, "Lieutenant Richmond P. Hobson," *Chautauguan* 27, 1898, p. 561.
14. "Lieut. Hobson's Promotion," *New York Times*, June 21, 1898, 1:4.
15. KJ to NT, December 6, 1897 [NTM].
16. Ibid., June 6, 1898.

17. NT, "Tesla's Latest Advances in Vacuum Tubes," *Electrical Review,* January 5, 1898, p. 9.

18. Cheiro (Louis Hamon), *Cheiro's Language of the Hand* (New York: Transatlantic Publishing Co., 1895).

19. Sphynx. Analysis of Tesla's palm. Private correspondence, August 1990.

20. KJ to NT, February 8, 1898 [NTM].

21. Ibid., March 12, 1898; March 25, 1898.

22. NK to KJ, March 12, 1898 [BLCU].

23. Ibid., December 3, 1898.

24. NK to KJ, November 3, 1898 [BLCU].

25. Ibid., March 9, 1899.

26. Marguerite Merrington papers, Museum of New York City; John O'Neill, *Prodigal Genius,* p. 302.

27. Virginia Cowles, *The Astors* (New York: Knopf, 1979), pp. 124–25.

28. Ibid., p. 135.

29. NT to JJA, December 2, 1898; January 6, 1899 [NTM].

30. Ibid., December 2, 1898 [NTM].

31. NT to JJA, January 6, 1899 [NTM].

32. Ibid., January 6, 1899; January 10, 1899; March 27, 1899 [NTM]. Whether Tesla actually received the full amount is unknown.

33. John O'Neill, *Prodigal Genius,* p. 176.

34. NK to KJ, November 3, 1898 [BLCU].

35. R. U. Johnson, *Remembered Yesterdays,* pp. 418–19.

36. "The Gentle Art of Kissing," *New York Times,* August 15, 1899, 6:2–4.

37. NT to RUJ, December 6, 1898 [BLCU].

38. Ibid., November 8, 1898.

39. "Lieut. Hobson's Career," *New York Times,* June 5, 1898, 2:4.

Chapter 25: Colorado Springs, pp. 214–219

1. Desire Stanton, "Nikola Tesla Experiments in the Mountains," *Mountain Sunshine,* Jul-Aug 1899, pp. 33–34.(Real name: Mrs. Gilbert McClurg.) Tesla's 1896 trip to Colorado was discovered by James Corum while researching articles at the Tesla Museum, Belgrade.

2. NT/Reginald Fessenden litigation, August 5, 1902 [BLCU].

3. NT. "Some Experiments in Tesla's Laboratory With Currents of High Potential and High Frequency," *Electrical Review,* March 29, 1899, pp. 193–97, 204.

4. T. Hunt and M. Draper, *Lightning,* p. 110.

5. Ibid.; NT, *On His Work in A.C.,* 1916/1992, p. 109.

6. T. Hunt and M. Draper, *Lightning,* p. 110.

7. Ibid.

8. NT/RF litigation, August 5, 1902, p. 12 [BLCU].

9. Drawings pertaining to the design of the Colorado Springs experimental station were created in 1896 and 1897. In the same manner, while at Colorado, Tesla also worked out plans for his next transmitter, which was erected on Long Island. NT, *My Inventions,* pp. 116–17.

10. T. Hunt and M. Draper, *Lightning,* p. 108.

11. Ibid.

12. Ibid.

13. According to present-day understanding, the ionosphere, or Kennelly-Heaviside layer, does not act as a carrier of the electrical waves, as Tesla hypothesized, but as a reflector, causing the energy "to bounce back and forth rapidly for long-distance transmission," and that is how it goes around the entire curve of the earth (Stanley Seifer, private correspondence, 1985).

14. NT/RF litigation, August 5, 1902, p. 51 [BLCU].

15. Alexander Marincic, "Research on Nikola Tesla in Long Island Laboratory." *Tesla Journal* 6, no. 7 (1988/89) pp. 25–28.

16. NT/RF litigation, August 5, 1902, [BLCU].

17. NT to George Scherff, June 22, 1899 [LC].

18. NT/RF litigation, August 5, 1902, p. 26 [BLCU].

19. The primary of the coil was a specially prepared cable spanning the inside perimeter of the building itself, and the secondary was a tubular shaped smaller coil in the center of the

structure which encircled a transmission tower that rose from a support column as a single spire. With a removable roof to augment the adjustment of the aerial, and a small bulb at its apex, the transmitter could be extended to a variable length that could reach a maximum of 200 feet from the ground. A. Marincic, *Colorado Springs Notes Commentary*, in Nikola Tesla, *Colorado Springs Notes*, A. Marincic, ed. (Belgrade, Yugoslavia: Nikola Tesla Museum, 1979).

20. Due to Tesla's extraordinary powers of eidetic imagery, a myth, perpetuated by O'Neill and Tesla's own autobiography, arose suggesting that the inventor worked out all designs and calculations solely in his mind. The original curators of the Tesla Museum therefore kept the Colorado notebook a secret, as they did not want to destroy this image of the inventor's extraordinary mental abilities. According to the present curator, Dr. Marincic, "The appearance of the Colorado notebook would show Tesla to be human, that he made mistakes, and so on." Marincic's position was totally different. He felt that the more people understood the real Tesla, the better would be the appreciation of his accomplishments. It was for this reason that Marincic prepared the notebook which was published by the musuem in 1979 (Tesla Museum, A. Marincic, Colorado Springs, August, 1990.)

21. NT/RF litigation, August 5, 1902, [BLCU].

22. NT, *My Inventions*, p. 86. See also *Colorado Springs Notes*, p. 174: "Now it was of importance to increase the magnifying factor...."

23. NT/RF litigation, August 5, 1902, p. 30 [BLCU].

24. NT, *Colorado Springs Notes*, pp. 28, 34.

25. NT to JJA, September 10, 1900 [NTM].

26. GS to NT, June 14, 1990 [LC].

27. GS to NT, June 22, 1899 [LC].

28. NT to GS, June 6, 1899 [LC].

29. A. Marincic, in *Colorado Springs Notes*, p. 15.

30. NT/RF litigation, Lowenstein testimony, August 5, 1902, pp. 99–101, 106 [BLCU].

31. NT, CSN, 1979, p. 37.

32. NT/RF litigation, Lowenstein testimony, August 5, 1902, pp. 106–8 [BLCU].

33. NT, *Colorado Springs Notes*, p. 61.

34. Ibid.

35. NT to GS, July 4 and 6, 1899 [LC].

Chapter 26: Contact, pp. 220–229

1. NT to RUJ, January 25, 1901 [BLCU].

2. NT to JH, December 8, 1899, in *Colorado Springs Notes*, p. 314.

3. NT, "Talking With the Planets," *Current Literature*, March 1901, p. 360.

4. *Pyramid Guide*, 1977 [LA].

5. NT, *Colorado Springs Notes*, pp. 109–110.

6. NT/RF litigation, August 5, 1902 [BLCU].

7. Ibid., pp. 127–33.

8. NT, "Talking With the Planets," February 9, 1901, *Colliers*, pp. 405–6; *Current Literature*, March 1901, pp. 429–31.

9. NT, "Interplanetary Communication," *Electrical World*, September 24, 1921, p. 620.

10. NT, "Signalling to Mars," *Harvard Illustrated*, March 1907, in *Tesla Said*, pp. 92–93.

11. GS to NT, July 1, 1899 [LC].

12. *New York Times* articles on wireles operators: D'Azar, September 3, 1899, 17:7; Marble November 7, 1899, 1:3; Riccia September 10, 1899, 10:4.

13. GS to NT, October 2, 1899 [LC].

14. NT to GS, September 27, 1899 [LC].

15. On July 28, in the *Colorado Springs Notes* Tesla also utilizes the word *feeble*. This same word appears in the 1901 article "Talking With the Planets." See also, Marc Seifer, 1979; 1984; 1986.

16. W. Jolly, *Marconi* (New York: Stein & Day, 1972), pp. 65–66.

17. Recent biographers, such as Hunt and Draper, attributed the impulses to "radio waves coming from the stars" or to pulsars. Tesla researcher Prof. James Corum suggests that he may have intercepted pulsed frequencies emanating from Jupiter or "the morning chorus," which are charged particles that "slosh back and forth between the North and South poles in the early morning." Additional possibilities include other natural phenomena associated with the lightning storms or telluric currents, faulty equipment, or self-delusion.

18. NT. Interplanetary communication. EW, September 9, 1921, p. 620.

19. R. Conot, *Streak of Luck: The Life Story of Edison* (N.Y.: Bantam Books, 1981), pp. 415–17.

20. Charles Batchelor, papers [TAE].

21. R. Conot, *Streak of Luck,* pp. 415–17.

22. Julian Hawthorne, "And How Will Tesla Respond to Those Signals From Mars?" *Philadelphia North American,* 1901 [BLCA].

23. Ibid.

24. Anonymous, "Mr. Tesla's Science," *Popular Science Monthly,* February 1901, pp. 436–37.The Tesla quotes are from NT, "The Problem of Increasing Human Energy," *Century,* June 1900.

25. NT to U.S. Navy, September 16, 1916 [NAR].

26. Francis J. Higginson to NT, May 11, 1899 [NAR].

27. NT to U.S. Navy, July 11, 1899 [NAR].

28. Ibid., August 20, 1899.

29. Ibid., September 14, 1899.

30. Ibid., September 27, 1899.

Chapter 27: Thor's Emissary, pp. 230–235

1. NT, "Tesla's reply to Edison," *English Mechanic & World Science,* July 14, 1905, p. 515, in *Tesla Said,* pp. 88–89.

2. Ibid., August 3, 1899.

3. Ibid., November 6, 1899.

4. John Ratzlaff and Fred Josst, *Dr. Nikola Tesla: English/Serbo-Croatian Diary Comparisons, Commentary and Tesla/Scherff Colorado Springs Correspondence.* (Millbrae, Calif.: Tesla Book Co., 1979), p. 73.

5. NT to GS, September 6, 1899 [LC].

6. NT to GS, September 22, 1899, in Ratzlaff and Jost, *Dr. Nikola Tesla,* p. 114.

7. Nancy Czito, Personal interview, November 1983, Inventor Commemoration Day, Washington, D.C.

8. NT, October 1919, p. 516; in *Tesla Said,* p. 216.

9. Leland Anderson, "John Stone on Nikola Tesla's Priority in Radio and Continuous-Wave Radiofrequency Apparatus," *Antique Wireless Association Review,* 1:1, 1986.

10. NT to GS, October 29, 1899 [LC].

11. Alexander Marincic, *Colorado Springs Notes,* 1979, p. 421.

12. NT, *Colorado Springs Notes,* 1979, p. 111.

13. John O'Neill, "Tesla Tries to Prevent World War II" (Originally unpublished chapter from Tesla biography), *Tesla Coil Builders Association,* July–August, 1988, pp. 13–14.

14. NT to RUJ, October 1, 1899 [BLCU].

15. NT, *Colorado Springs Notes,* 1979, p. 219.

16. O'Neill, 1988, p. 14.This work has been replicated by Professor James Corum by setting up two coils near each other, one with a low frequency (90 KH) and the other with a high frequency (200 KH). When exciting both coils, small fireballs sometimes appear. Placing a "thumbprint of carbon" on one of the coils also helps augment the process. It is possible, in this latter case, that the microparticles of carbon, when electrified, attract additional charges. Robert Golka, another Tesla researcher, has also produced fireballs. He suggests that rotational motion of a boundary layer of charges may be involved in the process. James Corum, "Cavity Resonator Developments," lecture before the International Tesla Society, Colorado Springs, August 1990.

17. NT, *Colorado Springs Notes,* 1979, p. 228.

18. NT, "Can Radio Ignite Balloons?" *Electrical Experimenter,* October 1919, pp. 516, 591–92.(Archives, Gernsback Publications, Farmingdale, NY).

19. As "the loss [of propagated waves] is proportional to the cube of the frequency…with waves 300 meters in length, economic transmission of energy is out of the question, the loss being too great. With wave-lengths of 12,000 meters [loss] becomes quite insignificant and on this fortunate fact rests the future of wireless transmission of energy." NT, "The Disturbing Influence of Solar Radiation on the Wireless Transmission of Energy," *Electrical Review and Western Electrician,* July 6, 1912; in *Tesla Said,* pp. 121–27.

20. NT, *Colorado Springs Notes,* 1979, p. 76.

21. H. Winfield Secor, "The Tesla High Frequency Oscillator," *Electrical Experimenter,* March 1916, pp. 614–15, 663.
22. NT, "Can Radio Ignite Balloons?" *Electrical Experimenter,* October 1919, p. 591.
23. Ibid.
24. John O'Neill, *Prodigal Genius,* p. 187; NT, *Colorado Springs Notes,* 1899/1979, p. 348.
25. KJ to NT, December 22, 1899 [NTM].

Chapter 28: The Hero's Return, pp. 236–244

1. RUJ to NT, July 7, 1900 [LC].
2. *Colorado Springs Gazette,* "Nikola Tesla to Come Here," October 30, 1903, 1:7; Tesla Sued for $180 by Electrical Co.," April 6, 1904, 3:1; "NT Says He Is Not Indebted to Duffner," September 6, 1905, 1:2.See also Ratzlaff and Anderson, pp. 79, 81, 86.
3. "Signor Marconi Arrival from Europe," *New York Times,* January 3, 1900, 1:3.
4. Dragislav Petkovich, "A Visit to Nikola Tesla," *Politika,* vol. XXIV, no. 6824, April 27, 1927 [LA].
5. Stanko Stoilovic, "Portrait of a Person, a Creator and a Friend," *Tesla Journal,* 4/5, 1986/87, pp. 26–29.
6. Pupin papers, patent no. 652,231, June 19, 1900 [BLCU].
7. Stanko, "Portrait," *Tesla Journal,* pp. 26–29.
8. U.S. patent letters to Pupin, June 30, 1896; July 25, 1896, Pupin papers [BLCU]; see also *Inventions, Researches, and Writings,* 1894, p. 292, and previous discussion in chapter 15.
9. NT, "Tesla's Wireless Torpedo," *New York Times,* March 20, 1907, 8:5, in *Tesla Said,* p. 96.
10. NT, "The Transmission of Electrical Energy Without Wires as a Means for Furthering Peace," *Electrical World & Engineer,* January 7, 1905, p. 22.
11. Admiral Higginson to NT, October 8, 1900 [NAR].
12. Vojin Popovic, "NT, true founder of radio communications," in *Nikola Tesla: Life and Work of a Genius* (Belgrade: Yugoslavia Society for the Promotion of Scientific Knowledge, 1976), V. Popovic, ed., p. 82.
13. The letter also makes reference to Tesla's continuing partnership with Peck and Brown, Tesla owning 4/9ths of all royalties on the invention. NT to GW, January 22, 1900 [LC].
14. Bernard A. Behrend, *The Induction Motor and Other Alternating Current Motors: Their Theory and Principles of Design* (New York: McGraw-Hill, 1921), pp. 261–62.
15. "The Tesla Patents," *Electrical Review,* September 19, 1900, pp. 288–92; see also discussions on priority of AC in earlier chapter.
16. GW to NT, September 5, 1900 [LC].
17. 685,012; 787,412; 725,605.
18. Swami Vivekananda to E. T. Sturdy, February 13, 1896, in *Letters of Swami Vivekananda* (Pithoragarth Himalayas: Advaita Ashrama May Avati, 1981), pp. 281–83.
19. RUJ to NT, March 6, 1900 [LC].
20. NT to RUJ, March 6, 1900 [LC].
21. NT, "The Problem of Increasing Human Energy," *Century,* June 1900, pp. 175–211.
22. NT to Corinne Robinson, [HL].
23. NT to JJA, May 2, 1900; March 30, 1900.
24. NT to RUJ, June 21, 1900; June 29, 1900 [LC].
25. "A Tesla Patent in Wireless Transmission," *Electrical World and Engineer,* March 26, 1900, p. 792.
26. NT to RUJ, June 15, 1900 [LC].
27. "Science and Fiction," *Popular Science Monthly,* July 1900, pp. 324–26.
28. NT to RUJ, July 12, 1900 [BLCU].
29. R. A. Fessenden, "Wireless Telegraphy," *Electrical World and Engineer,* January 26, 1901, pp. 165–66.
30. KJ to NT, August 2, 1900 [NTM].
31. NT to KJ, August 12, 1900 [BLCU].
32. JJA to NT, September 1900 [NTM].
33. NT to JJA, October 29, 1900 [NTM].

Chapter 29: The House of Morgan, pp. 245–255

1. NT, "Our Future Motive Power," *Everyday Science and Mechanics,* December 1931, pp. 78–81, 86.
2. Ibid.
3. H. Satterlee, *J. Pierpont Morgan, An Intimate Portrait,* p. 344.
4. NT to RUJ, January 29, 1900 [BLCU].
5. Werner Wolff, *Diagrams of the Unconscious* (New York: Grune & Stratton, 1948), p. 267.
6. H. Satterlee, *J. Pierpont Morgan,* Morgan, *An Intimate Portrait,* p. 344.
7. NT to JPM, November 26, 1900 [LC].
8. H. Satterlee and J. P. Morgan, *An Intimate Portrait,* p. 345.
9. Cass Canfield, *The Incredible Pierpont Morgan: Financier and Art Collector* (New York: Harper & Row, 1974).
10. A. Satterlee and J. P. Morgan, *An Intimate Portrait,* p. 343–44.
11. G. Wheeler, *Pierpont Morgan and Friends: Anatomy of a Myth* (Englewood Cliffs, N.J.: Prentice-Hall, 1973), pp. 61–62.
12. NT to JPM, October 13, 1904 [LC].
13. Note: All conversations between Tesla and Morgan have been recreated from their correspondence. Some literary license has been taken when in conversation form. Blocked quotes are verbatim. NT to JPM, November 26, 1900 [LC].
14. NT to JPM, December 10, 1900 [LC].
15. "Marconi's Signals," *New York Times,* April 8, 1899, in Jolly, p. 66.
16. "New Electric Inventions: Nikola Tesla's Remarkable Discoveries," *New York Recorder,* June 15, 1891.
17. "Besides, in this country, I have protected myself, though not quite so completely, in England, Victoria, New South Wales, Austria, Hungary, Germany, France, Italy, Belgium, Russia and Switzerland" NT to JPM, December 10, 1900 [LC].
18. Robert Hessen, *Steel Titan: The Life of Charles M. Schwab* (New York: Oxford University Press, 1975), pp. 116–17.
19. M. Josephson, *The Robber Barrons* (New York: J. J. Little, 1934), p. 426 and Satterlee, *J. Pierpont Morgan,* p. 347.
20. Satterlee, p. 348.
21. Wheeler, p. 233.
22. E. Hoyt, *The House of Morgan* (New York: Dodd, Mead, 1966), p. 245.
23. NT to JPM, December 10, 1900 [LC].
24. NT to JPM, September 7, 1902 [LC].
25. NT to JPM, March 5, 1901 [LC].
26. NT to JPM, October 13, 1904 [LC].
27. NT to JPM, December 10, 1900 [LC].
28. NT to JPM, October 13, 1904 (size calculated from point 8) [LC].
29. NT to JPM, October 13, 1904 [LC].
30. Ibid.
31. JPM to NT, February 15, 1901 [LC].
32. NT to JJA, January 3, 1901 [NTM].
33. JPM to NT, March 5, 1901 [LC].
34. NT to JPM, October 13, 1904 [LC].
35. NT to JPM, February 18, 1901 [LC].
36. NT to JJA, January 11, 1901 [NTM].
37. NT to JJA, January 22, 1901 [NTM].
38. One curious feature to this episode is that aside from lighting patents dating from 1890 to 1992, no circa-1900 Tesla patents have been uncovered which are specifically written up to describe fluorescent or neon lighting. This conclusion is corroborated by correspondence with other Tesla researchers (e.g., Leland Anderson and John Ratzlaff). If Tesla drafted patents on this invention, they were never filed in Washington. There may be copies in Morgan's archives or the Tesla Museum, or the invention might be somehow linked to other patents. A congressional investigation provides tangential evidence that Morgan purposely squelched this invention: "The introduction of fluorescent lighting in this country was slowed up by GE and Westinghouse, through control of patents, lest its efficiency cut too drastically the demand for current." (*Invention and the Patent System,* Report of Joint Economic Committee Congress of the United States, 88th Cong., 2d sess., December 1964, p. 100.)

39. NT to JPM, March 5, 1901 [LC].
40. NT to GW, March 13, 1901 [LC].

Chapter 30: World Telegraphy Center, pp. 256–265

1. NT to JPM, February 12, 1901 [LC].
2. NT to RUJ, March 8, 1900; March 9, 1900 [BLCU].
3. EH to NT, February 25, 1901 [LC].
4. NT to TCM, December 12, 1900 [NTM].
5. TCM to NT, December 13, 1900 [NTM].
6. TCM to NT, December 18, 1900; December 17, 1900 [NTM].
7. NT to Miss Emma C. Thursby, March 3, 1901 [NHS].
8. Julian Hawthorne, "Tesla's New Surprise," *Philadelphia North American* circa 1900 [BLCU].
9. NT to JH, January 16, 1901 [BLCU].
10. Paul Baker, *Stanny: The Gilded Life of Stanford White* (New York: Free Press, 1989), p. 15.
11. Ibid., p. 289.
12. Ibid., p. 321.
13. Literary license taken on conversation. Adapted from R. Fleischer, director, *The Girl in the Red Velvet Swing* (film), 1955; Michael Macdonald Mooney, *Evelyn Nesbit and Stanford White: Love and Death in the Gilded Age* (New York: Morrow, 1976), pp. 45–46.
14. NT to KJ, June 11, 1900 [BLCU].
15. J. Ratzlaff and L. Anderson, p. 70.
16. O'Neill.
17. NT to RUJ, January 1, 1901 [BLCU].
18. Interview with Mrs. Robert Underwood Johnson, July 1, 1990.
19. "Nikola Tesla Inventor," *Long Island Democrat,* August 27, 1901, 1:3.
20. O'Neill.
21. Historical Sketches of Northern Brookhaven Town: Shoreham, p. 68 [KSP].
22. "Mr. Tesla at Wardenclyffe, L.I." *Electrical World and Engineer,* September 28, 1901, pp. 509–10.
23. Ibid.: Warden's quote: "the ultimate number spoken of is 2000 to 2500 [workers]."
24. "When the Man Who Talked to Mars Came to Shoreham," *Port Jefferson Record,* March 25, 1971, p. 3.
25. W. Shadwell, *McKim, Mead & White: A Building List,* #818, NY, 1978.
26. SW to JPM, February 6, 1901; February 7, 1901 [SWP].
27. SW to NT, April 26, 1901 [SWP].
28. SW to NT, June 1, 1901 [SWP].
29. GS to NT, July 23, 1901 [LC].
30. SW to NT, January 1, 1901 [SWP].
31. G. Marconi, "Wireless Telegraphy and the Earth," *Electrical Review,* January 12, 1901; Recent Electrical Patents: "Marconi has been granted another patent on an improved receiver for electrical oscillations in his wireless telegraphy system.... *Electrical Review,* March 2, 1901; quotation in text is from "Syntonic Wireless Telegraphy," *Electrical Review,* part I, June 15, 1901, p. 755; part II, June 22, 1901, pp. 781–83.
32. NT to JPM, October 13, 1904 [LC].
33. NT, Wardenclyffe drawing and notes, May 29, 1901 [NTM].
34. Stephen Birmingham, *Our Crowd* (New York: Pocket Books, 1977). See also Satterlee, 1939 and Wheeler, 1973.
35. "Fear and Ruin in a Falling Market," *New York Times,* May 10, 1901, 1:6.
36. Edwin Hoyt, *The House of Morgan,* p. 251.
37. NT to JPM, October 13, 1904 [LC].

Chapter 31: Clash of the Titans, pp. 266–274

1. Thomas Edison, private notebook, March 18, 1902 [TAE, Reel M94].
2. O'Neill.
3. NT to RUJ, June 14, 1901 [BLCU].
4. SW to NT, June 1, 1901 [SWP].

5. "Long Island Automobiles," *Electrical World and Engineer,* January 26, 1901, p. 165.
6. Paul Baker, p. 318.
7. Lawrence Grant White letter to Kenneth Swezey, December 21, 1955 [KSP]. Lawrence had provided Swezey with three letters from Tesla, copies of which are in the Library of Congress and the Swezey Collection. He had asked for their return, but the originals are missing and copies do not exist in the Stanford White papers at the Avery Library.
8. The Tesla Museum has a photo of Tesla in one of these bill-board photos.
9. NT to KJ, August 8, 1901 [NTM].
10. The Johnsons went to Maine every August for a number of years. Tesla probably joined them during one of these sojourns.
11. Satterlee, p. 360.
12. *New York Times,* May 2, 1901, 7:1.
13. NT to JPM, February 8, 1903 [LC].
14. NT to JPM, October 13, 1904 [LC].
15. Sketch of Thomas F. Ryan (description of JPM). *New York World,* June 18, 1905, 1:3.
16. NT to JPM, August 8, 1901 [LC].
17. NT to SW, August 16, 1901 [SWP].
18. Paul Baker, *Stanny,* p. 326.
19. NT to SW, August 28, 1901 [LC].
20. NT to SW, August 30, 1901 [NTM].
21. Satterlee, p. 363.
22. NT to JPM, September 13, 1901 [LC].
23. NT to SW, September 13, 1901 [LC].
24. NT to SW, September 14, 1901 [NTM].
25. Shoreham, in *Historical Sketches of Northern Brookhaven Town,* pp. 69–70 [KSP].
26. Ibid.
27. NT to KJ, October 13, 1901 [BLCU].
28. NT to JPM, November 11, 1901 [LC].
29. NT to RUJ, November 28, 1901 [BLCU].
30. W. Jolly, *Marconi,* pp. 103–4.

Chapter 32: The Passing of the Torch, pp. 275–282

1. Lee DeForest, "Passage From Private Notebook," in *Father of Radio: An Autobiography* (Chicago: Wilcox and Follett, 1950).
2. NT, "Tesla on Marconi's Feat," April 13, 1930, *New York World.*
3. JAF to ET, January 11, 1927, in Abraham and Savin, p. 239. Fleming's connection with Tesla actually began a year earlier, when he wrote the inventor that "I have been charged with [your] description…on alternating currents of high frequency [and] am very anxious to repeat these in England." (JAF to NT, July 22, 1891, NTM).
4. NT, "Tesla on Marconi's Feat," April 13, 1930, *New York World.*
5. ET to Alba Johnson, January 29, 1930, in Abraham and Savin, p. 325; Jolly, p. 111.
6. "The Institute Annual Dinner and Mr. Marconi," *Electrical World and Engineer,* January 18, 1902, pp. 107–8, 124–26.
7. Charles Steinmetz, *Alternating Current Phenomena* (New York: McGraw-Hill, 1900), preface; see also preface of *Theoretical Elements of Electrical Engineering* (New York: McGraw-Hill, 1902).
8. Ronald Kline, "Professionalism and the Corporate Engineer: Charles P. Steinmetz and the AIEE," *IEEE Transactions on Education,* vol. E-23, *3,* August 1980.
9. *Electrical World and Engineer,* January 18, 1902, pp. 107–8, 124–26.
10. R. Conot, p. 413.
11. "Marconi Tells of His Wireless Tests," *New York Times,* January 14, 1902, p. 1.
12. *Electrical World and Engineer,* January 18, 1902, pp. 107–8, 124–26.
13. Ibid.
14. NT to JPM, January 9, 1902 [LC].
15. [KSP].
16. Lee DeForest, "A Quarter Century of Radio," *Electrical World,* September 20, 1924, pp. 579–80; D. McFarlane Moore quote from DeForest, 1950, p. 220.

17. Isaac Asimov, *Asimov's Biographical Encyclopedia of Science and Technology* (Garden City, N.Y.: Doubleday, 1964), pp. 464–65.
18. DeForest, *Electrical World*, September 20, 1924, p. 580.
19. R. Conot, pp. 413–14, 444.
20. Tesla–Fessenden U.S. Patent Interference Case, August, 1902, pp. 87, 97–98.
21. Ibid., pp. 99, 102.
22. NT to GS, August 9, 1902 [BLCU].

Chapter 33: Wardenclyffe, pp. 283–292

1. "Cloudborn Electric Wavelets to Encircle the Globe," *New York Times*, March 27, 1904 [condensed].
2. Alexander Marincic, "Research on Nikola Tesla in Long Island Laboratory," *Tesla Journal, 6/7*, 1988/89, pp. 25–28, 44–48.
3. P. Baker, p. 326; TCM to NT, March 21, 1895 [NTM].
4. RUJ to NT, June 19, 1902 [BLCU].
5. NT to JPM, July 3, 1903 [LC]; NT, *On His Work with AC*, 1916/1992, pp. 152, 169.
6. Arthlyn Ferguson, "When the Man Who Talked to Mars Came to Shoreham," *Port Jefferson Record*, March 25, 1971, p. 3; Natalie Stiefel to M. Seifer, April 10, 1997.
7. NT to JPM, September 5, 1902 [LC].
8. NT to JPM, September 7, 1902.Obviously, some people knew of Morgan's interest at this time (e.g., the Johnsons, Astor), and *The Echo*, August 1901, a local Port Jefferson paper, had revealed Morgan's interest, but his connection at this time was at the level of rumor. Details of Morgan's connection were never revealed until well after Tesla's death with the publication of the Hunt and Draper biography in 1967.
9. Mr. Steele (JPM) to NT, October 21, 1902.
10. NT to KJ, September 25, 1902 [BLCU]; interview with Mrs. R. U. Johnson Jr., New York City, 1886.
11. NT to Agnes, January 2, 1903 [BLCU].
12. The biography, coauthored with Frank Oyer, took ten years to complete.
13. T. C. Martin, "The Edison of To-day," *Harper's Weekly, 47*, Jan/Jun, 1903, p. 630.
14. NT to TCM, June 3, 1903 [NTM].
15. NT to RUJ, January 24, 1904 [BLCU].
16. NT to RUJ and GS corresp., March 14, 1905; January 10, 1909; March 24, 1909 [BLCU].
17. GS to NT, December 19, 1910; December 31, 1910 [BLCU].
18. NT to JPM, April 22, 1903; April 1, 1904 [LC].
19. "Cloudborn Electric Wavelets to Encircle the Globe," *New York Times*, March 27, 1904.
20. The workers included Mr. Hartman, Mr. Clark, Mr. Johannessen, Mr. Merckling, and Mr. Beers [GS to NT, April 14, 1903, BLCU].
21. These machines were probably hydraulically operated. Concerning the cupola, Tesla testified that one of his most important discoveries was that "any amount of electricity within reason could be stored provided [it was made] in a certain shape.... That construction enabled me to produce...the effect that could be produced by an ordinary plant of a hundred times the size," NT, *On His Work With AC*, pp. 170–77.
22. Mitchel Freedman, "Dig for Mystery Tunnel Ends With Scientist's Secret Intact," February 13, 1979, p. 24; "Famed Inventor, Mystery Tunnels Linked," March 10, 1979, p. 19. Both in *Newsday*. Also, personal interview with Edwin J. Binney, West Babylon, who as a boy, climbed down into the tunnels; personal inspection of site by author, 1984.Tesla was also conducting experiments with use of liquid nitrogen and energy transmission during his last days in Colorado.
23. KSP.
24. NT to GS, May 14, 1911 [BLCU].
25. A. Ferguson, op. cit., March 25, 1971.
26. Marnicic, 1989/90.
27. Strange Light at Tesla's Tower. *New York Herald Tribune*, July 19, 1903, 2:4.

Chapter 34: The Web, pp. 293–306

1. JPM to NT, July 16, 1903 [LC].

2. NT to GS, August 17, 1903 [BLCU].
3. NT to Dickson D. Alley, May 26, 1903 [BLCU].
4. Petar Mandic to NT, September 2, 1903, in *Tesla's Correspondence With Relatives*, p. 134 [NTM].
5. NT to JPM, September 13, 1903 [LC].
6. NT to GS, July 30, 1903 [BLCU].
7. NT to GS, August 17, 1903 [BLCU].
8. Virginia Cowles, *The Astors* (New York: Knopf, 1979), pp. 134–35.
9. JJA to NT, October 6, 1903 [NTM].
10. NT prospectus, January 1, 1904 [SWP].
11. NT to RUJ, September 22, 1903 [BLCU].
12. "NT Says We Will Be Soon Taking Around the World," *New York World*, July 14, 1905.
13. "The Reasons Why 5,000,000 Persons Demand that Higgins Investigate the Equitable," *New York World*, July 13, 1905, 1: 3–4.
14. Ibid.
15. "Eymard Seminary, Suffern, New York, Supported by Mrs. Ryan," *New York World*, July 1, 1905.
16. What John Skelton Williams Thinks of Thomas F. Ryan," *New York World*, June 18, 1905, Editorial Sec., p. 1.[Williams was critical of Ryan. This section of article was compiled by the editors of the newspaper.]
17. NT to TFR, December 20, 1905? [NTM].
18. NT to JPM, October 13, 1903 [LC].
19. R. U. Johnson, p. 482.
20. NT to RUJ, December 2, 1903 [BLCU].
21. G. Wheeler, p. 263.
22. Ibid.
23. "What J. Skelton William Thinks of T. F. Ryan," *New York World*, June 18, 1905, Editorial Sec., p. 1.
24. H. Satterlee, p. 426.
25. Stephen Birmingham, p. 328.
26. G. Wheeler, p. 266.
27. Marc Seifer and Howard Smukler, "The Tesla/Matthews Outer Space Connection: An Interview With Andrija Puharich," *Pyramid Guide*, Parts I & II, May and July, 1978.
28. Andrija Puharich, in *The Zenith Factor*, video by Sky Fabin, 1984.
29. *New York World*, Sunday sec., March 8, 1896.
30. Robert McCabe, personal correspondence, January 15, 1991, Flint, Michigan.
31. On a number of occasions, Tesla stated that Wardenclyffe was set up primarily for transmitting telephone conversations. Apparently his plan was to create identical magnifying transmitter-receivers at strategic points around the globe. These would be connected by means of wireless; however, individual subscribers could be linked to the central stations by means of conventional wires although a wireless connection to the local central station was also possible. So, for instance, a subscriber in Australia calling up America would make the wireless connection via the main intercontinental trunk line. Thus, the problem of providing free electricity was easily circumvented (*My Inventions*, p. 178).
32. Margaret Coit, *Mr. Baruch* (New York: Houghton Mifflin, 1957), p. 123.
33. "From 1905...to 1931 inclusive, the output was $2,871,300,000." John Hays Hammond Sr., *Autobiography* (New York: Farrar and Rinehart, 1935), p. 518.
34. Edwin Hoyt, *The Guggenheims and the American Dream* (United States: Funk and Wagnalls, 1967), p. 158.
35. NT to JPM, December 7, 1903 [LC].
36. R. Chernow, p. 140.
37. Ann Morgan to NT, April 26, 1928 [NTM]. Literary license taken on conversation. Adapted from John Kennedy, "When Woman Is Boss—An Interview with NT," *Colliers*, January 30, 1926.
38. Conversation recreated from NT to JPM December 11, 1903 and two undated communiques from the same period [LC].
39. Robert was publishing a paper by Madam Curie which Tesla was reading over. Tesla also conferred with Curie through the mail concerning her most recent discovery of radiant energy.

40. KJ to NT, December 20, 1903 [NTM].
41. NT to JPM, January 13, 1904 [LC].
42. NT to JPM, January 14, 1904 [LC].
43. NT to William Rankine, April 10, 1904 [Profiles in History Archives, Beverly Hills, Calif.].
44. NT, "The Transmission of Electric Energy Without Wires," *Electrical World and Engineer,* March 5, 1904, p. 429–31 [condensed].

Chapter 35: Dissolution, pp. 307–323

1. NT, "The House of Morgan," in *Tesla Said,* p. 243.
2. K. Mannheim, *The Sociology of Knowledge* (London: Routledge and Kegan, Paul, 1952).
3. J. Goethe, *Faust,* C. Brooks, ed./transl. (Boston: Ticknor and Fields, 1856).
4. NT, "Man's Greatest Wonder," circa 1930 [KSP].
5. NT to JPM, possibly not sent, circa 1903 [LC].
6. "Langley Airship Proves a Failure," *New York Herald,* January 8, 1903, 5:2.
7. NT, "Mr. Tesla Praises Professor Langley," *New York Herald,* October 9, 1903, 8:6.
8. Marincic, "Research on L. I. Laboratory," 1989/90, p. 26.
9. NT to GS, December 9, 1903 [BLCU].
10. Ibid.
11. P. Baker, p. 339.
12. NT to GS, March 21, 1904 [BLCU].
13. NT to John S. Barnes, April 20, 1904 [NYHS].
14. John Flynn, *God's Gold: The Story of Rockefeller and His Times* (New York: Harcourt Brace, 1932).
15. P. Baker, p. 313; literary license taken on conversations between White and Tesla.
16. Alfred Cowles, "Harnessing the Lightning," *The Cleveland Leader,* March 27, 1904.
17. NT to Kerr, Page and Cooper, April 8, 1904 [NYHS].
18. NT to GS, June 1, 1904 [BLCU].
19. GW announcement, October 28, 1958 [KSP].
20. NT to JPM, July 22, 1904 [LC].
21. NT to GS, June 1, 1904 [BLCU].
22. NT to JPM, September 9, 1904 [LC].
23. H. Satterlee, p. 413.
24. NT to JPM, October 13, 1904 [LC].
25. JPM to NT, October 15, 1904 [LC].
26. NT to JPM, October 17, 1904 [LC].
27. Ibid., December 16, 1904.
28. JPM to NT, December 17, 1904 [LC].
29. NT to JPM, December 19, 1904 [LC].
30. NT, "The Transmission of Electrical Energy Without Wires As a Means for Furthering Peace," *Electrical World and Engineer,* January 7, 1905, pp. 21–24, in *Tesla Said,* pp. 78–86.
31. NT to JPM, February 17, 1905 [LC].
32. NT to GS, January 23, 1905 [BLCU].
33. NT to RUJ, March 10, 1910 [BLCU].
34. Ibid., March 22, 1905; March 24, 1905; March 28, 1905.
35. Ibid., April 5, 1905; April 12, 1905.
36. Hobson to NT, May 1, 1905 [KSP].
37. Ginzelda Hull Hobson to K. Swezey, February 14, 1955 [KSP].
38. NT to GS, July 25, 1905 [BLCU].
39. NT to GS, November 13, 1905 [BLCU].
40. JPM to NT, December 14, 1905 [LC].
41. NT to JPM, December 15, 1905 [LC].
42. TCM to NT, December 24, 1905 [NTM].
43. NT to JPM, January 24, 1906 [LC].
44. KJ to Mrs. Hearst, March 15, 1906.

[Bancroft Library, Berkeley, Calif.]
45. Ibid., February 6, 1906.
46. NT to JPM, February 15, 1906 [LC].

47. GS to NT, April 10, 1906 [LC].
48. SW, April 24, 1906 [SWP].
49. B. Baker, 1989.
50. Marc Seifer, "Forty Years of the Handwriting of Nikola Tesla," Lecture before the National Society of Graphology, N.Y., 1979; "The Lost Wizard," in T. Grotz and E. Raucher, eds., *Tesla Centennial Symposium* (Colorado Springs, Colo.: International Tesla Society, 1984); *Nikola Tesla: Psychohistory of a Forgotten Inventor* (San Francisco: Saybrook Institute, 1986) doctoral diss.
51. NT to JPM, October 15, 1906 [LC].

Chapter 36: The Child of His Dreams, pp. 324–335

1. NT, "The People's Forum," *The World,* May 16, 1907.
2. "Tesla Tower to be Sold," *New York Times,* October 27, 1907, 6:4 (literary license on phone conversation).
3. NT to KJ, October 16, 1907 [LA].
4. "Miss Merrington, Long an Author," obituary column, *New York Times,* June 21, 1951.The play opened up in 1906 (literary license taken on related chain of events).
5. NT, "Sleep From Electricity," *New York Times,* October 16, 1907, 8:5.
6. NT, *Tribute,* p. D-11.
7. "Nikola Tesla Sued," *New York Times,* July 21, 1912, 7:2; "Syndicate Sues Nikola Tesla," *New York Sun,* July 21, 1912, 1:3; "Tesla Property May Go for Debt," *New York City Telegram,* April 17, 1922.Tesla's mortgage with the Waldorf Astoria was consummated in May 1908.
8. H. Satterlee, p. 456.
9. John Davis, *The Guggenheims: An American Epic* (New York: William Morrow, 1978), p. 106.
10. R. Chernow, pp. 123–26.
11. NT to GS, November 20, 1907 [BLCU].
12. NT to GS, April 1, 1907 [BLCU].
13. NT, "Tesla on the Peary North Pole Expedition," *New York Sun,* July 16, 1907, in *Tesla Said,* pp. 90–91.
14. NT, "Signalling to Mars," *Harvard Illustrated,* March 1907; in *Tesla Said,* pp. 92–93.
15. NT, "Tesla's Tidal Wave to Make War Impossible," *English Mechanic and World of Science,* May 3, 1907; in *Tesla Said,* pp. 98–102.
16. NT, "Tesla's Wireless Torpedo," *New York Times,* March 20, 1907, 8:5; in *Tesla Said,* pp. 96–97.
17. NT, *Tesla Said,* pp. 96–105.
18. NT, "Can Bridge Gap to Mars," *New York Times,* June 23, 1907, in *Tesla Said,* pp. 103–4.
19. O'Neill, 1944.
20. NT to J. O'Neill, February 26, 1916 [NTM]; "O'Neill Writes on Tesla's Life," *Nassau Daily Review Star,* 1944, p. 16; "Life of a Self-Made Superman," Book review of *Prodigal Genius, New York Times,* November 19, 1944 [KSP].
21. "Sheriff Takes Tesla Tower," *New York Sun,* June 13, 1907, 3:3.
22. NT to JJA, June 8, 1908 [NTM].
23. NT, "Nikola Tesla's Forecast for 1908," *New York World,* January 6, 1908; "Aerial Warships Coming, Tesla Tells," *New York Times,* March 11, 1908, 1:2; "Little Aeroplane Progress: So Says Nikola Tesla," *New York Times,* June 6, 1908, 6:5.
24. Ibid.
25. "Zeppelin Flies Over 24 Hours," *New York Times,* May 30, 1908, 1:1,2.
26. Carl Dienstback, "The Brucker Transatlantic Airship Expedition," *Scientific American,* January 21, 1911, pp. 1, 62.
27. R. U. Johnson, p. 580.
28. Dennis Eskow, "Silent Running," *Popular Mechanics,* July 1986, pp. 75–77.
29. Ibid.
30. Branimir, Jovanovic, *Tesla I Svet Vasduhoplovstva* (Belgrade: Tesla Museum, 1988), p. 42.
31. Frank Parker Stockbridge, "Tesla's New Monarch of Mechanics," *New York Herald Tribune,* October 15, 1911, p. 1.
32. *Illustrated World Encyclopedia* (Woodbury, N.Y.: Bobley Pub. Co., 1977).

33. *Encyclopedia of Science and Technology,* vol. 1, p. 288.
34. O. Chanute, "Progress in Aerial Navigation," *Engineering Magazine, 2,* 1891–92, pp. 1–15.
35. B. Jovanovic, *Tesla I Svet Vasduhoplovstva,* pp. 49–50.
36. John P. Campbell, "Vertical-Takeoff Aircraft," *Scientific American,* August 8, 1960, p. 48.
37. "A New Version of Space Shuttle," *Newsweek,* July 1, 1996, p. 69.
38. Wallace Cloud, "Vertical Takeoff Planes," *Popular Science,* August 1965, pp. 42–45; 176–77; *Wings,* [TV show], 1991.
39. *Encyclopedia of Science and Technology,* vol. 19, p. 203.
40. "The Allies' Firepower," *Newsweek,* February 18, 1991 (insert).
41. William Broad, "Flying on a Beam of Energy: New Kind of Aircraft Is on Horizon as Designers Try Microwave Power," *New York Times,* July 21, 1987, C1.
42. "Tesla Designs Weird Craft," *Brooklyn Eagle,* 8:2.
43. NT, "Method of Aerial Transportation," patent numbers 1,655,113 and 1,655,114, filed September 9, 1921, accepted January 3, 1928.
44. "The New Weapons," *Newsweek,* September 10, 1990, p. 28.

Chapter 37: Bladeless Turbines, pp. 336–342

1. NT to JJA, March 22, 1909 [NTM].
2. Literary license taken. "Messages From Dead Now Made Public; Sir Oliver Lodge Advertised," *New York Times,* September 15, 1908, 1:5; "Talk of Signals to Mars, Astronomers Gather in Paris," *New York Times,* April 21, 1909, 1:2; "How to Signal Mars, says N. Tesla," *New York Times,* May 23, 1909.
3. NT, "Tesla Predicts More Wonders," *New York Sun,* April 7, 1912, 1:3,4,5; 2:5.
4. O'Neill, 1944.
5. "French to Establish Wireless Station on Eiffel Tower," *New York Times,* January 26, 1908, 1:6.
6. "DeForest Tells of New Wireless," *New York Times,* February 14, 1909, 1:3.
7. "Dr. DeForest in Philadelphia. Wireless Telephone Soon," *New York Times,* March 23, 1909, 18:4.
8. "Steel Towers for Waldorf Wireless," *New York Times,* March 6, 1909, 14:2.
9. *Technical World Magazine,* circa 1911; in Jeffery Hayes, ed., *Tesla's Engine: A New Dimension For Power* (Milwaukee, WI: Tesla Engine Builders Assoc., 1994), p. 58.
10. NT, "New Inventions by Tesla," Address at NELA. *Electric Review and Western Electrician,* May 20, 1911, pp. 986–87.
11. F. P. Stockbridge, "Tesla's New Monarch of Mechanics," *New York Herald Tribune,* October 15, 1911, p. 1, in Hayes, ed., pp. 23–37.
12. Ibid.
13. GS to NT, January 10, 1909 [LC].
14. NT to GS, March 26, 1909 [LC].
15. GS to NT, November 11, 1909; November 4, 1910 [LC].
16. "Tesla Propulsion Company," *Electrical World,* May 27, 1909, p. 1263.
17. G. Freibott, "History and Uses of Tesla's Inventions in Medicine," Talk before International Tesla Society, Colorado Springs, Colo., 1984; M. Seifer, *Psychohistory,* p. 429; Frederick Sweet, Ming-Shian Kao, and Song-Chiau Lee, "Ozone Selectively Inhibits Growth of Human Cancer," *Science,* August 22, 1980, pp. 931–32.
18. NT to GS, February 22, 1910 [LC].
19. NT to GS, February 19, 1909; November 23, 1909; the company in Providence was Corliss NT to GW Co., January 12, 1909 [LC].
20. "Col. Astor Estate," *New York Times,* June 22, 1913, *V,* p. 2.
21. NT to Charles Scott, December 30, 1908[LC].

Chapter 38: The Hammond Connection, pp. 343–356

1. NT to JHH Jr., November 8, 1910 [LC].
2. John O'Neill, p. 175.
3. Hammond Jr. to Swezey, May 11, 1956 [KSP]. Note: No letters between NT and

Hammond Sr. were discovered at the Gloucester or Tesla Museums or in Hammond Sr.'s Yale papers.

4. John Hays Hammond, Sr., *Autobiography of John Hays Hammond* (New York: Farrar & Rhinehart, 1935).

5. M. Josephson, *Edison* (New York: McGraw-Hill, 1966), p. 292.

6. NT to JJA, January 6, 1899 [NTM].

7. John Hays Hammond, Jr., "The Future in Wireless," *National Press Reporter,* vol. XV, no. 110, May 1912.

8. John Hays Hammond Sr., *Literary Digest,* June 20, 1936, p. 27.

9. John Hays Hammond Sr., *Autobiography,* p. 129.

10. Ibid.

11. "Hammond's First Job," *New York Times,* December 31, 1915, 3:6.

12. "JHH Explains Why Is Ambitious to Become Vice President," *New York Times,* June 7, 1908, *v,* p. 11.

13. Nancy Rubin, *John Hays Hammond, Jr.: A Renaissance Man in the Twentieth Century* (Gloucester, Mass.: Hammond Museum, 1987), p. 4.

14. John Hays Hammond Sr., *Autobiography,* pp. 481–82.

15. Rubin, *John Hays Hammond, Jr.,* p. 8.

16. "John Hays Hammond, Jr.," Franklin Institute, April 15, 1959 [HC].

17. Mort Weisinger, "Hammond: Wizard of Patents," *Coronet,* May 1949, pp. 67–72.

18. Rubin, op. cit.; Weisinger, ibid., pp. 67–68.

19. JHH Jr. to NT, February 16, 1911 [NTM].

20. JHH Jr. to JHH Sr., September 17, 1909 [HC].

21. NT to JHH Jr. September 27, 1909; September 29, 1909 [NTM].

22. Rubin, op. cit., p. 8.

22. JHH Jr. to NT, November 10, 1910 [NTM].

24. NT to Harris Hammond, December 19, 1901 [NTM]. The Nobel Prize for achievements in wireless was shared with Carl F. Braun, an electrical inventor working for Marconi competitor Telefunken of Germany.

25. Rubin, *John Hays Hammond,* p. 4; guest book, John Hays Hammond Jr. Estate [HC].

26. John Hays Hammond Jr., "The Future of Wireless," *National Press Reporter,* May 1912 [HC].

27. NT to JHH Jr., November 12, 1910 [LC].

28. JHH Jr. to NT, November 10, 1910 [LC].

29. NT to JHH Jr., November 12, 1910 [LC].

30. John Hays Hammond Jr., op. cit., May 1912.

31. Ibid.

32. NT, "Possibilities of Wireless," *New York Times,* October 22, 1907, 8:6; in Ratzlaff, op. cit., p. 107.

33. Miessner to NT, November 8, 1915 [LC].

34. Swezey collection [KSP].

35. JHH Jr. to NT, February 16, 1911 [NTM].

36. NT to JHH Jr., February 18, 1911 [LC].

37. "New Inventions by Tesla," *Electrical Review and Western Electrician,* May 20, 1911, pp. 986–88.

38. "Tesla Tells of Wonders," *New York Times,* May 16, 1911, 22:5.

39. "Tesla's Plan for 'Wireless' Electric Lightning," *Electrical Review and Western Electrician,* January 8, 1910, p. 91.

40. NT to JHH Jr., February 28, 1911 [NTM].

41. Ibid., April 22, 1911; February 14, 1913.

42. Cleveland Moffett, "Steered by Wireless: The Triumph of a Man of Twenty-five," *McClure's Magazine,* March 1914, pp. 27–33.

43. JHH Jr. to NT, January 1, 1912 [LC]; Kaempffert, W., *A Popular History of American Invention* (New York: Scribners, 1924).

44. NT to JHH Jr., February 1913 [NTM].

45. Ibid., July 16, 1913.

46. Moffett, "Steered by Wireless," *McClure's,* March 1914.

47. JHH Jr. to JHH Sr., December 2, 1914 [HC].

48. JHH Jr. to Secretary of Navy, October 11, 1924 [NAR].

49. National Archives, Washington, D.C.

50. "World Court for Peace says John Hays Hammond," *New York Times,* March 22, 1915, 4:2; "To Test Hammond Torpedo," *New York Times,* August 29, 1916, 9:2; "Control Ships with Radio," *New York Times,* February 15, 1919, 3:8.

51. Rubin, *John Hays Hammond,* p. 12.

52. Hammond Jr., October 11, 1924.

53. NT, *On His Work in A.C.,* 1916/1992, pp. 19, 158.

54. Rubin, *John Hays Hammond,* p. 16.

55. Guest book [HC].

56. Andrija Puharich, *Beyond Telepathy* (Garden City, NY: Doubleday, 1962).

Chapter 39: J. P. Morgan Jr., pp. 357–367

1. Fritz Lowenstein to NT, April 18, 1912 [KSP].

2. "Electrified Schoolroom to Brighten Dull Pupils," *New York Times,* August 18, 1912. *5,* p. 1; "Tesla Predicts More Wonders," *New York Times,* April 7, 1912, *5,* 1:4–6.

3. "Marconi Lecture Before NY Electrical Soc." *Electrical World,* April 20, 1912, p. 835.

4. Erik Barnouw, *A Tower in Babel: A History of Broadcast United States* (New York: Oxford University Press, 1966), pp. 76–77.

5. Robert Sobel, *RCA* (New York: Stein & Day, 1986), pp. 19–20; Robert Harding, *George H. Clark Radiona Collection* (Washington, D.C.: Smithsonian Institution, 1990).

6. NT to GS, January 18, 1913 [LC].

7. W. Jolly, *Marconi,* p. 190.

8. NT to JPM Jr., March 19, 1914 [JPM].

9. NT to JPM Jr., July 23, 1913 (re: Girardeau, M. E. testimony), [JPM].

10. "Judgment Against NT," *New York Sun,* March 24, 1912, 1:3; "Syndicate [Stallo, Jacobash, Levy, and Sherwood Jr.] sues NT," *New York Sun,* July 21, 1912, 1:3; "NT Sued," July 21, 1912, *New York Times,* 7:2; "Wireless Litigation," *Electrical World,* June 28, 1913.

11. NT to JPM Jr., June 11, 1915 [LC].

12. NT to TAE, February 24, 1912 [TAE].

13. NT to GW, August 10, 1910; August 19, 1910; GW to NT, August 18, 1910 [LC].

14. Tesla received royalties of 5%, approximately $1200/month, until Tuckerton was seized in 1916, Tuckerton Radio Station to J. Daniels, Secretary of the Navy, July 3, 1916 [NA].

15. NT to JPM Jr., February 19, 1915 [JPM]; NT to Frank and NT to Pfund, circa 1912–1922 [NTM]; "19 More Taken as German Spies," *New York Times,* I, 1:3; "Find Radio Outfit in Manhattan Tower," *New York Times,* March 5, 1918, 4:4.

16. NT, *On His Work in A.C.,* 1916/1992, p. 133.

17. "Col. Astor Estate," *New York Times,* June 22, 1913, *5,* p. 2.

18. NT to JPM Jr., sympathy letter, March 31, 1913 [LC]; NT to Anne Morgan, March 31, 1913 [NTM].

19. NT to JPM Jr., May 19, 1913; May 20, 1913 [LC].

20. R. U. Johnson, *Remembered Yesterdays,* p. 142.

21. NT to GS, July 13, 1913 [LC].

22. NT to JPM Jr., June 6, 1913 [LC].

23. JPM Jr. to NT, June 11, 1913 [JPM].

24. NT, "Open Letter to His Grace," *Electrical Magazine,* March 18, 1912 [JPM].

25. NT to JPM Jr., June 15, 1913 [JPM].

26. NT to RUJ, December 24, 1914; December 27, 1914 [LC].

27. Josephson, *Edison,* 1959, p. 296.

28. JPM Jr. to NT, September 11, 1913 [JPM].

29. NT to JPM Jr., December 23, 1913 [LC].

30. There is some evidence that Tesla travelled to Krasnodar, Russia, east of the Black Sea, before the fall of the czar, circa 1914–16, where he lectured and gave demonstrations "at the circus building where the Kuban cinema is now," according to Semyon Kirlian (1896–1978). If he had actually seen Tesla (as opposed to a "Teslaic" demonstration by another engineer), this would mean that Tesla traveled to Europe probably right before World War I broke out. As he was negotiating with the king of Belgium, the kaiser of Germany, and engineers in Italy and Russia, it is possible that Tesla did indeed make a grand tour then, and if so, he would have most likely also visited his sisters in Croatia/Bosnia. Further evidence

would be needed to support this hypothesis. *Source:* Victor Adamenko, "In Memory of Semyon Kirlian." *MetaScience,* 4 (1980): pp. 99–103, unpublished.

31. NT to JPM Jr., December 29, 1913 [LC]; NT to JPM Jr., January 6, 1914 [JPM]; Ron Chernow, *The House of Morgan* (New York: *Atlantic Monthly Press,* 1990), p. 195.

32. NT to JPM Jr., March 14, 1914 [LC].

33. NT to RUJ, December 27, 1914 [LC].

34. Ron Chernow, *House of Morgan,* p. 190.

35. "Tribute of Former Associates for George Westinghouse." *Electrical World,* March 21, 1914, p. 637.

Chapter 40: Fifth Column, pp. 368–377

1. FDR re: NT and wireless priority, September 14, 1916 [NAR].

2. W. Jolly, *Marconi,* 1972.

3. Niel M. Johnson, *George Sylvester Viereck: German/American Propagandist* (Chicago: University of Illinois Press, 1972).

4. "Nation to Take Over Tuckertown Plant," *New York Times,* September 6, 1914, II, 14:1.

5. NT to JPM Jr., February 19, 1915 [JPM].

6. "Germans Treble Wireless Plant," *New York Times,* April 23, 1915, 1:6.

7. "Tesla Sues Marconi," *New York Times,* August 4, 1915, 8:1.

8. NT to JPM Jr., November 23, 1914; February 19, 1915 [JPM].

9. Erik Barnouw, *A Tower in Babel: A History of Broadcasting in the United States* (New York: Oxford University Press, 1966), vol. 1, pp. 35–36.

10. Hammond collection, National Archives.

11. Jolly, *Marconi,* p. 225.

12. J. Ratzlaff and L. Anderson, p. 100.

13. "Marconi Loses Navy Suit," *New York Sun,* October 3, 1914 [NT to JPM Jr. corresp., JPM].

14. "Prof. Pupin Now Claims Wireless His Invention," *Los Angeles Examiner,* May 13, 1915; R & A, p. 100.

15. "When Powerful High-Frequency Electrical Generators Replace the Spark-Gap," *New York Times,* October 6, 1912, VI, 4:1.

16. "Marconi Wireless vs. Atlantic Communications Co.," 1915 [LA].

17. NT, *On His Work in A.C.,* 1916/1992, p. 105.

18. Orin Dunlap, *Radio's 100 Men of Science* (New York: Harper & Brothers, 1944); Marconi Wireless vs. United States, *Cases Adjudged in the Supreme Court,* October 1942, v. 320, p. 17.This feature was obviously also part of Tesla's design, although the court eventually ruled Stone as the originator.

19. Leland Anderson, ed., "John Stone Stone on Nikola Tesla's Priority in Radio and Continuous-Wave Radiofrequency Apparatus," *The Antique Wireless Review,* vol. 1, 1986.

20. E. F. Sweet and FDR correspondence re: Tesla, September 14, 1916; September 16, 1916; September 26, 1916 [NAR].

21. Leland Anderson, "Priority in the Invention of the Radio: Tesla vs. Marconi," *The Tesla Journal,* vol. 2/3, 1982/83, pp. 17–20.

22. Lawrence Lessing. *Man of High Fidelity: Edwin Howard Armstrong* (New York: Lippincott, 1956), pp. 42–43.

23. Ibid., pp. 66–80.

24. [KSP].

25. Lloyd Scott, *Naval Consulting Board of the United States* (Washington, D.C.: Government Printing Office, 1920).

26. Johnson, *George Sylvester Viereck,* pp. 23, 34.

27. "Germany to Sink the Armenian. Navy May Seize Sayville Wireless," *New York Times,* July 1, 1915, 1:4–7.

28. NT to JPM Jr., July 1915 [LC].

29. "JP Morgan Shot by Man Who Set the Capitol Bomb," *New York Times,* July 3, 1915.

30. "Wireless Controls German Air Torpedo," *New York Times,* July 10, 1915, 3:6, 7.

31. NT, "Science and Discovery Are the Great Forces Which Will Lead to the Consummation of the War," *New York Sun,* December 20, 1914, in *Lectures, Patents, Articles,* pp. A-162–171.

32. "Federal Agents Raid Offices Once Occupied by Telefunken. Former Employee Richard Pfund Charged; No Arrests Made," *New York Times,* March 5, 1918, 4:4.

33. NT to GS, December 25, 1917 [LC].

34. Royalty check to NT for $1,567 from Hochfrequenz Maschienen Aktievgesell Schaft for drachlose Telegraphic, 1917 [Swezey Col.]. Tuckertown was still owned by the Germans, although seized by the U.S. Navy, and Tuckertown, with full knowledge of the "Director of Naval Communications," had agreed to pay Tesla royalties, see NT to GS, October 12, 1917 [LC].

35. Lloyd Scott, *Naval Consulting Board of the U.S.* (Washington, D.C.: Government Printing Office, 1920).

36. Interview with A. Puharich, 1984. According to Puharich, the Hammond/Tesla documents were removed from the Hammond Museum in Gloucester, Mass., after Hammond's death, and classified as top secret sometime in 1965.This author has read through many of these documents from the National Archives through the FOIA.

Chapter 41: The Invisible Audience, pp. 378–394

1. RUJ to NT, March 1916 [LC].
2. NT to JPM Jr. [JPM].
3. Hunt and Draper, 1964/77, pp. 170–71.
4. For literature: Romain Rolland, Hendrik Pontoppidan, Troeln Lund, and Verner von Heidenstam were announced; Theodor Svedberg was named for chemistry. Rolland was the only winner that year out of that group, with the others, except for Lund, eventually also winning.
5. Nobel nominations for 1915 and 1937 [Archives, Royal Swedish Academy of Sciences]; L. Anderson corresp., 1991.
6. The date of 1912 in the O'Neill book, and often echoed in various magazine articles, was probably a typographical error in the biography. O'Neill, 1944, p. 229.
7. NT to Light House Board, September 27, 1899 [NAR].
8. NT to RUJ, November 10, 1915 [BLCU].
9. Hunt and Draper, 1964/77, p. 167.
10. RUJ to NT, March 1916 [LC].
11. Probably Karl Braun who shared the 1909 Prize with Marconi; NT, *On His Work in A.C.*, p. 48.
12. "Tesla No Money; Wizard Swamped by Debts," *New York World*, March 16, 1916.
13. NT's Fountain," *Scientific American*, 1915.
14. "Can't Pay Taxes," *New York Tribune*, March 18, 1916; "Wardenclyffe Property Foreclosure Proceedings," *NY Supreme Court*, circa 1923 [L. Anderson files].
15. Abraham and Savin, 1971.
16. NT to GS, April 25, 1916 [LC].
17. NT, *My Inventions*, p. 103.
18. Leland Anderson, "Tesla Portrait by the Princess Vilma Lwoff-Parlaghy," *The Tesla Journal*, nos. 4/5, 1986/87, pp. 72–73.
19. Trbojevitch immigrated circa 1912. Interviews with William Terbo, 1984–1991.
20. John O'Neill to NT, February 23, 1916 (greatly condensed) [NTM].
21. NT to J. O'Neill, February 26, 1916 [NTM].
22. B. A. Behrend, "Edison Medal Award Speech, 1917," in *Tesla Said*, p. 180.
23. NT to Waldorf-Astoria mgt., July 12, 1917 [LA].
24. Minutes of the Annual Meeting of the AIEE, May 18, 1917, in *Tesla Said*.
25. O'Neill, 1944.
26. Minutes of the AIEE, May 18, 1917, in *Tesla Said*, pp. 189.
27. Lester S. Holmes was represented for the hotel as owner of said Tesla property. Baldwin and Hutchins to NT, July 13, 1917, from: *Wardenclyffe Property Foreclosure Proceedings*, New York Supreme Court, circa 1923 [LA].
28. Quoted in J. B. Smiley to Frank Hutchins, July 16, 1917 [LA].
29. John B. Smiley to Frank Hutchins, July 13, 1917 [LA].
30. NT to Waldorf-Astoria, July 12, 1917 [LA].
31. "Tesla's New Device Like Bolts of Thor," *New York Times*, December 8, 1915, 8:3.
32. NT to JPM Jr., April 8, 1916 [LA].
33. "Reason for Seizing Wireless," *New York Times*, February 9, 1917, 6:5.
34. "Spies on Ship Movements," *New York Times*, February 17, 1917, 8:2.
35. "19 More taken as German spies," *New York Times*, April 8, 1917, 1:3.

36. "Navy to Take Over All Radio Stations," *Enumeration*, April 7, 1917, 2:2.
37. F. J. Higginson to NT, May 11, 1899 [NAR].
38. F. Higgenson to NT, August 8, 1900 [NAR]. For the full correspondence of this event, see chapter 26; R. P. Hobson to NT, May 6, 1902 [LA].
39. L. S. Howeth, *History of Communications-Electronics in U.S. Navy* (Washington, D.C.: U.S. Government Printing Office, 1963), pp. 518–19; A. Hezlet, *Electronics & Sea Power* (New York: Stein & Day, 1975), p. 41.
40. Howeth, *History of Communications*, p. 64.
41. Hezlet, *Electronics & Sea Power*, pp. 41–42.
42. Robert Sobel, p. 43; Hezlet, p. 77.
43. Howeth, p. 256.
44. Ibid., pp. 375–76; Scherff to NT [LC]. U.S. Navy to Tuckerton Counsel, April 29, 1919 [NA].
45. Howeth, pp. 577–80.
46. NT, "Electric Drive for Battleships," *New York Herald*, February 25, 1917; in *Lectures, Patents, Articles*, p. A-185.
47. Patent no. 1,119,732, Apparatus for transmitting electrical energy, was applied for on January 16, 1902.The application was renewed on May 4, 1907 and granted on December 1, 1914. This patent, in essence, contains all of Tesla's key ideas behind the construction of Wardenclyffe.
48. KJ to Mrs. Hearst, circa 1917 [BLCU].
49. NT to GS, July 26, 1917 [LC].
50. GS to NT, August 20, 1917 [LC].
51. Howeth, p. 354.
52. The breakdown was as follows: GE 30%, Westinghouse 20%, AT&T 10%, United Fruit 4%, others 34%. Sobel, 1986, pp. 32–35.
53. Tesla would also be cut out of a secret agreement between GE and Westinghouse to hold back production of efficient fluorescent lighting equipment, as they did not want to undermine the highly profitable sale of normal Edison lightbulbs or "cut too drastically the demand for current" (S. C. Gilfillan, *Invention and the Patent System* (Washington, D.C.: U.S. Government Printing Office, 1964), p. 100.
54. "U.S. Blows Up Tesla Radio Tower," *Electrical Experimenter*, September 1917, p. 293.
55. "Destruction of Tesla's Tower at Shoreham, LI Hints of Spies," *New York Sun*, August 5, 1917.
56. Howeth, pp. 359–60.
57. Ibid., p. 361.
58. E. M. Herr to NT, November 16, 1920 [LC].
59. GW Corp. to NT, November 28, 1921 [LC].
60. NT to GW Corp., November 30, 1921 [LC].

Chapter 42: Transmutation, pp. 395–403

1. NT, "Edison Medal Speech," May 18, 1917, in *Tesla Said*, 181–82.
2. Hugo Gernsback, "Nikola Tesla and His Inventions," *Electrical Experimenter*, January 1919, pp. 614–15; R. Hugo Lowndes, *Gernsback: A Man With Vision, Radio Electronics*, August 1984, pp. 73–75.
3. Erik Barnouw, *A Tower in Babel: A History of Broadcast USA* (New York: Oxford University Press, 1966), pp. 28–30.
4. [KSP].
5. H. Gernsback, "NT: The man," *Electrical Experimenter*, February 1919, p. 697.
6. H. Winfield Secor, "The Tesla High Frequency Oscillator," *Electrical Experimenter*, March 1916, pp. 614–15, 663; NT, "Some Personal Reflections," *Scientific American*, June 5, 1915, pp. 537, 576–77.
7. Lester Del Ray, *Fantastic Science Fiction Art: 1926–1954* (New York: Ballantine, 1975).
8. NT to JPM Jr., June 13, 1917 [LC].
9. NT to GS, September 25, 1917 [LC].
10. Ibid., December 25, 1917.
11. Ibid.
12. O'Neill, pp. 222–28.
13. NT to GS, December 25, 1917.

14. NT to GS, June 11, 1918 [LC].
15. NT to GS, June 12, 1918 [LC].
16. NT to GS, May 15, 1918; June 22, 1918 [LC].
17. Leland Anderson, *Nikola Tesla's Residences, Laboratories and Offices* (Denver, Colo.: 1990).
18. GS to NT, March 29, 1918 [BLCU]; November 4, 1925 [LC].
19. Leland Anderson to M. Seifer, April 28, 1988; see *International Science and Tech.*, November 1963, pp. 44–52, 103.
20. Waltham advertisements, *New York Times*, June 8, 1921, 36:4, 5.
21. NT to GS, December 6, 1922 [LC].
22. NT to GS, October 18, 1918 and circa 1925 [LC].
23. RUJ to NT, December 30, 1919 [LC].
24. J. Abraham and R. Savin, *Elihu Thomson Correspondence* (New York: Academic Press, 1971), p. 400.
25. "Radio to Stars, Marconi's Hope," *New York Times*, January 19, 1919.
26. NT, "Interplanetary Communication," *Electrical World*, September 24, 1921, p. 620.
27. "Celestial Movies," February 3, 1919, 14:3.
28. H. Gernsback, "Nikola Tesla: The man," *Electrical Experimenter*, February 1919, p. 697.
29. Surmised in part from: "At Night and in Secret NT Lavishes Money and Love on Pigeons," *New York World*, November 21, 1926, Metropolitan Sec., p. 1.
30. O'Neill, 1944.
31. Ibid., pp. 224–26.
32. L. Anderson to M. Seifer, July 29, 1991.
33. C. R. Possell to Marc J. Seifer, phone interview and written correspondence, May 29, 1991; June 10, 1991; *Extraordinary Science*, IV, 2, 1992.
34. Jeffery Hayes, *Boundary-Layer Breakthrough* (Security, Colo.: High Energy Enterprise, 1990).
35. Interview with L. Anderson, July 29, 1988, Colorado Springs, Colo., as told to him and Inez Hunt. Also see James Caufield, "Radioed Light, Heat and Power Perfected by Tesla," *Harrisburg Telegraph*, March 22, 1924, Radio Sec., pp. 1–2: "The war was upon him and the gov't requested that [Wardenclyffe] come down. After the war Prof. Tesla again started to prove his theory, but this time he chose Colorado Springs as the location of his laboratory. It was while at the 'Springs' that he first demonstrated power transmission without the aid of wires."

Chapter 43: The Roaring Twenties, pp. 404–415

1. Tesla quote as told to John O'Neill and Bill Lawrence, O'Neill, pp. 316–17.
2. D. Wallechensky, November 1928, quoted in C. Cerf and V. Navasky, *The Experts Speak* (New York: Pantheon, 1984), p. 259.
3. NT to Hugo Gernsback, November 30, 1921 [KSP].
4. NT. "Views on Thought Transference," *Electrical Experimenter*, May 1911, p. 12.
5. Nikola Tesla v. George C. Bold Jr. Suffolk County Supreme Court, April 1921 [LA].
6. KJ to NT, April 24, 1920 [NTM].
7. Introduction compiled mostly from C. Daniel, ed., *Chronicle of the 20th Century* (Mt. Kisco, N.Y.: Chronicle, 1987); W. Langer, ed., *New Illustrated Encyclopedia of World History* (New York: Harry Abrams, 1975).
8. RUJ to NT, October 15, 1925 [LC].
9. W. Jolly, pp. 258–60.
10. Nancy Rubin, p. 25.
11. Dragislav Petkovich, "A Visit to Nikola Tesla," *Politika*, April 27, 1927, p. 6.
12. Adrian Potter, FBI report on Friends of Soviet Russia, 1921–1923 [FOIA].
13. George Seldes, *Witness to a Century* (New York: Ballantine, 1987), pp. 181–83.
14. Ronald Kline, "Professionalism and the Corporate Engineer: Charles P. Steinmetz," *IEEE Transactions*, August 1980, pp. 144–50.
15. L. Fischer, *The Life of Lenin* (New York: Harper & Row, 1964), p. 630.
16. There is a famous photo taken on April 3, 1921, during a trans-Atlantic wireless broadcast for RCA, GE, and AT&T. In a number of sources (M. Cheney, *Tesla: Man Out of Time* (Englewood Cliffs, N.J.: Prentice-Hall, 1981); M. Seifer, "The Inventor and the Corporation: Tesla, Armstrong & Jobs," *1986 Tesla Symposium Proceedings;* R. G. Williams,

Introducing Nikola Tesla Through Some of His Achievements (Mokelumne Hill, Calif.: Health Research, 1970) it has been suggested that the man standing between Albert Einstein and Charles Steinmetz was Nikola Tesla. Other people in the photo include Irving Langmuir and David Sarnoff. After reviewing the original caption and conferring with Tesla expert Leland Anderson, it has been ascertained that the man is not Tesla at all, but rather John Carson of AT&T. Coincidentally, this photo has been used by the GE public relations people on numerous occasions with all people but Einstein and Steinmetz airbrushed out. See *Life Magazine,* 1965 and B. Gorowitz, ed., *The Steinmetz Era: 1892–1923: The GE Story* (Schenectady, N.Y.: Schenectady Elfun Society, 1977), p. 39.

17. [KSP].

18. "Judgment [of $3,299] Filed Against Tesla by St. Regis," *New York Times,* May 25, 1924, 14:1.

19. "At Night and in Secret Nikola Tesla Lavishes Money and Love on Pigeons," *New York World,* Metropolitan Sec., November 21, 1926, 1:2–5.

20. M. Cheney, p. 84. Original source, Kenneth Swezey.

21. A. Nenadovic, "100 Years Since the Birth of Nikola Tesla," *Politika,* July 8, 1956 [KSP].

22. K. Swezey, "How Tesla Evolved Epoch-Making Discoveries," *Brooklyn Eagle,* April 4, 1926, pp. 8–9.

23. A. Nenadovic, July 8, 1956.

24. John B. Kennedy, "When Woman Is Boss," *Collier's,* January 30, 1926, pp. 17, 34.

25. Anne Morgan to NT, April 26, 1928 [NTM].

26. "At Night and in Secret Nikola Tesla Lavishes Money and Love on Pigeons," *New York World,* Metropolitan sec., November 21, 1926, p. 1. Other sick birds that he could not care for, Tesla took to animal hospitals.

27. C. Hedetniemi to OAP, January 29, 1943 [FOIA].

28. "Dr. Tesla Picks Tunney," *New York Herald Tribune,* September 27, 1927, 2:3.

29. Ginzelda Hull Hobson to K. Swezey, February 14, 1956 [KSP].

30. Petkovich, p. 4.

31. According to L. Anderson, Swezey said that the O'Neill stories of Tesla wiping his plates at the dinner table were untrue, the proof beign the pigeon link. "Meticulous," Tesla was clearly obsessed with avoiding contaminated water, phobic and fearful of germs, and so it seems likely to this researcher that he did, indeed, clean off his silverware and plates with napkins at eating establishments. Caring for pigeons, even in one's apartment (probably in a separate room where he kept a lab), is quite different than eating from unclean dishes.

32. NT to JPM Jr., November 21, 1924 [LC].

Chapter 44: Faster Than the Speed of Light, pp. 416–427

1. NT, 1960; translated from German by Edwin Gora.

2. NT to Flowers, 1917–1925 [NTM].

3. John B. Flowers, "Nikola Tesla's Wireless Power System and Its Application to the Propulsion of Airplanes," August 8, 1925 [Archives, Toby Grotz, Colorado Springs, Colo].

4. NT, "World System of Wireless Transmission of Energy," *Telegraph & Telephone Age,* October 16, 1927, pp. 457–60; in *Nikola Tesla,* 1981, pp. 83–86.

5. "In a solar eclipse the moon comes between the sun and the earth.... At a given moment, the shadow will just touch at a mathematical point, the earth, assuming it to be a sphere.... Owing to the enormous radius of the earth, [it] is nearly a plane. [Thus,] that point where the shadow falls will immediately, on the slightest motion of the shadow downward, enlarge the circle at a terrific rate, and it can be shown mathematically that this rate is infinite" (NT, *On His Work with AC,* pp. 137–39).

6. NT to Mrs. A. Trbojevic, November 20, 1928 [Wm. Terbo archives]; NT to Nikola Trbojevich, September 13, 1928, October 3, 1928; January 28, 1929; May 29, 1929, in *Correspondence with Relatives,* pp. 128, 135.

7. W. Terbo, Opening remarks, in S. Elswick, ed., *Proceedings of the 1988 Tesla Symposium,* Colorado Springs, Colo., 1988, pp. 8–11.

8. Myron Taylor, memorandum, September 28, 1931 [Archives, US Steel, USX Corp., Pittsburgh, PA]; Sava Kosanovic, August 30, 1952 [KSP].

9. Derek Ahlers, interview with Peter Savo, September 16, 1967 [Archives, Ralph Bergstresser].

10. "Beam to Kill Army at 200 Miles, Tesla's Claim on 78th Birthday," *New York Herald*

Tribune, July 11, 1934, 1:15; in *Solutions to Tesla's Secrets,* pp. 100–12.

11. Ronald Clark, *Einstein: The Life & Times* (New York: World Publishing—Times/Mirror, 1971).

12. Fritzof Capra, *The Tao of Physics* (Colo.: Shambhala, 1975), pp. 64, 208.

13. NT, New York paper on various theories, circa 1936 [KSP].

14. "Tesla Coil Used in Atom Splitting Machines," *N.Y. American,* November 3, 1929; O'Neill to NT, August 1, 1935 [NTM].

15. NT, "Tesla 'Harnesses' Cosmic Energy," *Philadelphia Public Ledger,* November 2, 1933; in *Solutions to Tesla's Secrets,* pp. 104–5.

16. "Tesla at 75," *Time,* July 20, 1931, pp. 27–28.

17. [KSP].

18. "Tesla, 79, Promises to Transmit Force," *New York Times,* July 11, 1935, 23:8.

19. "Sending Messages to Planets Predicted by Dr. Tesla on Birthday," *New York Times,* July 11, 1937, 13:2; in NT, *Solutions*..., pp. 132–34.

20. Nancy Czito, private corresp., Washington, D.C., 1983.

21. Patent numbers 685,957; 685,958, in NT, *Lectures, Patents, Aritcles,* 1956, pp. P-343–51.Robert Millikan coined the term "cosmic rays" in the mid-1920s. Tesla originally referred to the rays as a form of radiant energy. Dr. James Corum suggested that even if the rays did not travel faster than lightspeed, these statements by Tesla must have been based upon some real effect that had occurred (interview, 1992, Colorado Springs).

22. NT, "Dr. Tesla Writes of Various Phases of His Discovery," *New York Times,* February 6, 1932, 16:2; in *Tesla Said,* p. 237.

23. John O'Neill, "Tesla Cosmic Ray Motor," *Brooklyn Eagle,* July 10, 1932, in NT, *Solutions*..., pp. 95–96.This may be the description of a solar energy machine.

24. Carol Bird, "Tremendous New Power Soon to Be Unleashed," *Kansas City Journal-Post,* September 10, 1933; in *Solutions*..., pp. 101–2.

25. NT, "Tesla, 79, Promises...," *New York Times,* July 11, 1935, 23:8.

26. NT, "Expanding Sun Will Explode Some Day Tesla Predicts," *New York Herald Tribune,* August 18, 1935; in NT, *Solutions*..., pp. 130–32.

27. Joseph Alsop, "Beam to Kill Army at 200 Miles, Tesla's Claim," *New York Times,* July 11, 1934, pp. 1, 15; in *Solutions,* pp. 110–12; Walter Russell, *The Russell Cosmology: A New Concept of Light, Matter and Energy* (Waynesboro, Va.: The W. R. Foundation, 1953).

28. NT, "Expanding Sun," August 18, 1935.

29. NT, "Tesla, 79, Promises," July 11, 1935.

30. "Tesla at 78 Bares New 'Death-Beam'," *New York Times,* July 11, 1934, 18:1, 2.

31. H. Grindell-Mathews, "The Death Power of Diabolical Rays," *New York Times,* May 21, 1924, 1:2; 3:4, 5.

32. H. Grindell-Mathews, "Diabolical Rays," *Popular Radio,* August 8, 1924, pp. 149–54.

33. H. Gernsback, "The Diabolic Ray," *Practical Electrics,* August 1924, pp. 554–55, 601.

34. H. Grindell-Mathews, "Three Nations Seek Diabolical Ray," *New York Times,* May 28, 1924, 25:1,2.

35. Helen Welshimer, "Dr. Tesla Visions the End of Aircraft in War," *Everyday Week Magazine,* October 21, 1934, p. 3; in NT, *Solutions to Tesla's Secrets,* pp. 116–18.

36. L. Anderson, *NT's Residences, Labs & Offices* (Denver, Colo.: 1990). (Original source, a Dr. Watson of New York.)

37. Titus deBobula, Tesla tower blueprints, circa 1934 [NTM]; FBI archives [FOIA].

Chapter 45: Living on Credit, pp. 428–435

1. Hugo Gernsback, "NT and His Inventions," *Electrical Experimenter,* January 1919, p. 614.

2. NT to GS, July 11, 1935 [LC].

3. Branning, 1981, p. A-3.

4. NT to Carl Laemmle, July 15, 1937, Profiles in History Archives, Beverly Hills, Calif.; Neal Gabler, *An Empire of Their Own* (New York: Anchor Books, 1988), pp. 58, 205–6.

5. Mark Siegel, *Hugo Gernsback: Father of Modern Science Fiction* (San Bernardino, Calif.: The Borgo Press, 1988).

6. Lawrence Lessing, *Man of High Fidelity: Edwin Howard Armstrong* (New York: Lippincott, 1956).

7. "Tesla Is Provider of Pigeon Relief" [KSP].

8. Leland Anderson, "Nikola Tesla's Patron Saint," *American Srbobran,* August 14, 1991, p. 4; L. Anderson, *NT's Residences, Labs and Offices,* 1990.

9. OAP files [FOIA].

10. NT to GW Co. circa Jan–July, 1930, written from the Hotel Pennsylvania [KSP]; $2,000 debt, OAP files [FOIA].

11. O'Neill, p. 274.

12. Hugo Gernsback, Westinghouse recollections [KSP].

13. NT to GW Co., January 29, 1930; February 1, 1930; February 14, 1930; February 17, 1930; February 18, 1930; April 18, 1930 [LC].

14. A. W. Robertson, *About George Westinghouse and the Polyphase Electric Current* (New York: Newcomen Society, 1939), p. 28.

15. "Nikola Tesla," *Scientific Progress,* September 1934.

16. TdB to NT, November 1897; December 10, 1897 [NTM].

17. Ibid., July 26, 1901.

18. FBI deBobula files [FOIA].

19. TdB to NT, May 29, 1911; NT to TdB, May 31, 1911 [NTM].

20. Telephone interview with Robert Hessen, author of *The Steel Titan: The Life Story of Charles Schwab* (New York: Oxford University Press, 1975)—teaching at Stanford University; "Schwab Answers Suit of deBobula," *New York Times,* August 7, 1919, 15:6.

21. TdB to NT, July 11, 1935 [NTM].

22. NT to GS, June 17, 1937 [LC].

23. TdB to NT, November 25, 1935; July 6, 1936 [NTM].

24. deBobula FBI files, circa 1936–1949 [FOIA].

25. FBI deBobula files [FOIA]; "Tauscher Accuses Munitions Partner," *New York Times,* July 25, 1934, 36:4.

26. OAP NT files [FOIA].

27. Hugo Gernsback, Westinghouse recollections [KSP].

28. GW Co. to NT, January 2, 1934 [LC].

29. Mildred McDonald, December 1, 1952 [GWA].

Chapter 46: Loose Ends, pp. 436–445

1. Elmer Gertz, *Odyssey of a Barbarian: The Biography of G. S. Viereck* (Buffalo, NY: Prometheus Books, 1978).

2. M. Pupin to K. Swezey, May 29, 1931 [KSP].

3. D. Dunlap, *Radio's 100 Men of Science* (New York: Harper & Brothers, 1944), p. 124.

4. M. Pupin, *From Immigrant to Inventor* (New York: Scribners, 1930). Tesla is mentioned once on p. 285 within the phrase, "Tesla's AC motor and Bradley's rotary transformer..." (see chapter 10).

5. "Dr. Pupin Inspired," *New York Times,* 1927 [KSP].

6. Stanko Stoilkovic, "Portrait of a Person, a Creator and a Friend," *The Tesla Journal,* nos 4, 5, 1986/87, pp. 26–29.

7. NT to RUJ, circa 1929–1937 [LC]; Grizelda Hull Hobson to K. Swezey, February 14, 1955; Richmond P. Hobson Jr., in "Books of the Times," *New York Times,* Decemer 21, 1955 [KSP].

8. A brief excerpt from "The Haunted House" by G. S. Viereck, circa 1907, from Gertz, 1978.

9. Niel Johnson, *G. S. Viereck: German/American Propagandist* (Chicago: University of Illinois Press, 1972), p. 143.

10. Ibid., pp. 138–42.

11. NT corresp., March 2, 1942 [LA].

12. NT, as told to G. S. Viereck, "A Machine to End War," *Liberty,* February 1935, pp. 5–7.

13. Peter Viereck, phone interview, September 8, 1991.

14. Gertz, p. 24.

15. Elmer Gertz, June 1991 phone interview.

16. Cheney, p. 243.

17. NT to GSV, 1934 [from L. Anderson, "N. Tesla's Patron Saint," *American Srbobran,* August 14, 1991, p. 4]; NT to GSV, December 17, 1934 [in Cheney, p. 244].

18. "Sending of Messages to Planets Predicted by Dr. Tesla on Birthday," *New York Times,* July 11, 1937, 1:2–3; 2:2–3.

19. "Immigrant Society Makes Three Awards: Frankfurter, Martinelli and Tesla," *New York Times*, May 12, 1938, 26:1.

20. N. Johnson, pp. 204–10; GSV FBI files [FOIA].

21. "G. S. Viereck, 77, Pro-German Propagandist, Dies." *New York Times*, March 21, 1962.

22. NT, "Tesla and the Future," *Serbian Newsletter*, 1943.

23. O'Neill, 1944.

24. L. Anderson, August 14, 1991.

25. O'Neill, 1944.

26. "2000 Are Present at Tesla Funeral," *New York Times*, January 13, 1943.

27. Hugo Gernsback, "NT: Father of Wireless, 1857–1943," *Radio Craft*, February 1943, pp. 263–65, 307–10.

28. *Serbian Newsletter*, 1943, p. 5 [BLCU].

29. "NT Dead," editorial, *New York Sun*, January 1943.

Chapter 47: The FBI and the Tesla Papers, pp. 446–462

1. January 22, 1946, OAP report [FOIA].

2. J. Edgar Hoover, memorandum, January 21, 1943 [FBI, FOIA].

3. Cheney, p. 258.

4. M. Markovitch, personal interview, 1984.

5. As I rewrite this chapter in November 1995, Yugoslavia is in the midst of a civil war, with essentially all of the provinces having declared their independence. The most bellicose new nation is Serbia. It has attacked Bosnia repeatedly for over the last three years in attempts to drive out Croats and Muslims and capture as much land as possible. Many women have been raped, thousands of people have been killed, and over one million have had to flee their homes. At this point a solution appears to be futile.

6. Cheney, 1981, pp. 260–61.

7. J. Edgar Hoover, memorandum, January 21, 1943 [FBI, FOIA].

8. FBI, January 21, 1943 [FOIA].

9. "Floating Stretchers for Landings," *New York Times*, August 27, 1944, IV, 9:6; "Company Volunteers $1,500,000 Refund," *New York Times*, November 19, 1944, 1:3; "France's Honors Heaped on Spanel," *New York Times*, March 3, 1957, 26:5.

10. F. Cornels, January 9, 1943 [FBI, FOIA].

11. Fitzgerald to Tesla, March 8, 1939; December 20, 1942 [NTM]. MIT, however, had no record that Fitzgerald was a student in their school [M. Seifer to MIT, 1990]. Fitzgerald also met with Jack O'Neill to help on the biography. He also discussed with Jack the possibility of setting up a museum in the United States, perhaps with backing from Henry Ford.

12. J. O'Neill, "Tesla Tries to Prevent WWII," *TBCA News*, *7*, *3*, 8–9/1988, p. 15.

13. F. Cornels, FBI report, January 9, 1943 [FOIA].

14. L. Anderson, files from Cheney, 1981, p. 264.

15. D. E. Foxworth, FBI report, January 8, 1943 [FOIA].

16. D. E. Foxworth, FBI report, January 8, 1943; Donegan, FBI report, November 14, 1943 [FOIA].

17. Personal correspondence from Irving Jurow, Washington, D.C., July 5, 1993.

18. Werner Heisenberg, for instance, was in charge of the Nazis' version of the Manhattan Project. According to Heisenberg's autobiography, he knew that Germany did not have enough heavy water to construct an atom bomb, and he was just hoping that the war would end before such a device would be invented. Werner von Braun, of course, was also implementing the highly destructive V-2 rocket, which was yet another ultimate weapon.

19. Phone conversations and personal correspondence with Irving Jurow, June, July, 1993.

20. OAP memorandum, January 12, 1943; January 12, 1942 [1943 typographical error] [FOIA].

21. Cheney, p. 270.

22. W. Gorsuch, OAP report, January 13, 1943 [FOIA].

23. Trump resort, January 30, 1943 [LC]; C. Hedetneimi, OAP report, January 29, 1943 [FOIA]; interview with a guard from Manhattan Storage, FBI report, April 17, 1950 [FOIA].

24. C. Hedetniemi, OAP report, January 29, 1943 [FOIA].

25. Trump's conclusion, was that since the device was similar to the Van de Graaff electrostatic generator, Soviet engineers would find no ultimate value in it. This is somewhat

astonishing, as Trump also enclosed an article written by Tesla in 1934 in *Scientific American* where he states explicitly that his device was, operationally, completely unlike the Van de Graaff generator. As Trump worked with Van de Graaff at MIT, it would seem that his cavalier dismissal of the particle-beam weapon was based on professional jealousy. To Trump's credit, however, here we are, a half century later, and the Tesla weapon has yet to be perfected. (Trump report, FBI archives; N. Tesla, "Electrostatic Generators," *Scientific American,* March 1934, pp. 132–34; 163–65.)

26. Homer Jones to Lawrence Smith, February 4, 194[3] [OAP, FOIA].

27. NT, "The New Art of Projecting Concentrated Non-dispersive Energy Through the Natural Media" (1937), in E. Raucher and T. Grotz, eds., *1984 Tesla Centennial Symposium,* pp. 144–50.

28. According to a letter to Sava Kosanovic, Tesla was planning on selling eight particle beam weapons to Yugoslavia, three to Serbia, three to Croatia and two to Slovenia. NT to SK, March 1, 1941, *Correspondence with Relatives,* p. 183.

29. "$3,500,000 Payment by Amtorg Today." *New York Times,* November 15, 1932, 29:7; Amtorg and Bethlehem Steel, *New York Times,* April 30, 1935, 30:2; "To Catch a Spy," *Newsweek,* May 19, 1986, p. 7; etc., Amtorg Trading Corp. to M. Seifer, April 4, 1988.

30. J. Trump, letter quoted in Cheney, p. 276.

31. FBI NT memorandum, January 12, 1943 [FOIA].

32. J. Alsop, "Beam to Kill Army at 200 Miles," *New York Herald Tribune,* July 11, 1934, 1:15.

33. J. Corum and K. Corum, "A Physical Interpretation of the Colorado Springs Data," in E. Raucher and T. Grotz, eds., *The 1984 Tesla Centennial Proceedings,* pp. 50–58.

34. Alcoa Aluminum Co., private corresp., December 16, 1988.

35.NT, in *Tesla Said,* p. 278.

36. H. Welshimer, "Dr. Tesla Visions the End of Aircraft in War," *Every Week Magazine,* October 21, 1934, p. 3.

37. Charlotte Muzar, "The Tesla Papers," *The Tesla Journal,* 1982/83, pp. 39–42.

38. E. E. Conroy to J. Edgar Hoover, FBI, October 17, 1945 [FOIA].

39. Ralph Doty to OAP, January 22, 1946 [FBI, FOIA].

40. Cheney, p. 277.

M. Duffy to OAP, November 25, 1947 [FOIA]; FBI memorandum, April 17, 1950 [FOIA].

42. Andrija Puharich, phone interview, 1986; Ralph Bergstresser, phone interview, 1986.

43. "Are Soviets Testing Wireless Electric Power?" *Washington Star,* January 31, 1977, pp. 1, 5; "Russians Secretly Controlling World Climate," *Sunday Times,* Scranton, Penn., November 6.1977, pp. 14–15.

44. Tom Bearden, "Tesla's Secret and the Soviet Tesla Weapons," *Solutions to Tesla's Secrets,* John Ratzlaff, ed., 1981, pp. 1–35; Tom Bearden, "The Fundamental Concepts of Scalar Electromagnetics," *Tesla Conference Proceedings,* Steve Elswick, ed., 1986, pp. 7:1–20.

45. C. Robinson, "Soviet Push for a Beam Weapon," *Aviation Week,* May 2, 1977, pp. 16–27; N. Wade, "Charged Debate Over Russian Beam Weapons," *Science,* May 1977, pp. 957–59.

Chapter 48: The Wizard: Legacy, pp. 463–470

1. Lawrence P. Lessing, *Man of High Fidelity: Edwin Howard Armstrong* (New York: Lippincott, 1956, p. 286.

2. H. Fantel, "Armstsrong, Tragic Hero of Radio Music," *New York Times,* June 10, 1973, pp. 23–28; Lessing, 1956; Marc Seifer, "The Inventor and the Corporation: Case Studies of Innovators Nikola Tesla, Steven Jobs and Edwin Armstrong," *1986 Tesla Symposium,* S. Elswick, ed., pp. 53–74.

3. W. Whyte, *The Organization Man* (New York: Doubleday, 1956).

4. David Held, *Introduction to Critical Theory* (Berkeley, Calif.: University of California Press, 1980).

5. Bill Gates; interview, *Playboy,* September 1994, p. 64. In 1996, Jobs reemerged as an overnight billionaire with a highly successful stock offering of his new computer graphics company Pixar in 1996, and, in an amazing turnabout, Jobs was further resurrected as replacement CEO of Apple in 1997. Further, IBM has agreed to produce a Macintosh compatible computer.

6. Henry Aiken, corresp., phone interview, April 1986.
7. Ayn Rand, *Atlas Shrugged* (New York: 1957), p. 236.
8. M. Seifer and H. Smukler, "The Tesla/Matthews Outer Space Connection," *Pyramid Guide*, part I, May 1978, p. 5; part II, July 1978, p. 5 [FBI, FOIA].
9. "Tesla in Japan," *Tesla Memorial Society Newsletter*, Nicholas Kosanovich, ed., Fall-Winter 1995/96, pp. 2–3; David Kaplan and Andrew Marshall, "The Cult at the End of the World," *Wired*, July 1996, pp. 134–37, 176–84; Tom Bearden, "Tesla's Secret and the Soviet Tesla Weapons," *Solutions to Tesla's Secrets*, John Ratzlaff, ed., 1981, pp. 1–35.
10. P. O. Ouspensky, *New Model of the Universe* (New York: Vintage Books, 1971), pp. 29–31.
11. Dane Rudyar, *Occult Preparations for a New Age* (Wheaton, Ill.: Quest Books, 1975), p. 245; Robert Anton Wilson, *Cosmic Trigger: The Final Secret of the Illumenati* (New York: Pocket Books, 1975).

Appendix, pp. 471–476

1. J. R. Johler to Leland Anderson, August 15, 1959, in Anderson, "Nikola Tesla's Work in Wireless Power Transmission" (Denver, Colo., 1991, unpublished.
2. Eric Dollard, "Representations of Electric Induction: Nikola Tesla and the True Wireless," In S. Elswick, ed., *Proceedings of the 1986 Tesla Symposium* (Colorado Springs, Colo.: International Tesla Society, 1986), pp. 2-25–2-82.
3. NT, 1916/1992, p. 138.
4. James Corum and A-Hamid Aidinejad, "The Transient Propagation of ELF Pulses in the Earth-Ionosphere Cavity," *1986 International Tesla Symposium Proceedings*, pp. 3-1–3-12.
5. NT to KJ, April 19, 1907 [BLCU].
6. NT, "Terrestrial Night Light," *New York Herald Tribune*, June 5, 1935, p. 38.
7. NT, 1984, p. 225.
8. NT, *Colorado Springs Notes*, pp. 180–183; patent no. 649,621 in NT, 1956, p. P-293.
9. NT, May 16, 1900; patent no. 787,412, in NT, 1956, pp. P-332–33.
10. NT, "Tesla's New Discovery," 1901; in NT, 1984, p. 57.
11. NT, patent no. 685,012, in NT, 1956, pp. P-327–30. It is doubtful but possible that he considered using superconductivity, as this property of elements involving the expulsion of magnetism occurs at temperatures almost twice as cold. This effect, which is an abrupt and discontinuous transition from a magnetic state to a nonmagnetic state was officially discovered a decade later in 1911 by Kamerlingh Onnes (see J. Blatt, *Theories of Superconductivity* [New York: Academic Press, 1964]).
12. Discussions with Stanley Seifer, February 1991.
13. Tom Bearden, "Tesla's Secret," *Planetary Association for Clean Energy*, 3, pp. 12–24.
14. NT, 1897, in NT 1956, pp. P-293–94.
15. NT, 1956, p. P-293.
16. Ibid., p. P-328.

ACKNOWLEDGMENTS

This book began in earnest in the late 1970s, and has continued unabated until this point in 1996. Along the way, there were many individuals and institutions that helped me in my research. The first person to thank is my former partner in the field of consciousness research, Howard Smukler, who gave me the O'Neill book in 1976 along with the nutty text *Wall of Light: Nikola Tesla and the Venusian Space Ship*. Shortly thereafter, in 1977, I wrote my first article on the inventor. The second major hurdle was accomplished in 1979, after spending two years poring through the microfilm letters between Tesla and J. Pierpont Morgan, George Westinghouse, George Scherff, and Robert Underwood Johnson, which were obtained from the Library of Congress by Roberta Doren of the Interlibrary Loan Department at the University of Rhode Island (URI). After Roberta transferred to a different division, Vernice (Vicky) Burnett took over helping me, and she continued to do so unabated for another dozen years. I would like to thank Vicky for her resourcefulness and extraordinary efforts, and the rest of the staff at the URI library.

In 1980, I began a doctoral program at Saybrook Institute, San Francisco. The work resulted in a 725-page doctoral dissertation entitled *Nikola Tesla: Psychohistory of a Forgotten Inventor*. Stanley Krippner was not only my mentor; he was also a keen editor who corrected and criticized the entire treatise. It was a mammoth undertaking for him, and I am most appreciative. Other Saybrook committee members I wish to thank include Henry Alker, Octave Baker, Jurgen Kramer, Debra White, and the outside reader William Braud of the Mind Science Foundation in San Antonio, Texas. The dissertation was completed in 1986.

In 1987, I began to work full-time on a full-blown biography. Many entirely new avenues were revealed not covered in the doctoral dissertation. A number of key individuals, particularly Tesla experts, helped me enormously. From the start, Mike Markovitch of Long Island University provided me with important source material and translations; William

Terbo, Tesla's grandnephew, spent endless hours over many years with me discussing various details. In Belgrade, Alexander Marincic, director of the Tesla Museum and, in particular, his assistant, Branimir (Branko) Jovanovic, aided me in vital ways. And in the United States, I must also thank heartily Dr. Ljubo Vujovic, Jim and Ken Corum, and the patriarch of Tesla experts, Leland Anderson, whose cache of material, which, like the documents provided by the Tesla Museum, was indispensable in creating this treatise.

Other experts who helped include John Ratzlaff, of the original Tesla Book Company, John Pettibone of Hammond Castle, Paul Baker, Nick Basura, Tom Bearden, Ralph Bergstresser, Zoran Bobic, Nancy Czito, Steve Elswick, Uri Geller, Elmer Gertz, Robert Golka, Toby Grotz, James Hardetsky, Mrs. R. U. Johnson, Jr., John Karanfilovsky, Nicholas Kosanovich, John Langdon, J. W. McGinnis, Sanford Neuschatz, Nicholas Pribic, Dr. Andrija Puharich, Sid Romero, Lynn Sevigny, Richard Vangermeersch, J. T. Walsh, Tad Wise, and Japanese inventor extraordinaire Dr. Yoshiro NakaMats. Through their works, Hugo Gernsback, Kenneth Swezey, Inez Hunt and Wanetta Draper, Herbert Satterlee, and particularly Matthew Josephson and John O'Neill.

Important institutions included the Berkeley, Cornell, Harvard, MIT, Brown, and Yale University libraries, archives from the University of Prague, Columbia University Butler and Avery libraries, the New York Public Library and New York Historical Society, the Edison Menlo Park Archives, the J. Pierpont Morgan Library, Hammond Castle, the Westinghouse Corporation Archives, Hugo Gernsback Publications, the Library of Congress, the National Archives, the Smithsonian Institution, the FBI, the OAP and the instrument known as the Freedom of Information Act.

I would also like to thank my close friend Elliott Shriftman for his wisdom, great generosity, and continuing encouragement; the late Prof. Edwin Gora for his understanding of Tesla's link to theoretical physics, Roger Pearson, former dean of Providence College School of Continuing Education; and Raymond LaVertue, of Bristol Community College, for helping me put bread on my table; my sagacious agent John White, who has been with the project for over ten years; Allan Wilson (for believing in me) and Donald Davidson of Carol Publishing Group; and my loyal and altruistic partner in the screenplay *Tesla: The Lost Wizard,* Tim Eaton, visual effects editor for Industrial Light & Magic of Marin County.

The treatise is dedicated to my parents, Thelma and Stanley Seifer, my sister Meri Shardin, her husband, John Keithley and their children, Devin and Dara; my brother, Bruce Seifer, and his wife, Julie Davis; and my wonderful and understanding spouse, Lois Mary Pazienza, who has been with me throughout the entire twenty-year project.

This book is also dedicated to the Teslarians, who seek the truth from the past and a sane, ecologically minded technology for the future.

INDEX